Practical Sedimentology

Practical Sedimentology

Second Edition

Douglas W. Lewis
David McConchie

CHAPMAN & HALL
New York • London

An Apteryx Book

This edition published by
Chapman & Hall
One Penn Plaza
New York, NY 10119

Published in Great Britain by
Chapman & Hall
2-6 Boundary Row
London SE1 8HN

Printed in the United States of America

Library of Congress Cataloging in Publication Data

Lewis, D.W. (Douglas W.), 1937-

Practical sedimentology / D.W. Lewis and David McConchie. —2nd ed.
p. cm.

Includes bibliographical references and index.
ISBN 0-442-01217-9
1. Sedimentology. I. McConchie, David. II. Title.
QE471.L46 1993
552'.5—dc20 92-34768
CIP

British Library Cataloguing in Publication Data available

Please send your order for this or any Chapman & Hall book to **Chapman & Hall, 29 West 35th Street, New York, NY 10001, Attn: Customer Service Department.** You may also call our Order Department at 1-212-244-3336 or fax your purchase order to 1-800-248-4724.

For a complete listing of Chapman & Hall's titles, send your requests to **Chapman & Hall, Dept. BC, One Penn Plaza, New York, NY 10119.**

CONTENTS

PREFACE

Sedimentology has neither been adequately popularized nor commonly taught as an interdisciplinary subject, and many workers in the areas of modern environment studies have very limited knowledge of sedimentology. *Practical Sedimentology* (henceforth **PS**) is designed to provide an introduction and review of principles and interpretations related to sedimentary processes, environments, and deposits. Its companion volume, *Analytical Sedimentology* (henceforth **AS**), provides "cookbook recipes" for common analytical procedures dealing with sediments, and an introduction to the principles and reference sources for procedures that generally would be performed by specialist consultants or commercial laboratories. Specialist sedimentologists will find in them useful reviews, whereas scientists from other disciplines will find in them concepts and procedures that may contribute to an expanded knowledge of their working environment. Emphasis is given to summary figures and tables, and to literature sources that can provide the depth of understanding necessary to resolve particular problems; text discussion is minimized in the expectation that readers will seek detailed treatment in traditional texts as well as in the research papers cited. The selected bibliographies have been expanded relative to the first edition. They have been divided into sections according to pertinent subheadings; as a result, some of the items cited in the text that refer to several topics may have to be searched for in the bibliographies. We apologize for any slight inconvenience this situation may cause the reader.

There have been problems in segregating topics between the books. For example, discussion on interpretation of quantitative analytical results is coupled with discussion of principles of sediment textures (**PS**) rather than with quantitative analytical procedures (**AS**), because we think the potential results of a study need to be realized before the procedures for the study are considered. We hope that adequate cross-referencing between the books will minimize problems for readers in what may sometimes appear to be arbitrary segregations.

This book begins with a consideration of the complex end product of processes and materials, the sedimentary environment. It then proceeds to discuss the processes and materials themselves. The emphasis is on geological interpretations of ancient deposits, but most discussions are also relevant to modern sediments and can be used to predict environmental changes. A basic knowledge of geological jargon is anticipated for users of this book; we try to define most of the more esoteric terms in context, but if there are additional incomprehensible terms, refer to Bates and Jackson's *Glossary of Geology* (AGI, 1987).

ACKNOWLEDGMENTS

Many chapter drafts of **PS** were critically reviewed by Dr. M. Wizevich (at that time at Victoria University of Wellington); his helpful comments improved the book. Chapter 8 on carbonate sediments was reviewed by Dr. P. A. Scholle (Southern Methodist University, Dallas); Drs. J. Esterle (then of University of Canterbury) and A. A. Ekdale (University of Utah) contributed ideas and advice on several chapters. Fiona Davies-McConchie provided hours typing drafts and searching the literature. Marjorie Spencer, of Van Nostrand Reinhold Publishing Company, initially put much time and effort in the planning of the books. Bernice Pettinato (Beehive Production Services) capably and professionally handled the manuscript editing and production of the final presentation. Professors D. S. Gorsline, University of Southern California, and K. A. W. Crook, Australian National University, made general suggestions and encouraged the construction of the first edition of *Practical Sedimentology*. Glen Coates (at that time, University of Canterbury) provided some drawings; Lee Leonard drafted final copies of some of the illustrations. Several graduate students enjoyed their revenge by catching errors in the final draft.

1
Introduction

SCOPE OF SEDIMENTOLOGY

Sediments accumulate at the interface between the lithosphere and either the hydrosphere or the atmosphere; in settings such as river floodplains and intertidal flats, processes from all spheres influence sedimentation. The internal structures, the textures, and the composition of sediments may also reflect interactions with the biosphere, at scales from human to bacterial, and the lithosphere, wherein chemical and physical processes acting after deposition at temperatures less than about 300°C are considered to be within the realms of the sedimentary processes of weathering and diagenesis. Consequently, sedimentology is perhaps the most interdisciplinary subdiscipline of the earth sciences; it overlaps the fields of zoology, botany, agronomy, geography, hydrology, chemistry, physics, engineering, and other subdisciplines in geology such as geochemistry, geomorphology, and mineralogy.

Sedimentologists have many functions in the earth sciences. From the purely geological viewpoint, their fundamental job is to interpret the history of ancient deposits in the rock record, generally within the framework of basin analysis: the geographic and geologic setting for both derivation and accumulation of sediments (Fig. 1-1). The economic justification for these and more restricted studies is the discovery and exploitation of mineral deposits—both organic, such as oil and coal, and inorganic, such as metalliferous ores and many nonmetalliferous products (Figs. 1-2 and 1-3). To accomplish these tasks, sedimentologists must acquire data and understanding from studies of modern sedimentary deposits, environments, and processes (Fig. 1-4) following the philosophical tenets of the Principle of Uniformitarianism, or what is now commonly called Actualism.

Many sedimentologists exclusively focus on modern sediments and processes, and their work is of practical value to many other disciplines besides geology, such as agriculture, forestry, fisheries, fluvial and coastal erosion, land use planning, effluent dispersal and disposal, and many other facets of the natural and human-influenced terrestrial and marine environment (Table 1-1). Particularly in regard to guiding human developments for minimal environmental impact, there is a growing demand for the work of sedimentologists in multidisciplinary teams of specialists. Sedimentologists also are being requested more frequently to extrapolate actualistic principles into the future; by using their understanding of past trends and present process/product relationships, they can predict the nature and magni-

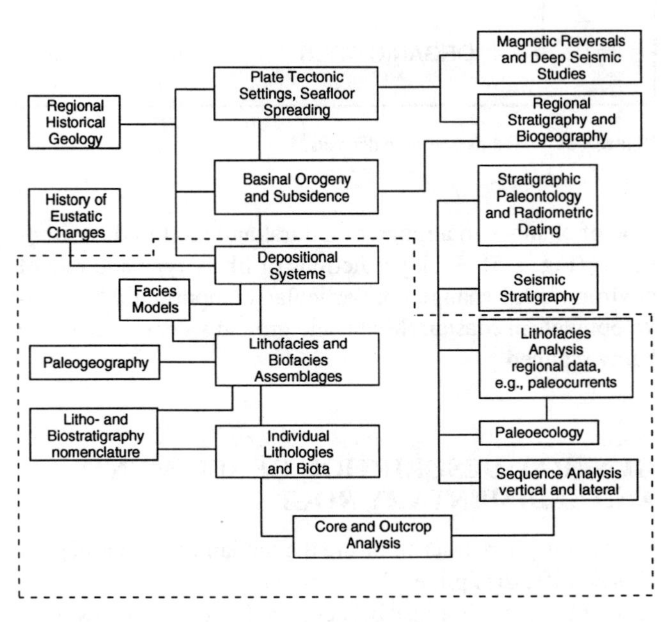

Figure 1-1. Example of the roles played by sedimentology in basin analysis. (After Miall 1990.)

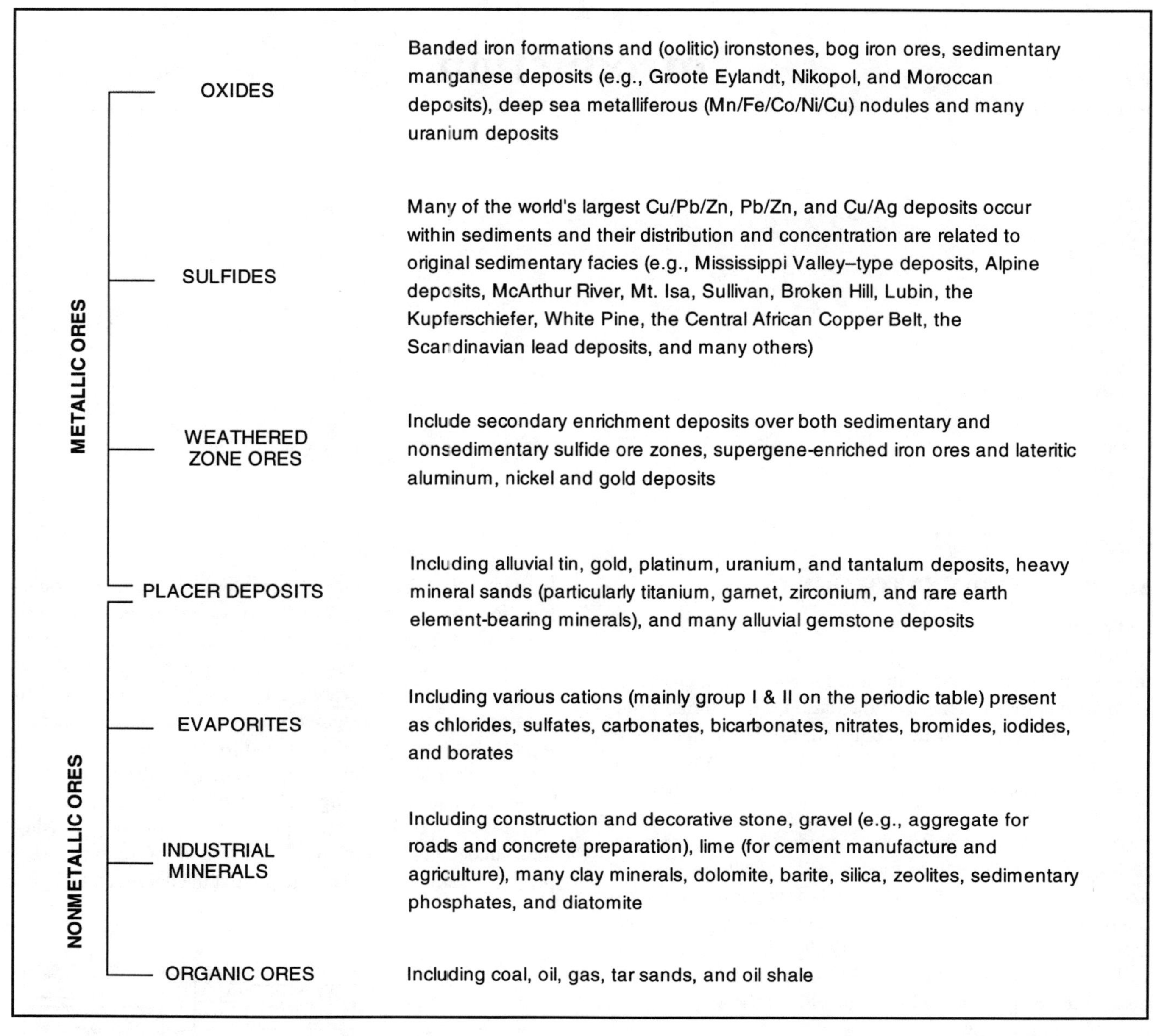

Figure 1-2. Sedimentary ore deposits.

tude of changes to a range of natural and modified environments (Fig. 1-5). Such predictions of likely type and rate of environmental change are particularly important where developments in coastal, fluvial, and groundwater regimens are being planned.

GENERAL DESCRIPTION OF SEDIMENTS AND SEDIMENTARY ROCKS

Observation and description are the fundamental first steps in all scientific disciplines. For sediments, three primary properties of materials must be described: structures, textures, and composition. *Sedimentary structures* are geometrical attributes given to grain assemblages by depositional and early postdepositional processes (e.g., current or slump processes); they are treated in Chapter 4. *Texture* refers to the size, shape, and group properties of the sedimentary particles in a deposit; they are treated mostly in Chapter 5 (for detrital sediments). *Composition* refers to the mineralogical or chemical characteristics of sedimentary particles and is treated in Chapter 6. Each property reflects, and therefore provides information about, different sets of processes acting in different settings; interrelationships are schematically shown in Fig. 1-6. Other properties, such as color and induration (if dealing with a cohesive sediment or rock) are also helpful distinguishing properties of sediments, but are secondary features.

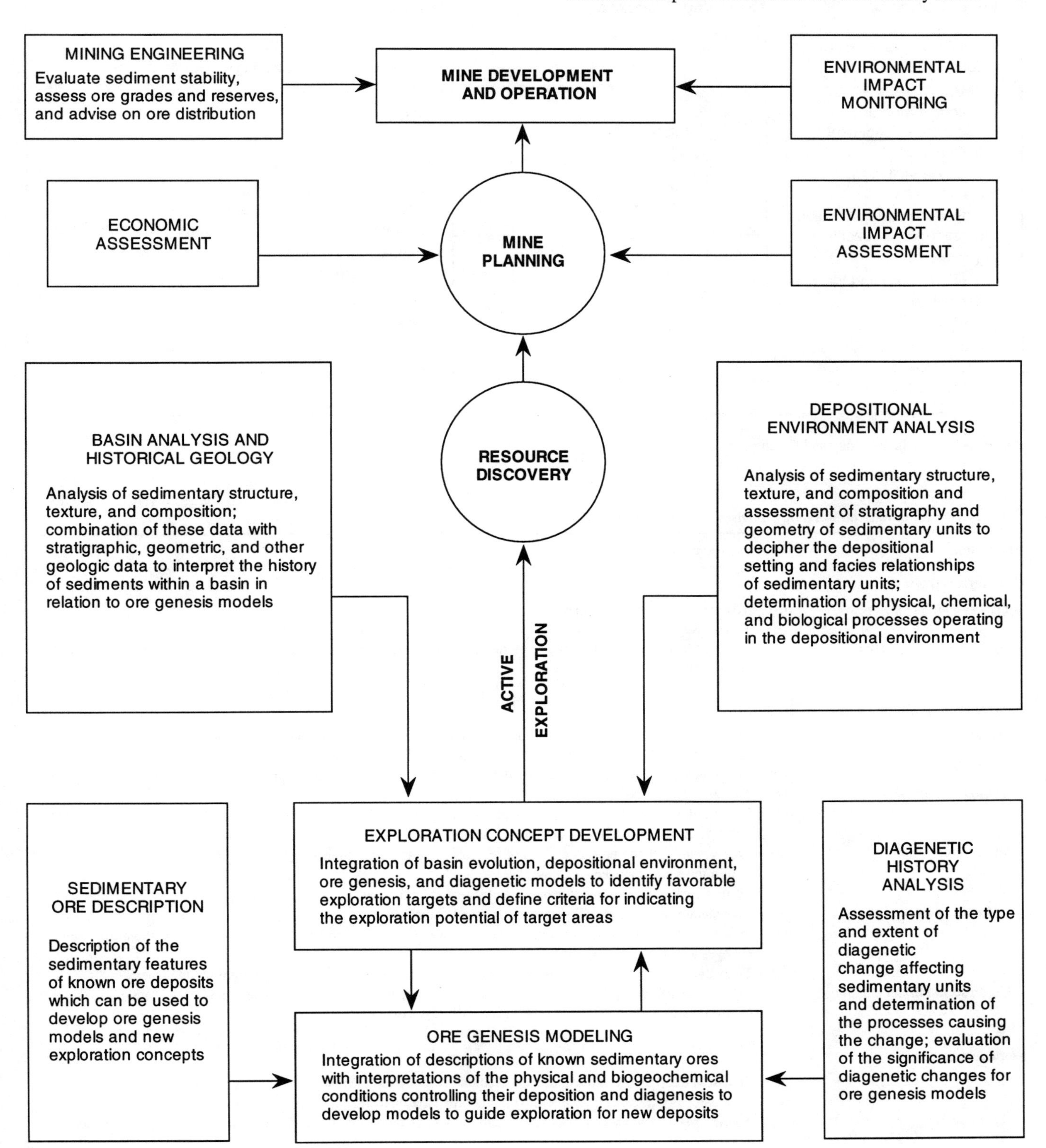

Figure 1-3. Roles of sedimentologists in mineral resource development. In the figure, "ore" refers to all economic mineral, coal, and hydrocarbon resources.

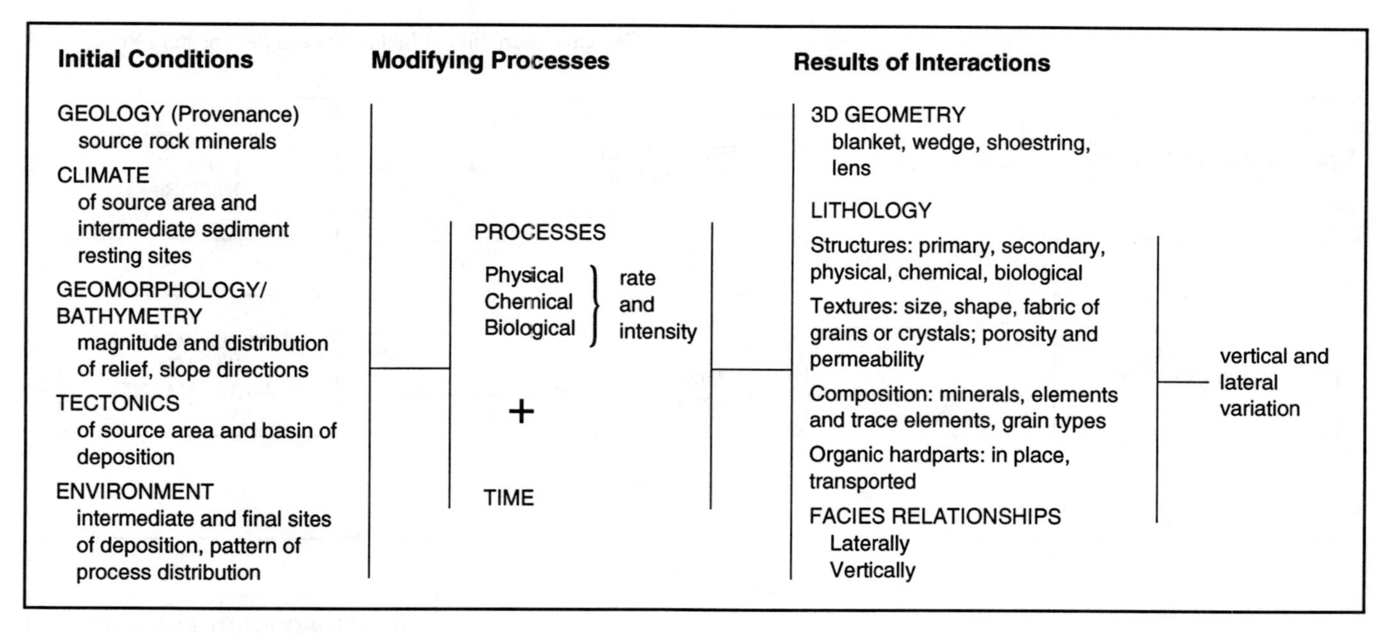

Figure 1-4. Sedimentology and the environment.

Table 1-1. Roles of Sedimentology in Disciplines Other Than Geology

Application		Principal Requirements
Primary Production		
Fisheries and aquaculture	(a)	To evaluate substrate sediment stability
	(b)	To assess the nature of chemical (nutrients and toxicants) transfers between sediment, water, and biota
Agriculture and forestry		To evaluate present conditions and predict the likely nature, rate, and extent of future changes in sediment/soil characteristics
Environmental Management		
Impact assessment	(a)	To describe the physical, chemical, and biological conditions in modern environments and predict the likely nature, rate, and extent of future natural and anthropogenic changes
	(b)	To collect and evaluate sedimentological data required for monitoring programs and environmental impact audits
Coastal management	(a)	To assess coastal sediment dynamics
	(b)	To describe physical, chemical, and biological conditions in modern environments and predict the likely nature, rate, and extent of future environmental changes
Land conservation	(a)	To evaluate past and present aspects of sediment/soil stability, mineralogy, and chemistry and propose procedures to minimize anthropogenic changes
	(b)	To recommend procedures to rehabilitate degraded land areas
Pollution control	(a)	To evaluate sediment stability, mineralogy, and chemistry
	(b)	To assess chemical transfers between sediment, water, and biota and predict the likely effects of anthropogenic changes to existing conditions
	(c)	To evaluate likely chemical transfers between sediment, water, and biota at potential waste disposal sites
Land use and urban planning		To describe present sedimentary environments, evaluate interactions between sediment, water, and biota in these environments, and predict the likely nature, rate, and extent of future changes
Engineering		
Civil	(a)	To assess the stability and engineering characteristics of sediment/soil
	(b)	To predict any changes in sediment/soil properties that may arise in response to proposed developments
Hydrological	(a)	To assess sediment dynamics in aquatic environments in relation to the distribution of sedimentary units and water flow conditions
	(b)	To describe the physical and chemical characteristics of subsurface sediments, evaluate the distribution and uniformity of sedimentary units, and determine their likely influence on groundwater flow
	(c)	To predict the likely effect on groundwater quantity and quality of future diagenetic changes in aquifer sediments
Resource Economics	(a)	To quantify the nature and extent of sedimentary mineral resources
	(b)	To assess the likely economic impact of sedimentary processes on proposed development programs
Archeology	(a)	To determine artifact provenance
	(b)	To describe the stratigraphy of archeological sites
	(c)	To interpret ancient environmental conditions at archeological sites

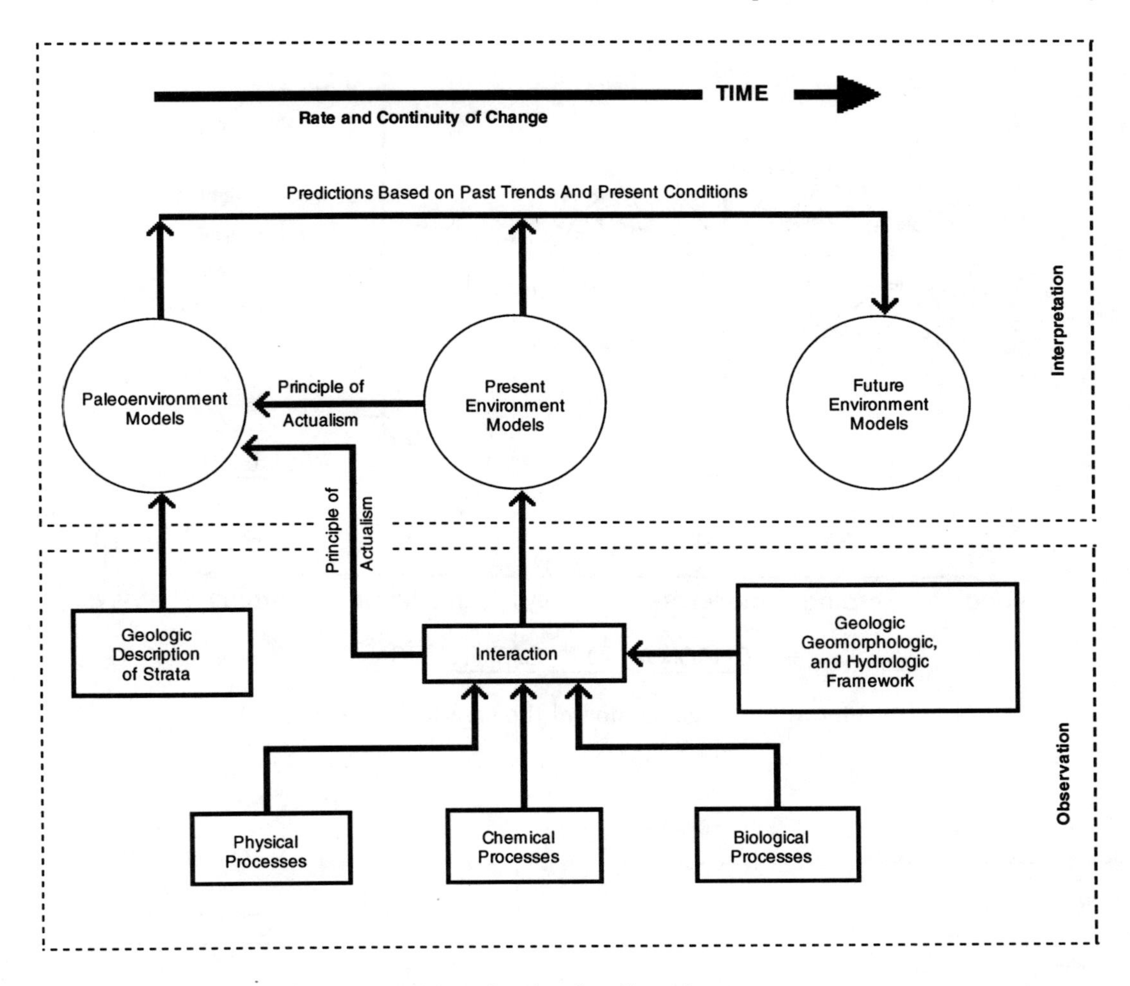

Figure 1-5. Fundamental controls, processes, and products in sedimentology.

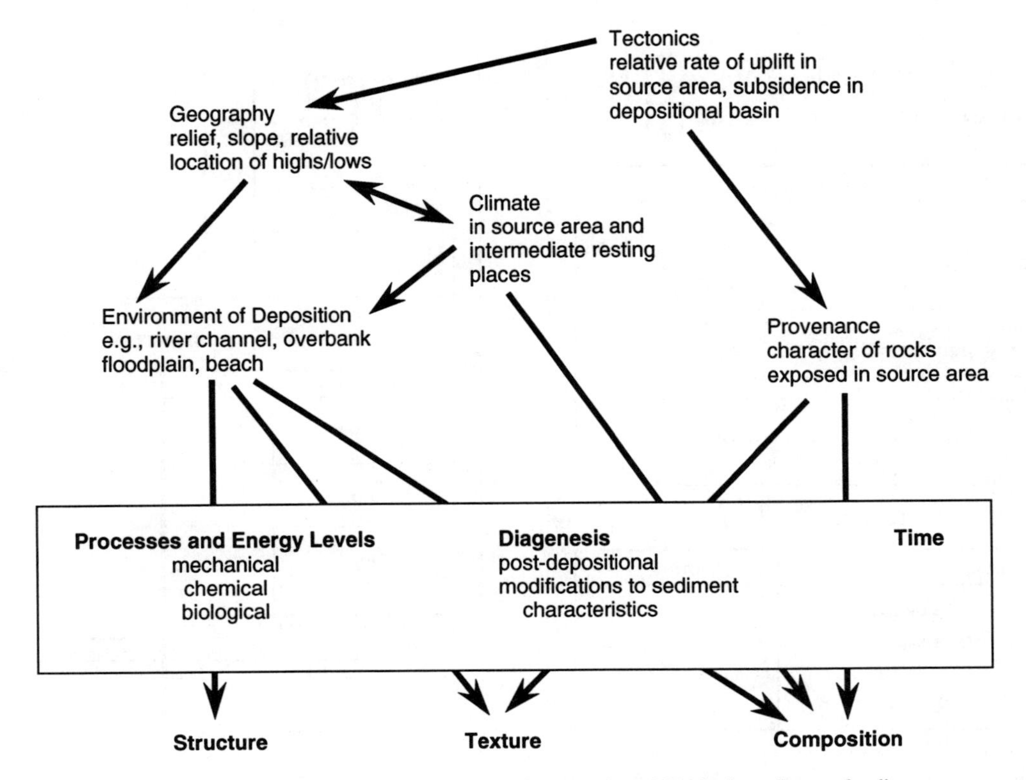

Figure 1-6. Schematic representation between the major controls and lithologic attributes of sediments.

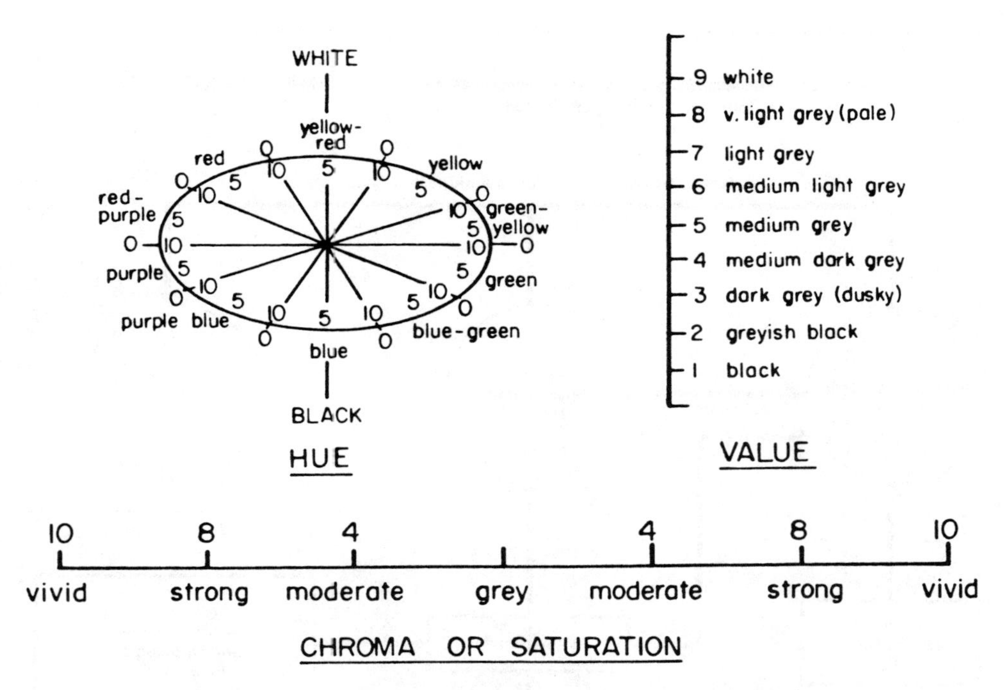

Figure 1. Organization of the standard rock color chart.

CARD SIDE 1

SAMPLE NUMBER .... DATE ....

GRID REF. .... MAP SHEET NO. ....

LOCATION: GEOG. ....

STRAT. ....

GENERAL DATA

COLOUR: FRESH .... WEATHERED ....

INDURATION: LOOSE ☐ FRIABLE ☐ INDURATED ☐

WELL INDURATED ☐ V. WELL INDURATED ☐

STRUCTURES	DESCRIPTIVE NOTES
BEDDING	....
CROSS-BEDDING	....
GRADED BEDDING	....
RIPPLES	....
TOOL MARKS	....
SCOUR, FLUTE AND GROOVE MARKS	....
SECONDARY PHYSICAL STRUCTURES	....
....	....
BIOGENIC STRUCTURES	....
....	....
OTHER	....
....	....

TEXTURE

GRAIN SIZE (ϕ)		OTHER
MAX.	% GRAVEL	SORTING
	% SAND	....
MIN.	% SILT	ROUNDING
MODE/S	% CLAY	....

CARD SIDE 2

MINERALOGY

MINERAL	%	COMMENT
QUARTZ		
FELDSPAR		
ROCK FRAGMENTS		
OTHER DETRITAL GRAINS		
....		
....		
....		
SHELL FRAGMENTS		
....		
GLAUCONITE		
OTHER NON-DETRITALS		
....		
....		
MATRIX		
....		
CEMENT		
....		

CHEMISTRY

MAJOR ELEMENT	%	TRACE ELEMENT	CONC. (ppm)
SiO_2			
Al_2O_3			
Fe_2O_3			
K_2O			
Na_2O			
MgO			
CaO			
MnO			
TiO_2			
P_2O_5			

PARTIAL DIGEST DATA DIGEST USED:

ELEMENT	CONC. (ppm)	ELEMENT	CONC. (ppm)

Figure 1-8. Example of a checklist for simplified recording of field or laboratory data on a note card.

Color

The human eye is capable of distinguishing between very subtle differences in color, and color is commonly a characteristic that can be used to distinguish between rock units. Color is best described by reference to the standard *Rock Color Chart* (Goddard et al. 1951). This chart is based on the Munsell color system, which is widely accepted in scientific disciplines. Colors are expressed by a shorthand notation that incorporates hue (given by a number) of red, yellow, green, blue or purple (given by a letter), value of lightness or darkness of the color, and chroma of the strength or saturation of the color (Fig. 1-7). For example, 5R 3/4 means hue of 5 red, value of 3, chroma of 4. If without a color chart, keep to primary colors: "chocolate brown" and "mauve" are visualized differently by people!

Wet and dry sediment usually have different colors; hence the same rock may look different on wet and dry days. In addition, unless one is looking only at fresh drill core or mine faces, color as seen in outcrop already has changed because of weathering; the same rock can have different colors depending on the type or duration of weathering conditions. Even in fresh exposures, different histories—early or late—may result in different colors between different parts of a single deposit. Each variety of color may need to be separately described.

In general, colors reflect the abundance of organic matter and the state of oxidation of iron and its compounds, and iron is commonly oxidized or reduced after deposition (e.g., Morad 1983). Black and dark gray suggest the presence of organic matter and reducing conditions since deposition (exceptions exist such as magnetite- or ilmenite-rich blacksands and some manganiferous deposits). Reds and yellow-browns indicate the presence of iron oxides or hydroxides, but intensity of coloration does not reflect abundance; even very small amounts (less than 1% of minerals such as hematite) can give strong colors. Yellowish colors often indicate that sulfides have oxidized to jarosite (an iron sulfate formed during weathering oxidation of sulfide compounds), whereas green colors often indicate the presence of the iron-rich phyllosilicates glauconite or chlorite.

Induration

Induration of sediments and rocks has a somewhat subjective scale, but the following are widely accepted: *loose; friable* (grains break free with finger pressure); *indurated* (broken easily with a hammer); *well indurated* (grains do not break when rock is broken); *very well indurated* (grains break when rock breaks—or hammer breaks!).

Other Properties

Quantitative measures used by soil scientists and engineers for rock and soil properties, such as for compressive and shear strength, swelling strain, penetrometer resistance or compaction indexes, which are not within the scope of our books.

Procedure

When describing any sediment or sedimentary rock, a consistent set and sequence of terms is required to effectively communicate comparable descriptions of salient features. A set that incorporates the most distinctive and important properties is as follows:

(color)
(induration)
(sedimentary structures)
(fabric term, such as sorting or matrix- vs. grain-supported)
(size term):
(cement term)
(compositional name, dependent on dominant sediment type).

Subsequent chapters in this book describe and discuss specific terminology and interpretations that can be made from these sediment characteristics.

Development of a consistent and thorough pattern of description, beginning with the above name and progressing to details and other properties, will aid the worker in the long run both in writing the final report and in avoiding the need to return to a locality to recollect information missed the first time. Data record cards (e.g., Fig. 1-8), which can be quickly designed and printed for each new project, save time and ensure both consistent and comprehensive collection of data. Cards of half-A4 size are suitable for cut-down clipboards and commonly are large enough to contain laboratory details; thus, data are not spread amongst several field and laboratory notebooks.

SELECTED GENERAL TEXTBOOKS AND REFERENCES

Ager, D. V., 1981, *The Nature of the Stratigraphical Record,* 2d ed. Macmillan, New York, 122p.

Bates, R. L., and J. A. Jackson (eds.), 1987, *Glossary of Geology,* 3d ed. American Geological Institute, Alexandria, Va., 749p.

Blatt, H., M. Middleton, and M. Murray, 1980, *Origin of Sedimentary Rocks,* 2d ed. Prentice-Hall, Englewood Cliffs, N.J., 782p.

Boggs, S., Jr., 1987, *Principles of Stratigraphy and Sedimentology.* Merrill Publishing Co., Columbus, Ohio, 784p.

Briggs, D. E. G., and P. R. Crowther, 1989, *Encyclopaedia of Palaeobiology.* Blackwell Scientific, in association with the Palaeontological Association, Oxford.

Davis, R. A., Jr., 1992, *Depositional Systems*, 2d ed. Prentice-Hall, Englewood Cliffs, N.J., 604p.

Dunbar, C. O., and J. Rogers, 1957, *Principles of Stratigraphy.* Wiley, New York, 358p.

Fairbridge, R. W. (ed.), 1972, *The Encyclopedia of Geochemistry and Environmental Sciences.* Van Nostrand Reinhold, New York, 1,321p.

Fairbridge, R. W., and J. Bourgeois (eds.), 1978, *The Encyclopedia of Sedimentology.* Dowden, Hutchinson & Ross, Stroudsburg, Pa., 828p.

Finkl, C. W. (ed.), 1984, *The Encyclopedia of Applied Geology.* Van Nostrand Reinhold, New York, 664p.

Folk, R. L., 1980, *Petrology of Sedimentary Rocks.* Hemphill, Austin, Tex., 182p.

Friedman, G. M., J. E. Sanders, and D.C. Kopaska-Merkel, 1992, *Principles of Sedimentary Deposits.* Macmillan, New York, 717p.

Fritz, W. J., and J. N. Moore, 1988, *Basics of Physical Stratigraphy and Sedimentology.* Wiley, New York, 371p.

Fuchtbauer, H., 1974, *Sediments and Sedimentary Rocks.* Halsted Press/Wiley, New York, 464p.

Garrels, R. M., and F. T. MacKenzie, 1971, *Evolution of Sedimentary Rocks.* Norton, New York, 397p.

Goddard, E. N., D. D. Trask, R. K. de Ford, O. N. Rove, J. T. Singlewald, and R. M. Overbeck, 1951, *Rock Color Chart.* Geological Society of America, New York.

Greensmith, J. T., E. H. Hatch, and R. H. Rastall, 1989, *Petrology of the Sedimentary Rocks*, 7th ed. Unwin Hyman, London, 502p.

Krumbein, W. C., and S. S. Sloss, 1963, *Stratigraphy and Sedimentation,* 2d ed. W. H. Freeman & Co., San Francisco, 660p.

Leeder, M. R., 1982, *Sedimentology.* Allen & Unwin, London, 344p.

Matthews, R. K., 1974, *Dynamic Stratigraphy.* Prentice-Hall, Englewood Cliffs, N.J., 370p.

Miall, A. D., 1990, *Principles of Sedimentary Basin Analysis,* 2d ed. Springer-Verlag, New York, 668p.

Pettijohn, F. J., 1975, *Sedimentary Rocks,* 3d ed. Harper & Row, New York, 628p.

Schwartz, M. L. (ed.), 1982, *The Encyclopedia of Beaches and Coastal Environments.* Hutchinson & Ross, Stroudsburg, Pa., 940p.

Selley, R. C., 1976, *An Introduction to Sedimentology.* Academic Press, New York, 408p.

Tucker, M. E., 1991, *Sedimentary Petrology,* 2d ed. Blackwell Scientific, Oxford, 260p.

von Engelhardt, W., 1977, *The Origin of Sediments and Sedimentary Rocks.* Halsted Press/Wiley, New York, 359p.

2
Environments of Sedimentation

The concept of sedimentary environment encompasses the entire complex of physical, chemical, and biological materials and processes involved from the time sediments are formed to the time they are examined. Thus, in addition to depositional setting, the tectonic setting, climate, provenance, postdepositional (diagenetic) conditions, and late-stage (postlithification) weathering effects are important and influential environmental factors. Furthermore, an understanding of the geochemistry of natural waters is important when interpreting many sedimentary deposits (e.g., Drever 1982). In this chapter, we focus on the more restricted depositional setting (e.g., Figs. 2-1, 2-2) in which sediments actually accumulate, and we provide a simple and generalized overview of the tectonic and climatic settings. Expanded discussion of processes acting in, and the characteristics of the sediments deposited in, these environments comprise the bulk of the remainder of this book.

TECTONICS, EUSTACY, CLIMATE: MEGACONTROLS OF ENVIRONMENT

The association of relative topographic/bathymetric highs and lows differentiates source areas, which shed detrital sediments, and basins, where detrital and nondetrital sediments accumulate. Sedimentary basins are not necessarily substantial troughs or depressions; they are merely areas in which

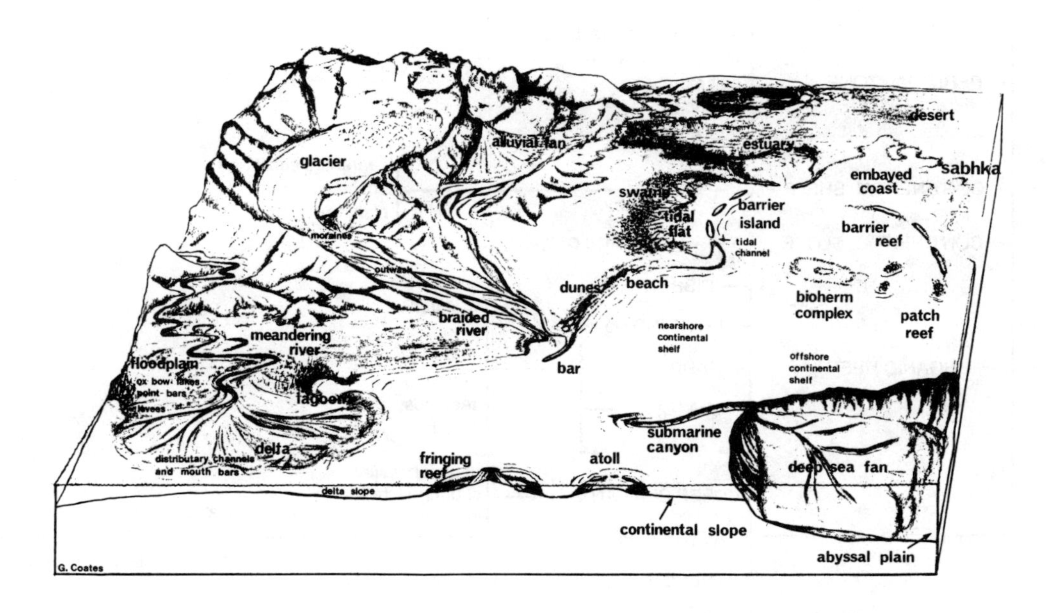

Figure 2-1. Schematic representation of major sedimentary environments. Relative positions of environments are realistic, but the overall association is for purposes of representation only. (Sketch by G. Coates.)

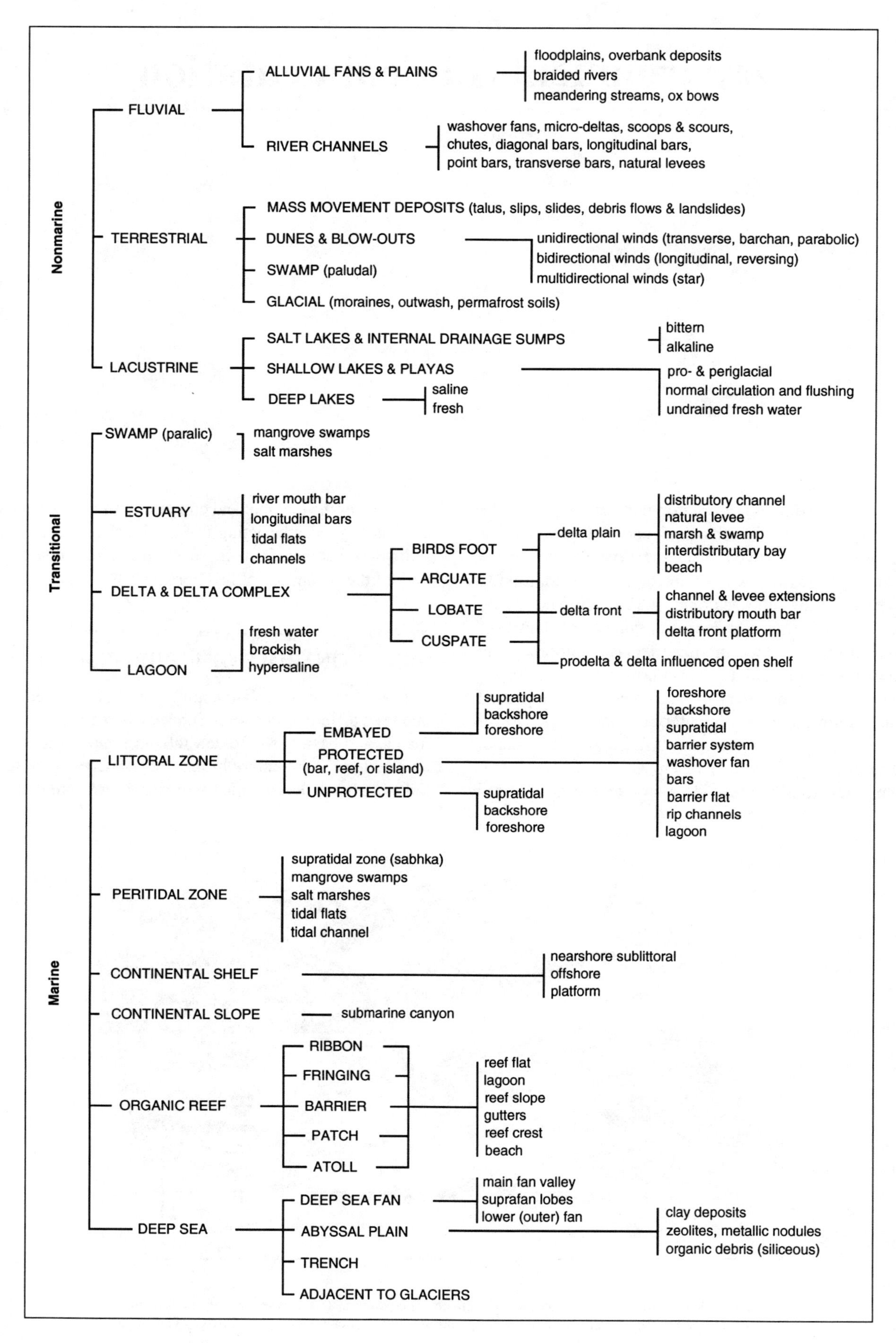

Figure 2-2. A variety of depositional environments.

Table 2-1. General Relationships between Plate Tectonics and Sedimentary Environments and Sequences

Basin Setting	Tectonism	Sedimentation
	Prerift arching	Thin alluvial fans on flanks of arch
	Rift valley grabens, and fault-angle depressions	Thick alluvial/deltaic wedges, lacustrine deposits (tuffs, evaporites, diatomites). Coarse marginal deposits interfinger with central finer sediments. Some volcaniclastics. Climatic factor important. Oceanic influence grows as rift opens. May have euxinic or evaporitic starved marine basins. However, generally sediments from continent provide detritus for stable to unstable shelf (miogeoclinal) sequences, tending coarse to fine both laterally and vertically, with carbonates often dominant distally or when relief is low. Eugeoclines of continent-derived detritus plus some volcanics supplied by seafloor spreading; submarine fans.
	Proto-Ocean	
	Narrow Ocean	
Rifted Continental Margins	Open ocean; basin separates into separate miogeoclines and eugeoclines	
	Failed rift valley	Commonly thick section with early volcanic phase (often rhyolitic), later arenites and/or carbonates. Finally filled by detritus from continent.
	Trench basin, in subduction zone	Melange on inner margin, basin fill of mixed pelagic/turbidite sequences scraped off the consumed seafloor. Submarine fans. Parts of ophiolites.
	Fore-arc basin, between trench and magmatic arc	Thick fluvial to deltaic to neritic to deep marine sequences (eugeoclinal), variable due to rate of compression and variable history
Convergent Margin Settings	Intra-arc basin, between continent and arc or two island arcs	Volcaniclastic sequences (eugeoclinal), with continental detritus in one case
	Back-arc basins Inter-arc, on oceanic crust behind arc	Volaniclastic (eugeoclinal) turbidite-rich sequences, minor pelagic sediments.
	Retro-arc, on continental crust behind arc	Alluvial-deltaic-shallow marine sequences to considerable thicknesses, derived from fold/thrust orogen.
	Suture belt, between crustal blocks; peripheral basins possible at junction	Complexly deformed sequences similar to those in retro-arc basins.
Intracontinental	Many may reflect crustal attenuation during early stages of rifting that subsequently failed. Others not easily related to plate tectonic theory. Commonly surrounded by stable continent. Alluvial to marine sequences; arenites-mudstones-carbonates common	
Oceanic	Rise crest, at midoceanic rift Rise flank Deep basin	Volcanic turbidites and pelagic carbonates; siliceous oozes

Source: After Dickinson (1974).

sediments can accumulate (depocenters), generally as a result of subsidence over a long period of time. The four-dimensional geometry of basins fundamentally dictates the size and shape of depositional environments. The main controls on the existence and geometry of sedimentary basins are *tectonism,* meaning differential earth movements, and *eustacy,* meaning sea-level changes. Broad-scale tectonic uplift or subsidence (*epeirogenic movements*) cannot effectively be distinguished from eustatic fall or rise. Concepts of tectonic processes, cycles, rates, and products have undergone major changes since the 1940s (e.g., Krynine 1942; Dickinson 1974; Dott 1978; Ingersoll 1988; Miall 1990). Earth movements on the broad scale are generally related to movements of giant plates of the earth's surface; these create the basins and dictate which areas will act as sources of detrital sediments (see Table 2-1, Fig. 2-3); Fig. 2-4 graphically depicts relationships of sedimentation and tectonics (some terms are defined in later chapters). Tectonism has a strong influence on the composition of detrital sediments produced in the various source areas, and compositional studies are the main indicators of the tectonic setting for deposition of ancient sediments (see Chapter 6). Eustatic changes also have received much attention since the 1980s, and large-scale patterns of sedimentation have been related to the consequential movements of shorelines and depositional environments (e.g., Vail, Mitchum, and Thompson 1977; Pittman 1979).

Whereas tectonics set the pattern of source areas and sites of deposition, a major influence on the environments of sedimentation and the nature of sediments accumulating in the

Rifted Continental Margin

(a) Early phase

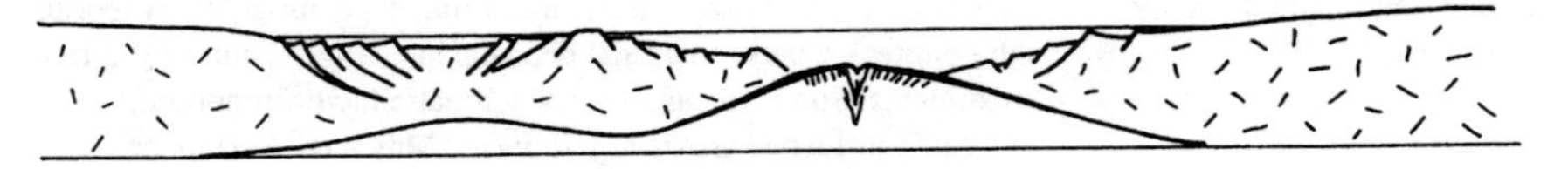

(b) After separation

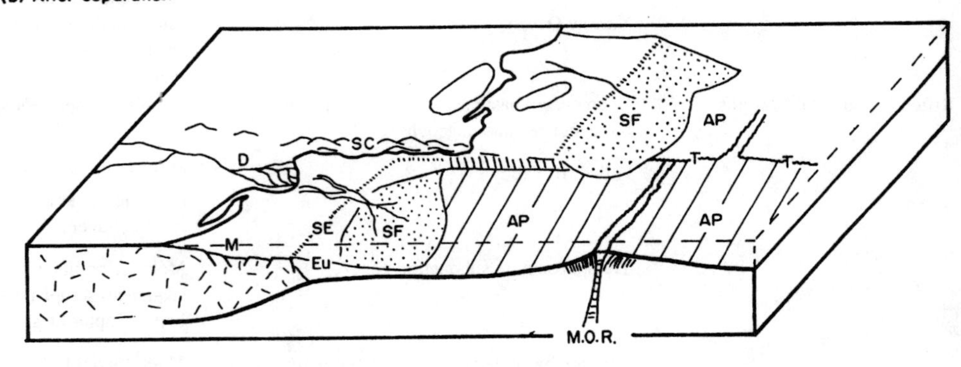

CONVERGENT MARGIN SETTING

(c)

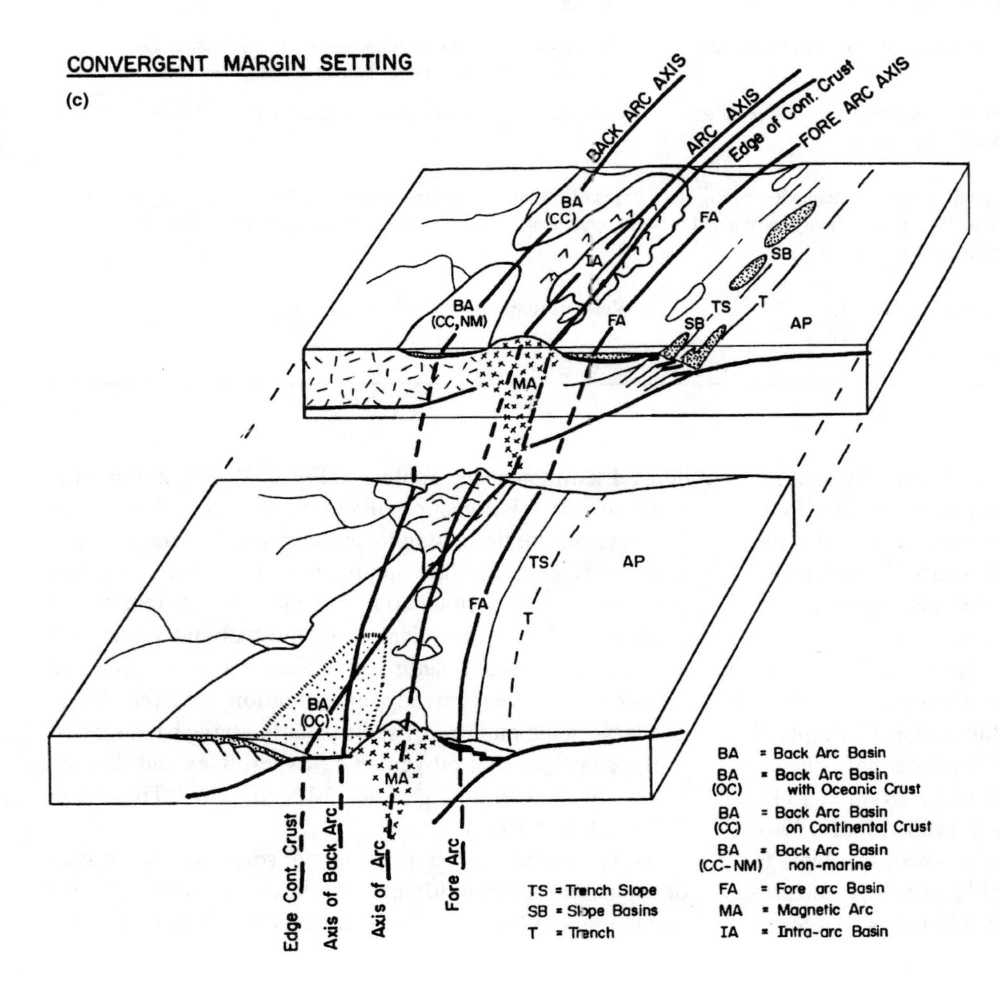

Figure 2-3. Schematic representation of major tectonic settings in relation to common sedimentary environments: a: Early phase of rifted continental margin; b: Late phase of rifted continental margin (this side of the continent is then termed the *passive margin* or *trailing edge*); c: Convergent margin setting (basins tend to be short-term and become highly deformed or evert as collision continues). (Sketches by J. Bradshaw.)

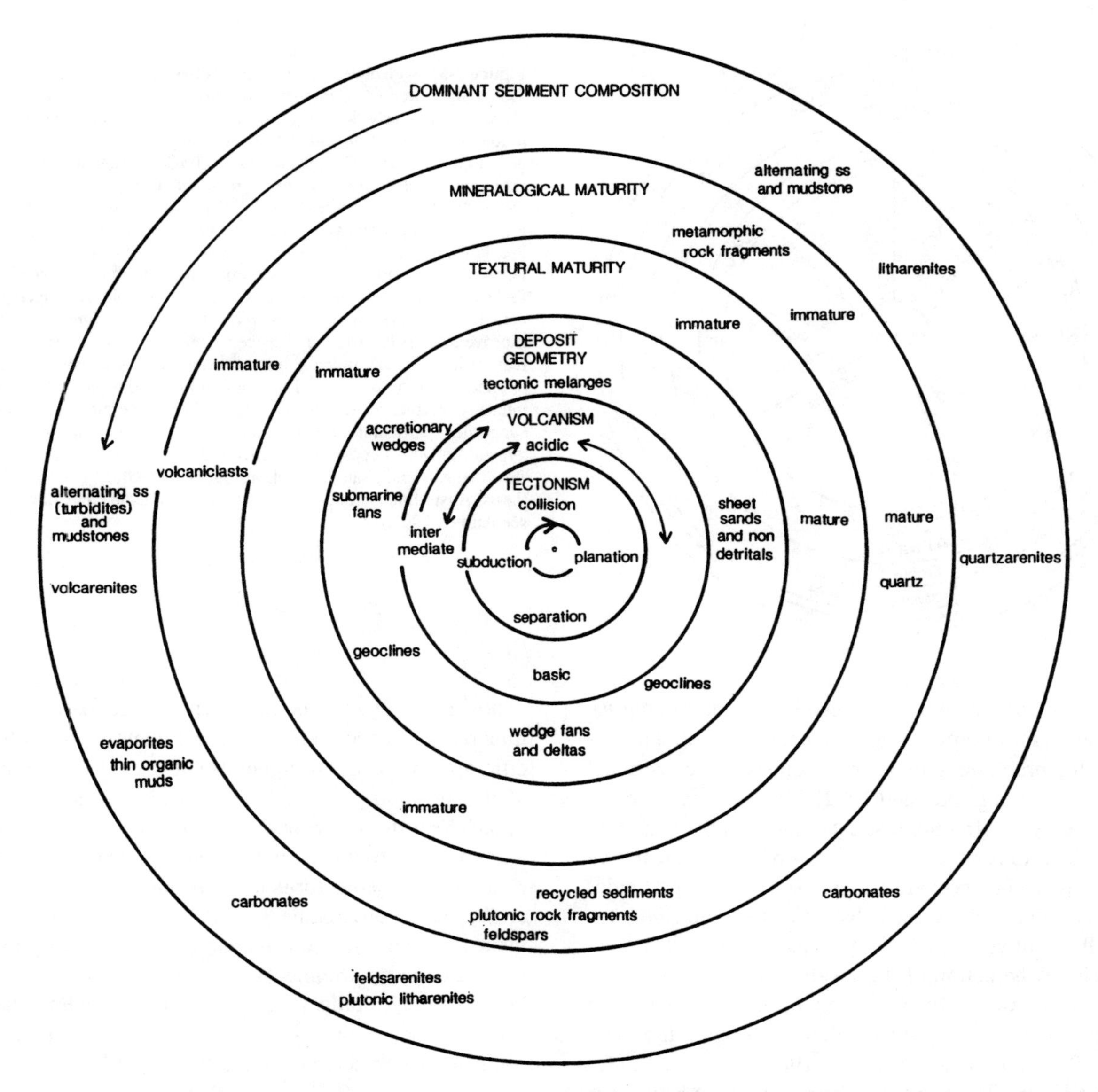

Figure 2-4. Schematic representation of the relationship between tectonics, environments, and sediment characteristics (Lewis Mark III version, 1991). Many exceptions exist, and a circular or cyclic development is not implied.

basins is climate (e.g., Fig. 2-5 and see Hollingsworth 1962). Climate determines the pattern of precipitation and thus controls the distribution of the dominant agents for both physical and chemical weathering, transportation, and deposition (see Chapter 3); its role with respect to weathering is enhanced through its influence on temperature and thus the rates of chemical and biochemical reactions. Climate-controlled weathering processes dictate the extent of alteration of detrital minerals, and the climate factor is important in determining what minerals may be precipitated by organic and inorganic processes (see Chapters 3 and 9). Climate also has an important influence on major oceanic current patterns and temperatures (major ocean currents result from warming and upwelling in the tropics, cooling and sinking in the cold polar regions).

Present-day climates are representative of the range of ancient climates, but the extent of the different climatic zones is similar to only the few and relatively short-term glaciation periods of the past; for most of geologic time there appear to have been no polar ice caps and warm paleoclimates were much more widespread (e.g., Bain 1963; Frakes 1979). Unfortunately, there are major problems in attempting to deduce paleoclimate from the sediment record (e.g., Beaty 1971; Barron 1983; Hecht 1985); biotic and mineralogical differences between laterally adjacent and vertically superimposed sequences may provide clues (e.g., Jacobs and Hays 1972; Singer 1980). The effects of climate insofar as it influences the composition of detrital sediments are further discussed in Chapter 6.

PROCESSES AND ENVIRONMENT

Many sedimentologists work exclusively with modern environments where settings and processes can be observed;

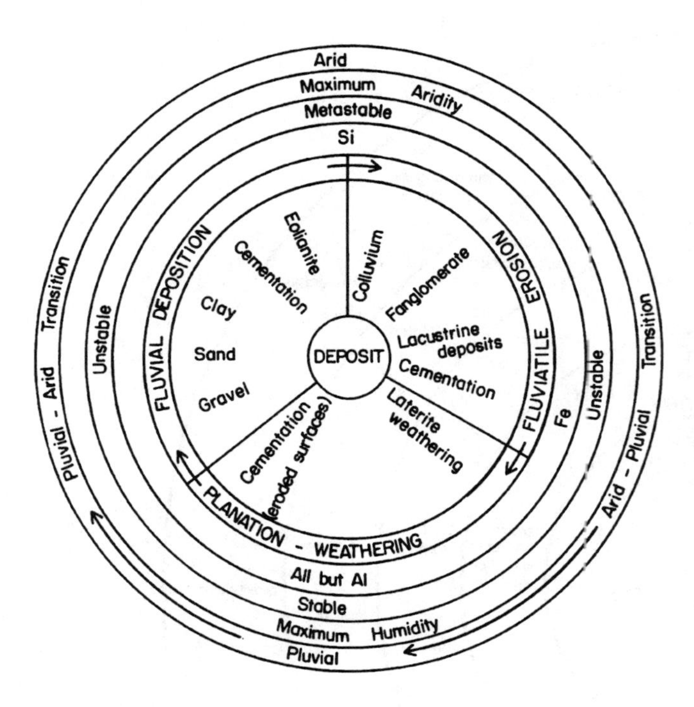

Figure 2-5. Idealized relationships between climate and continental sedimentary products. The major indicators of climate in sediments are: *Fossils,* particularly flora (e.g., palynology) and stenothermal biota. Main problems: redeposition and uncertainties in paleoecological relationships prior to the Tertiary. *Oxygen isotopes,* (O^{18}/O^{16}) ratios in carbonates. Problems: some organisms do not precipitate $CaCO_3$in equilibrium with water; diagenesis may alter the ratio. *Trace elements,* such as Sr/Ca and Mg/Ca ratios in carbonates, which can be temperature dependent. Problems: original shell mineralogy influences the ratios (e.g., aragonite inherently holds more Sr than calcite); diagenesis may change the ratios. *Sedimentary iron compounds* (see Chapter 9). Problems: diagenetic and epigenetic transformations. *Type and extent of detrital mineral alteration*: if provenance is known, the absence of minerals in the derived sediment may be indicative. Problems: imprecise and difficult to apply; a long time of exposure to mild chemical weathering may mimic short-time high-intensity weathering effects; diagenesis alters minerals also. *Miscellany*: evaporites (particularly the occurrence of those salts more soluble than gypsum); glacial deposits (problems of distinction from sediment gravity flow deposits and plant-rafted dropstones). (After M. J. Carr, B.Sc. thesis, Australian National University; derived from Chavaillon 1964—see Alimen 1965.)

these data and derivative process/response models comprise the foundation for interpretations of ancient depositional systems and for predictions of future changes in depositional systems (Fig. 1-3; e.g., Kennett 1982). However, the study of some processes requires sophisticated measuring equipment and/or long periods of observation to obtain representative data (e.g., the return period of major storms in a particular area may be many decades; substantial tectonic uplift requires millions of years); the study of ancient deposits is necessary to infer the action of these processes, and can be an effective means of discovering their effects and obtaining information that can be used predictively (e.g., Rigby and Hamblin 1972; Scholle and Spearing 1982; Legatt and Zuffa 1987). The accumulated and synthesized data can be used as models from which practical sedimentological investigations commonly can achieve short-term, cost-effective results for both modern and ancient settings; the models also can be very effective in economic exploration and development (e.g., Galloway and Hobday 1983).

The characteristics of sediments deposited in the various environments depend on the processes that have acted within each environment (Chapter 3). Sedimentary structures (Chapter 4) are the best indicators of processes. Textures (Chapter 5) also reflect the influence of depositional processes on sediment size distribution and fabric, but some shape and size characteristics are a function of the prior history of the particles (e.g., from aeolian and stream action before arrival in a lake, or as a consequence of the grain size distribution in the parent rock). Textural characteristics of a sediment may also be altered by postdepositional physical and chemical processes (e.g., the development of secondary grain overgrowths, chemical rounding, or development of preferred orientation of phyllosilicates during compaction). The composition of detrital sediments indicates nothing about the depositional environment, except where particular heavy minerals are concentrated because of their relatively high specific gravity, or where minerals show evidence of alteration after deposition (e.g., oxidation), in which case some aspects of the chemistry of the environment and/or residence time in a particular environment may be deduced. In contrast, the composition of minerals formed by chemical or biochemical processes in the environment (shells, authigenic minerals like glauconite, pore-filling cements) provides direct information about various environmental conditions (see Chapters 7, 8, 9). Unfortunately, similar processes act in most sedimentary environments: hence direct inference of ancient depositional environment from process interpretation is not feasible.

ORGANISMS AND ENVIRONMENT

Living organisms or fossils (particularly if they are in life position) and traces of organisms can be particularly useful in indicating both chemical and physical characteristics of the environment if we have sufficient information about their functional morphology, ecology, and paleoecology (e.g., Table 2-2, Fig. 2-6; see also Hedgepeth 1957; Drake 1968; Gould 1970; Schafer 1972; Dodd and Stanton 1981; and a multitude of other publications in the biological and paleontological literature). The distribution of organisms is influenced by temperature, light intensity, hydraulic energy, turbidity, substrate conditions, nutrient distribution, and interdependence; interpretations based on their presence must be related to these fundamental controls (e.g., Craig 1966). Unfortunately, most concentrations of hardparts reflect some degree of transport (in some cases it can be very extensive transport, e.g., Otvos 1976), and interpretations must consider hydrodynamic (and diagenetic) processes after the death of the organisms as well as the ecology of the living

Table 2-2. Some Common Ecological Terminology

Neritic: dwellers in relatively shallow waters (down to approx. 200 m or the edge of the continental crust)

Oceanic: dwellers of waters deeper than the neritic zone (living beyond the continental mass). (Sometimes used for forms living in upper or middle depths vs. bathypelagic forms of deep oceanic waters.)

Pelagic — living above the bottom
- nektic—swimmers (e.g., fish, squid, krill)
- planktic — floaters
  - phytoplankton: autotrophs (make their own food) (e.g., some hydrozoans, diatoms, dinoflagellates)
  - meroplankton: planktic only as juveniles (e.g., bivalve larvae)
  - zooplankton: heterotrophs (animals) (e.g., foraminifera, radiolaria)

Benthic — living on the bottom
- epifauna/flora: live on a substrate (epilithic if substrate is hard)
- infauna/flora: live within a substrate (endolithic if substrate is hard)

Vagile: organisms that can move (e.g., annelids, crustaceans, gastropods, some bivalves)
Sessile: organisms that cannot move (e.g., barnacles, bryozoa, oysters, crinoids, brachiopods)
Epibionts: attached to other organisms (e.g., some foraminifera or diatoms, or sponges)

Stenohaline: tolerant of a restricted range of salinity
Euryhaline: tolerant of a wide range of salinity
Stenothermal: tolerant of a restricted temperature range
Eurythermal: tolerant of a range of temperature
Photophilic: prefer to live in light
Sciatophilic: prefer to live in darkness

Assemblage: group of organisms or remains from a particular horizon.
Association: group of assemblages, with similar patterns of species.
Taphonomy: study of the history of organic remains, including subfield of biostratinomy—preservation history (pre- and postburial).

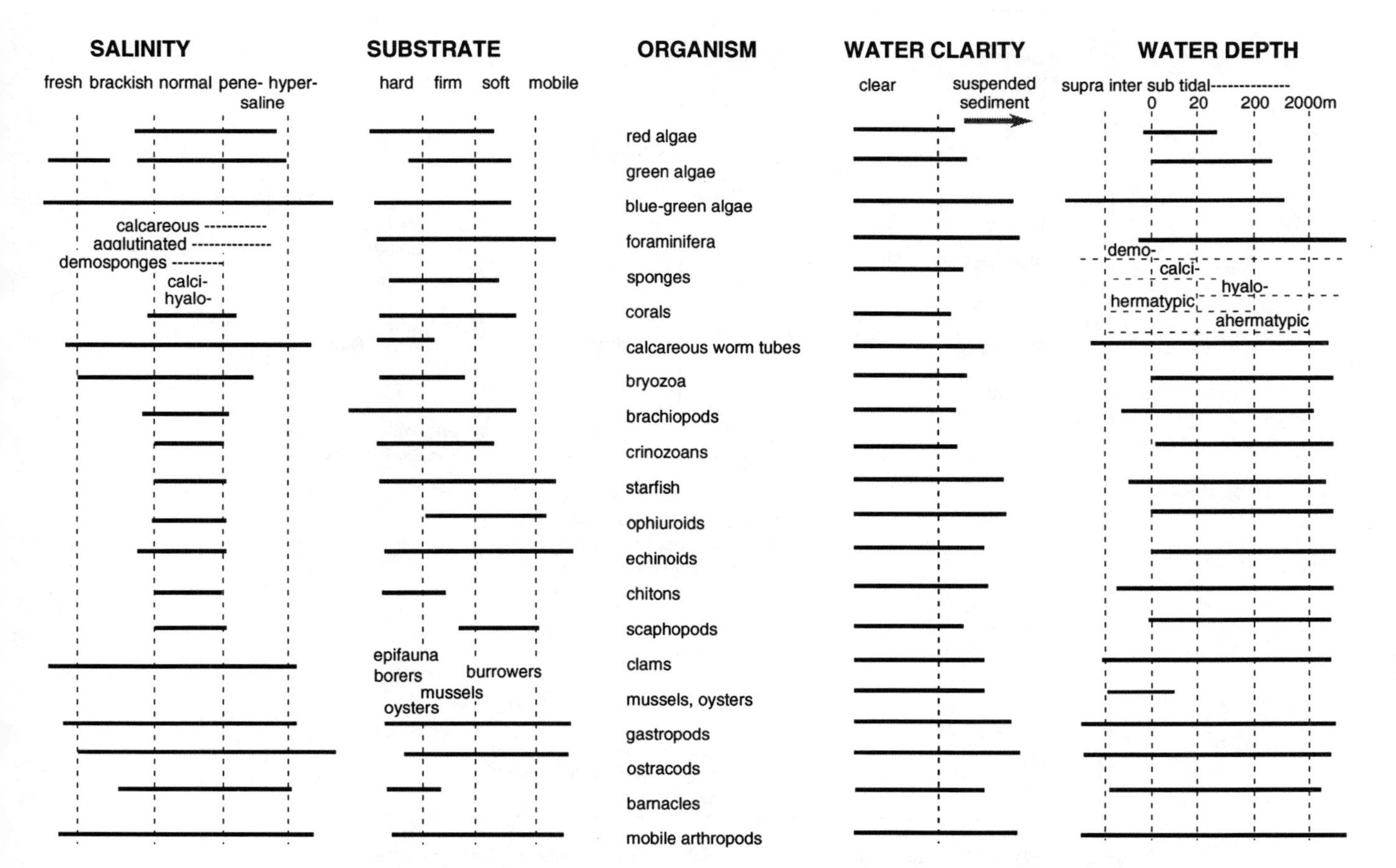

Figure 2-6. Paleoecological relationships between some groups of organisms and the environments in which they live. (After Heckel 1972, with advice from A. A. Ekdale and some modifications based on personal observation.)

Table 2-3. Descriptive Framework for Classification of Organic Hardpart Concentrations

Taxonomic Group		Packing	Geometry	Internal Structure	
				Simple	Complex
Monotypic		Matrix-supported	Stringer	X	
			Pavement	X	(X)
			Pod	X	
			Clump	X	
Polytypic		Shell-supported	Lens	X	(X)
			Wedge	(X)	X
			Bed	X	X
Indicative of:	Ecology (hydrodynamics)	Hydrodynamics (community and diagenesis)	Bathymetric relief and hydrodynamics (community)	Hydrodynamics (Community and ecological history)	

Source: Slightly modified from Kidwell, Fursich, and Aigner (1986).

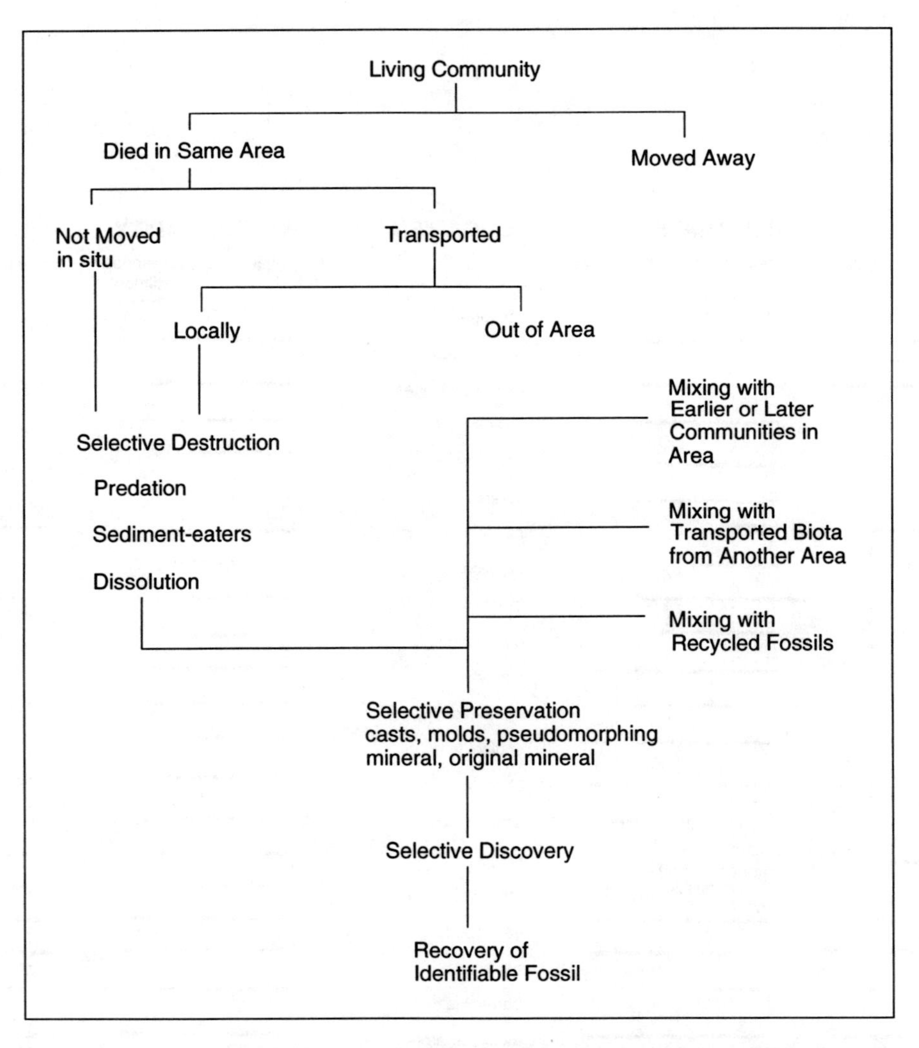

Figure 2-7. Possibilities that influence the occurrence of a fossil or other organic remains in sedimentary deposits. When trying to reconstruct communities, additional complications arise because quick-breeding organisms produce much more debris than slow-breeding ones, and there is likely to be selective destruction of members of the community (e.g., more delicate shell structure, shells with Mg-calcite or aragonite). (Adapted from Ager 1963.)

community (e.g., Table 2-3). With fossils, additional considerations as illustrated in Fig. 2-7 must be taken into account when interpreting their occurrence (see also Fagerstrom 1964; Lawrence 1968; Boucot 1981).

To complicate interpretations even further, some of the most effective biotic elements in environments (e.g., baffles to currents and agents for trapping and binding sediments) are plants such as algae, marine grasses, and mangroves, which have no preservable hardparts (e.g., Scoffin 1970; Brasier 1975; Gerdes, Krumbein, and Reineck 1985; Fonseca 1989). Changes in terrestrial vegetation through time have also had an important impact on nonmarine sedimentary environments (e.g., Schumm 1968). Despite the poor preservation potential of microbes such as cyanobacteria, their physical influence may result in the formation of stromatolites (thinly layered sediment masses of various morphologies), which are useful paleoenvironmental indicators and provide the earliest convincing evidence for life on Earth (e.g., Walter, Buick, and Dunlop 1980; Awramik 1986). Further discussion of biological processes and their influence in sedimentation is given in Chapter 3.

Not only is the observable relationship of biotic elements to the sediments useful in environmental interpretation, but in ancient sequences it is most commonly a knowledge of the age of fossils (particularly planktonic microfossils and pollen or spores) that provides a means for correlating between units and establishing a time-stratigraphic framework. Both during life and after death, organisms and organic matter in sediments can play an important role in water chemistry and metal-sediment-water interactions, during both deposition and diagenesis (e.g., Redfield 1958; Jackson, Jonasson, and Skippen 1978).

ENVIRONMENTAL MODELS

Any sedimentary deposit that can be characterized by a distinctive combination of features can be called a *facies* (e.g., Teichert 1958; Markevich 1960; Hallam 1981); facies analysis using the *Principle of Uniformitarianism* (or Actualism—the concept that physical, chemical, and biological laws acted in the past as they do today; e.g., Nairn 1965) comprises the fundamental philosophical basis for the reconstruction of sedimentary environments and the history of deposits. The Selected Bibliography at the end of this chapter provides an introduction to the literature on approaches to these kinds of studies.

When working with old deposits, in settings where environmental factors have changed and when dealing with sedimentary rocks generally, interpretation of processes alone cannot define the depositional environment because the same processes act in many environments. Interpretation requires synthesis of the three-dimensional geometry and lateral and vertical relationships of the various sedimentary units present with all available data indicative of process. Vertical stratigraphic sequences are among the most powerful aids to paleoenvironmental interpretation when there are no substantial erosional or nondepositional intervals (unconformities) in the succession. In such successions, the superposed lithologies represent originally laterally contiguous facies (an expression of Walther's Law, see Middleton 1973). For example, in a marine setting with progradation (outbuilding) or regression (relative sea-level fall), boundaries between different subenvironments tend to move seaward, and if there is net accumulation, the result is a vertical stacking of units of progressively nearer-shore deposits (e.g., Fig. 2-8). The re-

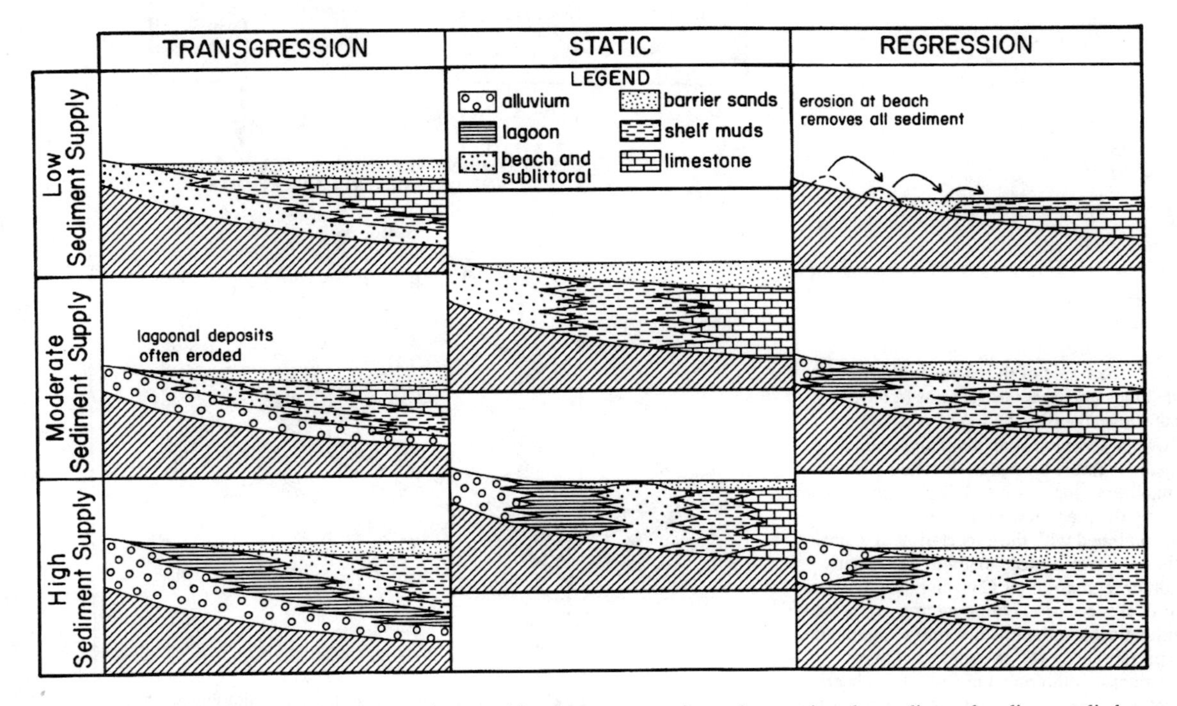

Figure 2-8. Idealized examples of successions resulting from transgression and regression along a linear shoreline supplied with detrital sediments. (After Selley 1970.)

verse occurs with transgression (relative sea-level rise) or retrogradation (which implies erosion as the shoreface retreats, thus much less likely preservation of depositional sequences). Similar progressive overlapping of subenvironments occurs without relation to shorelines, for example: (1) meandering river channels migrate across valley floors, and with net aggradation multistory sandy or gravelly deposits come to overlie finer levee and overbank deposits; (2) lobes of submarine fans grow radially outward from submarine canyons over the seafloor.

Sedimentary models known as *facies models* have been devised for many environments; some concentrate on processes operating in environments, others on the deposits at a single time, yet others on vertical successions that reflect deposition over time (e.g., Allen 1964; Spearing 1974; Reading 1987; Dott 1988; Walker 1990). Specific models are constructed either from observation in modern environments of processes and lateral relationships, or to explain stratigraphic sequences in the rock record (e.g., Figs. 2-9a and 2-9b). An example of the means of construction of a model from a vertical succession of facies is shown in Fig. 2-10a. Ideal models illustrate *expected* associations and relationships and are used as bases with which to compare the *actual* associations and relationships found in real sequences; real associations may differ substantially from the models. Models exist at various scales—from those accounting for sequences of features less than a meter thick to those that attempt to model sequences hundreds of meters thick. Models are not designed to be forced onto the data available, but to guide the worker during interpretation, including consideration of the causes of deviations from the ideal. Even widely used models, such as those for common graded-bed sequences, can be legitimately subject to different interpretations (e.g., Fig. 2-10b). In addition, it must always be borne in mind when dealing

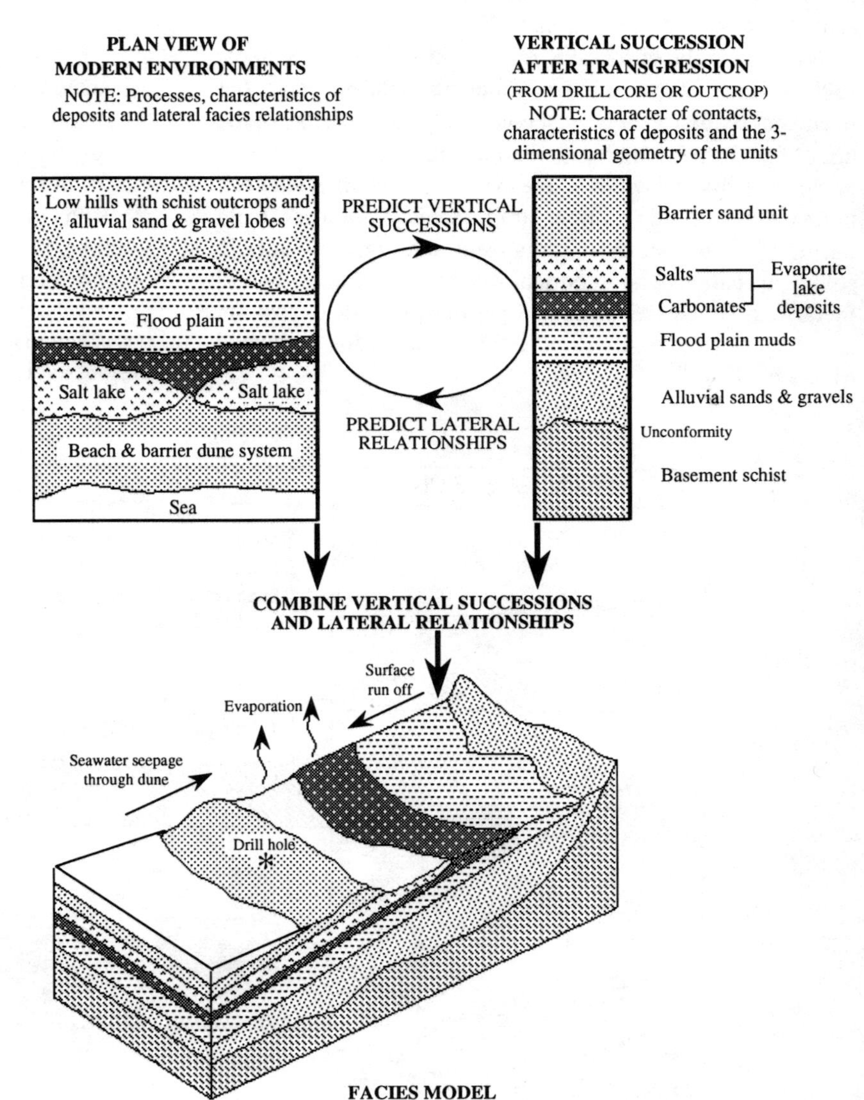

Figure 2-9a. Example of the routes to the construction of four-dimensional reconstructions (a block diagram that shows the distribution of rock units developed over a period of time; the magnitude of the time dimension is rarely known or depicted because of a general inability to locate time planes). Either route is valid: (1) using the plan view of an environmental association combined with the expectation of a vertical superposition of environments that will accompany transgression or regression; or (2) using the identification of a vertical succession (from outcrop or borehole studies) without substantial unconformity combined with the expectation that the environments that produced distinctive lithologic units will have been laterally contiguous at any time during their deposition.

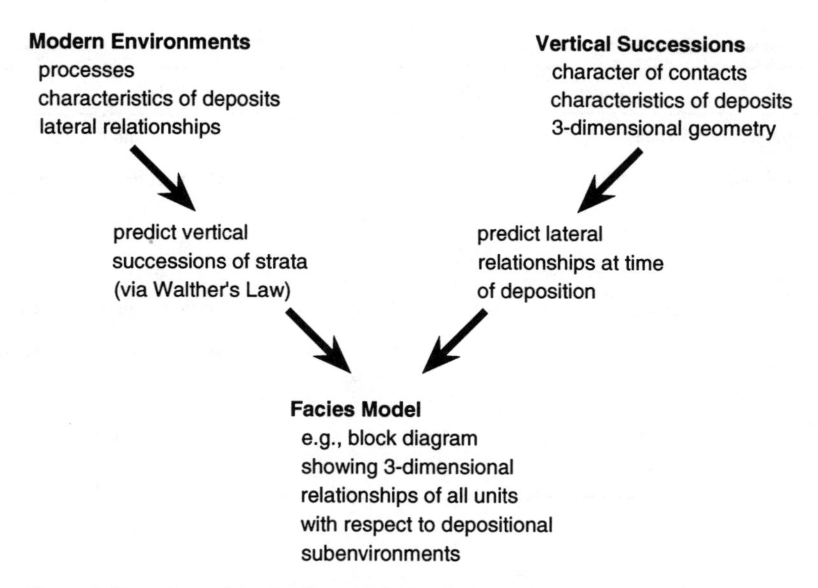

Figure 2-9b. Potential routes for establishing facies models.

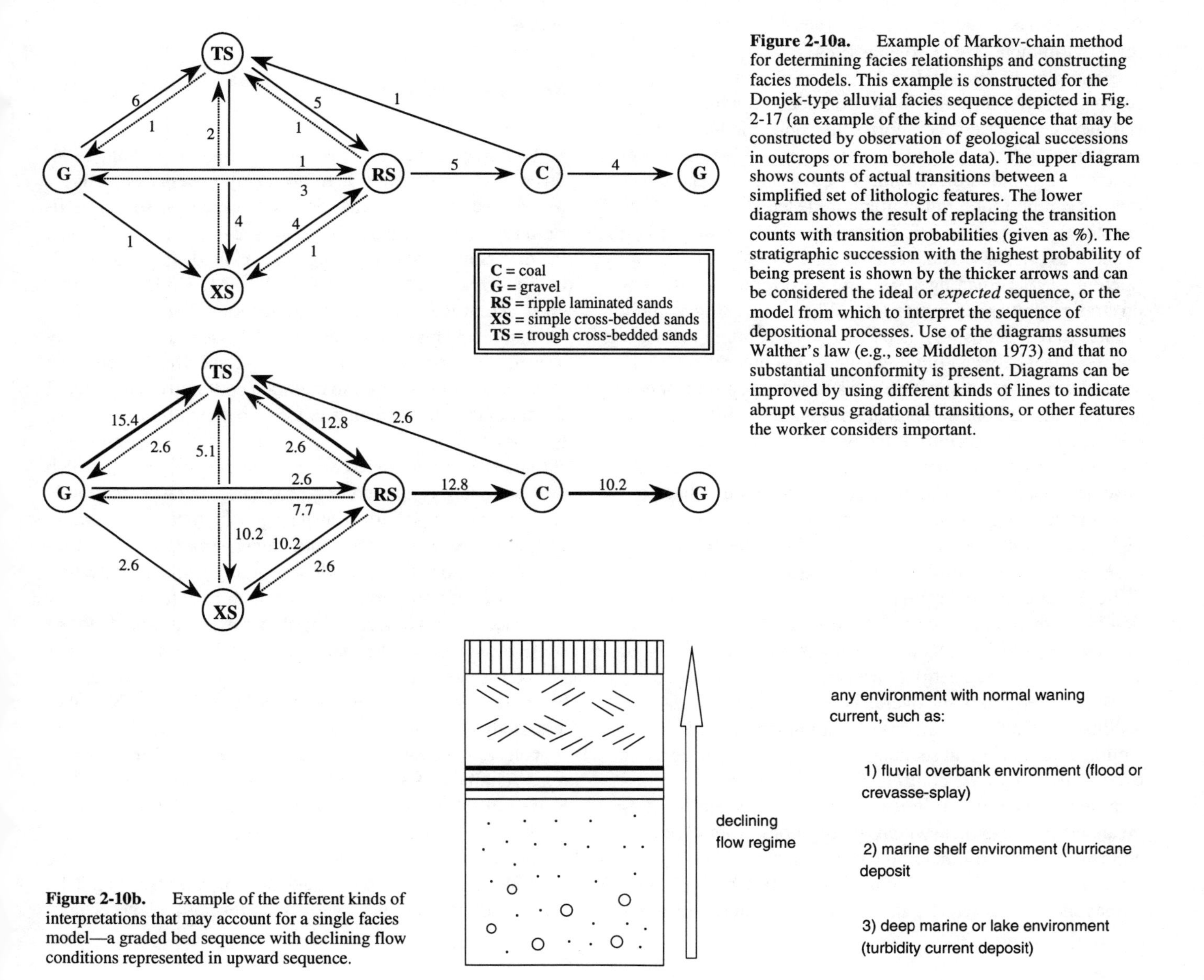

Figure 2-10a. Example of Markov-chain method for determining facies relationships and constructing facies models. This example is constructed for the Donjek-type alluvial facies sequence depicted in Fig. 2-17 (an example of the kind of sequence that may be constructed by observation of geological successions in outcrops or from borehole data). The upper diagram shows counts of actual transitions between a simplified set of lithologic features. The lower diagram shows the result of replacing the transition counts with transition probabilities (given as %). The stratigraphic succession with the highest probability of being present is shown by the thicker arrows and can be considered the ideal or *expected* sequence, or the model from which to interpret the sequence of depositional processes. Use of the diagrams assumes Walther's law (e.g., see Middleton 1973) and that no substantial unconformity is present. Diagrams can be improved by using different kinds of lines to indicate abrupt versus gradational transitions, or other features the worker considers important.

Figure 2-10b. Example of the different kinds of interpretations that may account for a single facies model—a graded bed sequence with declining flow conditions represented in upward sequence.

Table 2-4. An Example of a Depositional Hierarchy

Feature	Scale	Controls
Basin setting e.g., constructional passive continental margin	10^7 yr	Plate tectonics, eustacy
Depositional system e.g., river-dominated birdsfoot delta	10^6 yr	Climate, sediment supply, basin morphology, and others
		Allocyclic
		Autocyclic
Floodplain	10^4–10^5	Sediment character, relief, river discharge and variability
Channel belt	10–1000 yr	Gradient, water and sediment discharge and grain size
Point bar	Month–year	Meandering of talweg
Sand wave	Hour–week	Upper flow-regime flood
Ripple	Minute	Normal background processes of lower flow regime

Source: After Miall (1990).

with the stratigraphic record that only part of the sediment that was once in the area is preserved (much has been reworked and transported away), and preservation may be of the exceptional event (e.g., storm) rather than the norm for that environment (e.g., Crowley 1984).

Examples of physical and stratigraphic models representing specific large-scale environments in which detrital sediments accumulate comprise the bulk of this chapter; nondetrital deposits and environments are reviewed in later chapters. Expanded discussions of environments and facies models can be found in the literature listed at the end of the chapter (good starting points are Wolf 1973; Curtis 1978; Walker 1979, 1984; Reineck and Singh 1980; Scholle and Spearing 1982; Reading 1986; and most modern textbooks on stratigraphy or sedimentation). When attempting to apply models to real deposits, it is particularly important to ensure that a contact between lithologic units does not reflect a substantial time of nondeposition or a substantial episode of erosion (i.e., the lithologic units are not unconformities); if the contacts are unconformities, the record of the previous presence of one or more deposits (i.e., subenvironments) may have been removed and sediments from totally unrelated environments may be superposed across the contact (e.g., Miall 1990, chapter 4). Distinction of unconformities from short-term, normal-to-the-environment breaks (diastems) can be difficult (particularly in fluvial deposits) because sedimentation is nowhere continuous and erosional episodes occur in virtually all settings (e.g., as channels migrate laterally or storms increase the energy of transporting processes).

When interpreting the overall sequence (an entire stack of sediments, which may incorporate a number of superposed similar and/or different sequences), an attempt must be made to separate *autocyclic* controls, resulting from natural redistribution of energy within the depositional system, such as meandering or channel switching (avulsion), from *allocyclic* controls, generated outside the depositional system by tectonic, climatic, or eustatic changes (e.g., Beerbower 1964). Allocyclic effects will be more widespread and will influence greater thicknesses of sediment (e.g., see Miall 1990, chapter 8). Such considerations are particularly germane when explanations are sought for the different levels present in depositional hierarchies (e.g., Table 2-4).

Nonmarine Deposits

Soils

Soil environments are varied and complex (e.g., Foth 1984; Stace et al. 1972), and there are many genetic, objectively descriptive, and integrated descriptive-genetic soil classification schemes or models (e.g., Figs. 2-11, 2-12; Table 2-5). In the course of preparing the United Nations FAO/UNESCO "Soil Map of the World," FAO/UNESCO developed a comprehensive soil classification scheme, but to date most soil scientists continue to use the schemes previously used in their particular country (e.g., see Buol, Hole, and McCracken 1980), partly to maintain continuity and partly in the belief that the scheme is tailored to their country's needs. In a practical sense, the way in which soils are described, classified, and interpreted depends on the purpose for which they are being examined (e.g., for agricultural or economic use, engineering characteristics, geochemical exploration, conservation, or environmental management), and consequently, caution is needed when applying published interpretations of soils to purposes other than that for which they were originally intended. In applying existent models to the stratigraphic record, difficulties arise in that commonly the A horizon is not preserved, and in pre-Cretaceous time, soil biota were widely different (e.g., Schumm 1968). Nonetheless, study of ancient soils (*paleosols*) can provide information on paleoclimate as well as other environmental information (e.g., Retallack 1983; Wright 1986; Brown and Kraus 1987; Reinhardt and Sigleo 1988; Bronzer and Catt 1989).

For the sedimentologist, soils represent the complex products of diagenetic processes involving the atmosphere, the lithosphere, the hydrosphere, and the biosphere (see Fig. 3-1). Their study thus provides valuable information on such sub-

DRY HOT

WET COLD

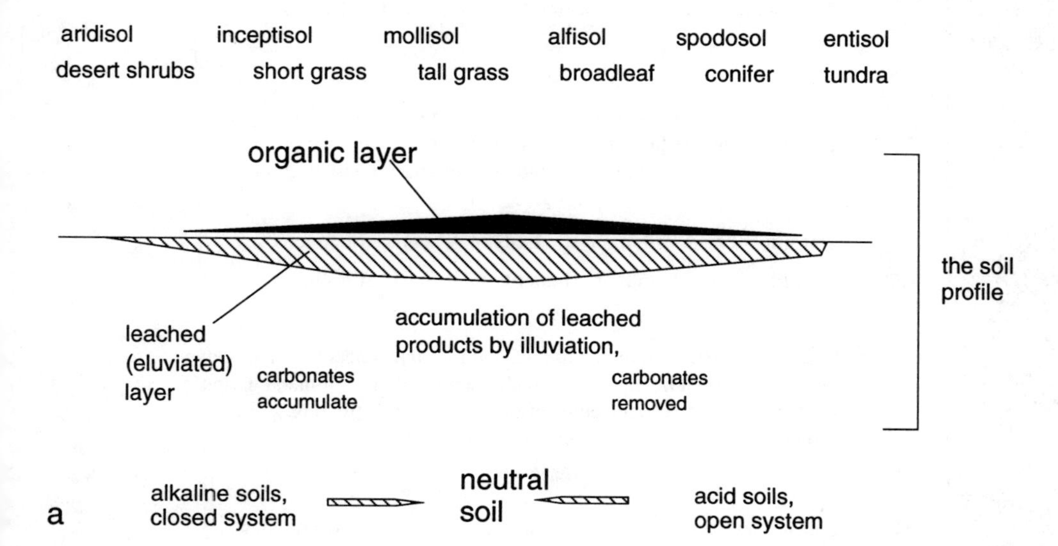

DRY COLD

WET COLD

entisol

spodosol

inceptisol

mollisol

alfisol

aridisol

ultisol

oxisol

b

DRY HOT

WET HOT

Figure 2-11. Climatic settings for common soil types: Fig. 2-11a and Fig. 2-11b show relative distributions of soil types defined in Table 2-5 (modified from Hunt 1972); Fig. 2-11c shows distribution of soils using the older standard terminology.

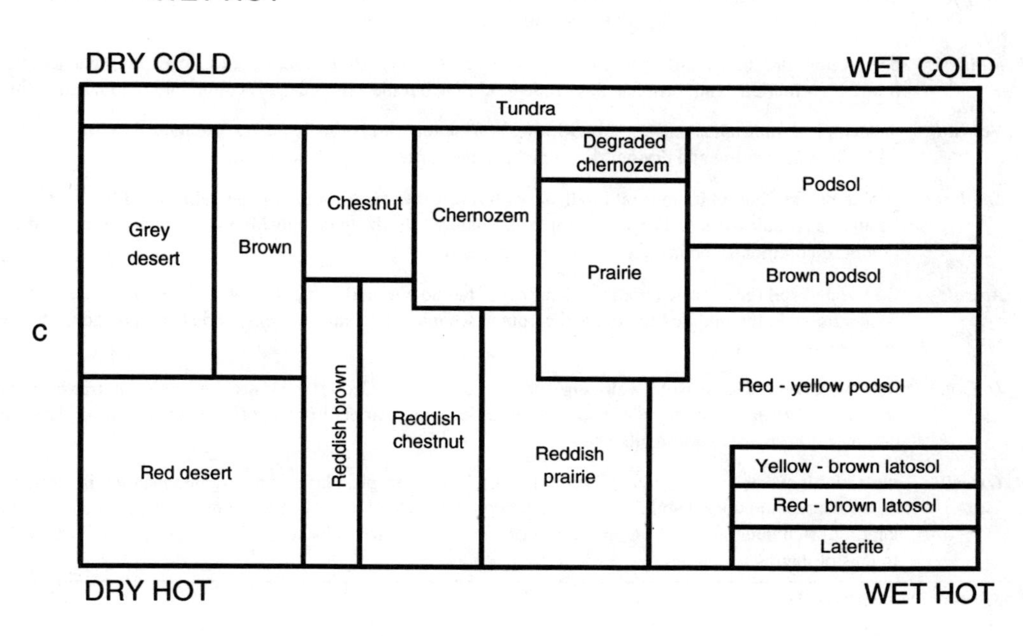

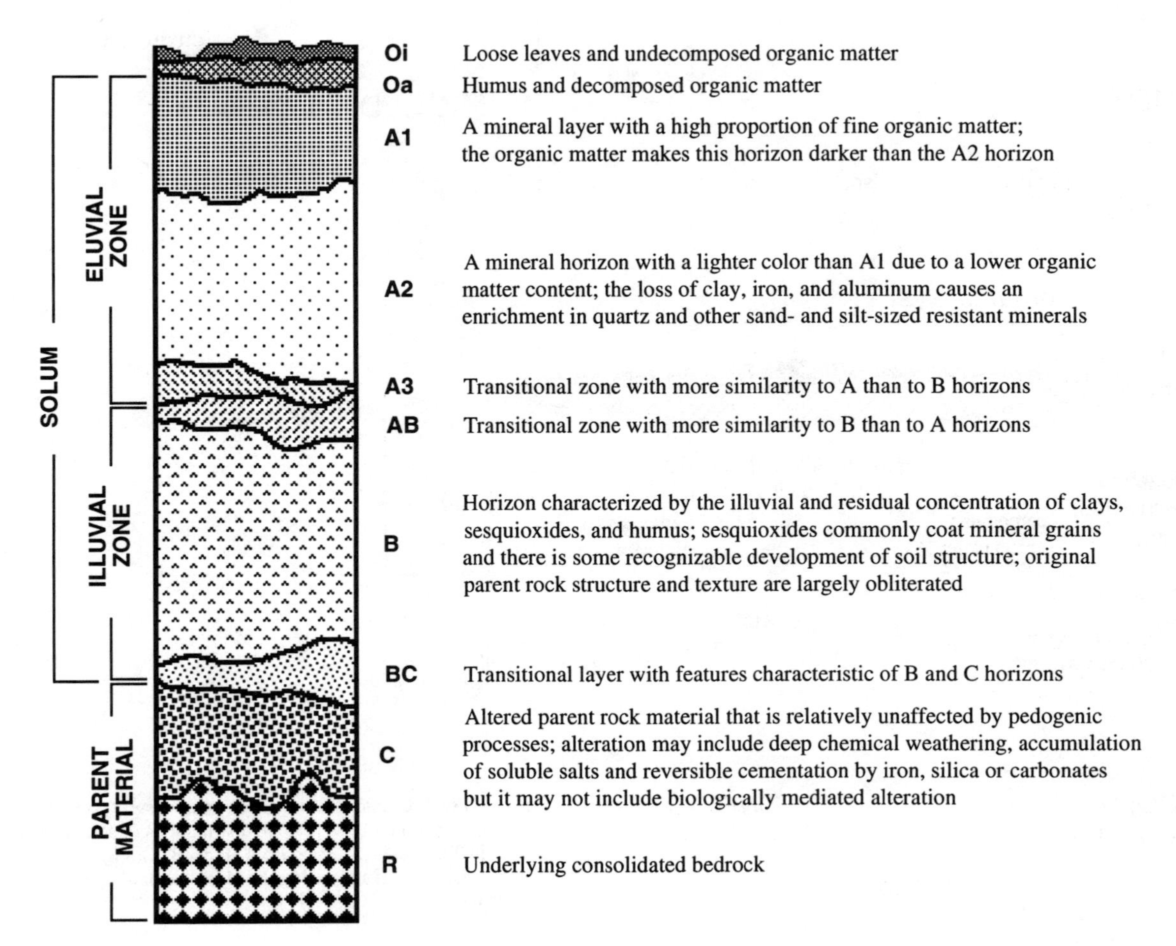

Figure 2-12. Common soil profile terms.

Table 2-5. One Example of Soil Terminology

Histosol:	organic matter soil, peats, bog and moor soils; any climate with water, which usually saturates them (some are periodically unsaturated).
Entisol:	poorly developed soils (azonal, tundra, or humic gley), of any climate, with little or no middle horizon if any horizon at all, and properties largely of parent material; little chemical weathering, highly variable composition, commonly with illite and smectite clays.
Inceptisol:	humid region soils with zonation beginning but not well developed (leaching in upper horizon but unstable minerals still present); wide variety of climatic conditions and compositions; smectites common together with allophanes, interlayered clays and illites.
Aridisol:	desert soils with leached (eluviated) A horizon, (illuviated) B horizon enriched in carbonates and other soluble salts and may be locally silicified; commonly illite in A, smectite in B, and mixed-layer clays in both.
Mollisol:	dark A horizon, lighter carbonate-rich B horizon grassland soil (chernozem, prarie, chestnut) that commonly borders desert regions and reflects limited rainfall/leaching; original clays little modified, but the assemblage is commonly dominated by illite and smectite.
Vertisol:	dark tropical-subtropical clayey soils with weak zonation, little organic matter; dominated by smectite clays that swell and shrink repeatedly with alternate wetting and drying (deep cracks in the surface occur at some time during the year). Some kaolinite where leaching occurs.
Spodosol:	cool temperate-humid region soil (podzol), with four zones: dark (organic) upper, bleached (eluviated), reddish to black illuviated (amorphous Fe and Al sesquioxides and organo-complexes), sandy C horizon; composition varied, with complex mixture of illite and chlorite alteration products, common smectite, vermiculite, and mixed-layer clays.
Alfisol:	well-zoned soil (gray-brown podzol, planosol) of humid alternating wet-to-dry climates, with clay sizes eluviated from the upper and concentrated in the middle horizon, and common sesquioxide concentrations; varied composition, kaolinite to smectite clays, vermiculite common (from illite).
Ultisol:	more weathered than alfisols, well-zoned soils (red-yellow podzolic) in climates where evapotranspiration exceeds precipitation in the dry season and precipitation in the wet season causes leaching; variety of compositions with common kaolinite, halloysite, vermiculite, and commonly sesquioxides and gibbsite.
Oxisol:	most highly weathered, little-zoned soils (laterites) of wet tropics; little or no organic material in soils, no soluble materials or unstable primary minerals; sesquioxides, kaolinite and gibbsite dominant. Goethite and hematite pisolites common near surface, underlain by gibbsite-rich then kaolinite-rich intervals; gibbsite may dominate near the surface as bauxite. The weathered zone may be heavily ferruginized for a thickness of up to tens of metres.

Source: After USDA (1975).

jects as the composition of the Precambrian atmosphere (e.g., Holland 1984), as well as on the climate and chemical factors dictating the difference between laterites (aluminous end products of intensive wet tropical weathering) and silcretes (semi-arid soils in which silica has been concentrated in the B horizon; e.g., Stephens 1971). Many detrital sediments that accumulate elsewhere have spent time in soils as intermediate resting places, and thus have been influenced by pedogenetic processes, a selection of which are listed in Table 2-6. Changes to the soil regime, either naturally or artificially, can have dramatic results: for example, if soils formed on pyrite-bearing muds (common in organic-rich intertidal environments, which may become supratidal because of sea-level fall, tectonic or reclamation processes) are drained, oxidation of the pyrite may cause the soil to develop a very high acidity (pH values below 3 are common in acid sulfate soils), which can lead to environmental problems such as fish mortality in nearby waterways, loss of agricultural productivity, or serious corrosion of concrete foundations, pipes, and underground service lines (e.g., see Bear 1964 for some chemical processes and effects in soils). Soil processes are important to the mining industry, insofar as they are involved in the formation of lateritic aluminium (bauxite), nickel, and gold ores (e.g., Butt 1989; Mann 1984). In geochemical exploration, knowledge of the influence of soil composition and soil processes on element mobilities (Table 2-7) is essential for interpreting soil surveys (Joyce 1984; Hoffman 1986).

Table 2-6. Common Soil Processes and Their Effects

Process	Effect
Accumulation	Physical addition of solid material to the surface of a soil
Acidification	Increase in the hydrogen ion concentration in soils due to processes such as ferrolysis, oxidization of sulfides, or addition of humic acids
Alkalinization	Accumulation of sodium ions on exchange sites in a soil
Braunification	Release of iron from primary minerals, its dispersion within the soil, and oxidization to enhance brown to red-brown soil coloration
Calcification	Addition (usually as a cement) of calcium carbonate to a soil
Calcretization	Cementation of a soil by calcium carbonate (often with traces of other carbonates)
Decomposition	Chemical, physical, and biological breakdown of mineral or organic soil components
Desalinization	Removal of soluble salts from saline soil horizons
Eluviation	Movement of material out of a particular part of a soil profile
Enrichment	Addition of new material to a soil
Erosion	Physical removal of solid material from the surface of a soil
Ferrolysis	Reaction of ferrous iron with water and oxygen to produce ferric hydroxides/oxides and hydrogen ions in oxidizing soil profiles
Gleization	Reduction of iron (and manganese) under anaerobic (waterlogged) conditions to produce a bluish to greenish gray soil matrix
Humification	Transformation of complex organic matter into humus in the soil profile
Hydrolysis	Reaction of soil with water; involves the uptake of hydrogen and hydroxide ions by soil constituents and their release of ions originally present
Illuviation	Movement of material into a portion of a soil profile
Kaolinization	Leaching and alteration of soil material (under acidic conditions) to produce kaolinite
Lateritization	Decomposition of silicates and the selective removal of silica from a soil to leave it strongly enriched in iron and aluminium
Leaching	Removal in solution (eluviation) of soluble material from a soil
Lessivage	Mechanical migration of small mineral particles from the A to the B horizon in a soil, causing clay enrichment in the B horizon
Oxidization	Oxidization of reduced inorganic and organic material in the soil primarily involving the production of ferric hydroxides and oxides and the loss of organic matter
Pedoturbation	Reworking of soil material (often resulting in loss of horizon definition) by biological or physical processes (e.g., freeze-thaw action)
Podzolinization	Chemical removal of iron, aluminum, and/or humic material from a soil horizon to leave it enriched in silica
Reduction	Reduction of oxidized soil material under anaerobic conditions; reduction is usually aided by microbial activity and primarily involves the loss of ferric hydroxides and oxides
Ripening	Physical, chemical, or biological changes occurring after air penetrates previously waterlogged organic soil
Salinization	Accumulation of soluble salts such as chlorides and sulfates of sodium, magnesium, calcium, and potassium in a soil
Silcretization	Cementation of a soil by silica
Silicification	Addition (usually as a cement) of silica to a soil
Synthesis	Formation of new mineral or organic material in the soil

Table 2-7. Mobility of Some Common Trace Elements in Soils

Element	pH Conditions			Additional Factors that Reduce Metal Mobility		
	< 5.5	5.5–7.0	> 7.0	Fe/Mn Oxides	Organic Matter	Other
Antomony	L	L	L	yes		sulfide, -ve Eh
Arsenic	M	M	M	yes		sulfide, clays
Barium	L	L	L			sulfate, carbonate, clay, –ve Eh
Bismuth	L	L	L	yes		–ve Eh
Boron	VH	VH	VH			
Cadmium	M	M	M			–ve Eh
Chromium	VL	VL	VL			
Cobalt	H	M/L	VL			sulfide, adsorption
Copper	H	M/L	VL	yes	yes	sulfide, adsorption
Fluorine	H	H	H			calcium, adsorption
Gold	I	I	I			
Ferric iron	VL/H	VL	VL	yes		
Ferrous iron	H	M/L	VL	yes		+ve Eh
Lead	L	L	L			carbonate, sulfate
Manganese	H	H	H/VL	yes		clays
Mercury (aq)	M	L	L	yes		sulfide
Molybdenum (Eh)	L	M	H	yes		sulfide, carbonate, adsorption, –ve
Nickel	H	M/L	VL			sulfide, adsorption, silicates
Selenium	H	H	VH	yes		–ve Eh, adsorption
Silver	H	M/L	VL	yes	yes	–ve Eh, sulfide, chromate, arsenate
Thorium	VL	VL	VL			adsorption on clay/Al hydroxides
Tin	I	I	I			
Uranium	L/M	H	VH	yes	yes	–ve Eh, adsorption
Vanadium	H	H	VH			–ve Eh, adsorption, silicates
Zinc	H	M/H	L/VL	yes	yes	sulfide, carbonate, phosphate

Notes: Negative Eh (–ve Eh) generally indicates reducing conditions in natural environments whereas a positive Eh (+ve Eh) indicates oxidizing conditions.
Element mobility rating: I = immobile/very insoluble, VL = very low mobility, L = low mobility, M = moderate mobility, H = high mobility, VH = very high mobility

Alluvial Deposits

Considerable attention has been devoted to alluvial facies models, which are generally separated into *braided* (including most alluvial fan) and *meandering* end members (e.g., Fig. 2-13); however, there is a continuous gradation between these end members and some recognize other ideal members (e.g., *anastomosing* streams, which have major meandering channels, each of which is braided, that do not shift laterally with time; e.g., Smith and Smith 1980; Rust 1981). Both intra- and extra-basinal controls play very important roles in the development and history of all alluvial systems (e.g., Gordon and Bridge 1987), and the architectural analysis approach to depositional systems has been particularly necessary and effective in analyzing alluvial systems (e.g., Miall 1985 and brief discussion in Chapter 4).

Alluvial fans occur in close proximity to highlands and tend to be the coarsest end of the fluvial continuum and have the steepest slopes (hence appreciable initial dips in cross section); depending on climate, debris flows may prove as significant in deposition as water-driven processes (e.g., Blissenbach 1954; Bull, in Rigby and Hamblin 1972). Models are varied and overlap those of braided streams (see below).

The meandering stream has a relatively simple and well-understood ideal model (Fig. 2-14a), but individual settings can show considerable complexities (e.g., Fig. 2-14b). Meandering streams are favored by relatively low slopes and a high suspended:bed load ratio. Deposition is dominated by *lateral accretion* of relatively coarse (mainly sand) point bars that build out from the convex face of the meanders toward the erosive concave face, and by overlying *vertical accretion* deposits of fine sediment, mud, and organic matter, resulting from overbank flooding and vegetal growth away from the stream channel (e.g., Allen 1970; Jackson 1981). Interruptions to normal development take place by *avulsion,* when the stream dramatically changes its course during flood stage then maintains a new channel belt after the flood wanes (e.g., Smith et al. 1989).

In practice, interpreting the lateral migration of meanders and associated point bars is further complicated by the gradual downstream migration of the meander loops and by the fact that a single major flood can radically modify the geometry of the meander belt. High-energy, gravelly meandering river patterns (e.g., Carson 1986), together with other properties, give rise to transitional systems between the me-

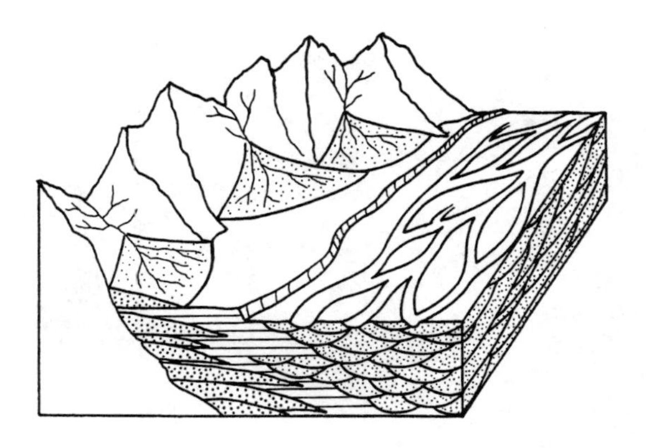

Figure 2-13a. Simplified models illustrating settings and geometrical attributes of common alluvial facies (stippled patterns indicate gravel or sand-dominated units; clear patterns represent fine sediment). Upper sketch depicts alluvial fans in a piedmont setting and braided streams in a channel belt some distance from the valley sides; characteristics of both tend to be few muds (mainly confined to the edges of the active fluvial systems) and vertically amalgamated channeled sequences of coarse sediment. Lower sketch depicts high- and low-sinuosity meandering systems, which have many more fines than braided systems; multistory sand layers (right side) may result from meandering of the channel across the floodplain if the overall rate of aggradation is high (perhaps the most common expectation would be for fewer and thinner sands in overbank muds). (After Allen 1965.)

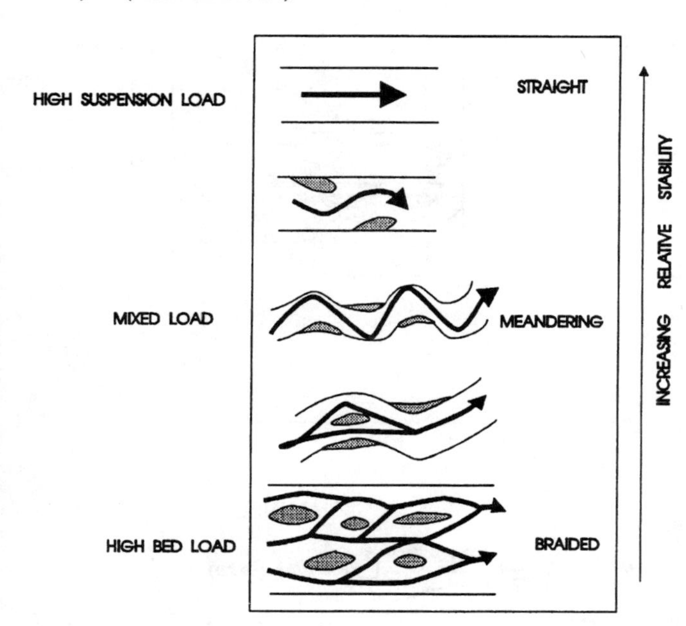

Figure 2-13b. Schematic representation of the relationships between river type and sediment load. (After Schumm and Meyer 1979.)

andering and braided ideals (e.g., Shelton and Noble 1974; Cant and Walker 1976). Coal accumulations are characteristically associated with meandering stream deposits, but there are plentiful exceptions. The fines and organic matter in the system concentrate in cohesive and resistant clay plugs as they fill abandoned meander loops (e.g., oxbow lakes); these plugs act to confine the channel system to the central tract of the channel belt. Marked similarities to meandering stream geometry, mechanics, and deposition occur in tidal channels in both detrital and carbonate systems (in some broad, low-gradient high-tidal-range tidal flat systems such as the Ganges/Brahmaputra/Meghna delta complex of Bangladesh, it is virtually impossible to morphologically distinguish the end of meandering river channels from the beginning of tidal channels).

Braided streams are dominated by complex shifting patterns of channels and various types of bars (e.g., Fig. 2-15); the bars are stable during low flow conditions but may exhibit rapid changes in position and morphology during flood stages (e.g., Williams and Rust 1969). They generally develop where sediments have a low suspension load relative to bed load (therefore erosion-resistant clay plugs are characteristically lacking, although in some systems peat bogs may play a similar role). Braided drainage is favored by abundant sediment supply and irregular stream discharge, particularly in areas that are normally characterized by low flow velocities but experience frequent flood conditions. They commonly form in the steeper gradient parts of the fluvial system and tend to have abundant gravels. However, braided streams may exist in systems where there are few or no gravels (e.g., Cant and Walker 1976), particularly in relatively low-gradient, broad braid plains, which have a higher preservation potential than the steeper parts of fluvial systems.

Deposition occurs primarily in various types of bar, which form in or migrate into channels and are individually highly ephemeral, being destroyed and reconstructed elsewhere as the braided channels change their position. Many different varieties of braided stream have been documented (e.g., Fig. 2-16 and Table 2-8, which lists commonly recognized lithofacies) and there is no single or simple model for them (see Smith 1970; Rust 1972; Miall 1977, 1978; Rust and Jones 1987). Despite the complexities associated with interpreting facies distributions in ancient braided stream systems, considerable attention is focused on their study by the mining industry because these facies host major gold deposits and some economic placer deposits of uraninite, platinum, tin, and some gemstones (e.g., Smith and Minter 1980; Pretorius 1981); they can also make good petroleum reservoir rocks.

Eolian Deposits

Eolian deposits with dunes and blowouts are common features of coastal and arid landscapes (e.g., Bowler 1976; Glennie 1987; see Fig. 2-17), but because they are highly mobile subaerial deposits they are not commonly preserved in the geologic record (many exceptions exist, as in the Me-

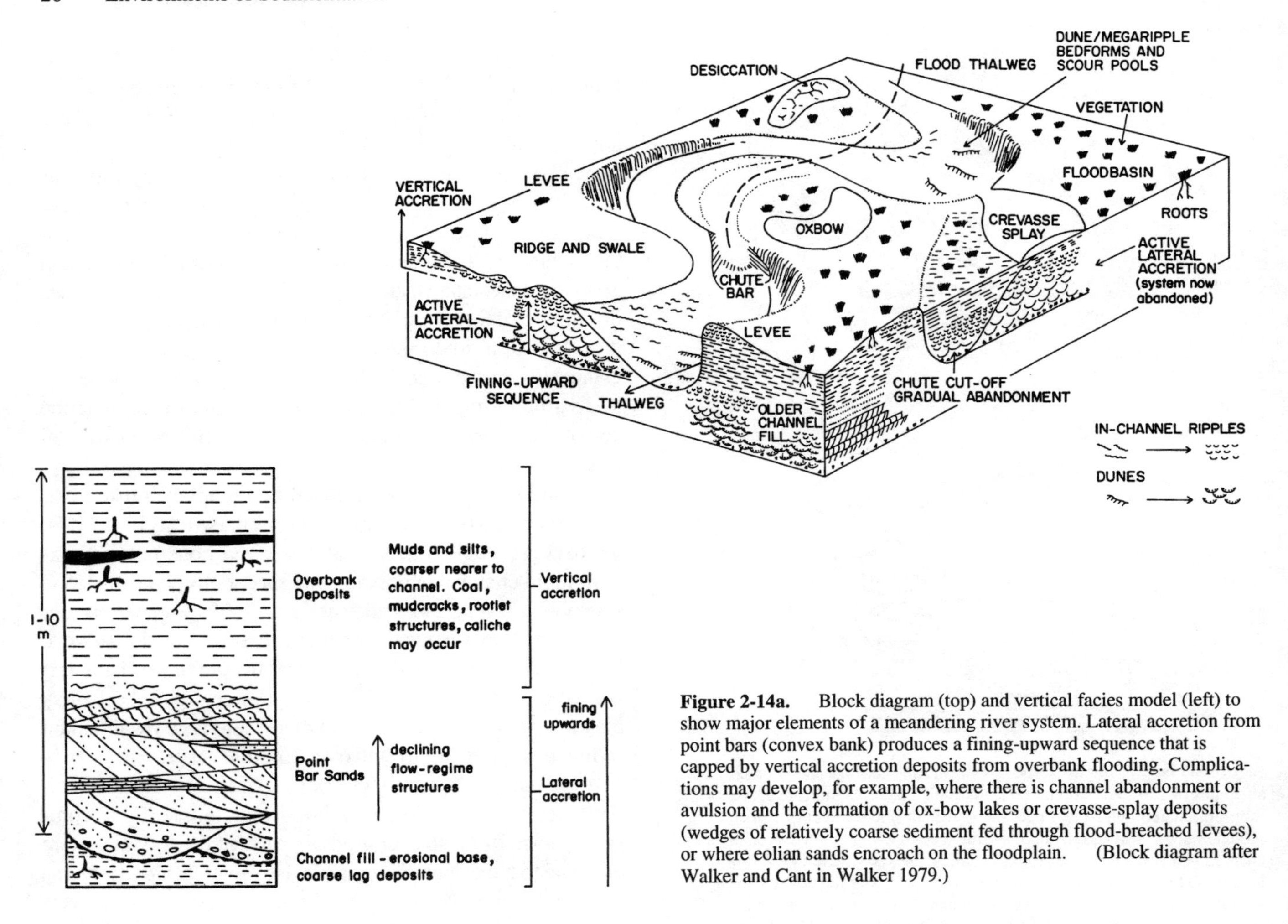

Figure 2-14a. Block diagram (top) and vertical facies model (left) to show major elements of a meandering river system. Lateral accretion from point bars (convex bank) produces a fining-upward sequence that is capped by vertical accretion deposits from overbank flooding. Complications may develop, for example, where there is channel abandonment or avulsion and the formation of ox-bow lakes or crevasse-splay deposits (wedges of relatively coarse sediment fed through flood-breached levees), or where eolian sands encroach on the floodplain. (Block diagram after Walker and Cant in Walker 1979.)

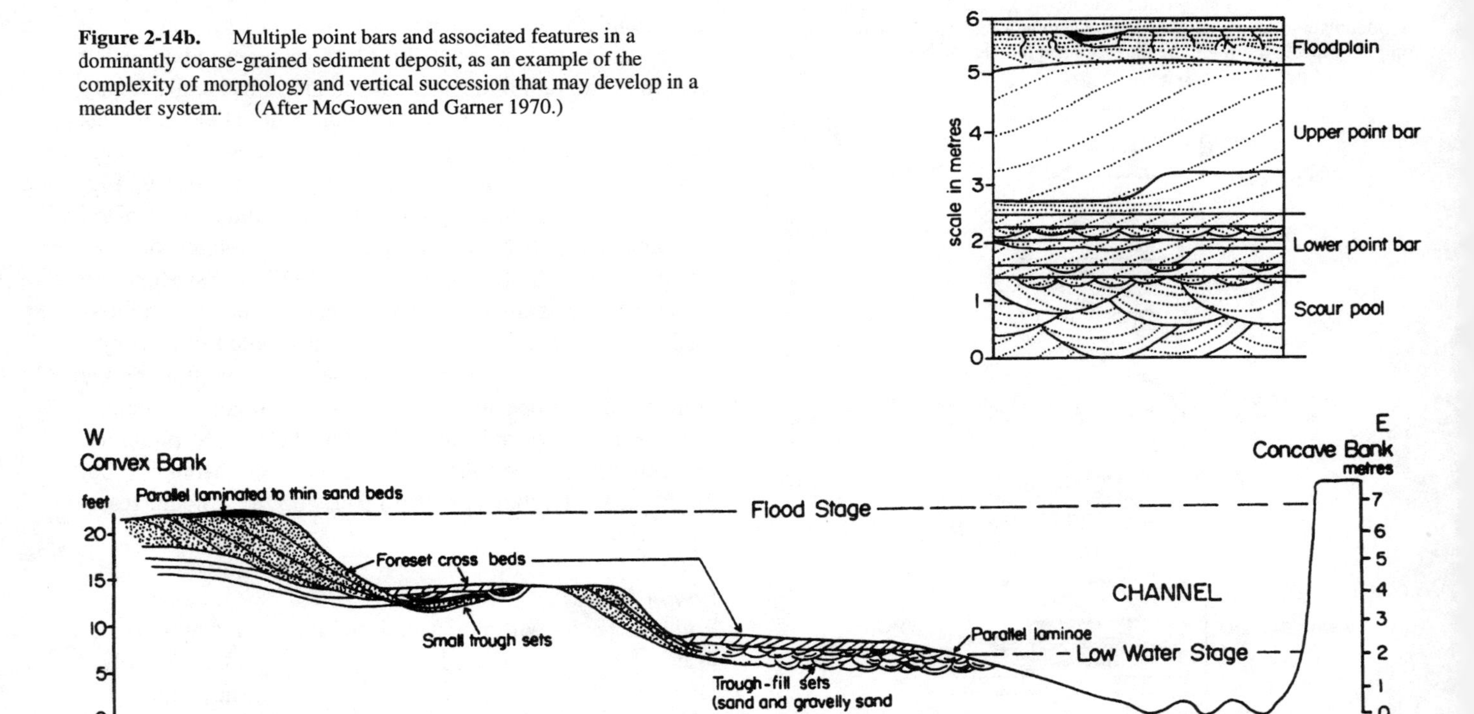

Figure 2-14b. Multiple point bars and associated features in a dominantly coarse-grained sediment deposit, as an example of the complexity of morphology and vertical succession that may develop in a meander system. (After McGowen and Garner 1970.)

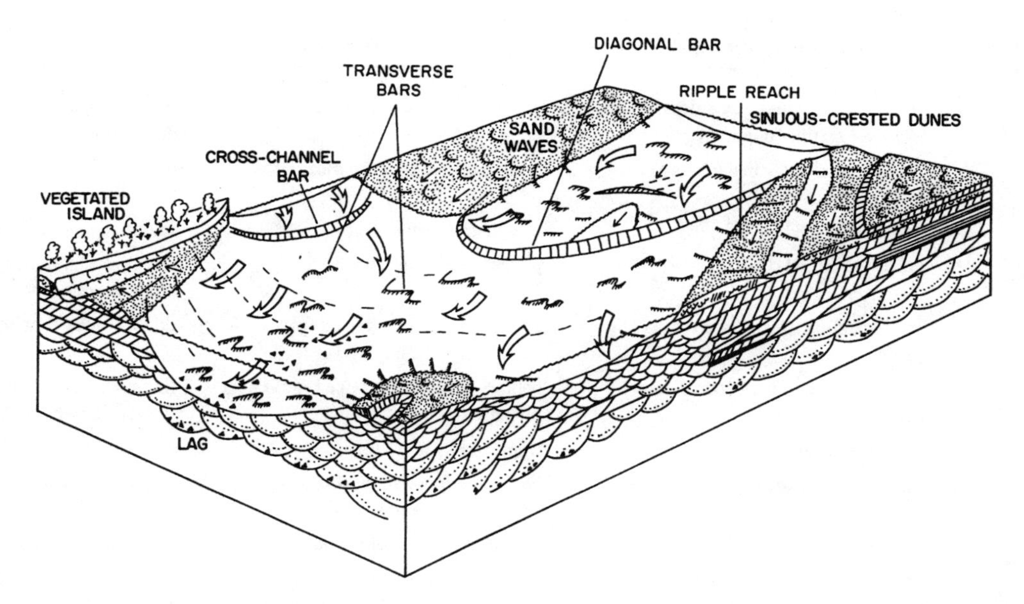

Figure 2-15. Generalized block diagram for a braided stream system, showing major morphological elements; see Fig. 2-17 for various types of vertical facies sequences that may result. (After Walker and Cant in Walker 1979.)

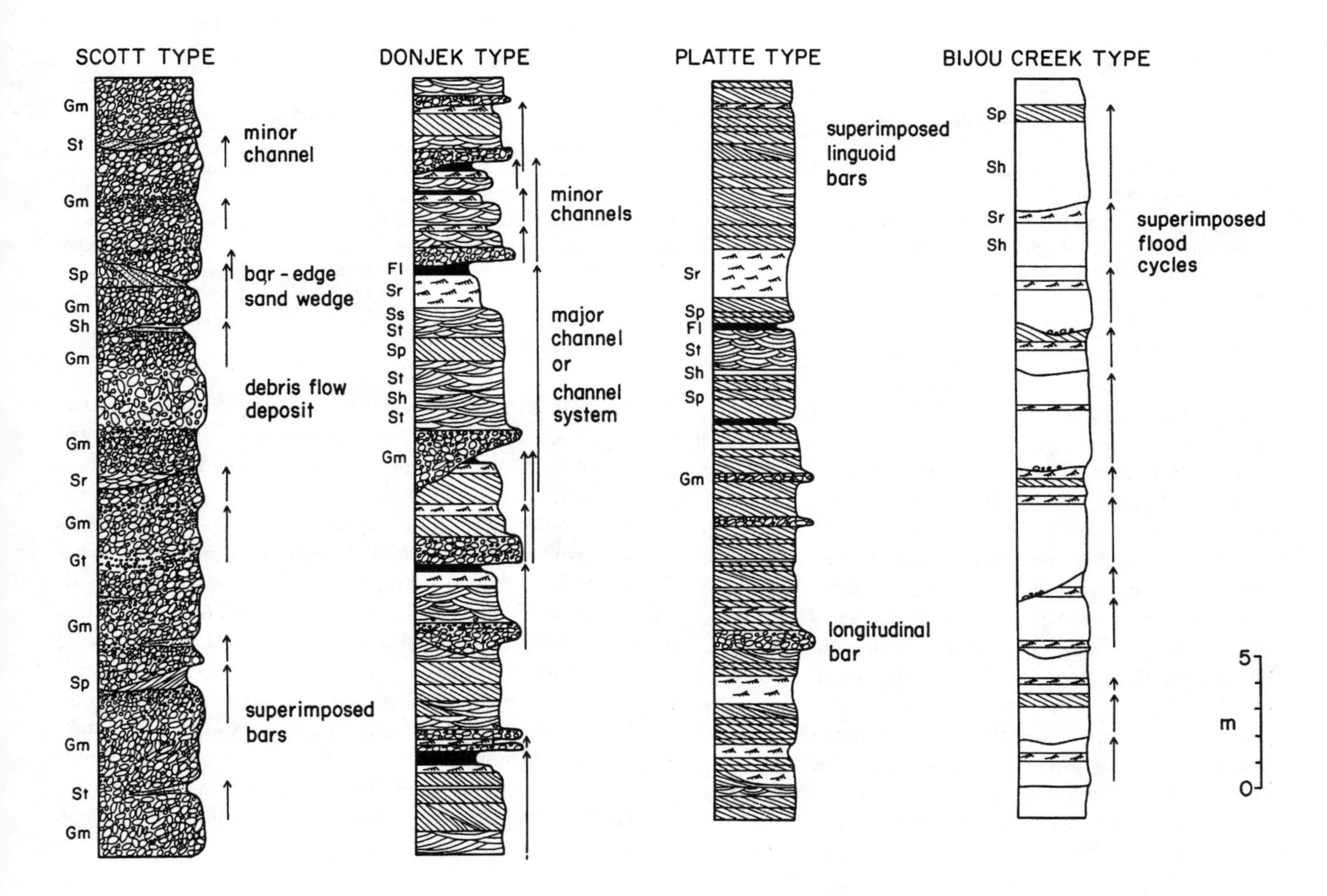

Figure 2-16. A few varieties from a wide spectrum of vertical facies sequences that may result from aggradation in braided stream environments; environmental conditions are so variable in braided alluvial systems that no facies sequence is generally representative. See Table 2-8 for the shorthand notations for individual facies types. (After Miall, in Miall 1978.)

Table 2-8. Lithofacies and Interpretation for Principal Facies Assemblages of Braided Stream Deposits Depicted in Fig. 2-16

Facies Code	Lithofacies	Sedimentary Structures	Interpretation
Gms	massive, matrix-supported gravel	none	debris flow deposits
Gm	massive or crudely bedded gravel	horizontal bedding, imbrication	longitudinal bars, lag deposits, sieve deposits
Gt	gravel, stratified	trough cross-beds	minor channel fills
Gp	gravel, stratified	planar cross-beds	linguoid bars or deltaic growths from older bar remnants
St	sand, medium to very coarse, may be pebbly	solitary (theta) or grouped (pi) trough cross-beds	dunes (lower flow regime)
Sp	sand, medium to very coarse, may be pebbly	solitary (alpha) or grouped (omikron) planar cross-beds	linguoid, transverse bars, sand waves (lower flow regime)
Sr	sand, very fine to coarse	ripple marks of all types	ripples (lower flow regime)
Sh	sand, very fine to very coarse, may be pebbly	horizontal lamination, parting or streaming lineation	planar bed flow (lower and upper flow regime)
Sl	sand, fine	low-angle (<10) cross-beds	scour fills, crevasse splays, antidunes
Se	erosional scours with intraclasts	crude cross-bedding	scour fills
Ss	sand, fine to coarse, may be pebbly	broad, shallow scours including eta cross-stratification	scour fills
Sse, She, Spe	sand	analogous to Ss, Sh, Sp	eolian deposits
Fl	sand, silt, mud	fine lamination, very small ripples	overbank or waning flood deposits
Fsc	silt, mud	laminated to massive	backswamp deposits
Fcf	mud	massive, with freshwater molluscs	backswamp pond deposit
Fm	mud, silt	massive, desiccation cracks	overbank or drape deposits
Fr	silt, mud	rootlets	seat earth
C	coal, carbonaceous mud	plants, mud films	swamp deposits
P	carbonate	pedogenic features	soil

Source: After Miall (1977).

Note: In relation to Fig. 2-16, note that characterizing facies are *Scott type:* proximal rivers; predominantly stream flows and including alluvial fans. Main facies: Gm. Also Gp, Gt, Sp, St, Sr, Fl, Fm. *Donjek type:* distal gravelly rivers; cyclic deposits. Main facies: Gm, Gt, St. Also Gp, Sh, Sr, Sp, Fl, Fm. *Platte type:* Sandy braided rivers; virtually noncyclic. Main facies: St, Sp. Also Sh, Sr, Ss, Gm, Fl, Fm. *Bijou Creek type:* ephemeral or perennial rivers subject to flash floods. Main facies: Sh, Sl. Also Sp, Sr.

sozoic of the western U.S., e.g., Walker and Harms 1972; McKee 1979; Sanderson 1974; Porter 1987; Langford and Chan 1989). Cementation, commonly by calcium carbonate or iron oxides precipitated from groundwaters that have moved toward the surface by capillary attraction, can result in stabilization and resistance to erosion. Even where dunal deposits are preserved, they may be difficult to distinguish from subaqueous deposits (e.g., Kocurek and Dott 1981), and individual units vary rapidly in thickness and are laterally discontinuous (see Chapter 4 for further discussion of eolian structures). High-angle cross-bedding with a variety of dip azimuths is a traditional criterion for their recognition, but many eolian deposits do not show this association. Processes and stratification sequences are commonly complex (e.g., McKee 1966; Clemmensen and Hegner 1978; Fryberger, Schenk, and Krystinik 1988; Sweet et al. 1988).

Sands are generally well sorted, but because the major deposits with which they can be confused are beach sands, this also is not a good recognition feature. Grains are commonly frosted (difficult to determine in consolidated rocks), may have a thin desert varnish of oxides (commonly lost in diagenesis), and gravels can be sandblasted into multifaceted shapes (e.g., *dreikanter*). In semi-arid and arid regions, interdune and blowout depressions commonly reach the groundwater table and ephemeral to semi-permanent lakes can form, in which substantial evaporite deposits may accumulate (e.g., Arakel 1980); evaporitic lake deposits may alternate with dune sands in the rock record (e.g., Ahlbrandt and Fryberger in Ethridge and Flores 1981). In subrecent dune fields, blowouts may concentrate archeological artifacts in the lag deposits. Because human activities are commonly concentrated along coasts, study of dune migration and stabilization is an important aspect of modern environmental evaluation. In addition, eolian deposits have high initial porosities and permeabilities; thus, they make good aquifers and petroleum reservoir beds.

Lacustrine Deposits

There are few general models for lake deposits, which have a low preservation potential compared to marine deposits in

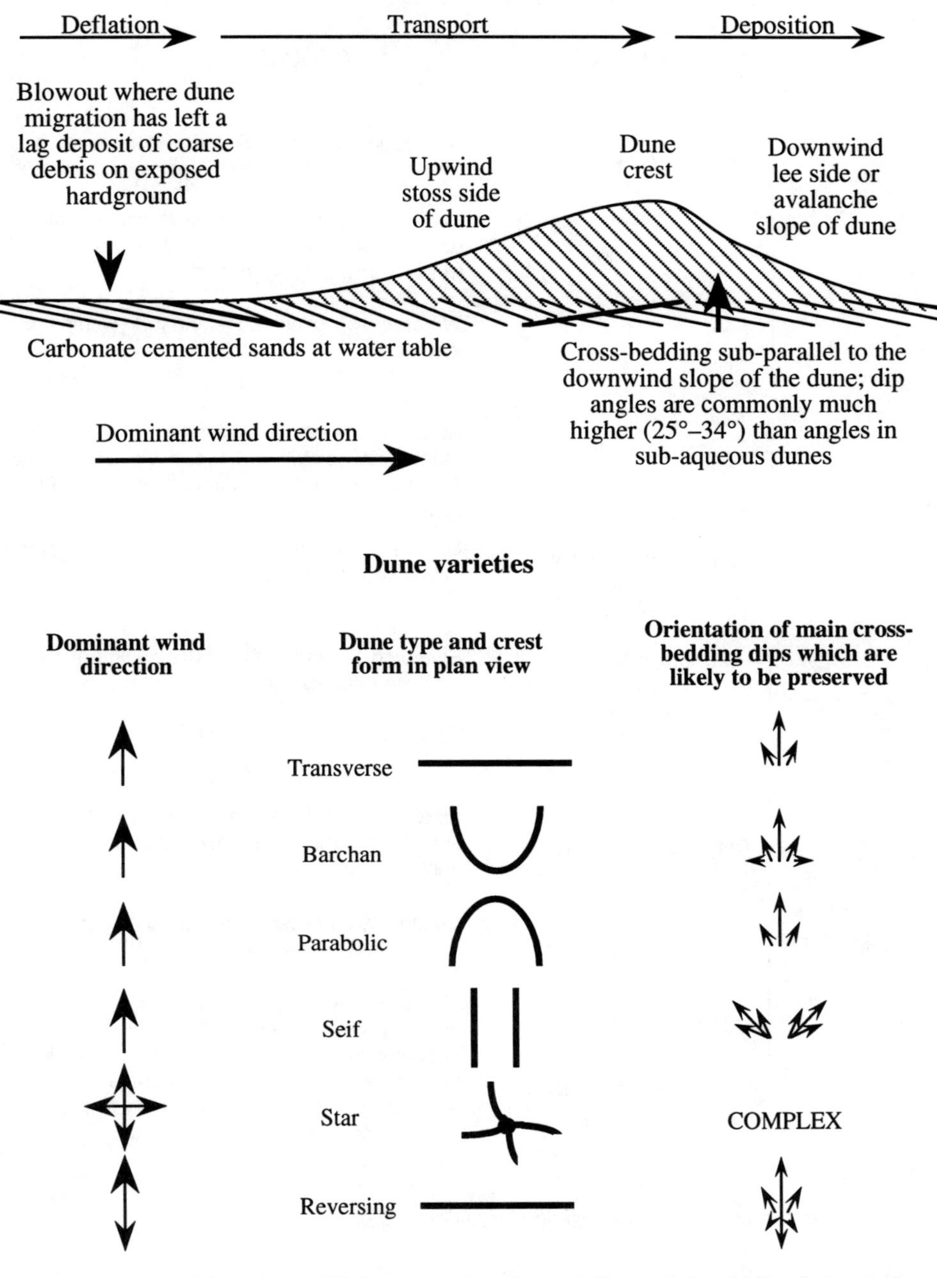

Figure 2-17. Schematic simplified representation of general characteristics of eolian dune systems.

the rock record, but they may cover large areas and they occasionally leave thick sequences (e.g., Reeves 1968; Picard and High 1972; Matter and Tucker 1978). The sedimentology and geochemistry of lakes can be complex and can differ markedly (Wetzel 1976), depending on local tectonics, source rocks, climatic factors, biotic elements present, bathymetric configuration, water supply to the lakes, and presence and size of any outlet. Deposits may be primarily detrital, biogenic, evaporitic, or mixed. Criteria for distinguishing lacustrine and fluvial deposits are sparse, particularly since they tend to occur together and with deltas (e.g., Monroe 1981; Yuretiach et al. 1984); in such interfingering sequences, the lake sediments are commonly regarded as the monotonous fine-grained sequences, but because the same range of processes operates in large lakes as in the oceans (Fig. 2-18 and Chapter 3), this type of generalization is a misleading simplification. There are economic deposits of importance in some ancient lacustrine sequences (e.g., oil shales, Eugster and Surdam 1973), and in modern lake sediments (e.g., varied salts from Australian hypersaline lakes, Arakel and McConchie 1982; Jankowski and Jacobson 1989; Arakel and Cohen 1991); they can also be good hydrocarbon source rocks. Fig. 2-19 illustrates the depth relationships between some chemical variables in lakes, and Fig. 2-20 depicts an example of the way in which evaporitic lake brines may interact with groundwaters and sediment beneath a lake.

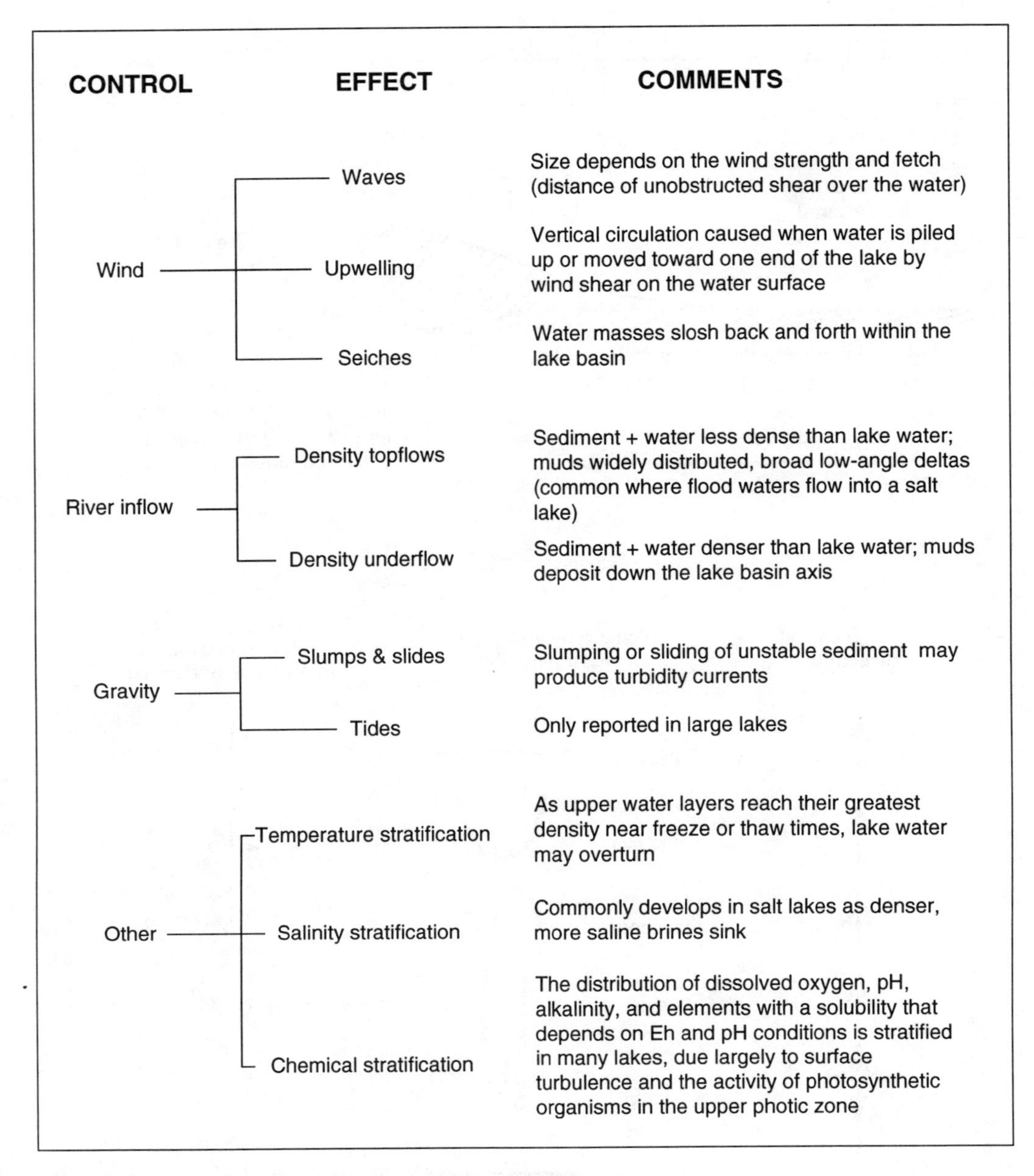

Figure 2-18. Summary of dominant processes operating in lakes.

Transitional Deposits

Shoreline (beach and shallow nearshore) facies are critical in interpretations of basin analysis because they can generally be recognized in the sediment record and because they mark the boundary between the marine and nonmarine realms (e.g., Howard 1972; Archer and Maples 1984; Cant 1984); their movement with time displays transgressions or progradational episodes, and regressions or retrogradational episodes (e.g., Curray 1964). Without such a marker unit, these large-scale trends can be difficult to identify. These environments, transitional between the open ocean and terrestrial realms, can contain a very complex array of processes and subenvironments (e.g., Curray 1969). In this book we distinguish transitional environments as those with a mix of both marine and freshwater influences; nearshore water tends to be brackish (to hypersaline in arid regions) and has a weak to strong salt wedge that migrates across much of it during the tidal cycle.

We consider the beach, tidal flats, and the closely related shallow nearshore deposits to be open marine (see below), but it is a moot point whether they can be separated clearly from the environments discussed in this section, since beaches are also present in all of them, tidal flats in many, and the bulk of the sediments in large deltas are deposited under fully marine conditions. The shoreline is also the locus of major human settlements; hence, original environmental conditions are commonly changed substantially with various effects that are only sometimes anticipated; engineering, geomorphic, geochemical, geographic, and geologic studies are thus abundant (e.g., for the range of topics, see Hails and

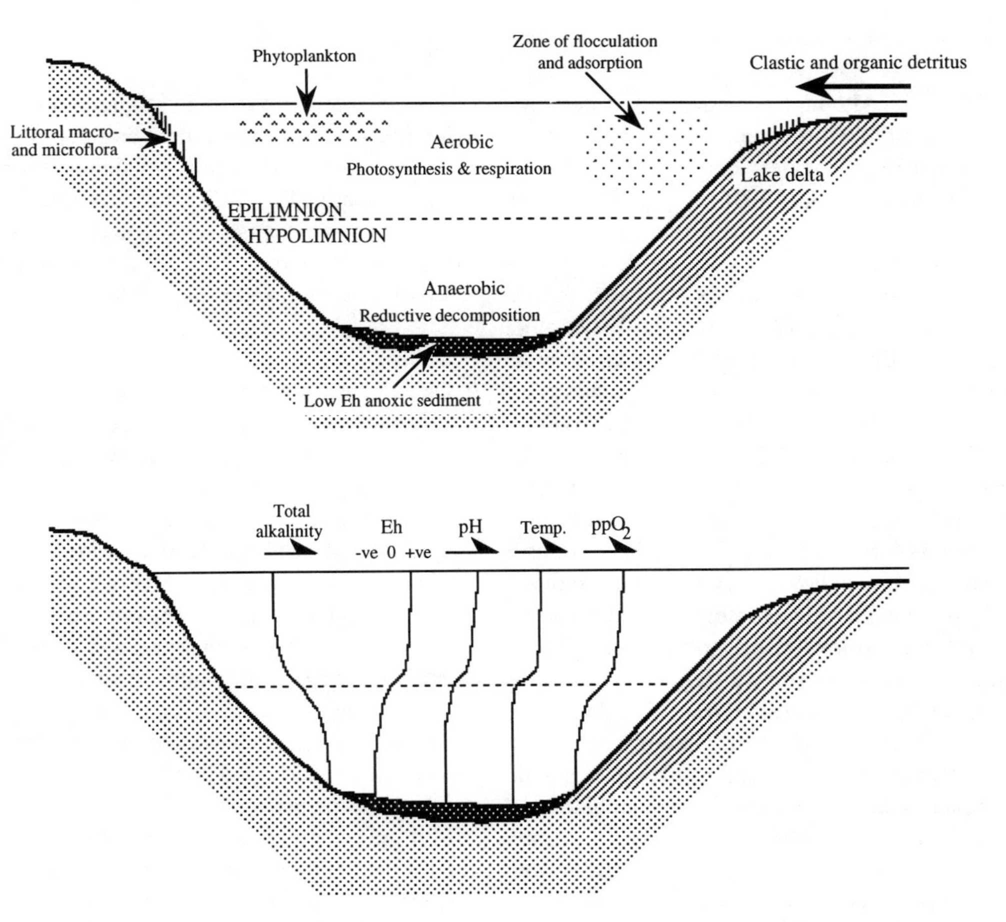

Figure 2-19. Schematic representation of important factors in lake sedimentary environments. The *epilimnion* is the warmer, slightly less dense surface layer of water that floats on the denser *hypolimnion* when thermal stratification occurs in lakes.

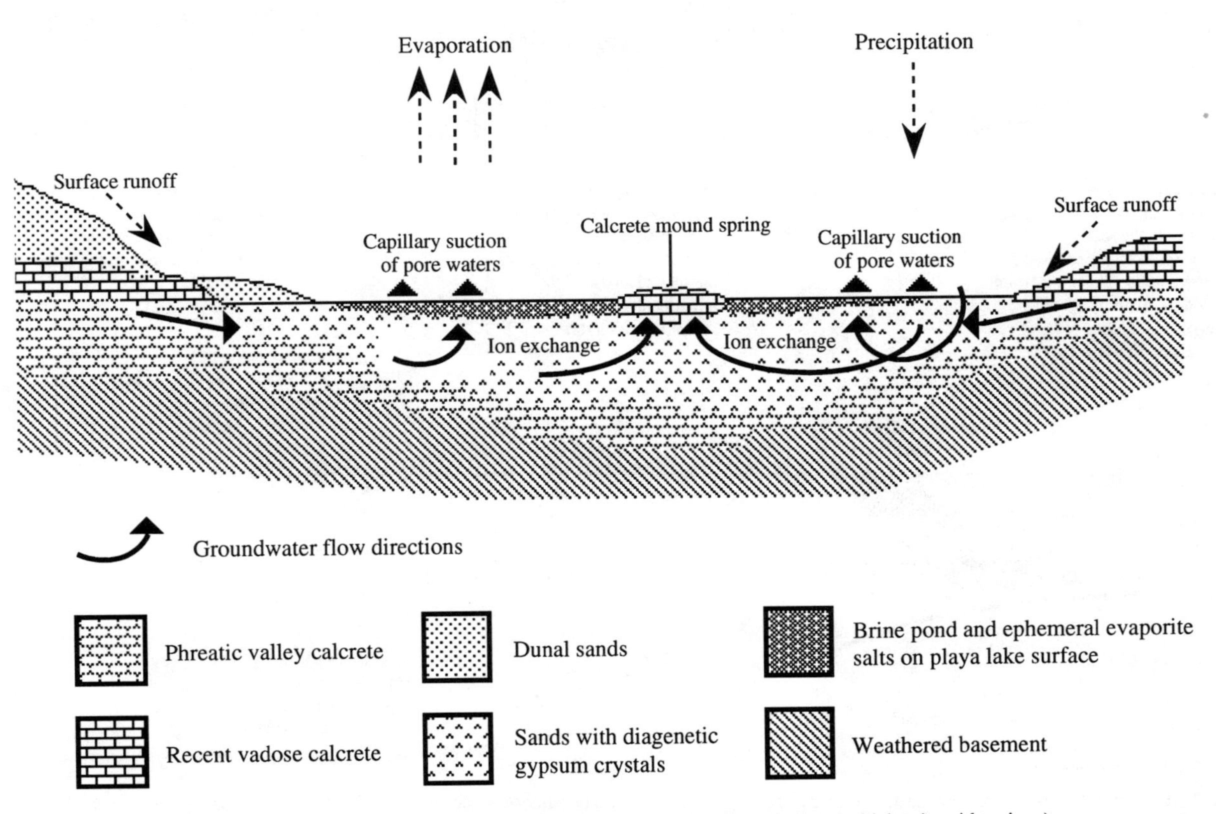

Figure 2-20. Schematic representation of important processes in playa lake settings (ephemeral lakes in arid regions).

Carr 1975; Davis 1985). Although we concentrate here on detrital sediment accumulations, geometry and physical processes are very similar in carbonate sequences (e.g., Shinn, Lloyd, and Ginsburg 1969; Evans 1970).

Estuaries

The estuarine environment, at the mouths of rivers, is the smallest transitional environment in area, but may leave substantial deposits in the sedimentary record. This environment is of major concern to modern sedimentologists, since it is perhaps the one most likely to suffer from human activities (e.g., Figs. 2-21, 2-22; Nelson 1972; Howard and Frey 1975; Frey and Howard 1986; Leckie and Singh 1991). Although the morphology of some estuaries is controlled by tectonic processes or by barriers formed by longshore sediment transport, most are drowned river valleys. They are more common in modern settings than most of the geologic record because of the Recent postglacial sea-level rise. Deposits can be difficult to recognize because transgression frequently reworks estuarine sediments into other shoreline/nearshore deposits. However, because of their origin and morphology, most estuarine subenvironments accumulate sediments rapidly, and if tectonic subsidence is sufficiently rapid, substantial thicknesses of sediment may accumulate, such as *paralic* coal measure sequences (*paludal* deposits of lake shorelines can be very similar). Marked lateral and vertical facies changes occur in estuarine deposits, particularly where river discharge was intermittent. In progradational sequences, estuaries fill with fluvial sediment and deltas eventually form on the seaward side; estuarine deltas such as the Ganges/Brahmaputra (Fig. 2-23) show complete gradation between these two environments. Seagrass and mangrove communities may be particularly influential in sedimentation (e.g., Ward, Boynton, and Kemp 1984).

Deltas

Deltas are complex mixes of terrestrial to fully marine subenvironments created by a prograding deposit of sediment at the boundary between an alluvial system and the sea; they are commonly called marine deltas to distinguish them from less complex, smaller-scale lacustrine deltas. Precise definitions are difficult because of the wide scope and relative scale of subenvironments (e.g., Moore and Asquith 1971; Nemec 1990). Marine deltas are sites of major sediment accumulations in terms of both thickness and areal extent (e.g., Morgan and Shaver 1970); substantial quantities of sediment accumulate each year (Table 2-9), and because they are progradational features and are often in areas of high subsidence, they have a high preservation potential in the rock record. The diverse subenvironments shift their geographic position markedly over time spans of just a few hundred years; major changes in river/distributary routes (*avulsions*) can take place virtually overnight (except where humans create artificial levees, in which case avulsion can be delayed temporarily).

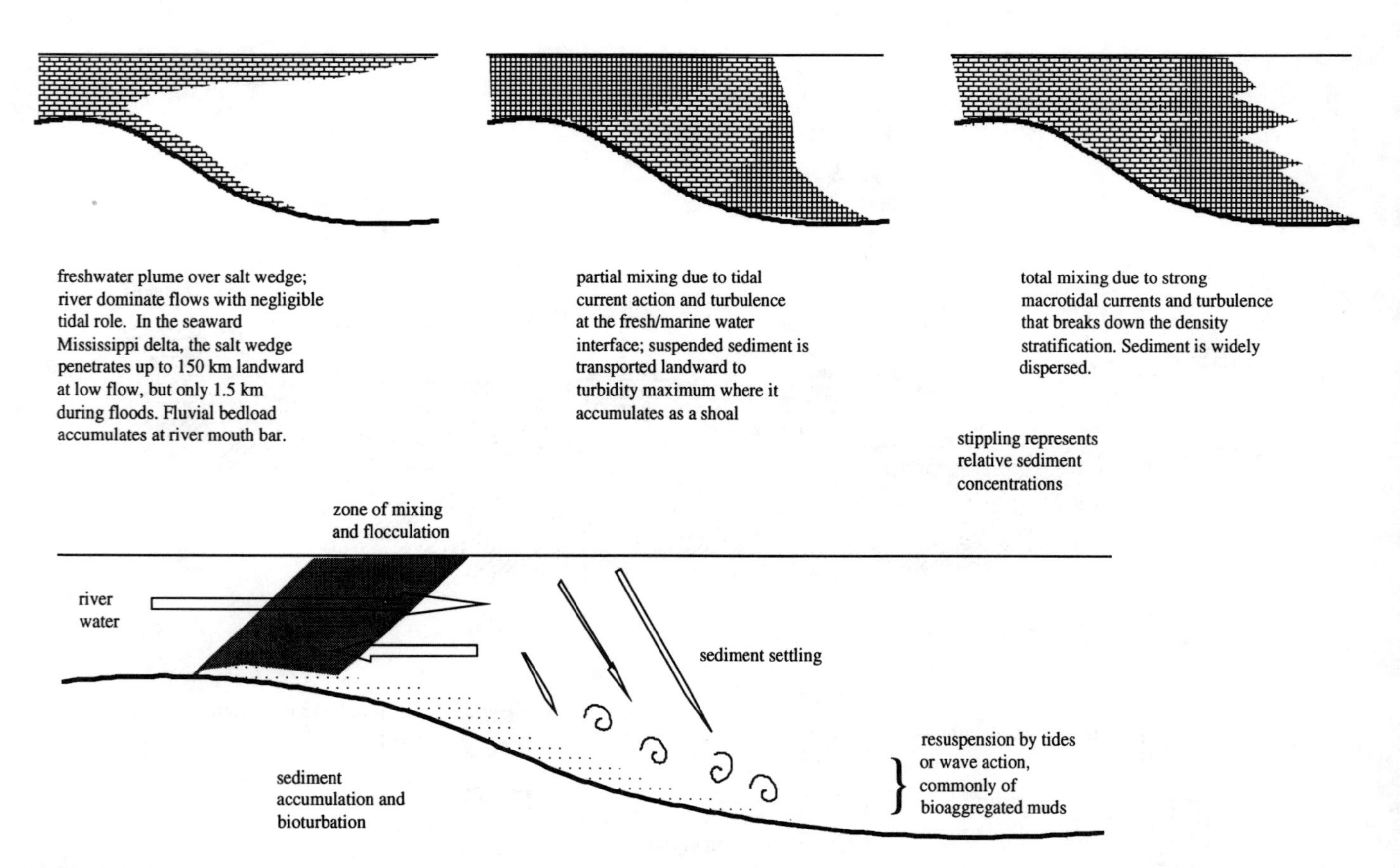

Figure 2-21. Relative sediment concentration in different types of estuarine circulation systems. (After Nichols and Biggs, in Davis 1985.)

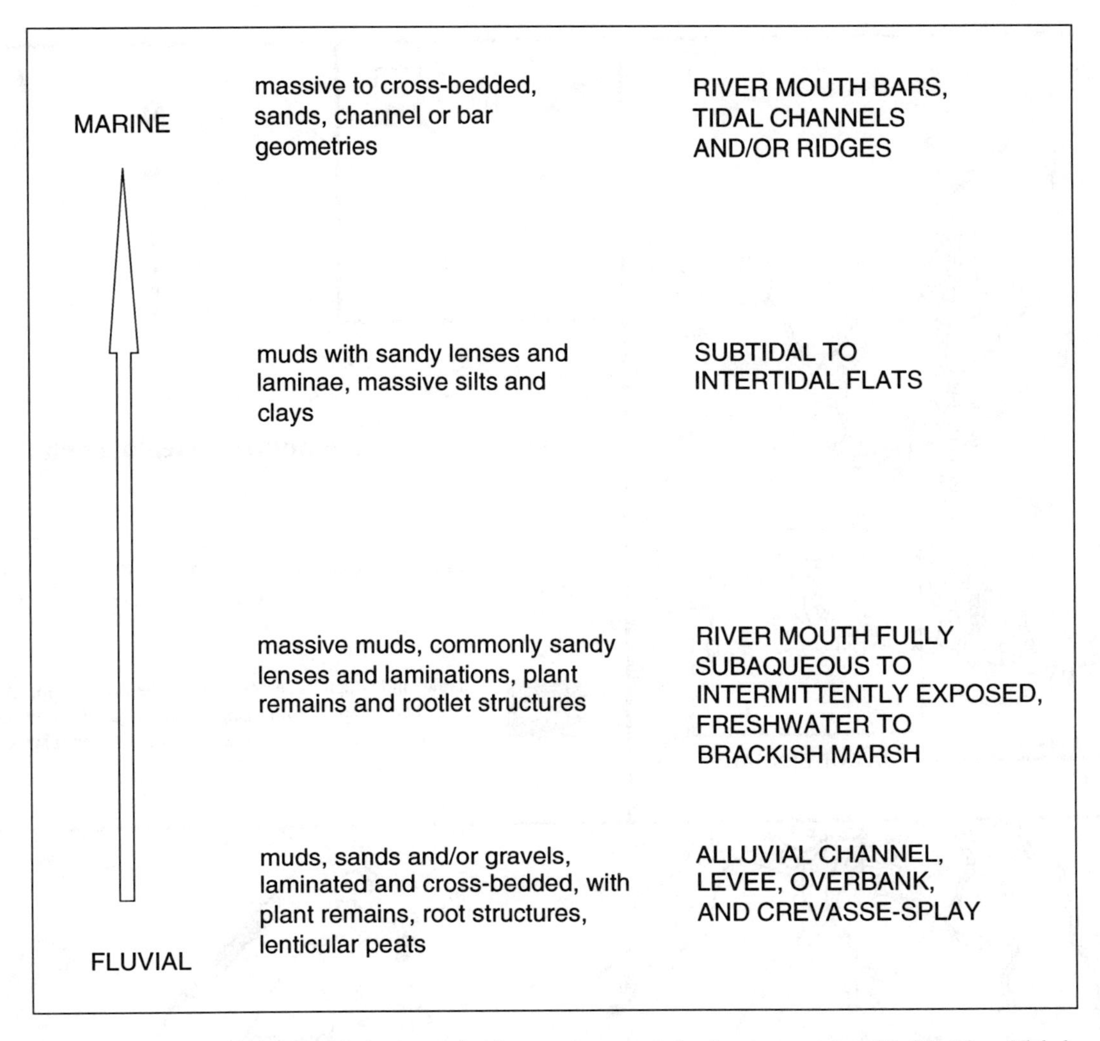

Figure 2-22. Simplified vertical facies succession for estuarine aggradation in a transgression. (Modified from Nichols and Banks, in Davis 1985.)

The bulk of most delta deposits are fine muddy sediments; hence, the coarser permeable units are commonly isolated and comprise excellent potential petroleum traps. Organic matter is abundant with the muds; hence, there is a close association of potential source and reservoir sediments (e.g., papers in Shirley and Ragsdale 1966; Fisher et al. 1969; Coleman 1982). However, some deltas are dominated by coarse sediment (e.g., Colella and Prior 1990). Because many urban centers have developed in rapidly evolving deltaic settings, an understanding of sediment dynamics in deltas is particularly important for land-use planners, and there is a high demand for data from sedimentologists and geomorphologists. For example, most of the present land area of Bangladesh consists of deltaic sediments that attain thicknesses exceeding 5 km (R. Khandoker, Dhaka University, pers. comm.) and extend offshore for over 100 km along much of the Bengali coastline; since the 1970s, new subaerial land has been prograding into the Bay of Bengal at an average rate of c. 35 km^2/y! (Because the new land is of such low relief, yet is settled by families in desperate need of space, the results are disastrous during storm-caused floods and temporary rises in sea level.)

Deltas are commonly classed on a triangular diagram, each apex of which represents the dominance of one of the major processes influencing delta morphologies and subenvironments (Fig. 2-24). Deltas dominated by river processes are *high-constructive* deltas because they build out into the large marine (or lake) basin despite the activity of tidal and wave processes. Where basinal processes dictate the shape and characteristics of the deltaic deposits supplied by the rivers, *high-destructive* wave or tidal deltas result (e.g., Figs. 2-25, 2-26). In cross-section, the general delta geometry is commonly represented by the profile shown in Fig. 2-27, but that kind of sketch vastly exaggerates the vertical scale; whereas in *Gilbert-type* deltas (characteristic of some lake deltas where topset, foreset, and bottomset beds are distinguishable even on outcrop scale) the foresets can dip over 15°, in most large deltas the foresets of the delta front and upper pro-delta environments rarely exceed 1–2°. Generalized vertical facies models of the deposits resulting

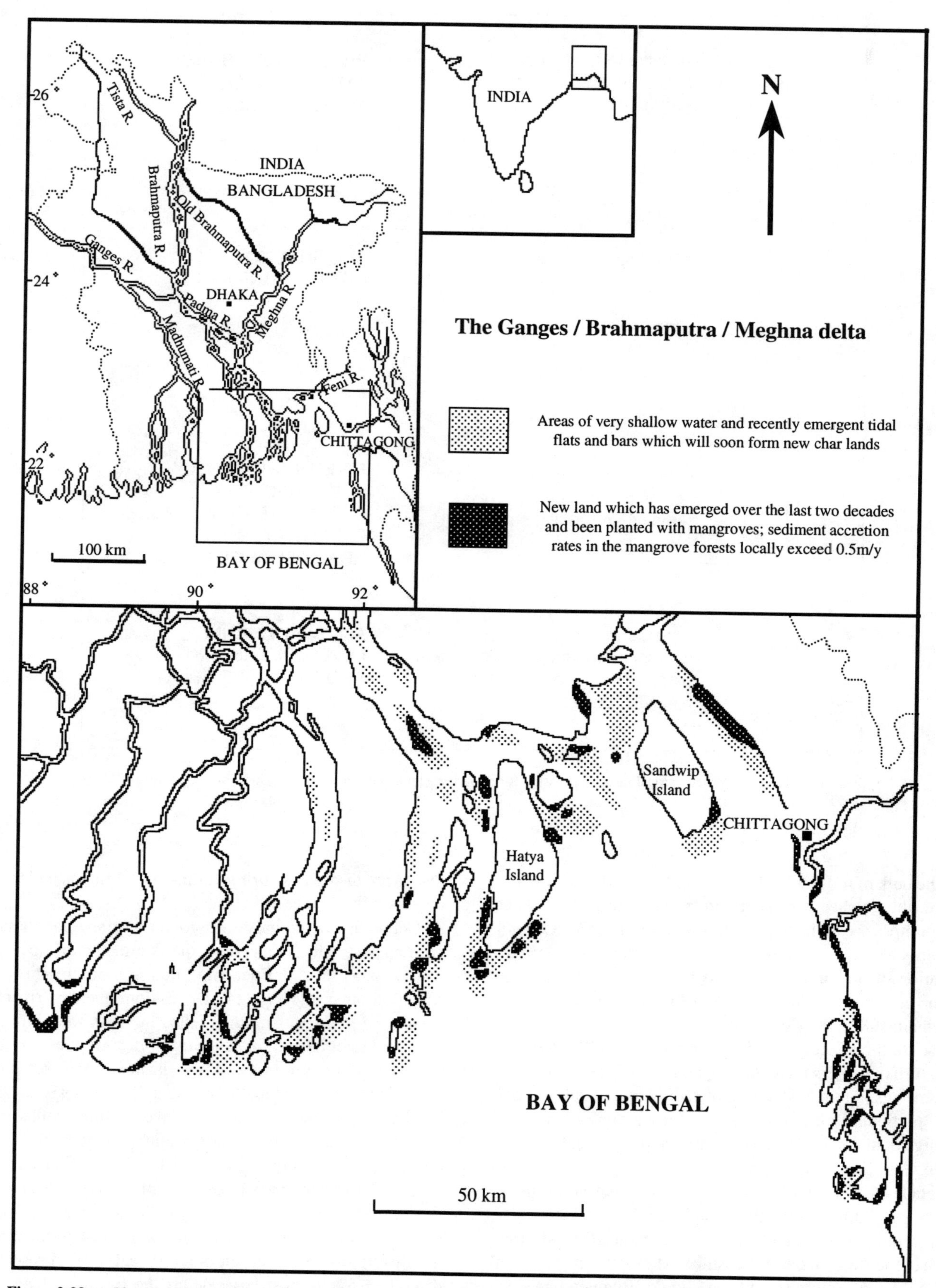

Figure 2-23. Plan-view morphology of part of the Ganges/Brahmaputra/Meghna delta. Morphologically, the delta has many of the characteristics of a fractal: as scale changes from large map views to detailed field observations of the intertidal zones, finer detail is revealed, but the pattern of the channels remains remarkably constant.

Table 2-9. Characteristics of Some of the World's Longest Rivers

River	Catchment Area 10^3 km^2	Mean Water Discharge 10^3 m^3·s^{-1}	Suspended Sediment Discharge 10^6 tons·y^{-1}
Ganges/Brahmaputra/Meghna	1048	30.8	2240
Yellow (Huangho)	770	1.6	1080
Amazon	6150	199.8	900
Yangtze	1940	28.5	478
Irrawaddy	430	13.6	285
Mississippi	3269	18.4	210
Orinoco	990	34.9	210
Mekong	795	14.9	160
Indus	969	7.5	100
Danube	810	6.5	67

Source: Ganges/Brahmaputra/Meghna data from R. Khandoker, Dhaka University (pers. comm.), other data after Milliman and Meade (1983).

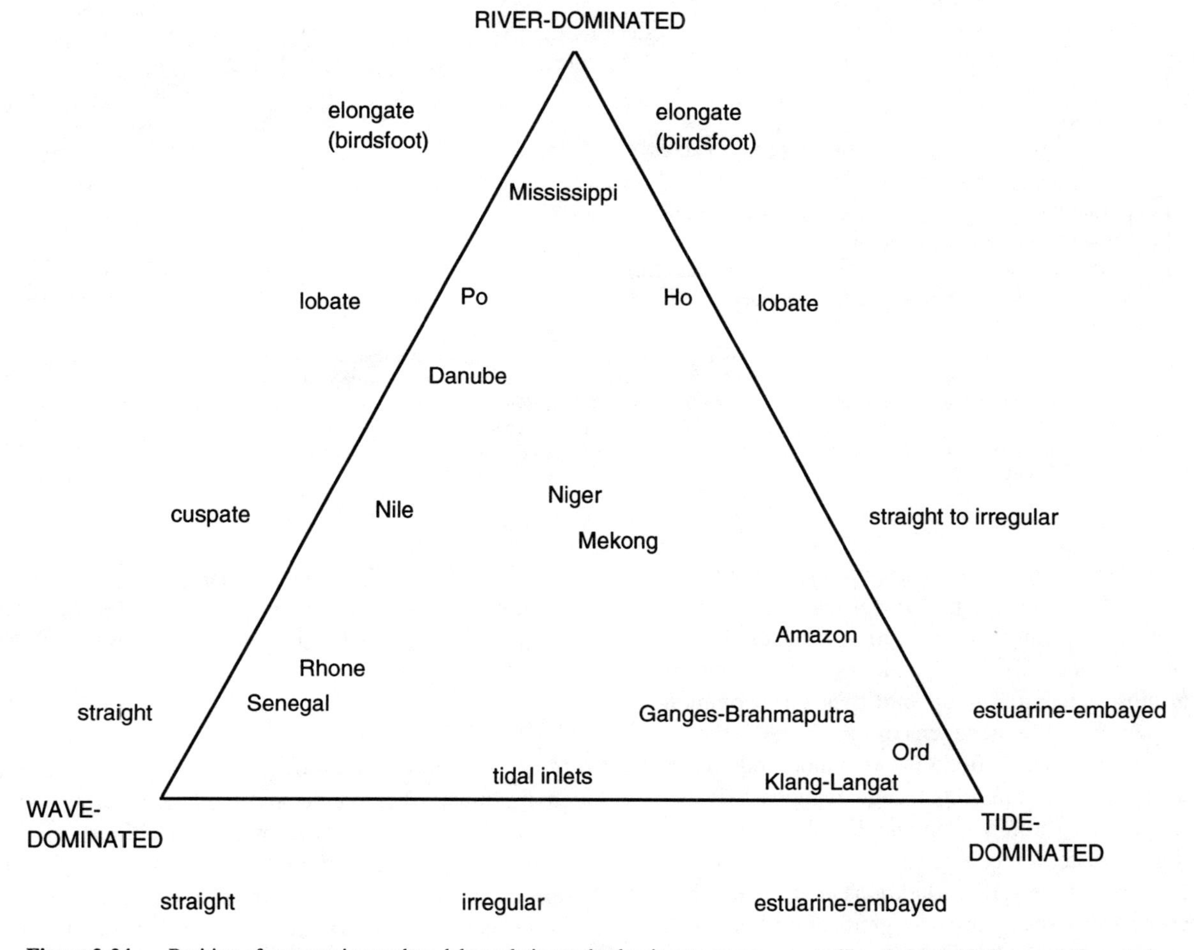

Figure 2-24. Position of some major modern deltas relative to the dominant processes controlling their morphology and characteristics. (Modified from Wright, in Davis 1985.)

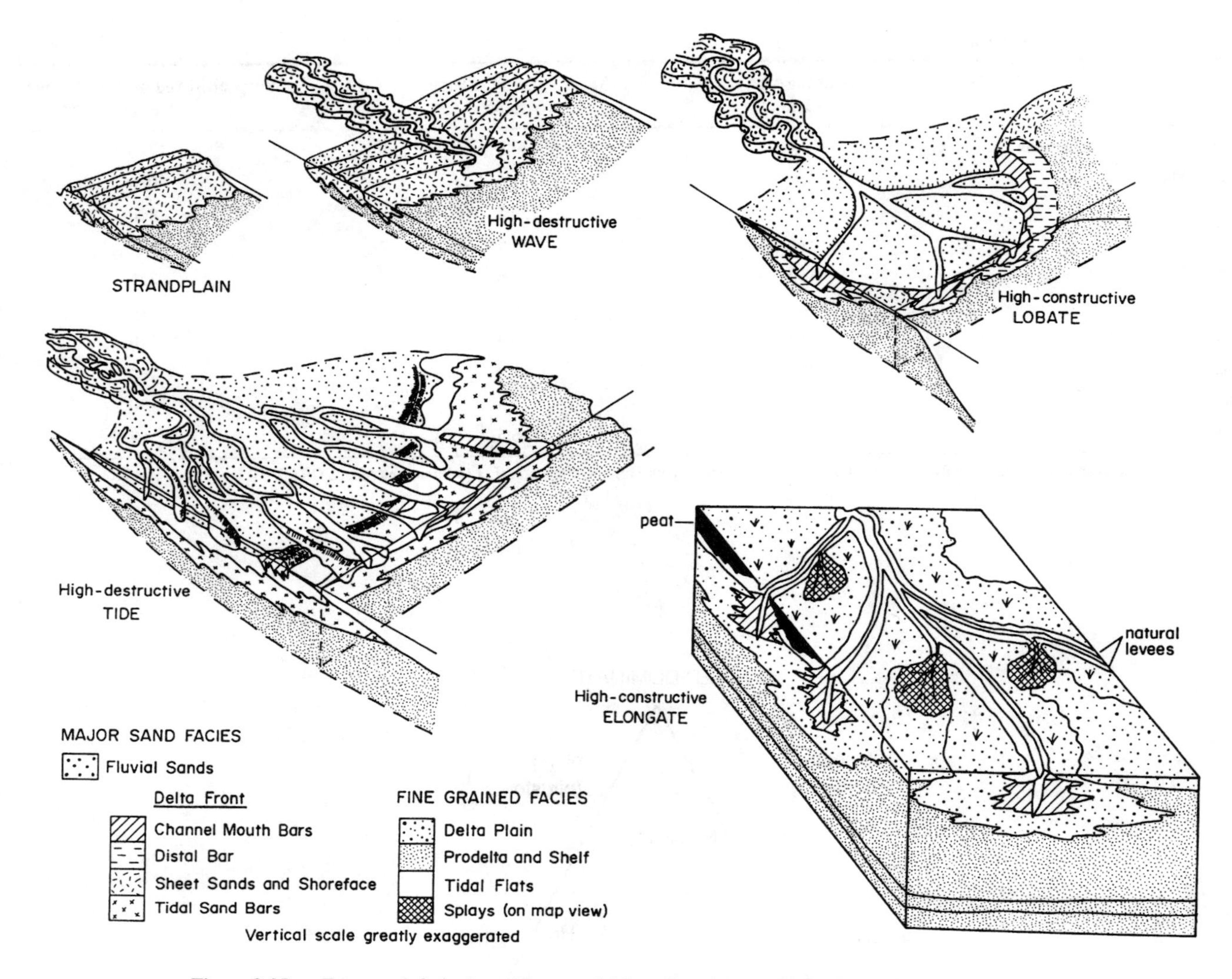

Figure 2-25. Framework facies in major types of deltas. *Strandplain*, without delta and dominated by wave action, shown for comparison with delta coastlines. *High-destructive wave delta*, composed primarily of shoreface and associated fluvial sands. *High-destructive tidal delta*, with extensive tidal shoal or sandflat facies. *High-constructive lobate delta*, with associated fluvial sands, channel mouth bars and delta front sheet sands. *High-constructive elongate delta*, with thick channel mouth bars or bar fingers. (Modified after A. J. Scott, in Fisher et al. 1969.)

from each major type of delta construction are shown in Fig. 2-28. Because deltas are progradational features, there is an overall coarsening vertical trend in their deposits; however, because of the variety of subenvironments and the abundance of mud in most deltas, fining-upward trends are common in internal sequences that may be tens of meters thick. Figure 2-29 contrasts processes and effects in two major modern deltas.

A special class of *fan delta* is recognized where rivers deposit their sediment directly from an alluvial fan into a deep-water basin; no general model is valid in which this situation develops because of the varied conditions of sediment character and supply, onshore and offshore gradient, and energy levels of the wave, tidal, and longshore regimes. However, since the 1980s, much attention has been devoted to both modern and ancient fan deltas, and numerous local models have been established (e.g., Galloway 1976; Chan and Dott 1983; Nemec and Steel 1988). Most contain coarse sediment derived from the fluvial system, but the associated marine facies may be dominated by fine sediment (e.g., Lewis and Ekdale 1991).

Marine Deposits

Most sediments in the rock record were deposited in marine environments (Figs. 2-30 and 2-31 summarize general aspects), and consequently, most depositional environment models focus on this setting. Because many of the most developed areas of the world are within 50 km of the coast, and because coastal environments are the most dynamic and changeable geomorphic settings on earth, resource managers, environmental planners, and engineers also have major interests in coastal process models developed by sedimentologists and geomorphologists. Of particular interest are the nature

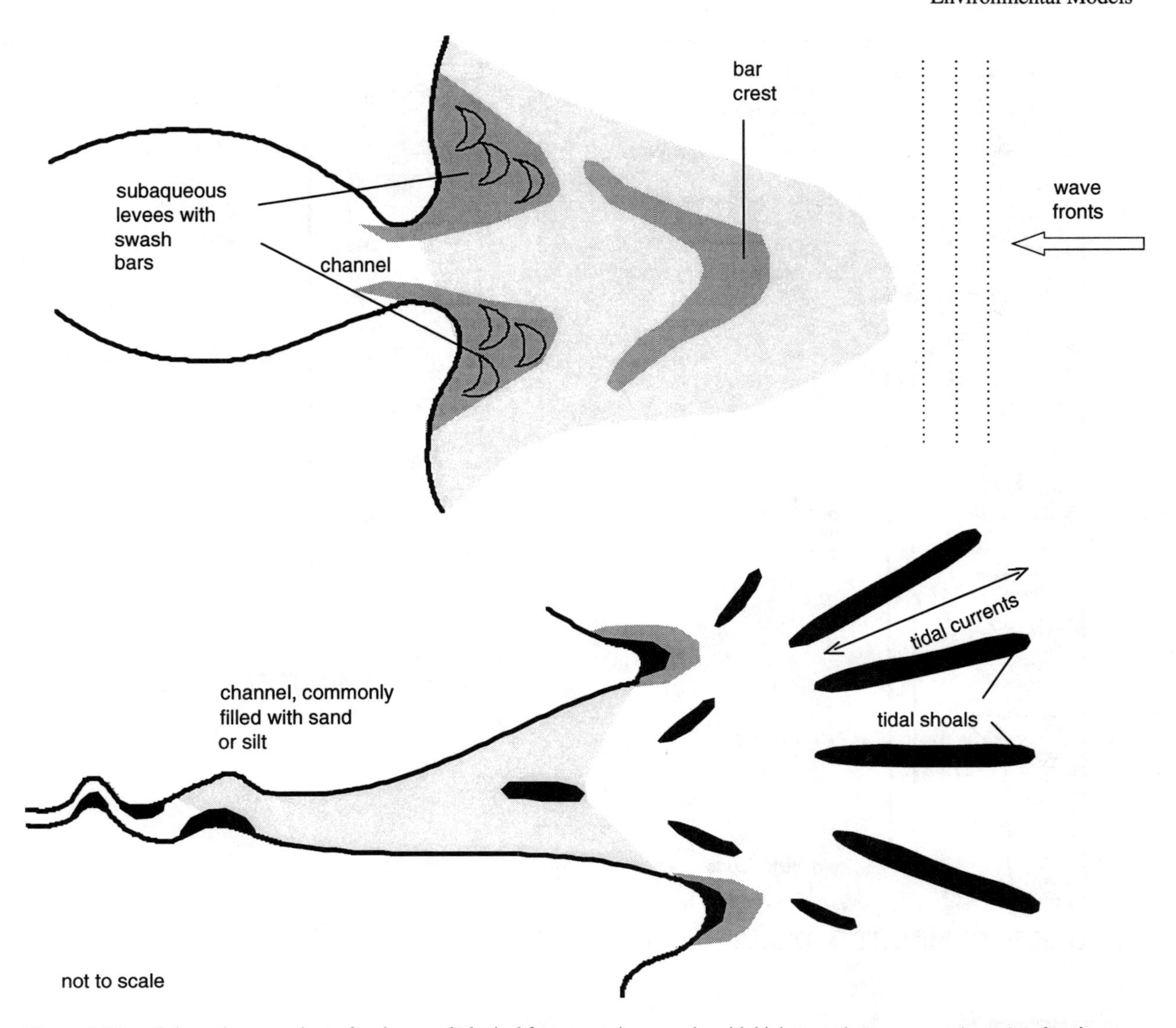

Figure 2-26. Schematic comparison of major morphological features at river mouths with high normal wave energy (upper) and at river mouths with high macrotidal energy. (After Wright 1977.)

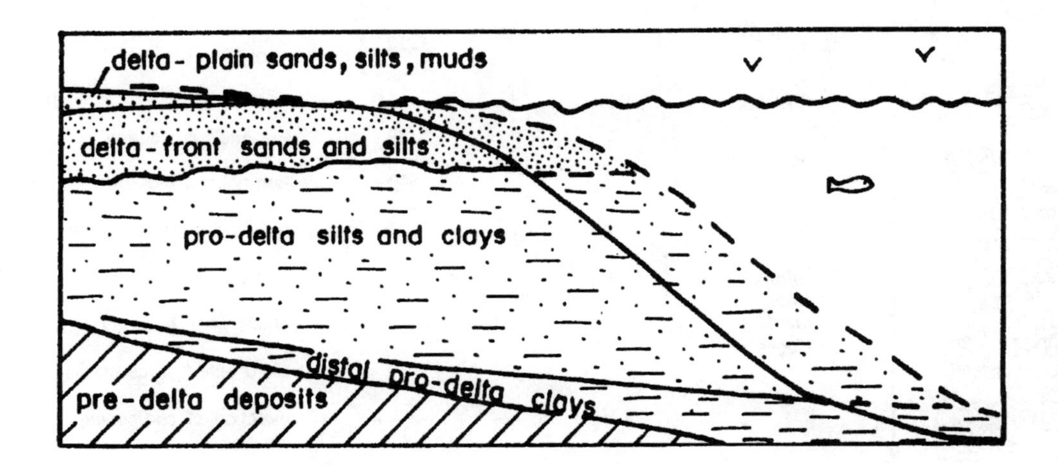

Figure 2-27. Idealized longitudinal cross-section through a high-constructive delta to show major named zones; dashed lines represent configuration of the delta profile after further progradation. In lakes and some fan deltas, bedding within (*Gilbert-type*) deltas is subparallel to the delta profile; the relatively horizontal coarse upper beds are then called *topsets*, the steeply dipping beds are *foresets*, and the relatively horizontal, finest lowermost beds are *bottomsets*.

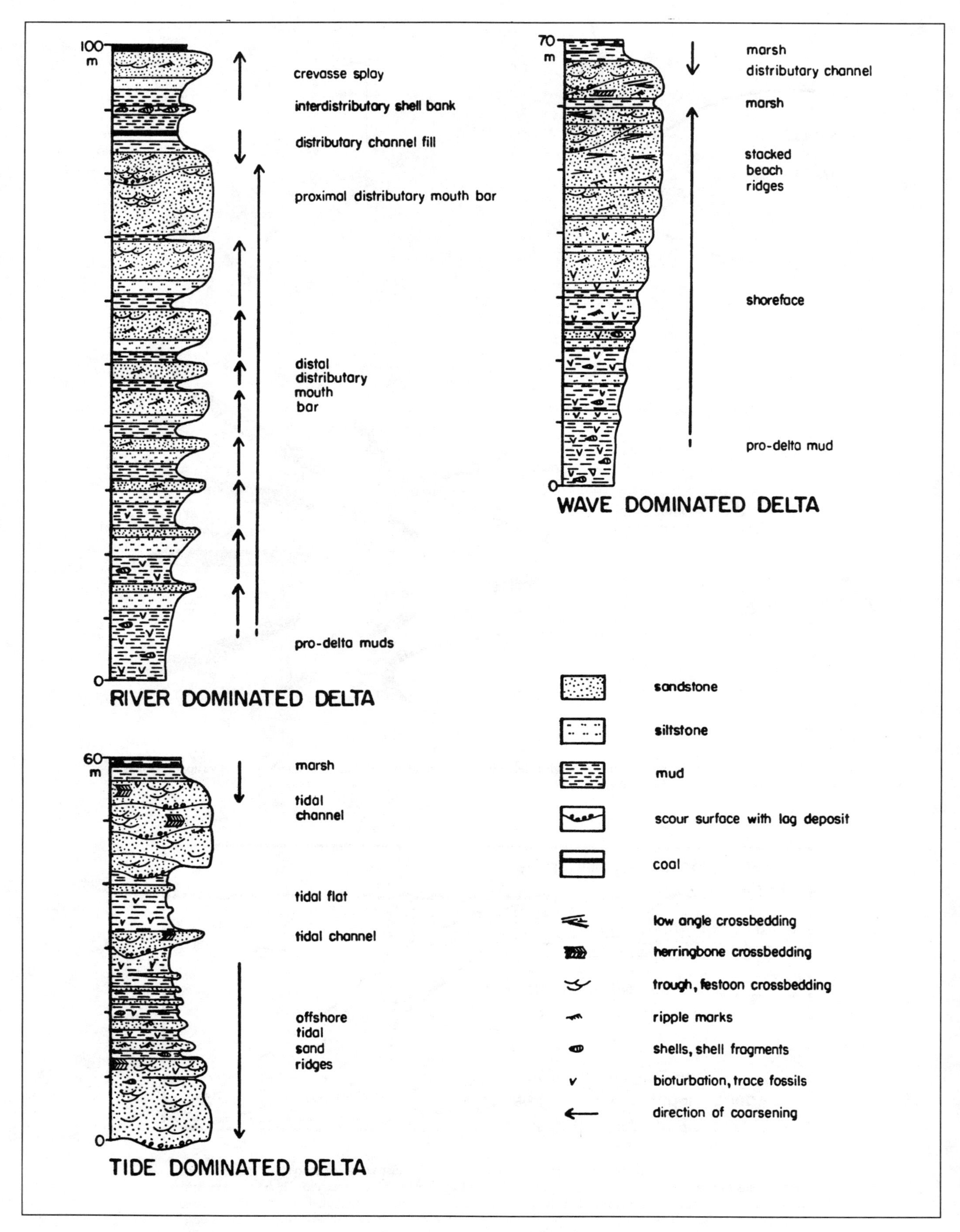

Figure 2-28. Examples of deltaic vertical facies sequences. (Redrawn from Miall, in Walker 1979.)

Influence			Result	Ganges-Brahmaputra Delta	Mississippi Delta
River discharge	Flood stage	sediment load	Transported sediment load rises during floods	Very high (Monsoonal floods)	High
River discharge	Flood stage	grain size	Mean grain size of transported sediment increases during floods	Silts and clays	Silts and clays
River discharge	Low flow stage	sediment load	Transported sediment load much lower than during floods	Moderate	Moderate
River discharge	Low flow stage	grain size	Mean grain size of transported sediment lower (dominantly clays and silts) than during floods	Clays and silts	Clays and silts
Coastal processes	Wave energy		High wave energy leads to erosion, reworking and winnowing of deltaic sediments	Low to moderate	Low
Coastal processes	Tidal range		High tidal range distributes wave energy over a large area and produces currents that rework and winnow deltaic sediments	>3 m	<1 m
Coastal processes	Current velocity		Strong littoral currents generated by waves and tides redistribute sediment alongshore, offshore, and inshore	Mostly high but low velocity areas are common	Low
Structural stability	Stable area		Rigid basement precludes subsidence and causes delta sediments to build upward and seaward	Rapid subsidence due to faulting and sediment compaction leads to >200 m of sediment in the modern delta	Rapid subsidence due to regional downwarping and sediment compaction has produced a deltaic sequence about 120 m thick
Structural stability	Subsiding area		Downwarping and sediment compaction allow thick deltaic sequences to accumulate, overlapping progradational lobes develop		
Structural stability	Rising area		Uplift of land (or fall in sea level) causes river distributaries to rework older deltaic sediments		
Climatic conditions	Wet area	Hot / warm	Warm wet climate leads to dense vegetation that traps sediment transported by river and tidal currents	Dense vegetation covers much of the delta plain, vast mangrove forests follow the coast and all major tidal channels, salt-tolerant grasses cover large areas of tidal flats	Dense vegetation covers much of the delta plain, minor mangrove forests and salt-tolerant grasses along the coast
Climatic conditions	Wet area	Cool / cold	Seasonal plant growth traps some sediment; plant debris accumulates during winter, forming delta peats		
Climatic conditions	Dry area	Hot / warm	Sparse vegetation limits sediment trapping; aeolian processes have a strong influence on the delta plain		
Climatic conditions	Dry area	Cool / cold	Sparse vegetation limits sediment trapping; winter ice interrupts fluvial processes; spring floods strongly influence sediment transport and deposition		

Figure 2-29. Factors that influence sedimentation in deltaic complexes. Some data are from personal observations of the Ganges-Brahmaputra Delta over large areas by D. McConchie who measured deposition rates between 0.5 and 1 m per year, and by R. Khandoker (Geology Department, Dhaka University) who noted thickness values considerably greater than 200 m resulting from compaction- and fault-caused subsidence.

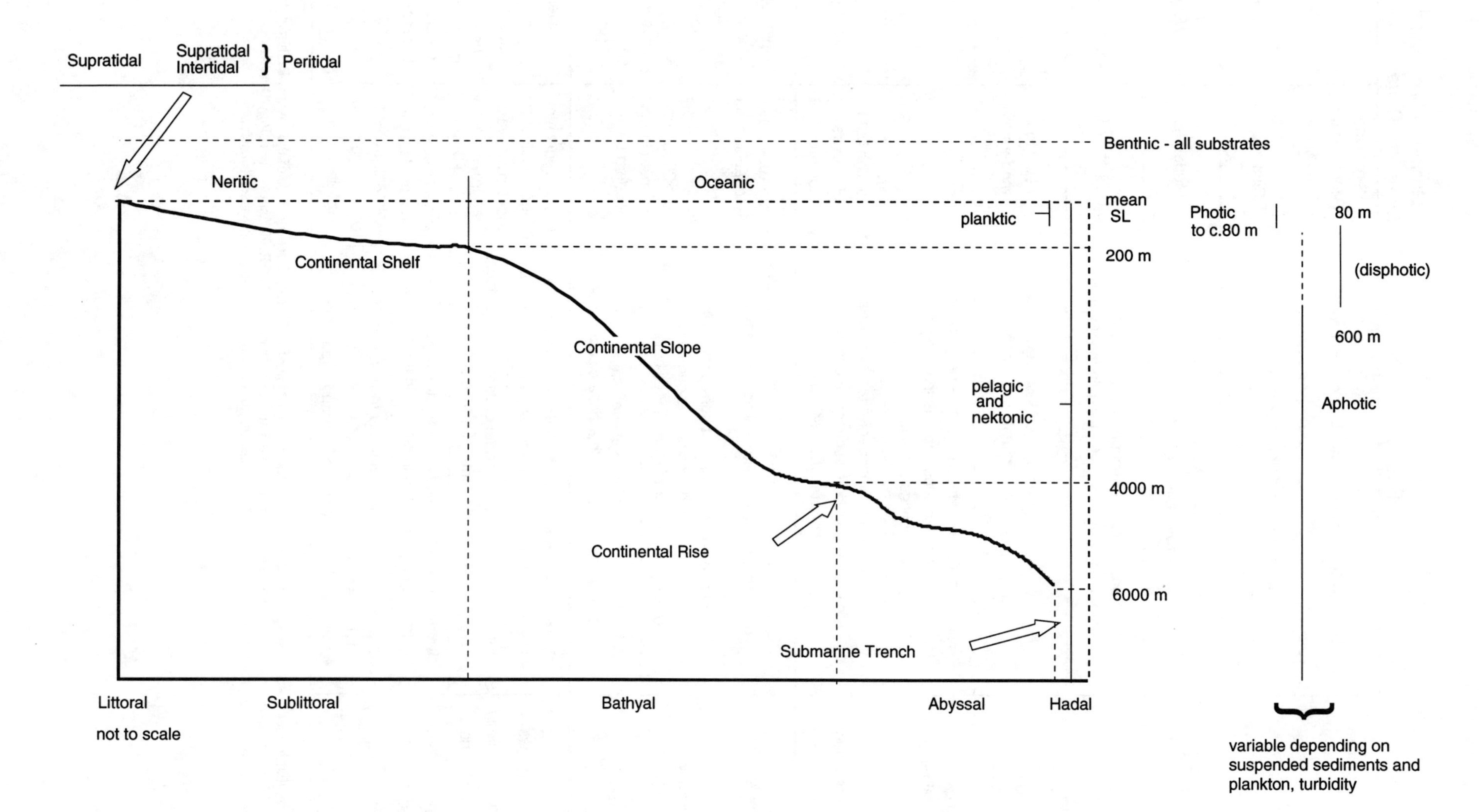

Figure 2-30. Common terminology for marine environments.

Depth Zone		Littoral	Shelf or Neritic (0–200m)	Bathyal (200–1000m)	Abyssal (1000 + m)
	Light	may be restricted by suspended matter; in clear tropical waters intensity falls to 35% at 20m, 20% at 40m, 13% at 60m, 5% at 100m; different wavelengths with penetration depths		surface waters as neritic Nil on substrate	
	Temperature	highly variable	varies with latitude, weather, surficial currents	thermocline-rapid decrease with depth	stenothermal ca. 4°C
	Salinity	highly variable	varies with latitude, weather, surficial currents	stenohaline	
	Turbulence	high or low	variable, storm waves effective	low; rare turbidity currents	
Physical Factors	Textures	gravels, sands muds on some tidal flats	variable-sand, mud; well to poorly sorted (storm reworked or bioturbated)	muds with sandy turbidites; gravelly mass flow deposits near submarine canyons calcareous ooze	siliceous ooze
	Structures	beach lamination, dune cross-bedding, flaser/lenticular bedding ripple mark, mud cracks	variable-planar cross and graded bedding, hummocky stratification, bioturbation	regular planar bedding, graded bedding; chaotic/contorted slide masses at base of slope	
	Authigenic/Perigenic components	beachrock cement			
			glauconite		
				. . . .Mn nodules	
			. . . . phosphates		
		. . . . .pyrite			
		. . . . .evaporites			
			oolites		
		broken shells			
		dolomite		sparse rhombs	
	Flora	mangroves . . . . sea grasses . . . . blue green algae (mats and stromatolites) rhodoliths calcified chlorophytes . . . . boring algae and fungi			
Benthic Biota	Fauna	barnacles, bivalves, corals, foraminifera, bryozoa, cephalopods, echinoderms, annelids, crustacea (representatives in virtually all depth zones except abyssal)			
	Traces	tracks . . . . dwelling traces . . . . .feeding traces . . . . .grazing traces . . . . (maximum diversity)			

Figure 2-31. Characteristics of the marine environment.

and rate of change in coastal morphology resulting from human developments and periodic high-energy natural events. (Although coastal environments could also be classed as transitional, whereas marine processes are dominant, most sedimentologists consider them within the marine realm.)

Beaches

Despite abundant studies, problems remain even in the identification of the controls on modern beach morphology (e.g., Williams 1973). Wind, waves, tidal action, longshore currents, and gravity-induced movements all combine in different proportions to produce the varied beaches that can be encountered along the coastline of any landmass. Beach or open (*nonbarred*) marine shoreline morphology is sketched in Fig. 2-32, common processes in Fig. 2-33, and some relationships between normal background (fair-weather) and extreme (storm) processes and deposits in Figs. 2-34 and 2-35. In general, continuous wave winnowing along the beach gives rise to well-sorted sands, or sorted, segregated-by-shape and imbricated gravels, but energy levels and the effectiveness of the winnowing process vary with factors such as offshore gradient and aspect of the coast (how open it is to long-fetch waves and prevailing storms and the orientation of the beach relative to storm winds; e.g., Kumar and Sanders 1976; Heward 1981; Hart and Plint 1989).

Compared to open shorelines, *barred* coasts, where an emergent offshore sand or gravel bar or spit creates a more or less enclosed lagoon, have a more complex geometry, range of energy levels in depositional processes, and series of subenvironments. The lagoon may accumulate sediments that are identical to either estuarine or tidal-flat deposits (see discussions above and below); the setting is distinguished by the presence of the bar (e.g., Figs. 2-36, 2-37; Davis, Ethridge, and Berg 1971; Davidson-Arnott and Greenwood 1976; Hunter, Clifton, and Phillips 1979; Kraft and John 1979). A variety of relatively small-scale features may be distinguished within the lagoon and bar, such as storm washovers, tidal channels and deltas, subaqueous and subaerial dune systems, salt and/or freshwater marshes (e.g., Schwartz 1982).

Tidal Flats

Tidal flats can occupy a broad or narrow band, depending on the gradient of the coast and on the tidal range; they may show a range of lithologic and biogenic characteristics (e.g., Fig. 2-38; Reineck in Rigby and Hamblin 1972; Ginsburg 1975; Klein 1977; Frey, Howard, and Park 1989), reflecting a range of processes (Fig. 2-39; e.g., Yeo and Risk 1981; Gerdes, Krumbein, and Reineck 1985). The tidal flat environment includes both the true tidal flats and the higher-en-

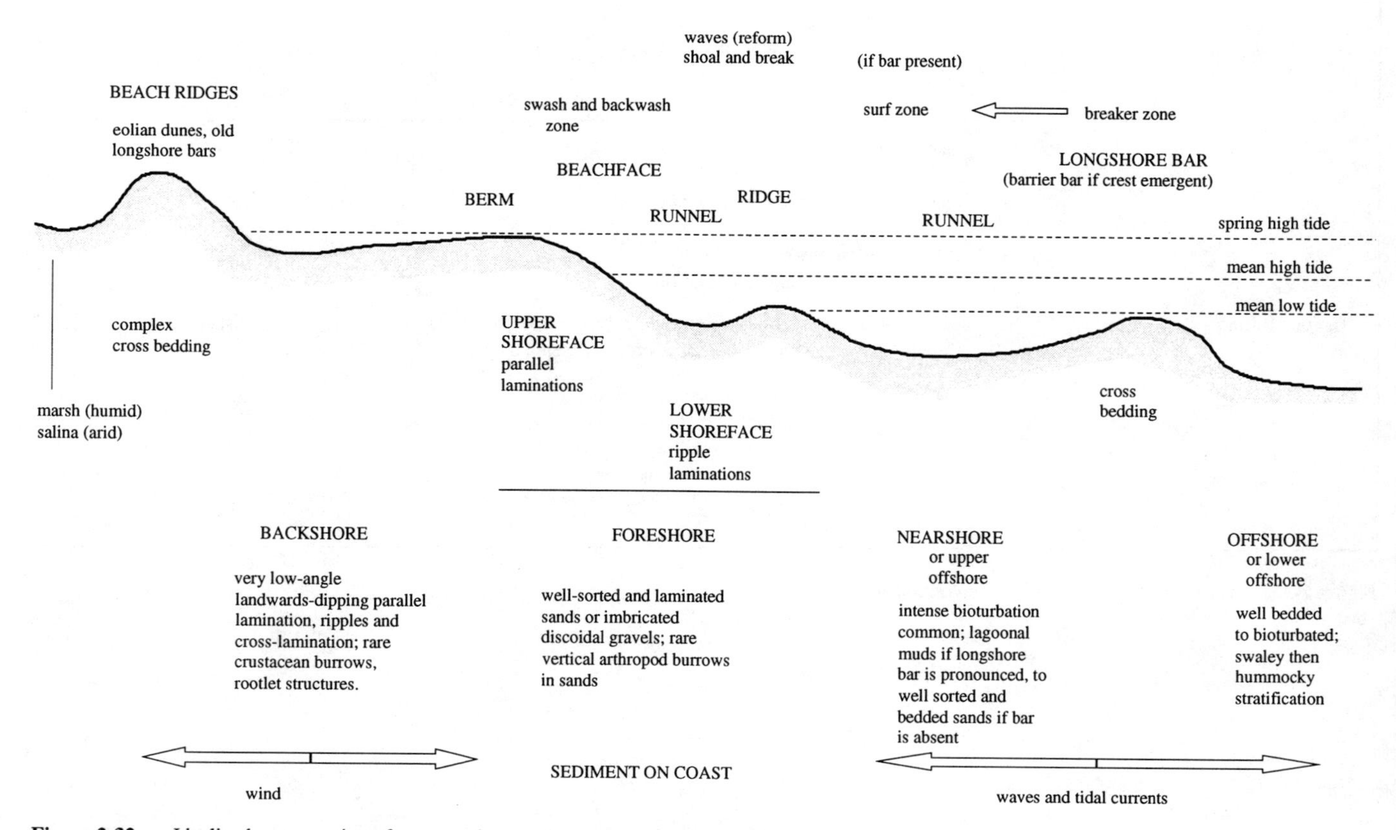

Figure 2-32. Idealized cross-section of open sandy coastline to show typical morphology, processes, and deposits.

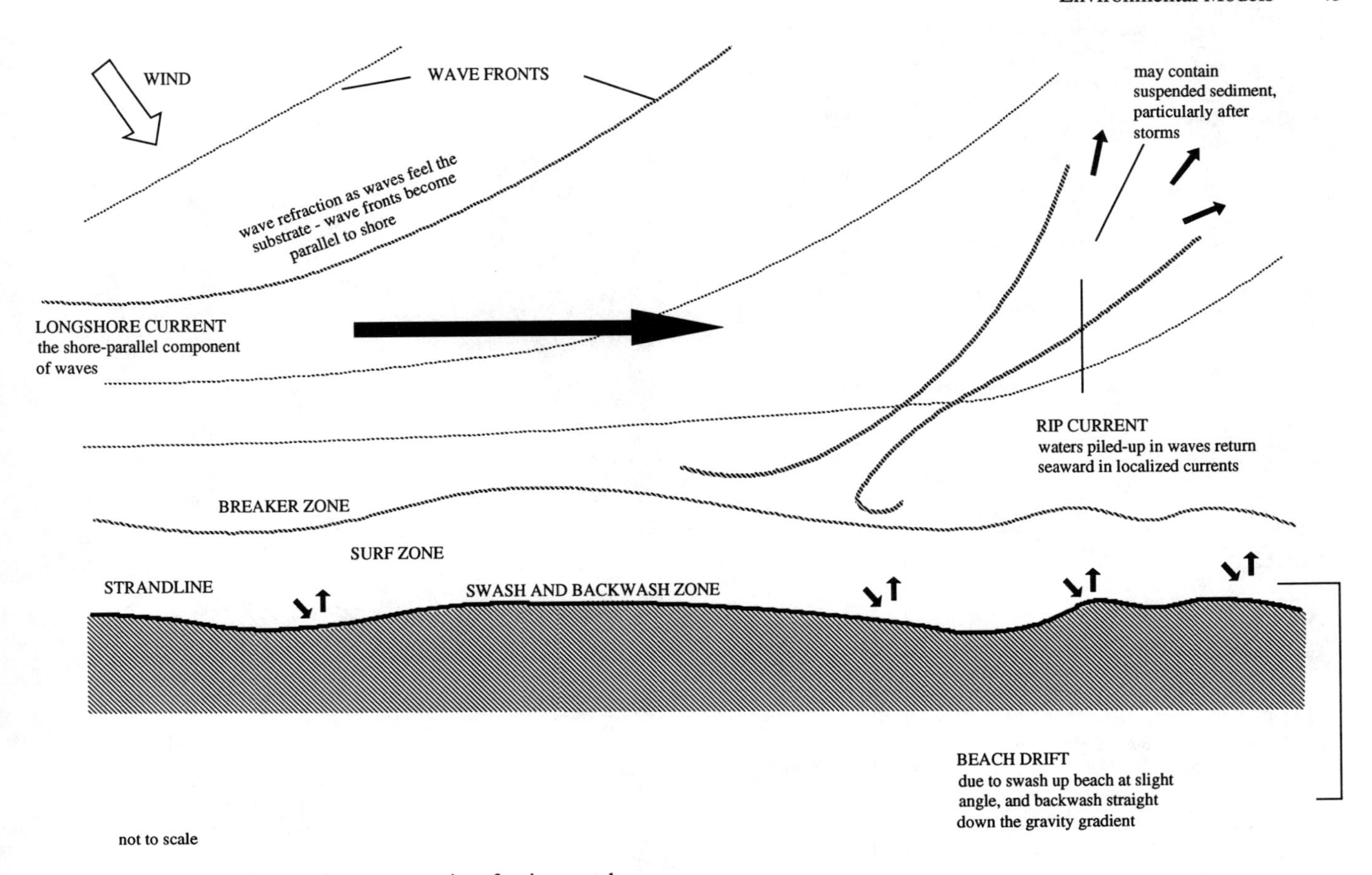

Figure 2-33. Schematic plan-view representation of major coastal processes.

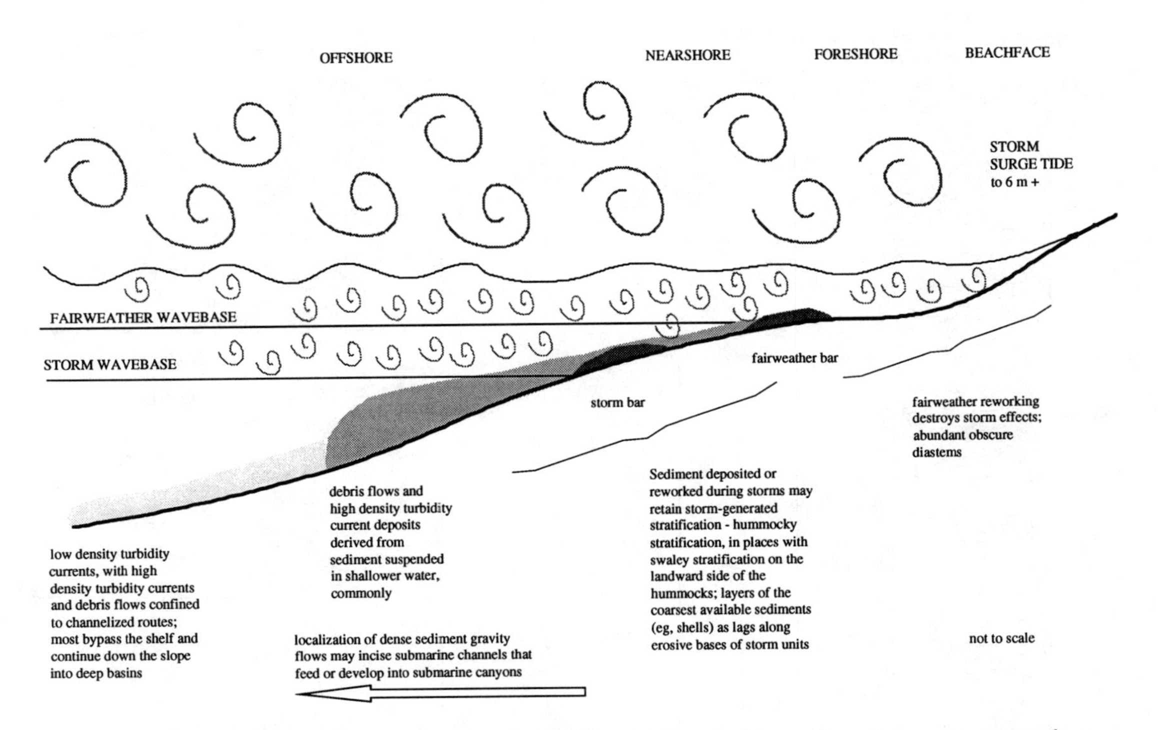

Figure 2-34. Schematic cross-sectional representation of differences between fair-weather and storm processes and products on an open coastline.

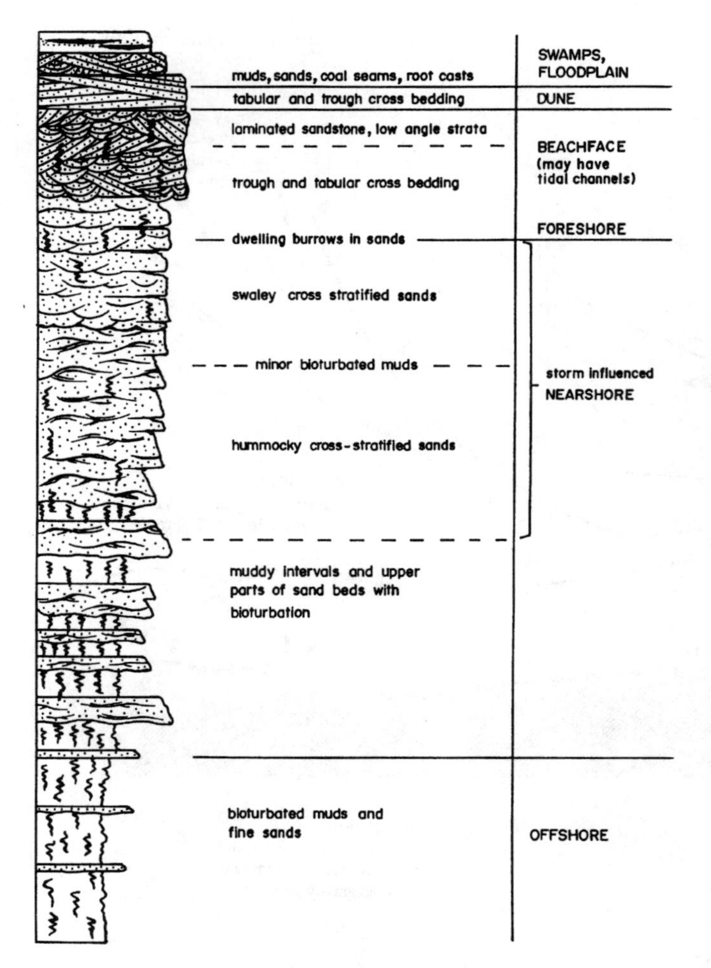

Figure 2-35. Idealized offshore-onshore vertical facies sequence for an open sandy coast.

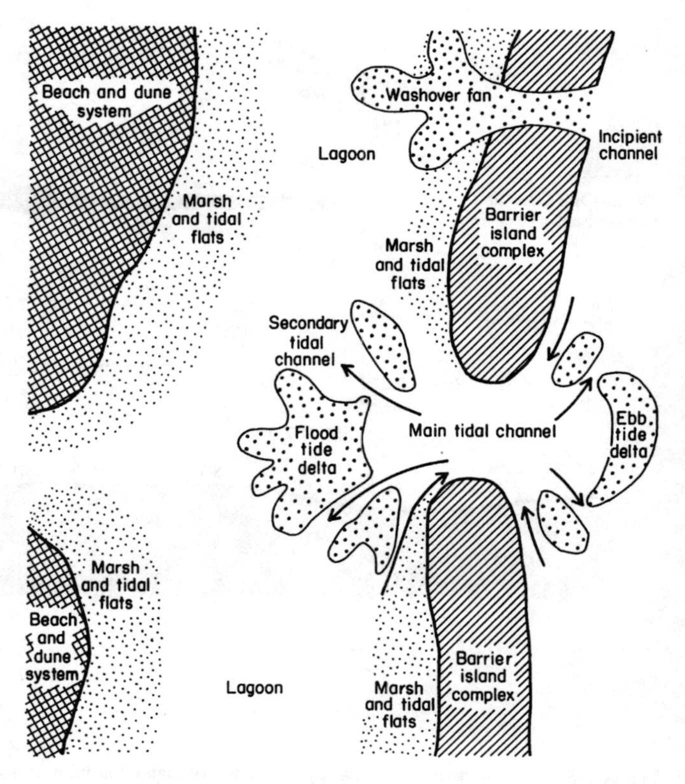

Figure 2-36. Plan-view representation of morphological features on a barred coast.

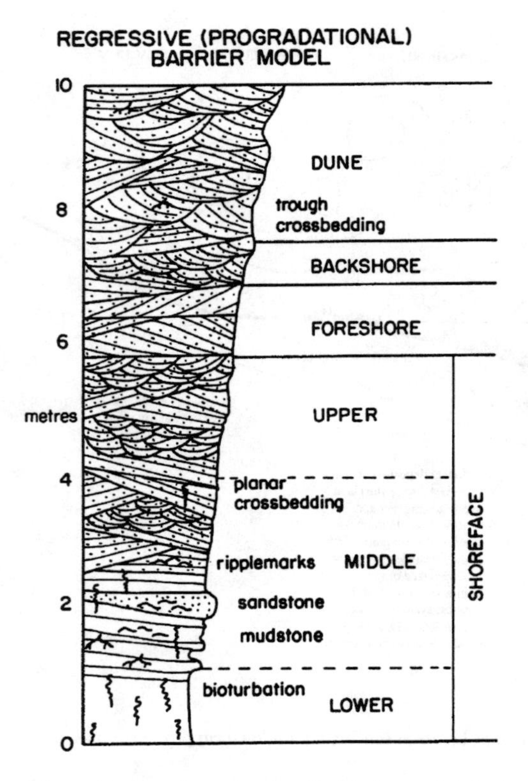

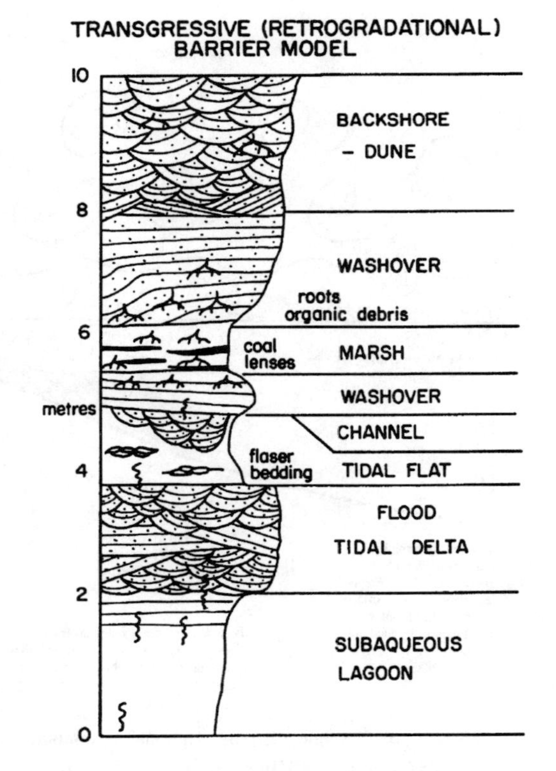

Figure 2-37. Generalized vertical facies models for two types of barred coast. (After Reinson, in Walker 1979.)

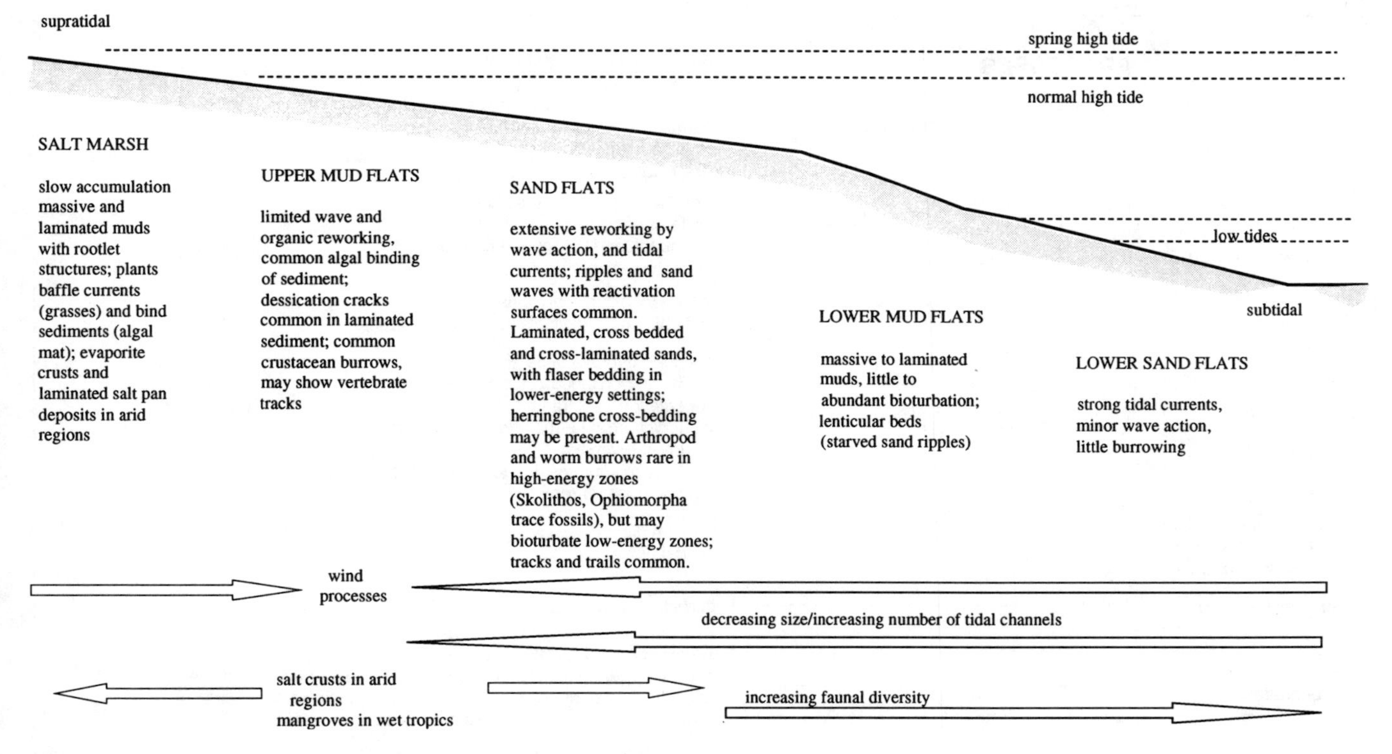

Figure 2-38. Schematic cross-section to show common subenvironments of intertidal flats.

ergy tidal channels that cut across them. Tidal channels form dendritic patterns, from numerous but small channels near mean high-water level to fewer but deeper and wider ones below mean low-water level. The smaller channels slowly shift their position, reworking the tidal flat sediments and introducing coarser sediment into the facies (e.g., Kumar and Sanders 1974). The rate of channel migration is largely controlled by the slope of the flats, the tidal range, and the rate of sediment supply, but periodic high-energy events can also play a major role. In areas such as the Bay of Bengal, where a high tidal range is combined with a high sediment influx and a low coastal gradient, the tidal flat environment may be several kilometers wide and only the largest tidal channels remain stable for more than one year. Intertidal sand bars play a dominant role in sedimentation of many sand-dominated tidal flats (e.g., Klein 1970).

Tidal flat environments differ markedly depending on climatic conditions. In arid areas, the upper part of the mud flats and the salt marsh, which form the supratidal zone (Fig. 2-38), are commonly characterized by accumulations of gypsum, anhydrite, and carbonate minerals with an ephemeral halite crust; vegetation is limited to sparse halophytic shrubs, grasses, and algal mats. In the supratidal zone of wet tropical areas, evaporite minerals are rare and the area is usually characterized by extensive mangrove forests, halophytic grasses, and algal mats (e.g., Chapman 1977; Hutchings and Saenger 1987). In more temperate regions, where neither mangroves nor evaporite deposits are common, the supratidal zone is dominated by halophytic grasses and algal mats. As a consequence of their texture, composition, depositional setting, and common association with sulfate-reducing bacteria, many ancient tidal flat sediments are host to economically important ore deposits (particularly copper; e.g., Maynard 1983). Interest in the sedimentology and stability of modern tidal flat deposits is also increasing as new value is found in them for mangrove afforestation in countries with timber demands that exceed the natural resource base (e.g., in Bangladesh, over 120,000 hectares of mangroves have been planted since the 1970s; McConchie 1990; Saenger and Siddiqi 1993; McConchie and Saenger 1991).

Shelves

Definition of the *shelf* is controversial (e.g., Lewis 1974; Burk and Drake 1974). "From the seaward limit of the shoreline" begs the question since many argue about what marks the outer limit of the transitional, peritidal, and littoral zones! "To the continental break" (where the offshore gradient increases significantly) or "to the 200-m isobath" (which is the approximate average depth of the modern continental break) involve problems in that it is uncertain whether in pre-Holocene times either of these features were as distinct as they are consequential to the Pleistocene sea-level changes. Although the concept of a relatively shallow marine environment is clearly in the minds of all workers, precise boundaries cannot generally be defined because there are so few criteria for depth in marine settings (e.g., see Hallam 1967; Anon 1988).

Shelf environments are the most complex of all marine environments because of their diversity of processes and

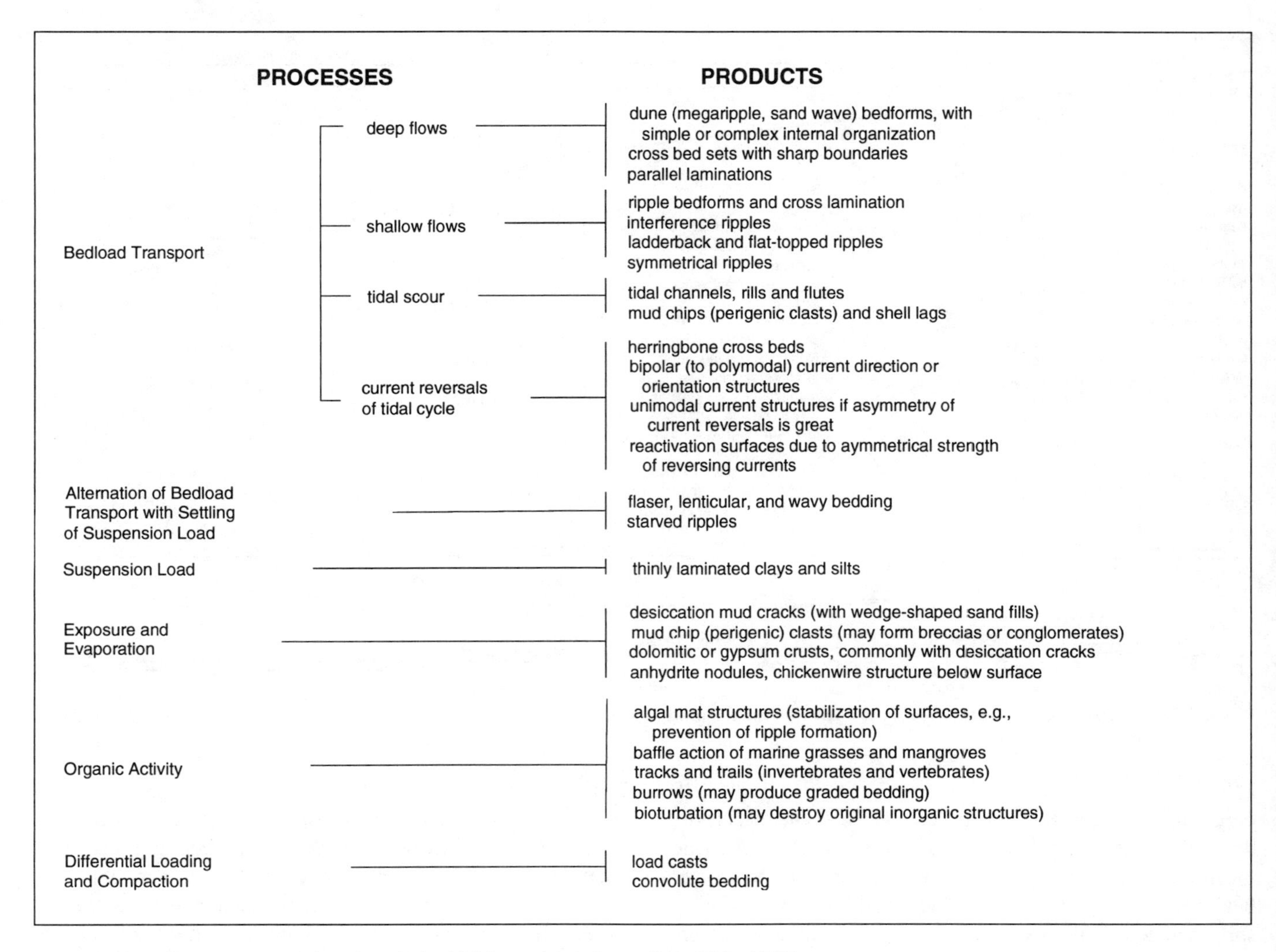

Figure 2-39. Some processes and products in the tidal flat environment. (After Klein, 1977.)

bathymetry; modern shelf settings are particularly complex because of the Holocene variations in sea level (e.g., discussion and review by Walker in Walker 1984). Consequently, whereas much is known about individual areas in terms of processes and responses, there are few or no general models of wide applicability (cf. Johnson 1956; Goldring and Bridges 1973; Bouma et al. 1980; Tillman and Siemers 1984; Knight and Maclean 1986). Both muddy and sandy shelves exist, and well-sorted sands can be deposited far from the shoreline by relatively deep sand waves (e.g., McCave 1971). Attempts at distinguishing between inner and outer shelf subenvironments are commonly feasible on the basis of facies differences (the inner shelf is above storm wave base and thus shows much more evidence of higher-energy processes), but are not always successful and are certainly not depth-dependent (e.g., gradient and aspect can determine storm wave base; transgressions and regressions leave relict sediments with out-of-phase characteristics; sediment supplied may be restricted in size; thus, similar deposits will accumulate over the entire shelf).

A major indicator of the inner shelf is the presence of *hummocky stratification*—a structure produced in association with major storms and considered to be characteristic of shallow shelf environments between normal fair-weather and storm wave base (Figs. 2-34, 2-35, and see discussion and Fig. 4-11 in Chapter 4). Most shelf deposits are muddy, but longshore currents and offshore channels may produce *stringers* of coarser detrital sediment, and some shelves exposed to common storms (e.g., Drake, Cacchione, and Karl 1985) or supplied with little clay are dominantly sandy. Many modern shelf areas also contain substantially coarser sediment (sands and gravels) deposited by fluvial processes during Pleistocene low-sea-level stands; both modern and ancient shelves also may contain channels filled with coarse sediment that was funneling toward the deeper oceanic repositories such as submarine fans (e.g., Lewis 1982). In general, shelf deposits tend to be bioturbated, but exceptions exist in which storms rework the sediment and may redeposit thicknesses greater than those penetrated by the infauna (alternatively, bioturbation may destroy most storm effects; e.g., Gagan, Johnson, and Carter 1988).

Shelf and submarine platform sediments are commonly

dominated by biogenic debris, mainly carbonate shell fragments but also silicious hardparts; a variety of authigenic minerals (such as glauconite, zeolites, and phosphates, and during the Precambrian, banded iron formation, see Chapter 9) are also found in areas where the influx of detrital and biogenic sediment is slow. These sediments are discussed more fully in Chapters 7, 8, and 9. However, because of the major environmental role played by reef systems, it is appropriate to mention them in this overview. In modern tropical regions, particularly on the western side of the oceans where equatorial currents swing poleward, coral reefs rise above the shelf surface to form substantial barriers to currents and waves (they thus influence greater areas than are occupied by the reefs themselves). Although initiated on a variety of substrates (such as volcanic mounds), modern reefs are largely developed on top of older reefs and reflect in their growth history the complex sea-level changes of the Holocene (e.g., Orme et al. 1978, Grimes, Searle, and Palmieri 1984). Nearshore fringing reefs surround many islands and may follow parts of continental coastlines, but most of these are small scale in contrast to the large barrier (and some patch) reefs of the mid- to outer shelf. Large coralgal reef systems are ecologically diverse and include a variety of depositional subenvironments with individual combinations of physical, biological, and chemical conditions (e.g., Fig. 2-40; see also Fig. 8-6). Because of the common high permeability and porosity of parts of reef systems, they are particularly suitable reservoir rocks for petroleum and are also hosts to major hydrothermal metalliferous ores. Modern coral reef ecosystems have a high conservation and economic importance to fisheries, recreation, biomedical research, coastal engineering, and in the preservation of genetic diversity of the biomass.

Deep Marine Settings

Deep marine environments mainly accumulate the finest detritus such as hemipelagic muds composed of planktic and pelagic organic hard parts (organic soft parts are generally decomposed before reaching the substrate in deep oceans and calcareous skeletal material may also dissolve where water depths exceed the carbonate compensation depth), fine authigenic minerals (e.g., some clays and zeolites), eolian detritus (frequently volcanic ash), and locally dominant other materials including ice-rafted detritus and extraterrestrial debris (e.g., Lizitzin 1972; Hsu and Jenkyns 1974; Pickering, Hiscott, and Hein 1989). The environments for accumulation of sediment include (1) the continental slope (e.g., Doyle and Pilkey 1979), which is an environment that accumulates pelagic sediment like the bathyal and abyssal seafloor but also collects spillover sediment from the shelf and funnels it down to the abyssal plain; and (2) the bathyal to abyssal basin plain (Fig. 2-30), in which accumulate the finest pelagic sediment settling from suspension, and occasional sediment gravity flows such as turbidites (see Chapter 3). *Submarine*

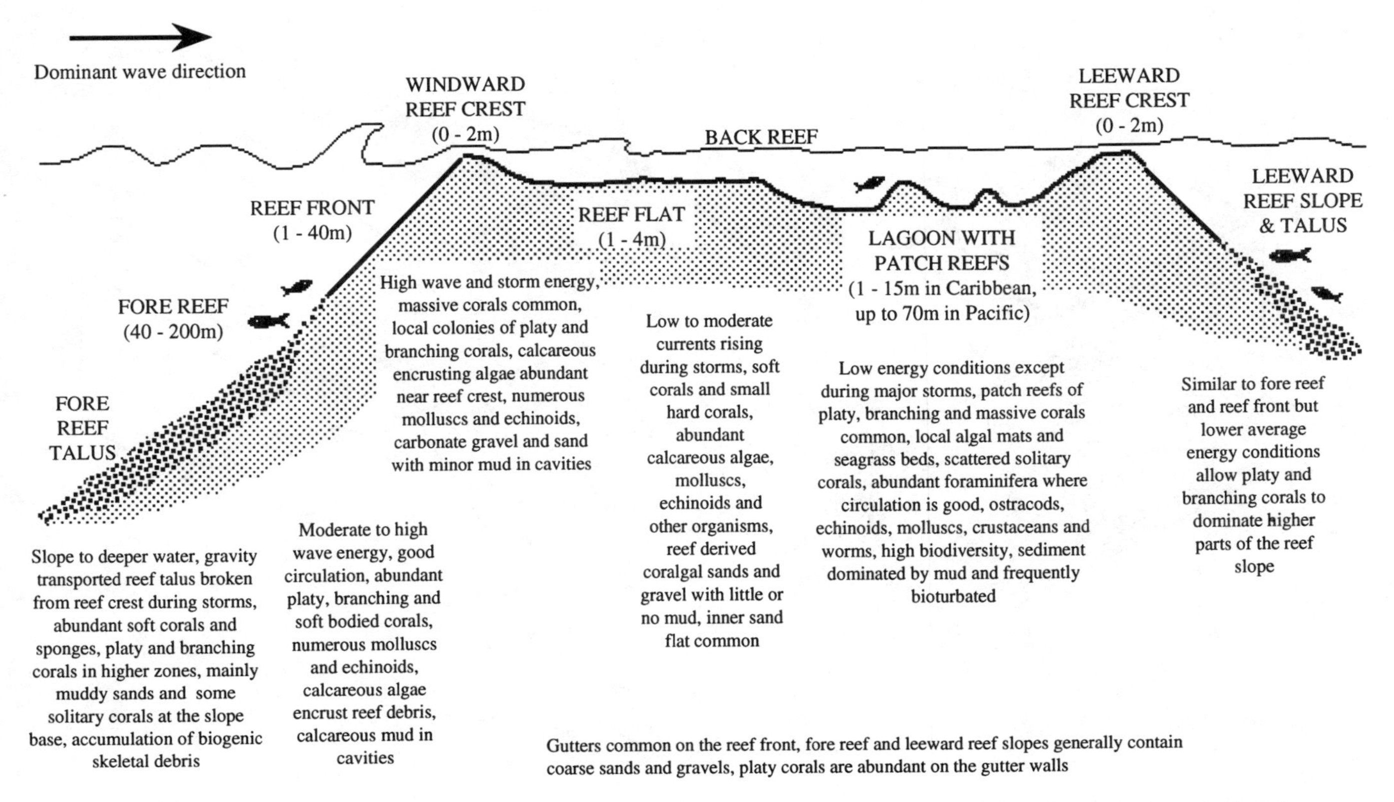

Figure 2-40. Schematic cross-section showing the varied subenvironments of a coral reef atoll (typical water depths are indicated, but the diagram is not to scale).

fans debouch from submarine canyons of various sizes onto the abyssal plains (e.g., Whitaker 1976; Stanley and Kelling 1978). Studies of modern deposits are carried out by seismic, sonar, and limited drilling (e.g., Normark and Gutmacher 1988; DSDP and JOIDES Reports). Facies models treating these environments are concentrated on settings where more varied sequences result from the intermittent injection of turbidity currents and other sediment gravity flows carrying coarser sediment from shallow marine depths (see discussion of gravity-flow processes in Chapter 3; eg, Middleton and Bouma 1973; Macdonald 1986; Pickering et al. 1986).

Most attention has focused on developing models for submarine fans (e.g., Figs. 2-41, 2-42; e.g., Normark 1970, 1978), many of which have been recognized in the rock record (e.g., Dott and Shaver 1974; Mutti and Ricci-Lucchi 1978; Walker 1978; Shanmugam and Moiola 1988). Despite a large number of papers on, and interpretations involving, the submarine fan and its characteristic suite of turbidite deposits, relatively few have been adequately recorded and mapped, and problems remain in producing an accordance between the features of modern and ancient examples (see Mutti and Normark in Zuffa 1987). Submarine fans, together with marine deltas and some alluvial fans, are widespread sites for the accumulation of thick and varied sediments; Table 2-10 shows similarities and differences between their deposits. Attention has also focused on authigenic minerals in deep ocean sediment, primarily the pelagic ferromanganese nodules (e.g., Glasby and Read 1976; Heath 1981). As a result of the immense size of some deep sea ferromanganese nodule deposits and developments in recovery technologies, they are being viewed by economic geologists as an important future metal resource; consequently, an improved understanding of deep sea sedimentation and geochemistry can be expected from the associated exploration work.

SELECTED BIBLIOGRAPHY

General

Allen, P., 1964, Sedimentological models. *Journal of Sedimentary Petrology* 34:289–93.

Anon., 1988, Determining paleobathymetry. *Palaios* 3(Special Issue/5):454–536.

Beerbower, J. R., 1964, Cyclothems and cyclic depositional mechanisms in alluvial plain sedimentation. In D. F. Merriam (ed.), *Symposium on Cyclic Sedimentation. Kansas Geological Survey Bulletin* 169:31–42.

Coneybeare, C. E. B., 1979, *Lithostratigraphic Analysis of Sedimentary Basins*. Academic Press, New York, 555p.

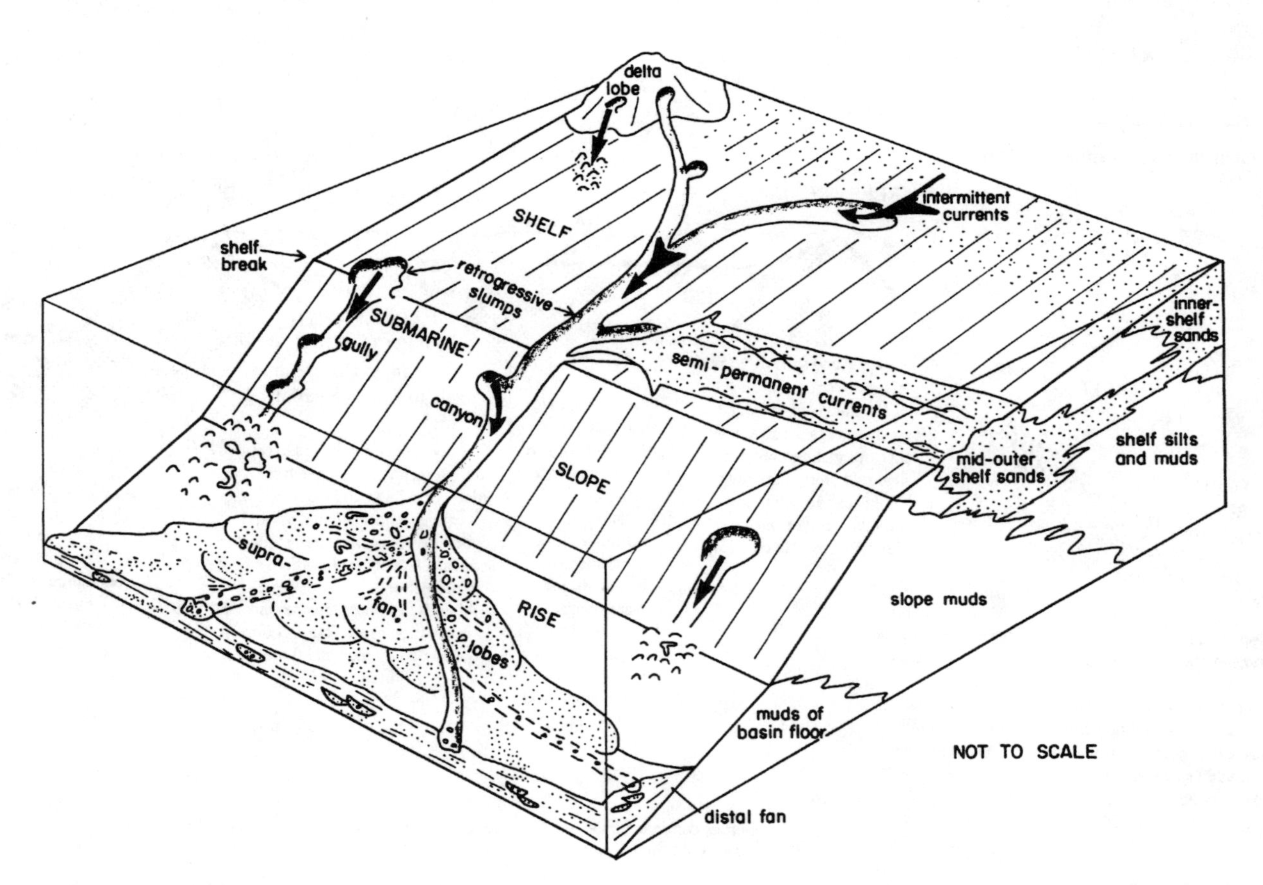

Figure 2-41. Schematic block diagram of a deep-water submarine fan setting and sediment supply system. (From Lewis 1982, *New Zealand Journal of Geology and Geophysics.*)

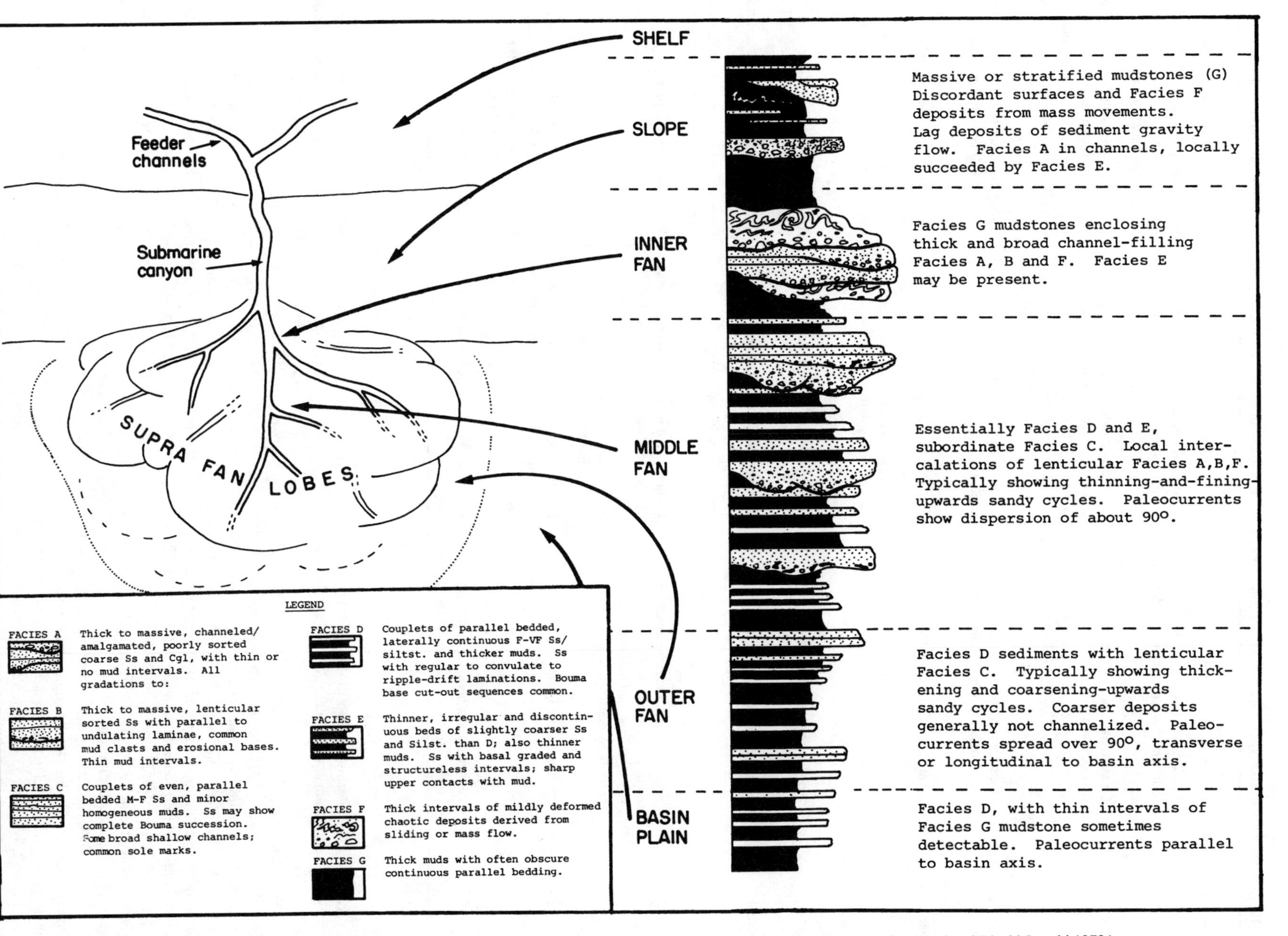

Figure 2-42. Vertical facies sequence in relation to plan-view morphology of an idealized deep-water submarine fan system. (Column after Mutti and Ricchi-Lucchi 1978.)

Table 2-10. Comparison of General Characteristics of Alluvial Fan, Delta, and Submarine Fan Deposits, with Emphasis on Differences Between Them

	Alluvial Fan	Delta	Submarine Fan
Bedding	Thick massive beds of similar sediment character; sparse thin lenses of fines	Well-bedded to massive sequences of similar sediment character (more variety on delta plain and front)	Well-bedded sequences of alternating sands and muds away from submarine canyon mouth and channels
Geometry	Characteristic channel structures and lensoidal bed geometry on outcrop scale	Channels on delta plain, bar bedforms and some channels on delta front; in large deltas, bed geometry not apparent at outcrop scale	Channels in proximal fan or suprafan areas, sheetlike geometry in distal units; amalgamation of units common; channeling may be visible on outcrop scale
Mass wasting	Occasional slump scarps, but rarely recognizable slump/slide deposits	Common slump and slide scarps and deposits	Proximal slump/slide deposits, related to base-of-slope position; some in channels
Grain size	Mainly gravel and sand; sparse muds; rare clasts of locally derived sediment	Rare gravel; mainly sand and mud in delta plain and delta front, silt and mud in prodelta; common clasts and rare rafts of locally derived sediment	Gravel in channels proximal to submarine fan, sand and mud more distal deposits; common rafts and clasts of locally derived sediment
Sorting	Poor sorting of particles	Sorted sand fraction	Sorting good to poor
Roundness	Angular to subround detritus	Subround to well-rounded detritus	Rounded detritus to angular carbonate clasts
Bioturbation	No bioturbation	Bioturbation common but may not be obvious, and in some areas of delta front, may be absent	Rare tracks, trails, grazing traces; rare intervals with more bioturbation probably reflecting temporary activity of displaced shallower-water fauna
Flora	Sparse terrestrial macro-plant remains, mainly as fragments of logs moved during floods; rare buried forests (mainly by debris flows)	Common obvious terrestrial plant debris, mainly as concentrations of leaves and twigs on bedding planes; common ash-rich peat/coal deposits and rare buried forests on delta plain	Very rare terrestrial plant debris
Fauna	No organic hardparts	Locally transported or in life position macro-invertebrate shells and shelly layers on delta front; dispersed macro- and micro-organism shells on prodelta	Rare and dispersed, displaced macro-invertebrate shells in proximal coarse facies; micro-organism shells in proximal and distal fines
Weathering	Pedogenic weathering horizons (soil profiles, rootlet or caliche horizons)	Pedogenic weathering on delta plain only	No pedogenic weathering

Crowley, K. D., 1984, Filtering of depositional events and the completeness of sedimentary sequences. *Journal of Sedimentary Petrology* 54:127–36.

Curtis, D.M. (ed.), 1978, *Depositional Environments and Paleoecology: Environmental Models in Ancient Sediments.* Society for Sedimentary Geology Reprint Series 6, Tulsa, Okla., 240p.

Degens, E. T., E. G. Williams, and M. L. Keith, 1957, Geochemical criteria for differentiating marine and fresh-water shale. *American Association of Petroleum Geologists Bulletin* 41:2427–55.

Dickinson, W. R. (ed.), 1974, *Tectonics and Sedimentation.* Society for Sedimentary Geology Special Publication 22, Tulsa, Okla. (See esp. pp. 1–27 by Dickinson.)

Dott, R. H., Jr., 1988, Something old, something new, something borrowed, something blue—a hindsight and foresight of sedimentary geology. *Journal of Sedimentary Petrology* 58:358–64.

Drever, J. I., 1982, *The Geochemistry of Natural Waters.* Prentice–Hall, Englewood Cliffs, N.J., 388p.

Einsele, G. (ed.), 1992, *Sedimentary Basins: Evolution, Facies, and Sediment Budget.* Springer–Verlag, Berlin, 550p.

Galloway, W. E., and D. K. Hobday, 1983, *Terrigenous Clastic Depositional Systems: Applications to Petroleum and Uranium Exploration.* Springer–Verlag, New York, 423p.

Hallam, A., 1967, Depth indicators in marine sedimentary environments. *Marine Geology* 5:329–567.

Hallam, A., 1981, *Facies Interpretation and the Stratigraphic Record.* W. H. Freeman & Co., San Francisco, 291p.

Kennett, J. P., 1982, *Marine Geology.* Prentice–Hall, Englewood Cliffs, N.J., 813p.

Leggett, J. K., and G. G. Zuffa (eds.), 1987, *Marine Clastic Sedimentology: Concepts and Case Studies.* Graham & Trotman Ltd., London.

Markevich, V. P., 1960, The concept of facies. *International Geology Review* 2:376–9, 498–507, 582–604.

Maynard, J. B., 1983, *Geochemistry of Sedimentary Ore Deposits.* Springer–Verlag, New York, 305p.

Miall, A. D., 1988, Facies architecture in clastic sedimentary basins. In K. Kleinspehn and C. Paola (eds.), *New Perspectives in Basin Analysis,* Springer–Verlag, New York, pp. 63–81.

Miall, A. D., 1990, *Principles of Sedimentary Basin Analysis,* 2d ed. Springer–Verlag, New York, 668p.

Middleton, G. V., 1973, Johannes Walther's Law of the correlation of facies. *Geological Society of America Bulletin* 84:979–88.

Nairn, A. E. M., 1965, Uniformitarianism and environment. *Palaeogeography Palaeoclimatology Palaeoecology* 1:5–11.

Reading, H. G. (ed.), 1986, *Sedimentary Environments and Facies,* 2d ed. Blackwell Scientific Publications, Oxford, 615p.

Reading, H. G., 1987, Fashions and models in sedimentology: A personal perspective. *Sedimentology* 34:3–9.

Reineck, H.-E., and I. B. Singh, 1980, *Depositional Sedimentary Environments,* 2d ed. Springer–Verlag, New York, 439p.

Rigby, J. K., and W. K. Hamblin (eds.), 1972, *Recognition of Ancient Sedimentary Environments.* Society for Sedimentary Geology Special Publication 16, Tulsa, Okla., 340p.

Scholle, P. A., and D. Spearing (eds.), 1982, *Sandstone Depositional Environments.* American Association of Petroleum Geologists Memoir 31, Tulsa, Okla., 410p.

Selley, R. C., 1970/1985, *Ancient Sedimentary Environments.* Cornell University Press, Ithaca, N.Y., 237/317p.

Shimp, N. F., J. Witters, P. E. Potter, and J. A. Schleicher, 1969, Distinguishing marine and freshwater muds. *Journal of Geology* 77:566–80.

Spearing, D. R., 1974, Summary sheets of sedimentary deposits. Geological Society of America Maps and Charts M-8, New York, 6 sheets.

Teichert, C., 1958, Concepts of facies. *American Association of Petroleum Geologists Bulletin* 42:2718–44.

Vail, P. R., R. M. Mitchum, Jr., and S. Thompson, III, 1977, Global cycles of relative changes of sea level. In C. E. Paynton (ed.), *Interregional Unconformities and Hydrocarbon Accumulation.* American Association of Petroleum Geologists Memoir 36, Tulsa, Okla., pp. 83–97.

Walker, R. G. (ed.), 1979, *Facies Models.* Geoscience Canada Reprint Series 1, Geological Association of Canada, Toronto, Ontario, 211p.

Walker, R. G. (ed.), 1984, *Facies Models,* 2d ed. Geoscience Canada, Reprint Series 1, Geological Association of Canada, Toronto, Ontario, 317p.

Walker, R. G., 1990, Facies modeling and sequence stratigraphy. *Journal of Sedimentary Petrology* 60:777–86.

Wolf, K. H., 1973a, Conceptual models, 1: Examples in sedimentary petrology environmental and stratigraphic reconstructions and soil, reef, chemical and placer sedimentary ore deposits. *Sedimentary Geology* 9:153–93.

Wolfe, K. H., 1973b, Conceptual models, 2: Fluviatile-alluvial, glacial, lacustrine, desert and shorezone [bar-beach-dune-chenier] milieus. *Sedimentary Geology* 9:261–81.

Organisms and Environment

Ager, D. V., 1963, *Principles of Palaeoecology.* McGraw Hill, New York, 371p.

Awramik, S. M., 1986, New fossil finds in old rocks. *Nature* 319:446–7.

Boucot, A. J., 1981, *Principles of Marine Benthic Paleoecology.* Academic Pess, New York, 463p.

Brasier, M. D., 1975, An outline history of seagrass communities. *Palaeontology* 18:681–702.

Chapman, V. J. (ed.), 1977, *Wet Coastal Ecosystems.* Elsevier, Amsterdam.

Craig, G. Y., 1966, Concepts in paleoecology: Importance of ecological approach to study of fossil faunas. *Earth-Science Reviews* 2:127–55.

Curtis, D. M. (ed.), 1978, *Depositional Environments and Paleoecology: Environmental Models in Ancient Sediments.* Society for Sedimentary Geology Reprint Series 6, Tulsa, Okla., 240p.

Dodd, J. R., and R. J. Stanton, 1981, *Paleoecology, Concepts and Applications.* Wiley–Interscience, New York, 559p.

Drake, E. T. (ed.), 1968, *Evolution and Environment.* Peabody Museum Centennial Symposium, Yale University, 1966; Yale University Press, New Haven, 470p.

Fagerstrom, J. A., 1964, Fossil communities in paleoecology. *Geological Society of America Bulletin* 75:1197.

Fonseca, M. S., 1989. Sediment stabilization by *Halophila decipiens* in comparison to other seagrasses. *Estuarine, Coastal and Shelf Science* 29:501–7.

Gerdes, G., W. E. Krumbein, and H.-E. Reineck, 1985, The depositional record of sandy, versicolored tidal flats (Mellum Island, southern North Sea). *Journal of Sedimentary Petrology* 55: 265–78.

Gould, S. J., 1970, Evolutionary paleontology and the science of form. *Earth-Science Reviews* 6:77–119.

Grimes, K. G., D. E. Searle, and V. Palmieri, 1984, The geological setting of the Capricornia section, Great Barrier Reef Marine Park. In W. T. Ward and P. Saenger (eds.), *The Capricornia Section of the Great Barrier Reef, Past, Present, and Future.* Royal Society Queensland and Australian Coral Reef Society, pp. 5–17.

Heckel, P. H., 1972, Recognition of ancient shallow marine environments. In J. K. Rigby, and W. K. Hamblin (eds.), *Recognition of Ancient Sedimentary Environments.* Society for Sedimentary Geology Special Publication 16, Tulsa, Okla., pp. 226–86.

Hedgpeth, J. W. (ed.), 1957, *Treatise on Marine Ecology and Paleoecology.* Geological Society of America Memoir 67 (vol. 1, Ecology; vol. 2, Paleoecology), New York.

Hutchings, P., and P. Saenger, 1987, *Ecology of Mangroves.* University of Queensland Press, Brisbane, Australia, 388p.

Jackson, K. S., I. R. Jonasson, and G. B. Skippen, 1978, The nature of metals–sediment–water interactions in freshwater bodies, with emphasis on the role of organic matter. *Earth-Science Reviews* 14:97–146.

Kidwell, S. M., F. T. Fursich, and T. Aigner, 1986, Conceptual framework for the analysis and classification of fossil concentrations. *Palaios* 1:228–38.

Lawrence, D. R., 1968, Taphonomy and information losses in fossil communities. *Geological Society of America Bulletin* 79:1315–30.

McConchie, D. M., and P. Saenger, 1991, Mangrove forests as an alternative to civil engineering works in coastal environments of Bangladesh: Lessons for Australia. In A. V. Arakel (ed.), *Proceedings of the 1990 Workshop on Coastal Zone Management* (Yeppoon, Queensland). Geoprocessors Ltd., Brisbane, Australia, pp. 220–33.

Redfield, A. C., 1958, The biological control of chemical factors in the environment. *American Scientist* 46:205–21.

Rigby, J. K., and N. D. Newell (eds.), 1971, Reef organisms through time. In *Symposium volume—Proceedings of the North American Paleontological Convention, Part J.* Allen Press, Lawrence, Kans.

Saenger, P., and N. A. Siddiqi, 1993, Land from the sea: The mangrove afforestation program of Bangladesh. *Ocean and Coastal Management* 20(1):23–39.

Schafer, W., 1972, *Ecology and Palaoecology of Marine Environments.* Oliver & Boyd, Edinburgh, 568p.

Schumm, S. A., 1968, Speculations concerning paleohydrologic controls of terrestrial sedimentation. *Geological Society of America Bulletin* 79:1573–88.

Scoffin, T. P., 1970, The trapping and binding of subtidal carbonate sediments by marine vegetation in Bimini Lagoon, Bahamas. *Journal of Sedimentary Petrology* 40:249–73.

Walter, M. R., R. Buick, and J. S. R. Dunlop, 1980, Stromatolites 3,400–3,500 Myr. old from the North Pole area, Western Australia. *Nature* 284:443–5.

Tectonics and Environment

Dickinson, W. R. (ed.), 1974, *Tectonics and Sedimentation.* Society for Sedimentary Geology Special Publication 22, Tulsa, Okla., 204p.

Dott, R. H., Jr., 1978, Tectonics and sedimentation a century later. *Earth-Science Reviews* 14:1–34.

Douglas, R. G., I. P. Gorsline, and D. S. Colburn, 1981, *Depositional Systems of Active Continental Margin Basins.* Short Course Notes, Society for Sedimentary Geology Pacific Section, Los Angeles, Calif., 165p.

Ingersoll, R. V., 1988, Tectonics of sedimentary basins. *Geological Society of America Bulletin* 100:1704–19.

Krynine, P. D., 1942, Differential sedimentation and its products during one complete geosynclinal cycle. *First Pan American Congress of Mining and Engineering Geology Annals,* Mexico, vol. 2, part 1, pp. 537–61.

Pitman, W. C., III, 1978. Relationship between eustacy and stratigraphic sequences of passive margins. *Geological Society of America Bulletin* 89:1389–1403.

Climate and Environment

Alimen, H., 1965, *The Quaternary Era in the Northwest Sahara.* Geological Society of America Special Paper 84.

Bain, G. W., 1963, Climatic zones throughout the ages. In *Polar Wandering and Continental Drift,* Society for Sedimentary Geology Special Publication 10, Tulsa, Okla., pp. 100–30.

Barron, E. J., 1983, A warm, equable Cretaceous: The nature of the problem. *Earth-Science Reviews* 19:305–38.

Beaty, C. B., 1971, Climatic changes: Some doubts. *Geological Society of America Bulletin* 82:1395–8.

Crook, K. A. W., 1967, Tectonics, climate, and sedimentation. In *7th Sedimentological Congress Abstracts,* Reading, England.

Dickens, J.M., 1993, Climate and the Late Devonian to Triassic, *Palaeogeography, Paleoclimatology, Paleoecology* 100:89–94.

Frakes, L. A., 1979, *Climates Throughout Geologic Time.* Elsevier, Amsterdam, 310p.

Hecht, A. D., 1985, *Paleoclimate Analysis Modeling.* Wiley–Interscience, New York, 445p.

Hollingsworth, S. H., 1962, The climatic factor in the geological record. *Quarterly Journal Geological Society of London* 118:1–21.

Jacobs, M. B., and J. D. Hays, 1972, Paleo-climatic events indicated by mineralogical changes in deep-sea sediments. *Journal of Sedimentary Petrology* 42:889–98.

Singer, A., 1980, The paleoclimatic interpretation of clay minerals and weathering profiles. *Earth-Science Reviews* 15:303–26.

Nonmarine Environments

Allen, J. R. L., 1965, A review of the origin and characteristics of recent alluvial sediments. *Sedimentology* 5:89–191.

Allen, J. R. L., 1970, Studies in fluviatile sedimentation: A comparison of fining-upwards cyclothems, with special reference to coarse-member composition and interpretation. *Journal of Sedimentary Petrology* 40:298–323.

Arakel, A. V., 1980, Genesis and diagenesis of Holocene evaporitic sediments in Hutt and Leeman lagoons, Western Australia. *Journal of Sedimentary Petrology* 50:1305–26.

Arakel, A. V., and A. Cohen, 1991, Deposition and early diagenesis of playa glauberite in the Karinga Creek drainage system, Northern Territory, Australia. *Sedimentary Geology* 70:41–59.

Arakel, A. V., and D. M. McConchie, 1982, Classification and genesis of calcrete and gypsite lithofacies in paleodrainage systems of inland Australia and their relationship to carnotite mineralization. *Journal of Sedimentary Petrology* 52:1149–70.

Bear, F. E., 1964, *Chemistry of the Soil.* Reinhold, New York, 515p.

Blissenbach, E., 1954, Geology of alluvial fans in semi-arid regions. *Geological Society of America Bulletin* 65:175–89.

Bowler, J. M., 1976, Aridity in Australia: Age, origins and expression in aeolian landforms and sediments. *Earth-Science Reviews* 12:279–310.

Bronzer, A., and J. A. Catt (eds.), 1989, *Paleopedology—Nature and Application of Paleosols.* Catena Supplement 16, Catena Verlag, Cremlingen-Dested, Germany.

Brown, T. M., and M. J. Kraus, 1987, Integration of channel and floodplain suites, 1: Developmental sequence and lateral relations of alluvial paleosols. *Journal of Sedimentary Petrology* 57:578–601.

Bull, W. B., 1977, The alluvial fan environment. *Progress in Physical Geography* 1:222–70.

Buol, S. W., F. D. Hole, and R. J. McCracken, 1980, *Soil Genesis and Classification,* 2d ed. Iowa State University Press, Ames, 406p.

Butt, C. R. M., 1989, Genesis of supergene gold deposits in the lateritic weathering regolith of the Yilgarn Block, Western Australia. In R. R. Keays, W. R. H. Ramsay, and D. I. Groves (eds.), *The Geology of Gold Deposits: The Perspective in 1988.* Economic Geology Monograph 6, Economic Geology Publishing Co., El Paso, Texas, pp. 460–70.

Cant, D. J., and R. G. Walker, 1976, Development of a braided-fluvial facies model for the Devonian Battery Point Sandstone, Quebec. *Canadian Journal of Earth Sciences* 13:102–19.

Cant, D. J., and R. G. Walker, 1978, Fluvial processes and resulting facies sequences in the sandy braided South Saskatchewan River. *Sedimentology* 25:625–48.

Carson, M. A., 1986, Characteristics of high-energy "meandering" rivers: The Canterbury Plains, New Zealand. *Geological Society of America Bulletin* 97:886–95.

Clemmensen, L. B., and J. Hegner, 1991, Eolian sequence and erg dynamics: The Permian Corrie Sandstone, Scotland. *Journal of Sedimentary Petrology* 61:768–74.

Ethridge, F. G., and R. M. Flores (eds.), 1981, *Recent and Ancient Nonmarine Depositional Environments: Models for Exploration.* Society for Sedimentary Geology Special Publication 31, Tulsa, Okla., 349p.

Eugster, H. P., and R. C. Surdam, 1973, Depositional environment of the Green River Formation of Wyoming. *Geological Society of America Bulletin* 84:1115–20.

Foth, H. D., 1984. *Fundamentals of Soil Science,* 7th ed. John Wiley & Sons, New York, 435p.

Fryberger, S. G., C. J. Schenk, and L. F. Krystinik, 1988, Stokes surfaces and the effects of near-surface groundwater-table on aeolian deposition. *Sedimentology* 35:21–41.

Glennie, K. W., 1987, Desert sedimentary environments, present and past—a summary. *Sedimentary Geology* 50:135–65.

Gordon, E. A., and J. S. Bridge, 1987, Evolution of Catskill (upper Devonian) river systems: Intra- and extra-basinal controls. *Journal of Sedimentary Petrology* 57:234–49.

Hoffman, S. J., 1986, Soil sampling. In J. M. Robertson (series ed.), Exploration Geochemistry Design and Interpretation of Soil Surveys. *Reviews in Economic Geology* 3:39–77.

Holland, H. D., 1984, *The Chemical Evolution of the Atmosphere and Oceans.* Princeton University Press, Princeton, 582p.

Hunt, C. B., 1972, *Geology of Soils, Their Evolution, Classification and Uses.* W. H. Freeman & Co., San Francisco, 344p.

Jackson, R. G., II, 1981, Sedimentology of muddy fine-grained channel deposits in meandering streams of the American Middle West. *Journal of Sedimentary Petrology* 51:1169–92.

Jankowski, J., and G. Jacobson, 1989, Hydrochemical evolution of regional groundwaters to playa brines in Central Australia. *Journal of Hydrology* 108:123–73.

Joyce, A. S., 1984, *Geochemical Exploration.* Australian Mineral Foundation, Adelaide, 183p.

Katz, B. J. , 1990, *Lacustrine Basin Exploration—Case Studies and Modern Analogies.* American Association of Petrolium Geologists Memior 50, AAPG Bookstore, Tulsa Okla. 340p.

Kocurek, G., and R. H. Dott, Jr., 1981, Distinctions and uses of stratification types in the interpretation of eolian sand. *Journal of Sedimentary Petrology* 51:579–95.

Langford, R. P., and M. A. Chan, 1989, Fluvial–aeolian interactions: Part II, ancient systems. *Sedimentology* 36:1037–51.

McGowen, J. H., and L. E. Garner, 1970, Physiographic features and stratification types of coarse-grained point bars: Modern and ancient examples. *Sedimentology* 14:77–111.

McKee, E. D., 1966, Structure of dunes at White Sands National Monument, New Mexico (and a comparison with structures of dunes from other selected areas). *Sedimentology* 7:1–70.

McKee, E. D. (ed.), 1979, Continental arid climate lithogenesis. *Sedimentary Geology* 22.

McKee, E. D., E. J. Crosby, and H. L. Berryhill, Jr., 1967, Flood deposits, Bijou Creek, Colorado, June 1965. *Journal of Sedimentary Petrology* 37:829–51.

Mann, A. W., 1984, Mobility of gold and silver in lateritic weathering profiles: Some observations from Western Australia. *Economic Geology* 79:38–49.

Matter, A., and M. E. Tucker (eds.), 1978, *Modern and Ancient Lake Sediments.* International Association of Sedimentologists Special Publication 2, Blackwell Scientific Publications, Oxford, 290p.

Miall, A. D., 1977, A review of the braided-river depositional environment. *Earth-Science Reviews* 13:1–62.

Miall, A. D., (ed.), 1978, *Fluvial Sedimentology.* Canadian Society of Petroleum Geologists Memoir 5, Calgary, Canada, 859p.

Miall, A. D., 1985, Architectural-element analysis: A new method of facies analysis applied to fluvial deposits. *Earth-Science Reviews* 22:261–308.

Monroe, S., 1981, Late Oligocene–early Miocene facies and lacustrine sedimentation, upper Ruby River Basin, southwestern Montana. *Journal of Sedimentary Petrology* 51:939–51.

Picard, M. D., and L. H. High, Jr., 1972, Criteria for recognizing lacustrine rocks. In J. K. Rigby and W. K. Hamblin (eds.), *Recognition of Ancient Sedimentary Environments,* Society for Sedimentary Geology Special Publication 16, Tulsa, Okla., pp. 108–45.

Porter, M. L., 1987, Sedimentology of an ancient erg margin: The Lower Jurassic Aztec Sandstone, southern Nevada and southern California. *Sedimentology* 34:661–80.

Pretorius, D. A., 1981. Gold and uranium in quartz–pebble conglomerates. *Economic Geology,* 75th anniv. vol., pp. 117–38.

Reeves, C. C., Jr., 1986, *Introduction to Paleolimnology.* Developments in Sedimentology 11, Elsevier, Amsterdam, 225p.

Reinhardt, J., and E. R. Sigleo (eds.), 1988, *Paleosols and Weathering Through Geologic Time: Principles and Applications.* Geological Society of America Special Paper 216.

Retallack, G. J., 1983, A paleopedological approach to the interpretation of terrestrial sedimentary rocks: The mid-Tertiary fossil soils of Badlands National Park, South Dakota. *Geological Society of America Bulletin* 94:823–40.

Rust, B. R., 1972, Structure and process in a braided river. *Sedimentology* 18:221–46.

Rust, B. R., 1981, Sedimentation in an arid-zone anastomosing fluvial system: Coopers Creek, central Australia. *Journal of Sedimentary Petrology* 51:745–55.

Rust, B. R., and B. G. Jones, 1987, The Hawkesbury Sandstone south of Sydney, Australia: Triassic analogue for the deposit of a large, braided river. *Journal of Sedimentary Petrology* 57:222–33.

Sanderson, I. D., 1974, Sedimentary structures and their environmental significance in the Navajo Sandstone, San Raphael Swell, Utah. *Brigham Young University Geology Studies* 21:215–46.

Schumm, S. A., and D. F. Meyer, 1979, Morphology of alluvial rivers of the Great Plains. *Great Plains Agricultural Council Publication* 91:9–14.

Shelton, J. W., and R. L. Noble, 1974, Depositional features of braided-meandering stream. *American Association of Petroleum Geologists Bulletin* 58:742–9.

Smith, D. G., and N. D. Smith, 1980, Sedimentation in anastomosed river systems: Examples from alluvial valleys near Banff, Alberta. *Journal of Sedimentary Petrology* 50:157–64.

Smith, N. D., 1970, The braided stream depositional environment: Comparison of the Platte River with some Silurian clastic rocks, north-central Appalachians. *Geological Society of America Bulletin* 81:2993–3014.

Smith, N. D., and W. E. L. Minter, 1980, Sedimentologic controls on gold and uranium in two Witwatersrand paleoplacers. *Economic Geology* 75:1–26.

Smith, N. D., T. A. Cross, J. P. Dufficy, and S. R. Clough, 1989, Anatomy of an avulsion. *Sedimentology* 36:1–24.

Stace, H. C. T., G. D. Hubble, R. Brewer, K. H. Northcote, J. R. Sleeman, M. J. Mulcahy, and E. G. Hallsworth, 1972. *A Handbook of Australian Soils.* Rellim, Glenside, S. Australia, 435p.

Stephens, C. G., 1971, Laterite and silcrete in Australia: A study of the genetic relationships of laterite and silcrete and their companion materials, and their collective significance in the formation of the weathered mantle, soils, relief and drainage of the Australian continent. *Geoderma* 5:5–52.

Sweet, M. L., J. Nielson, K. Havholm, and J. Farrelley, 1988, Algodones dune field of southeastern California: Case history of a migrating modern dune field. *Sedimentology* 35:939–52.

van Houten, F. B. (ed.), 1977, *Ancient Continental Deposits.* Benchmark Papers in Geology 43, Dowden, Hutchinson & Ross, Inc., Stroudsburg, Pa.

Walker, T. R., and J. C. Harms, 1972, Eolian origin of flagstone beds, Lyons Sandstone (Permian), type area, Boulder County, Colorado. *Mountain Geologist* 9:279–88.

Wetzel, R. G., 1975, *Limnology.* Saunders, Philadelphia.

Williams, P. F., and B. R. Rust, 1969, The sedimentology of a braided river. *Journal of Sedimentary Petrology* 39:649–79.

Wright, V. P. (ed.), 1986, *Paleosols—Their Recognition and Interpretation.* Blackwell Scientific Publications, Oxford, 315p.

Yuretiach, R. F., L. J. Hickey, B. P. Gregson, and Y.-L. Hsia, 1984, Lacustrine deposits in the Paleocene Fort Union Formation, northern Bighorn Mountains, Montana. *Journal of Sedimentary Petrology* 54:836–52.

Transitional Environments

Archer, A. W., and C. G. Maples, 1984, Trace-fossil distribution across a marine-to-nonmarine gradient in the Pennsylvanian of southwestern Indiana. *Journal of Paleontology* 58:448–66.

Cant, D. J., 1984, Development of shoreline-shelf sand bodies in a Cretaceous epeiric sea deposit. *Journal of Sedimentary Petrology* 54:541–56.

Chan, M. A., and R. H. Dott, Jr., 1983, Shelf and deep-sea sedimentation in Eocene forearc basin, western Oregon—fan or nonfan? *American Association of Petroleum Geologists Bulletin* 67:2100–116.

Cole, R. B., and P. G. DeCelles, 1991, Subaerial to submarine transitions in early Miocene pyroclastic flow deposits, southern San Joaquin Basin, California. *Geological Society of America Bulletin* 103:221–35.

Colella, A., and D. B. Prior (eds.), 1990, *Coarse-Grained Deltas.* International Association of Sedimentologists Special Publication 10, Blackwell Scientific Publications, Oxford, 357p.

Coleman, J. M., 1982, *Deltas: Processes of Deposition and Models for Exploration,* 2d ed. International Human Resources Development Corp., Boston, 124p.

Curray, J. R., 1976, Transgression and regressions. In R. L. Miller (ed.), *Papers in Marine Geology.* Macmillan, New York, pp. 175–203.

Curray, J. R., 1969, Estuaries, lagoons, tidal flats and deltas. In D. J. Stanley (ed.), *The New Concepts of Continental Margin Sedimentation.* American Geological Institute Short Course Lecture Notes, pp. JC3-1–3-30.

Davidson-Arnott, R. G. D., and B. Greenwood, 1976, Facies relationships on a barred coast, Kouchibouguac Bay, New Brunswick, Canada. In R. A. Davis and R. L. Ethington (eds.), *Beach and Nearshore Sedimentation.* Society for Sedimentary Geology Special Publication 24, Tulsa, Okla., pp. 149–68.

Davis, D. K., F. G. Ethridge, and R. R. Berg, 1971, Recognition of barrier environments. *American Association of Petroleum Geologists Bulletin* 55:550–65.

Davis, R. A., Jr. (ed.), 1985, *Coastal Sedimentary Environments,* 2d ed. Springer–Verlag, New York, 716p.

Evans, G., 1970, Coastal and nearshore sedimentation: A comparison of clastic and carbonate deposition. *Geological Association of London Proceedings* 81:493–508.

Fisher, W. L., L. F. Brown, Jr., A. J. Scott, and J. H. McGowan, 1969, *Delta Systems in the Exploration for Oil and Gas.* Bureau of Economic Geology, University of Texas at Austin, 78p.

Frey, R. W., and J. D. Howard, 1986, Mesotidal estuarine sequences: A perspective from the Georgia Bight. *Journal of Sedimentary Petrology* 56:911–24.

Frey, R. W., J. D. Howard, S.-J. Han, and B.-K. Park, 1989, Sediments and sedimentary sequences on a modern macrotidal flat, Inchon, Korea. *Journal of Sedimentary Petrology* 59:28–44.

Galloway, W. E., 1976, Sediments and stratigraphic framework of the Copper River fan-delta, Alaska. *Journal of Sedimentary Petrology* 46:726–37.

Gerdes, G., W. E. Krumbein, and H.-E. Reineck, 1985, The depositional record of sandy, versicolored tidal flats (Mellum Island, southern North Sea). *Journal of Sedimentary Petrology* 55:265–78.

Ginsburg, R. N. (ed.), 1975, *Tidal Deposits: A Casebook of Recent Examples and Fossil Counterparts.* Springer–Verlag, New York, 428p.

Hails, J., and A. Carr (eds.), 1975, *Nearshore Sediment Dynamics and Sedimentation.* J. H. Wiley, London, 316p.

Hart, B. S., and A. G. Plint, 1989, Gravelly shoreface deposits: A comparison of modern and ancient facies sequences. *Sedimentology* 36:551–7.

Heward, A. P., 1981, A review of wave-dominated clastic shoreline deposits. *Earth-Science Reviews* 17:223–76.

Howard, J. D., 1972, Trace fossils as criteria for recognizing shorelines in the stratigraphic record. In J. K. Rigby and W. K. Hamblin (eds.), *Recognition of Ancient Sedimentary Environments.* Society for Sedimentary Geology Special Publication 16, Tulsa, Okla., pp. 215–25.

Howard, J. D., and R. W. Frey (eds.), 1975, Estuaries of the Georgia Coast, U.S.A.: Sedimentology and biology. *Senkenbergiana Maritima* 7:1–307.

Hoyt, J. H., 1969, Chenier versus barrier, genetic and stratigraphical distinction. *American Association of Petroleum Geologists Bulletin* 53:299–306.

Hunter, R. E., H. E. Clifton, and R. L. Phillips, 1979, Depositional processes, sedimentary structures and predicted vertical sequences in barred nearshore systems, southern Oregon coast. *Journal of Sedimentary Petrology* 49:711–26.

Klein, G. D. V., 1970, Depositional and dispersal dynamics of intertidal sand bars. *Journal of Sedimentary Petrology* 40:1095–127.

Klein, G. D. V., 1977, *Clastic Tidal Facies.* CEPCO Champaign–Urbana, Ill., 149p.

Kraft, J. C., and C. J. John, 1979, Lateral and vertical facies relations of transgressive barrier. *American Association of Petroleum Geologists Bulletin* 63:2145–63.

Kumar, N., and J. E. Sanders, 1974, Inlet sequences: A vertical succession of sedimentary structures and textures created by lateral migration of tidal inlets. *Sedimentology* 21:291–323.

Kumar, N., and J. E. Sanders, 1976, Characteristics of shoreface storm deposits: Modern and ancient analogues. *Journal of Sedimentary Petrology* 46:145–62.

Leckie, D. A., and C. Singh, 1991, Estuarine deposits of the Albian Paddy Member (Peace River Formation) and lowermost Shaftsbury Formation, Alberta, Canada. *Journal of Sedimentary Petrology* 61:825–49.

Lewis, D. W., and A. A. Ekdale, 1991, Lithofacies relationships in a late Quaternary gravel and loess fan delta complex, New Zealand. *Palaeogeography, Palaeoclimatology, Palaeoecology* 81:221–51.

McConchie, D. M., 1990. Report on land stability problems affecting coastal plantations in Bangladesh. U.N., FAO Working Paper 25, 31p.

McPherson, J. G., G. Shanmugam, and R. J. Moiola, 1987, Fan deltas and braid deltas: Varieties of coarse-grained deltas. *Geological Society of America Bulletin* 99:331–40.

Milliman, J. D., and R. H. Meade, 1983, World-wide delivery of river sediments to the oceans. *Journal of Geology* 91:1–21.

Moore, G. T., and D. O. Asquith, 1971, Delta: Term and concept. *Geological Society of America Bulletin* 82:2563–8.

Morgan, J. P., and R. H. Shaver (eds.), 1970, *Deltaic Sedimentation, Modern and Ancient.* Society for Sedimentary Geology Special Publication 15, Tulsa, Okla., 312p.

Nelson, B. W. (ed.), 1972, *Environmental Framework of Coastal Plain Estuaries.* Geological Society of America Memoir 133, New York.

Nemec, W., 1990, Deltas—remarks on terminology and classification. In A. Colella and D. B. Prior (eds.), *Coarse-Grained Deltas.* International Association of Sedimentologists Special Publication 10, Blackwell Scientific Publications, Oxford, pp. 3–12.

Nemec, W., and R. J. Steel (eds.), 1988, *Fan Deltas: Sedimentology and Tectonic Settings.* Blackie & Son, Ltd., Glasgow, 464p.

Nummedal, D., O. H. Pilkey, and J. D. Howard (eds.), 1987, Sea-level fluctuation and coastal evolution. Society for Sedimentary Geology Special Publication 41, Tulsa, Okla., 267p.

Schwartz, R. K., 1982, Bedform and stratification characteristics of some modern small-scale washover sand bodies. *Sedimentology* 29:835–49.

Shinn, E. A., R. M. Lloyd, and R. N. Ginsburg, 1969, Anatomy of a modern tidal flat, Andros Island, Bahamas. *Journal of Sedimentary Petrology* 39:1202–28.

Shirley, M. L., and J. A. Ragsdale (eds.), 1966, *Deltas in Their Geologic Framework.* Houston Geological Society, Houston, Tex., 251p.

Ward, L. G., W. R. Boynton, and W. M. Kemp, 1984, The influence of waves and seagrass communities on suspended particulates in an estuarine embayment. *Marine Geology* 59:85–103.

Williams, A. T., 1973, The problem of beach-cusp development. *Journal of Sedimentary Petrology* 43:857–66.

Wright, L. D., 1977, Sediment transport and deposition at river mouths: A synthesis. *Geological Society of America Bulletin* 88:857–68.

Yeo, R. K., and M. J. Risk, 1981, The sedimentology, stratigraphy, and preservation of intertidal deposits in the Minas Basin, Bay of Fundy. *Journal of Sedimentary Petrology* 51:245–60.

Shelf Environments

Bouma, A. H., D. S. Gorsline, C. Monty, and G. P. Allen (eds.), 1980, Shallow marine processes and products. *Sedimentary Geology* (Special Issue) 26:1–279.

Burk, C. H., and C. L. Drake (eds.), 1974, *The Geology of Continental Margins.* Springer–Verlag, New York, 1,009p.

Drake, D. E., D. A. Cacchione, and H. A. Karl, 1985, Bottom currents and sediment transport on San Pedro shelf, California. *Journal of Sedimentary Petrology* 55:15–28.

Gagan, M. K., D. P. Johnson, and R. M. Carter, 1988, The Cyclone Winifred storm bed, central Great Barrier Reef shelf, Australia. *Journal of Sedimentary Petrology* 58:845–6.

Goldring, R., and P. Bridges, 1973, Sublittoral shelf sandstones. *Journal of Sedimentary Petrology* 43:736–47.

Johnson, J. W., 1956, Dynamics of nearshore sediment movement. *American Association of Petroleum Geologists Bulletin* 40:2211–32.

Knight, R. J., and J. R. Maclean (eds.), 1986, *Shelf Sands and Sandstones.* Canadian Society of Petroleum Geologists Memoir 11, Calgary, Canada, 347p.

Lewis, D. W., 1982, Channels across continental shelves: Corequisites of canyon-fan systems and potential petroleum conduits. *New Zealand Journal of Geology and Geophysics* 25:209–55.

Lewis, K. B., 1974, The continental terrace. *Earth-Science Reviews* 10:37–71.

McCave, I. N., 1976, Sand waves in the North Sea off the coast of Holland. *Marine Geology* 10:199–225.

Orme, G. R., J. P. Webb, N. C. Kelland, and G. E. G. Sargent, 1978, Aspects of the geological history and structure of the northern Great Barrier Reef Province. *Philosophical Transactions of the Royal Society of London* A291:23–35.

Tillman, R. W., and C. T. Siemers (eds.), 1984, *Siliciclastic Shelf Sediments.* Society for Sedimentary Geology Special Publication 34, Tulsa, Okla., 268p.

Deep Marine Environments

Bouma, A. H., W. E. Normark, and N. E. Barnes (eds.), 1985, *Submarine Fans and Related Turbidite Systems.* Springer–Verlag, New York, 351p.

Dott, R. H., Jr., and R. H. Shaver (eds.), 1974, *Modern and Ancient Geosynclinal Sedimentation.* Society for Sedimentary Geology Special Publication 19, Tulsa, Okla.

Doyle, L. J., and O. H. Pilkey (eds.), 1979, *Geology of Continental Slopes.* Society for Sedimentary Geology Special Publication 27, Tulsa, Okla., 374p.

Glasby, G. P., and A. J. Read, 1976, Deep-sea manganese nodules. In K. H. Wolf (ed.), *Handbook of Strata-Bound and Stratiform Ore Deposits.* Elsevier, Amsterdam, pp. 295–340.

Heath, G. R., 1981, Ferromanganese nodules of the deep-sea. *Economic Geology,* 75th anniv. vol., pp. 736–65.

Hedberg, H. D., 1970, Continental margins from viewpoint of the petroleum geologist. *American Association of Petroleum Geologists Bulletin* 54:3–43.

Hsu, K. J., and H. C. Jenkyns, 1974, *Plagic Sediments: On Land and Under the Sea.* International Association of Sedimentologists Special Publication 1, Blackwell Scientific Publications, Oxford, 447p.

Lizitzin, A. P., 1972, *Sedimentation in the World Ocean.* Society for Sedimentary Geology Special Publication 17, Tulsa, Okla., 218p.

Macdonald, D. I. M., 1986, Proximal to distal sedimentological variation in a linear turbidite trough: Implications for the fan model. *Sedimentology* 33:243–59.

Middleton, G. V., and A. H. Bouma (eds.), 1973, *Turbidites and Deep-Water Sedimentation.* Short Course Lecture Notes, Society for Sedimentary Geology Pacific Section, Anaheim, Calif., 157p.

Mutti, E., and F. Ricci-Lucci, 1978, Turbidites of the northern Appenines: Introduction to facies analysis. *International Geology Review* 20:125–66.

Normark, W. R., 1970, Growth patterns of deep-sea fans. *American Association of Petroleum Geologists Bulletin* 54:2170–95.

Normark, W. R., 1978, Fan valleys, channels and depositional lobes on modern submarine fans: Characters for recognition of sandy turbidite environments. *American Association of Petroleum Geologists Bulletin* 62:912–31.

Normark, W. R., and C. E. Gutmacher, 1988, Sur submarine slide, Monterey Fan, central California. *Sedimentology* 35:629–47.

Pickering, K. T., R. N. Hiscott, and F. J. Hein, 1989, *Deep Marine Environments: Clastic Sedimentation and Tectonics.* Uwin–Hyman, Boston, 416p.

Pickering, K. T., D. A. V. Stow, M. P. Watson, and R. N. Hiscott, 1986, Deep-water facies, processes and models: A review and classification scheme for modern and ancient sediments. *Earth-Science Reviews* 23:75–174.

Shanmugam, G., and R. J. Moiola, 1988, Submarine fans: Characteristics, models, classification, and reservoir potential. *Earth-Science Reviews* 24:383–428.

Stanley, D. J., and G. Kelling (eds.), 1978, *Sedimentation in Submarine Canyons, Fans, and Trenches.* Dowden, Hutchinson & Ross, Stroudsburg, Pa.

Walker, R. G., 1978, Deep water sandstone facies and ancient submarine fans: Models for exploration for stratigraphic traps. *American Association of Petroleum Geologists Bulletin* 62:932–66. (Cf. Discussion by T. H. Nilsen and Reply, 1980, *American Association of Petroleum Geologists Bulletin* 64:1094–112, 1101–8.)

Whitaker, J. H. McD., 1976, *Submarine Canyons and Deep Sea Fans.* Benchmark Papers in Geology, Dowden, Hutchinson & Ross, Stroudsburg, Pa.

3
Processes in Sedimentation

An appreciation of the chemical, biological, and physical processes acting at or near the surface of the earth is integral to understanding and interpreting many characteristics of sediments, in particular sedimentary structure (Chapter 4), texture (Chapter 5), and composition, both with respect to the alteration of detrital minerals and with regard to the origin of nondetrital sediments (Chapters 7, 8, and 9). These processes are very closely interrelated and influence the history of most sediments. Interactions are particularly important at boundaries such as the sediment/water interface and the sediment/air interface, where gradients in physical, chemical, and biological processes are greatest (Santschi et al. 1990). For clarity, the different types of processes are treated separately here, but it is important to remember that although their relative influence may differ spatially and temporally, they almost always act in unison.

The intimacy of interaction between chemical and both physical and biological processes is illustrated in the formation of soils (Fig. 3-1). In soils, physical processes move products of chemical decay and (bio)chemical precipitates, and the thickness of a soil profile reflects a balance between in situ chemical weathering and physical removal and physical delivery of weathered products. In dry or cold mountainous areas, physical erosion dominates over chemical weathering, and unstable minerals (see discussion in Chapter 6) may be eroded and transported; soils may barely be formed. In contrast, where physical erosion is slow and water supply abundant, heavy chemical alteration may produce soil profiles many tens of meters thick. Physical processes commonly accelerate chemical alteration by mechanically breaking mineral grains into finer particles, which have a greater surface area/unit volume and thus are more readily subject to alteration by chemical processes; positive feedback occurs because chemical processes accelerate erosion by weakening the minerals that comprise rocks. The extent to which chemical reactions can progress in soils and sediments also depends on the efficiency with which physical processes (such as circulating solutions) can supply reactants and remove products; inadequate supply of a reactant (e.g., water for hydrolysis reactions) or sluggish removal of reaction by-products (i.e., residues) can slow or stop a particular reaction.

SURFICIAL CHEMICAL PROCESSES

Chemical (and biochemical) processes involving interactions between the lithosphere and any combination of the hydrosphere, atmosphere, and biosphere fundamentally influence the production, transport, deposition, and early diagenesis of most sediments. Rocks, minerals, and primary organic material are modified on or near the surface by a combination of chemical and biochemical processes to produce a variety of products (Fig. 3-2). An understanding of their nature and effects is also essential in modern environmental studies, because these various processes govern mobilization, dispersion, and chemical speciation of potential pollutants (e.g., Förstner and Wittman 1981; Kabata-Pendias and Pendias 1992). Unfortunately, although basic chemical principles are known, quantification of the influence of many of these processes is limited because there are too many intimately related variables in nature and chemical equilibrium is seldom attained in natural environments (e.g., Holland 1972); few natural geological settings at the earth's surface even approach closed systems. Despite these constraints, the descriptive and predictive utility of an examination of surficial chemical processes is remarkably high. Hence we treat fundamental geochemical principles more fully than most readers will initially think warranted. Further details

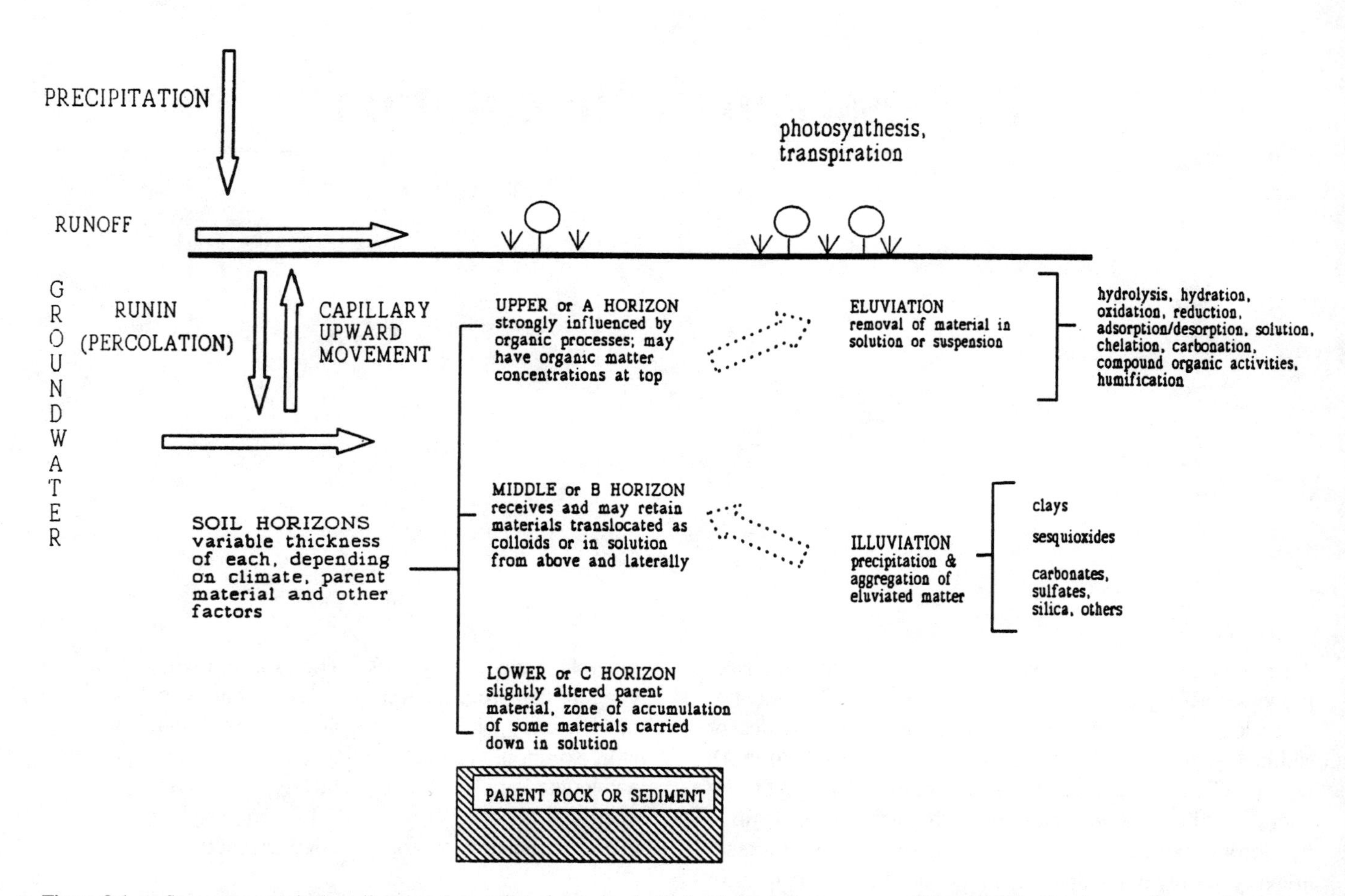

Figure 3-1. Some processes in the soil system. Inorganic and organic, physical and chemical processes are intimately interrelated; e.g., rainwater physically displaces solids and solutes, and is concurrently involved in chemical and biochemical reactions. Compounds are taken into solution through chemical reactions (e.g., dissolution via acid reactions), which may be dependent on biotic activity (e.g., production of humic acids) and require physical removal of the products from the site for reactions to have a substantial effect on the parent material. It is simplistic to consider any of the processes in isolation within natural systems.

concerning geochemical processes discussed below can be obtained from Garrels and Christ (1965); Berner (1971); Mueller and Saxena (1977); Holland (1978); Iler (1979); Krauskopf (1979); Yariv and Cross (1979); Anderson and Rubin (1981); Stumm and Morgan (1981); Aiken et al. (1985); Drever (1988); Kabata-Pendias and Pendias (1992).

Dissolution and Precipitation

Water is the fundamental agent in most chemical processes. Because the H_2O molecule has two hydrogen atoms 104° apart on the edges of a larger oxygen atom, its charge distribution is asymmetrical and it behaves as a dipole, with one positively and one negatively charged end. The attraction of cations to the negative end of the water dipole, and of anions to the positive end, weakens bonding within minerals and promotes dissolution, hydration (addition of water molecules to compounds), and hydrolysis (addition of H^+ or OH^- ions to compounds), e.g.:

$$(Mg,Fe)_2SiO_4 \text{ [olivine]} + 2H_2O + CO_2 \Rightarrow Fe(OH)_3 + Mg^{2+} + HCO_3^- + SiO_2 + e^-$$

In solutions, H_2O dipoles orient about dissolved ions, further enhancing the solvent power of the water. Water circulation promotes reactions by supplying new reactants and removing reaction products that might otherwise accumulate and retard further reaction. Water circulation is particularly influential in chemical reactions where postdepositional compaction causes lateral expulsion of fluids containing various ion mixes from muds that are generally organic-rich. Although water is an excellent solvent, it does not have an unlimited capacity to dissolve salts or other materials; the amount of a solid that can be dissolved under equilibrium conditions is given by its solubility constant (K_{sp}; values for 25°C and 1 atm can be found in most chemical data tables). This constant is used in calculations, such as for $CaCl_2$ (the common road salt of icy northern climes):

$$K_{sp} = \frac{[Ca^{2+}] \cdot [Cl^-]^2}{CaCl_{2(s)}}$$

where $[Ca^{2+}]$ and $[Cl^-]$ are the concentrations in moles per litre of the calcium and chloride ions, and $CaCl_{2(s)}$ is the concentration of the solid that will equal 1. (The concentration of the solid phase is 1 for all compounds.)

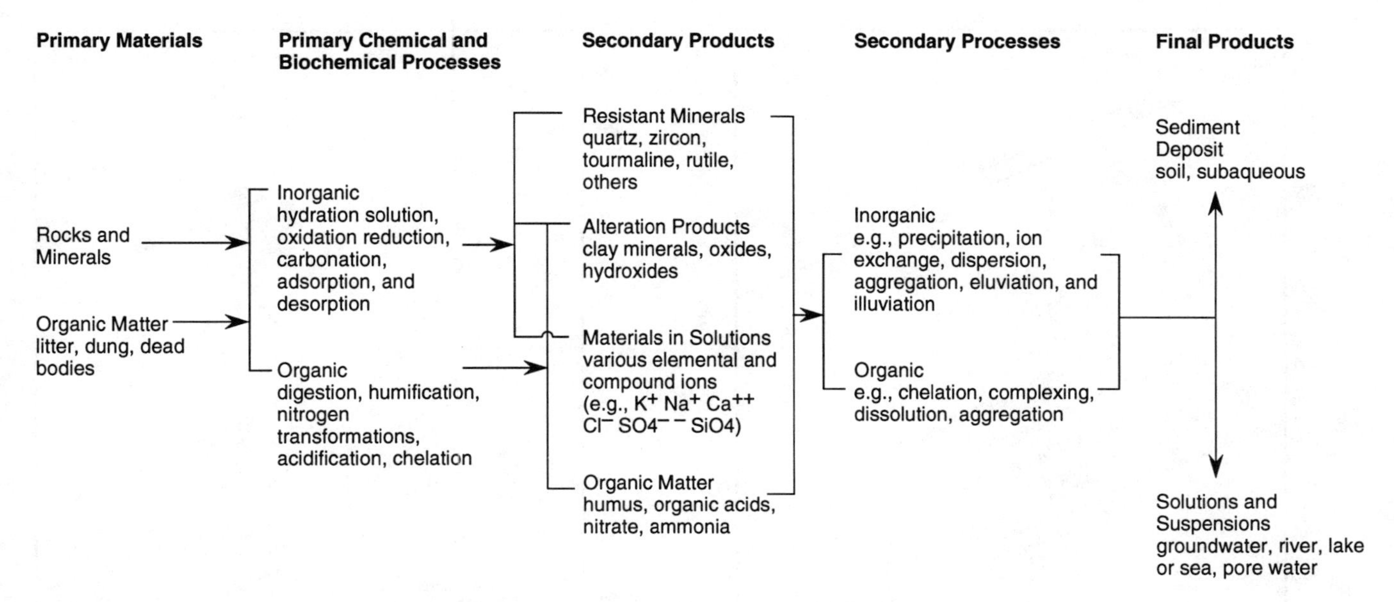

Figure 3-2. The chemical and biochemical weathering route from primary materials to secondary products.

The smaller the value of K_{sp}, the less of the substance can be dissolved. When the concentrations of ions from a particular substance reach the limit determined by the K_{sp} for a particular temperature (the solubility of solids increases with increasing temperature), the solution is saturated. In a saturated solution, if water evaporates or the temperature falls, the solution will become supersaturated and precipitation of a solid is likely (see extended discussion on evaporite sediments in Chapter 9). Precipitation of solids may also result if another solution with a common ion is added (*the common ion effect:* see Dean and Schreiber 1978; Krauskopf 1979 for further details), e.g., gypsum may precipitate if sulfate-rich waters from oxidizing sulfides (e.g., in mine tailings) reach a salt lake that is nearly saturated with respect to calcium sulfate. Precipitation can also be caused, even from undersaturated solutions, by the precipitation of another solid (coprecipitation); e.g., if iron oxyhydroxides are precipitated from a solution that also contains a significant (but less than saturation) amount of dissolved silica, silica may also precipitate (e.g., Harder and Flemhig 1970; Iler 1979).

Numerous other factors can influence the solubility of solids in natural waters, but the most important are *pH*, *Eh* (see separate subsections that follow), and the formation of *complexes* and *chelates*. Complexes, formed by bonding a central ion (commonly a metal) to surrounding ligand species, can cause a significant increase in solubility; e.g., AgCl has a very low solubility ($K_{sp} = 1.78 \times 10^{-10}$), but in the presence of high chloride concentrations $AgCl_2^-$, $AgCl_3^{2-}$, and $AgCl_4^{3-}$ can form, and these are much more soluble. Chelates, formed by bonding central ions (such as metals) to surrounding organic molecules, act in the same way to increase the solvent ability of waters derived from, or that have passed through, organic-rich sediment (e.g., swamps).

The total amount of dissolved salts in a solution is commonly referred to as the *salinity* and is expressed in terms of the weight of solids dissolved in 1 kg (≅1 L) of water; salinity is usually expressed in ppt (parts per thousand = g/L) for saline waters, or ppm (parts per million = mg/L) for fresh waters. The salinity of normal marine waters is about 35 ppt, hypersaline waters may exceed 200 ppt, brackish waters are commonly around 10 ppt, and freshwater averages are about 100 ppm (±10 ppm). In seawater, the relative abundance of the major cations that contribute to the salinity is $Na^+ \gg Mg^{2+} > Ca^{2+} > K^+$ and for the anions it is $Cl^- \gg SO_4^{2-} > HCO_3^-$. In fresh water the relative abundance of the major ions is much more strongly influenced by catchment lithologies, climate, and vegetation, and therefore varies widely from place to place; but the relative cation abundance is usually $Ca^{2+} \gg Na^+ > Mg^{2+} > K^+$ and the relative anion abundance is $HCO_3^- \gg SO_4^{2-} > Cl^-$. Salinities and ion ratios in interstitial waters in sediments vary widely as solutions migrate, mix with other solutions, and interact with the sediment.

pH

The measure of acidity of a solution is pH: the negative logarithm of the hydrogen ion (H^+ or hydronium ion H_3O^+) concentration in solution. In pure water at 25°C and 1 atm pressure, water dissociates to give an H^+ concentration (equal to the OH^- concentration) of 10^{-7} moles/L (i.e., pH = 7); the dissociation constant $K_d = [H^+][OH^-] = 10^{-14}$. The hydrogen ion concentration increases or decreases by an order of magnitude for each change of a full pH unit. At 25°C, pH values <7 indicate acidic water and pH values >7 indicate basic conditions. The temperature dependence of pH is important because, for example, at 120°C (often attained in diagenesis) the pH of neutral pure water is about 6 rather than 7, whereupon reactions can take place that would not at pH 7. The pH

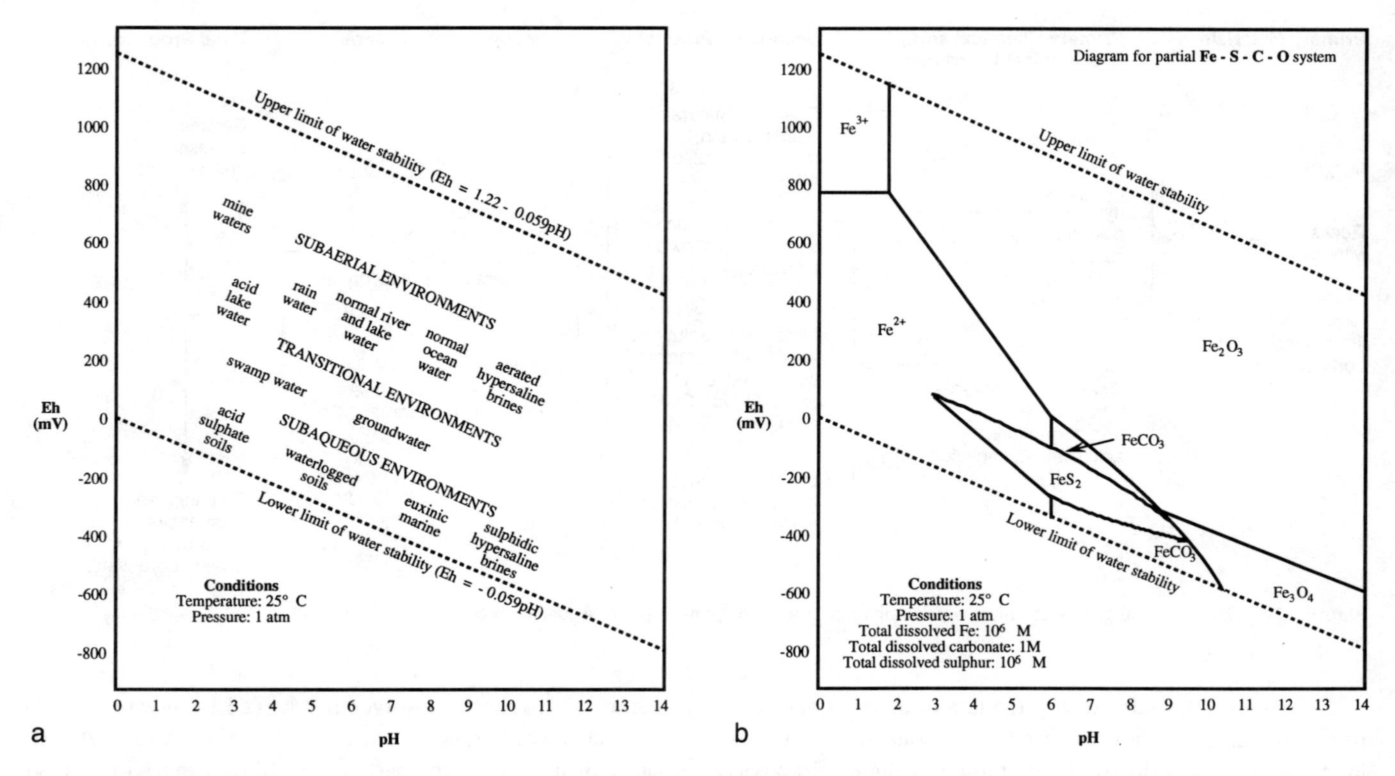

Figure 3-3. Examples of phase diagrams to show the range and importance of Eh and pH in sedimentary systems. *a:* Eh/pH "fence" showing the common range of values for sedimentary environments. *b:* Example of an Eh/pH phase diagram in a system with considerable relevance to both modern depositional systems (particularly those influenced by humans) and to diagenetic systems in which iron-bearing minerals are present. The diagram shows the relationship of common iron compounds to Eh and pH at the stated conditions; fields change in size as those conditions change. (After Brookins 1988.)

of natural waters ranges from near zero in some hot springs in volcanic regions to 10 or greater in some highly alkaline sea or lake waters, but most natural waters have a pH between 4 and 8.5; open marine waters have a pH of 7.5–8.5 (see Fig. 3-3a).

The importance of pH in surficial chemical processes (e.g., see Krauskopf 1979 and tables and figs. in Rosler and Lange 1972; Brookins 1988) lies in the fact that it indicates the availability of H^+ and OH^- ions for replacing other cations and anions in minerals and for interacting with other ions in solution, e.g.,

$$2NaAlSi_3O_8 + 2H^+ + 9H_2O \Rightarrow Al_2Si_2O_5(OH)_4 + 4H_4SiO_4 + 2Na^+$$

(albite) (kaolin)

Thus, the solubility of many substances is strongly pH-dependent (e.g., Fig. 3-4); for example, a small rise in pH (at constant Eh) may cause substantial precipitation of $Fe(OH)_3$ from a solution containing Fe^{2+} (Fig. 3-3b). Many chemical reactions involving minerals also have a feedback effect on fluid chemistry and may increase or decrease the solution pH; for example, experiments show that originally neutral water will develop a pH of 9–11 when amphiboles or pyroxenes are abraded in it, and a pH of 5–7 with abrasion of pure quartz; hydrolyzing reactions at the feldspar/water interface are thought to produce a pH>11 at the interface (e.g., Stevens and Carron 1948).

Oxidation and Reduction

Oxidation and reduction (or redox) reactions are among the most important chemical processes in sedimentology because both oxidizing (e.g., well-drained soil) and reducing (e.g., marine mud) environments are common, and as conditions change or sediments are transported from one condition to the other, compositional changes frequently take place as equilibrium is reestablished under the new set of conditions (e.g., Fig. 3-3a). For example, insoluble iron oxides eroded from lateritic soils and deposited with reducing muds in an

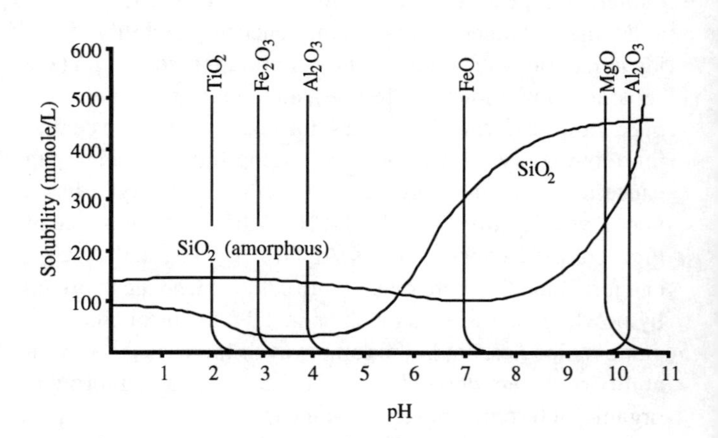

Figure 3-4. Variation in solubility of some common oxides as a function of pH. (After Rosler and Lange 1972.)

estuary are likely to be reduced, releasing soluble ferrous ions; these are then free to react with sulfide ions, produced in the sediment by bacteria, to form iron monosulfides and pyrite. Conversely, when reduced marine muds are exposed by a relative fall in sea level (e.g., some forms of land reclamation), sulfide minerals may be oxidized to form ferric iron oxyhydroxides and jarosite (an iron sulfate), with a large consequential release of hydrogen ions in the process. The excess of H^+ causes acid sulfate soils, which may create environmental and engineering problems in many low-lying coastal areas. Elements within sediments that are transported from an environment where one set of redox conditions prevails to another with different redox conditions can be strongly influenced by the change; e.g., cadmium is generally more mobile under oxidizing conditions than under reducing conditions, whereas iron is more mobile under reducing conditions (Fig. 3-3b, Table 2-7, and see Krauskopf 1979; Brookins 1988).

Just as all acid/base reactions involve proton (H^+) transfer, all redox reactions involve electron transfer. Oxidation involves the loss of one or more electrons, whereas reduction involves a gain of one or more electrons. Because redox reactions involve electron transfer between two reactants, the oxidation/reduction potential of the system can be readily measured in volts (or mV). All redox reactions involve both an oxidation and a reduction: one *half-cell* reaction releases and the other consumes electrons; the measured potential is for the two reactions combined (tables of half-cell electrode potentials can be found in many geochemistry and thermodynamic data books). The half-cell reaction that has been selected as a standard reference is:

$$2H^+ + 2e^- \Leftrightarrow H_2 \qquad E^\circ = 0.0 \text{ V}$$

E° is the potential measured at standard conditions of temperature = 25°C, pressure = 1 atm and unit activity (similar to concentration) for each chemical species present. Such ideal conditions are virtually never attained in natural systems, so the practical measure *Eh* is used: the potential developed relative to this reference half cell after correction for all departures from those standard conditions. The Eh thus varies with temperature and with concentration of the reacting substances.

An example of the measured potential for two reactions is:

$$Fe^{2+} + 2e^- \Leftrightarrow Fe \qquad E^\circ = -\,0.41 \text{ V}$$

$$Cu^{2+} + 2e^- \Leftrightarrow Cu \qquad E^\circ = +\,0.34 \text{ V}$$

which combine as:

$$Fe + Cu^{2+} \Leftrightarrow Fe^{2+} + Cu \qquad E^\circ = +\,0.75 \text{ V}$$

Note: These half-cell reactions are written so that the more oxidized species is on the left-hand side as recommended by the International Union of Pure and Applied Chemists, but the opposite approach is also used by geochemists; either way the sign of the standard electrode potentials will remain the same. The convention of placing the more oxidized species on the left is helpful in interpreting the meaning of the formula: the more positive the value of E°, the greater the likelihood that the reaction will proceed to the right. A positive value for the overall reaction (obtained by combining the two half reactions, with the oxidized species in the more positive half reaction positioned to the left) means that the reaction will tend to proceed to the right spontaneously. This particular reaction can be used as a form of copper mining, although *Eh* rather than E° potentials will apply; if Cu^{++} solutions (e.g., water drained through sediments or mine tailings containing copper-bearing minerals) are passed over iron compounds (e.g., waste tin cans), the copper will exchange for the iron (almost pure copper cans can be collected later).

Oxidation/reduction potentials are also commonly expressed in some sciences as pE (= –log[electron activity], analogous to pH = –log[proton activity, similar to concentration]), but the use of *Eh* is preferred by geochemists because it is directly measurable whereas pE is not. The relationship betwen pE and Eh is given by:

$$pE = \frac{F}{2.303\ RT}\ \text{Eh} = 16.9 \text{ Eh at } 25°\text{C}$$

where *F* is the Faraday constant (96.42 kJ/volt gram equivalent or 23.061 cal/volt gram equivalent), *R* is the gas constant (0.008314 kJ/degree or 0.001987 kcal/degree), *T* is the absolute temperature (298.15 at the standard temperature of 25°C), and 2.303 is the conversion from natural to base 10 logs.

The relationship between Eh, E°, the activities of the reactants (the more oxidized species in the reaction), products (reduced or less oxidized species), and temperature for the reaction ($aA + bB \Leftrightarrow cC + dD$) is given by the Nernst equation:

$$\text{Eh} = E^\circ - \frac{RT}{nF} \ln \frac{[C]^c \cdot [D]^d}{[A]^a \cdot [B]^b} = E^\circ - \frac{0.059}{n} \log \frac{[C]^c \cdot [D]^d}{[A]^a \cdot [B]^b}$$

where *F* is the Faraday constant, *R* is the gas constant, *T* is the absolute temperature and *n* is the number of electrons involved in the reaction. The activity of a dissolved species will approximately equal its concentration in dilute solutions, but activity decreases relative to concentration as the concentration increases (see Krauskopf 1979); a pure substance such as a solid has unit activity. The Nernst equation can be written with the position of the reactant (more oxidized species) and product (reduced or less oxidized species) reversed, but when this is done the minus sign must be altered to a plus.

Relationships involving Eh and the Nernst equation are the basis for constructing the Eh/pH diagrams (e.g., Fig. 3-3), which are widely used in sedimentology and geochemistry. They can also be used to calculate ion ratios such as the Fe^{2+}/Fe^{3+} ratio for environments where the Eh is known, to predict the effects of mixing different water bodies, and to predict the effect of a wide range of sediment/water interactions. For example, to determine the Fe^{+2}/Fe^{+3} ratio in interstitial waters of acid sulfate soils (pH commonly as low as 2) where the Eh is +0.5 V: by using the value of E° = +0.77 V for the reaction $Fe^{+3} + e^- \Leftrightarrow Fe^{+2}$ and assuming a temperature of 20°C, the Nernst equation states that the Fe^{+2} concentration will be about 38,000 times the Fe^{+3} concentration. Iron will therefore be mobile through that soil system; iron oxyhydroxides will not precipitate unless mixing with other waters or reaction with other solids causes the pH to rise, or exposure to air (e.g., at the soil surface) increases the availability of free oxygen.

Oxygen is the strongest oxidizing agent in natural environments—a stronger oxidant would break down water, consequently releasing oxygen—but redox reactions need not involve molecular oxygen. In natural environments, Eh generally reflects the abundance of free oxygen; however, many redox reactions on the earth's surface are slow and require biological mediation (the slow rate at which many natural redox reactions attain equilibrium is fortunate—if it were otherwise, all organic matter, including ourselves, would spontaneously combust!). A positive Eh usually gives rise to oxidizing reactions because of the availability of oxygen, but a negative Eh does not always result in reducing reactions—a *reducing capacity* (ability to cause reduction reactions) is also necessary; which in sediments is generally provided by organic matter. Where organic matter is lacking, negative Eh values may not result in reduction of materials like hematite (the iron oxide that colors sediments red).

The Eh = 0 boundary may be above, at, or below the sediment–water interface in sedimentary systems. Its position depends on the oxygen content of the water above the sediment, the chemical oxygen demand of the sediment (generally equals the quantity of organic matter present), and the extent to which physical and biological reworking of the sediment allows oxygen from the water to mix with the sediment. For further discussion of redox reactions in natural environments and the procedures for constructing Eh/pH diagrams, see Garrels and Christ (1965), Krauskopf (1979), and Stumm and Morgan (1981).

Adsorption, Desorption, and Ion-Exchange Reactions

The smallest particles (primarily clay minerals and colloids; see Chapter 7) have a very high surface area to volume ratio and many carry a surface electric charge (e.g., Grim 1968; Yariv and Cross 1979). The high surface area/volume ratio and high charge/mass ratio give these particles a high capacity to adsorb ions from solution. Adsorbed ions are not tightly bound, and either can be desorbed if conditions change or can exchange with other ions in surrounding solutions. Ion-exchange capacities of some fine-grained sedimentary materials are shown in Table 3-1. Adsorption, desorption, and ion-exchange reactions in soils play an important role in plant nutrition; through sediment/water interactions, they can have a major influence in other sedimentary environments both on water chemistry and on sediment composition.

Table 3-1. Ion-Exchange Capacities of Some Common Sedimentary Materials

Material	Exchange Capacity meq/100g
Kaolinite	3–15
Illite	10–30
Chlorite (nonswelling)	20–40
Chlorite	40–100
Montmorillonite	80–150
Colloidal iron hydroxide	5–30
Humic acids in soils	150–700

Ions adsorbed onto fine sediment particles near the sediment/water interface may play a major role in later diagenetic reactions involving both minerals and pore fluids.

Which ions will be adsorbed, desorbed, or exchanged in any sediment will be controlled by a large number of factors, including the composition of the minerals providing the exchange sites, the type, size, concentration, and charge of the ions involved, the total ionic strength and pH of the solution, temperature, and time. Chemical species (e.g., microbially produced sulfide anions) that form low-solubility precipitates with some potentially exchangeable ions (e.g., some metal cations) can remove those ions both from the solution acting as the exchange medium and from the mineral exchange sites. On clay minerals, the surface charge is mostly negative; the surface charge of humic acids is also usually negative. Therefore, it is mainly cations that are involved in the exchanges.

For colloids, the sign and magnitude of the charge depend on the composition of the particle, the pH of the solution, and the type and concentration of other ions present. Exactly how the charge develops is not well understood, but it is known to be strongly pH-dependent; most colloidal particles have a net positive charge when the pH is below the isoelectric point (point of no net charge; Fig. 3-5, Table 3-2), but they have a net negative charge when the pH is above this point (e.g., Yariv and Cross 1979). In sedimentary environments the change in charge density and sign can be very important when some colloidal materials are transported from an environment characterized by one pH to another environment where the prevailing pH is on the other side of the isoelectric point. This effect is well illustrated by colloidal hematite (e.g., McConchie and Lawrance 1991), which has an isoelectric point (IEP) between pH = 7.5 and pH = 8.0 (Fig. 3-5): it is negatively charged in normal marine waters where the prevailing pH is about 8.2, but positively charged in nonmarine systems where the pH is usually <7. In these circumstances, anions bound to colloidal hematite in a river will be released when the river reaches the sea, and cations will take their place. The pH at which the charge on the colloidal particles is zero is also important because this is the pH at which the particles will be most likely to flocculate and settle out. IEP values for a selection of colloid-forming materials are shown in Table 3-2.

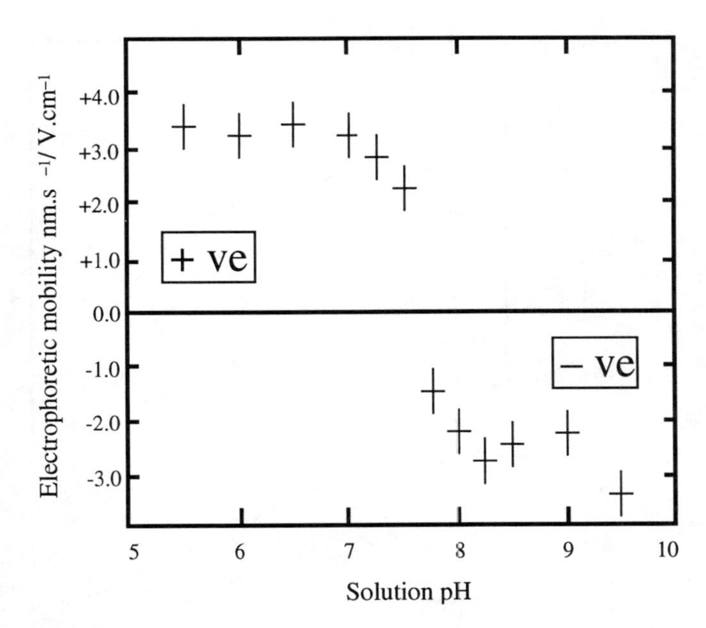

Figure 3-5. Relationship between electrophoretic mobility of colloidal hematite and pH of the solution.

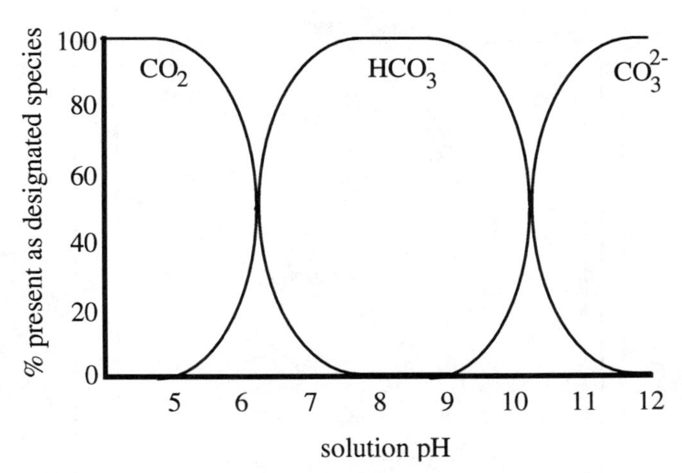

Figure 3-6. Species distribution in the $CO_2/HCO_3^-/CO_3^{2-}$ aqueous system in relation to pH.

Reactions Involving Carbon Dioxide

Carbon dioxide will combine with water to form bicarbonate ions (HCO_3^-) and hydrogen ions over most of the pH range encountered in natural environments, i.e.:

$$H_2O + CO_2 \Leftrightarrow HCO_3^- + H^+ \qquad K_{d1} = 4.45 \times 10^{-7}$$

$$HCO_3^- \Leftrightarrow CO_3^{2-} + H^+ \qquad K_{d2} = 4.69 \times 10^{-11}$$

The species distribution for this system is shown in Fig. 3-6.

Bicarbonate ions produced in the first dissociation are the dominant anionic species in river water (the hydrogen ions released in the process contribute to the lower-than-neutral pH of rain and most river waters) and can react with calcium ions to form calcite, i.e.:

$$Ca^{2+} + 2HCO_3^- \Leftrightarrow CaCO_3 + H_2O + CO_2$$

Calcite will also react readily with hydrogen ions under acidic conditions, i.e.:

$$CaCO_3 + 2H^+ \Leftrightarrow Ca^{2+} + CO_2 + H_2O \text{ (in low pH solutions)}$$

or

$$CaCO_3 + CO_2 + H_2O \Leftrightarrow Ca^{2+} + 2HCO_3^- \text{ (in less acidic solutions)}$$

These reactions and the formation of solid carbonates provide an important buffering system for both ocean waters and the carbon dioxide partial pressure in the atmosphere (vital in issues such as the *greenhouse effect*). The same suite of reactions provides much of the explanation for the formation of calcretes, beachrock, cave limestone, and for dissolution and reprecipitation reactions during carbonate diagenesis and carbonate cementation of detrital sediments (see also Chapters 8 and 9). A simplified version of the carbon cycle which involves these reactions is shown in Fig. 3-7. Because cold waters hold more CO_2 than warm waters, $CaCO_3$ is more soluble at depth in the ocean (colder waters are denser than warmer waters). For much of the Precambrian, when the carbon dioxide content of the atmosphere was substantially (perhaps 100 times) higher than it is today (Holland 1984; McConchie 1987), these reactions indicate that the mean pH of the world's oceans was lower than it is today, with substantial consequences with respect to nondetrital sedimentation (e.g., Walker 1983 and see Chapters 7, 8, and 9).

Carbon dioxide also plays an indirect role in rock weathering, which is particularly important in soil formation, e.g.:

$$\underset{\text{(orthoclase)}}{2KAlSi_3O_8} + H_2CO_3 + H_2O \Leftrightarrow 2K^+ + CO_3^{2-} + \underset{\text{(kaolin)}}{Al_2Si_2O_5(OH)_4} + 4SiO_2$$

Table 3-2. Isoelectric Points (IEP) for Common Colloidal Particles

Compound	IEP
$Al(OH)_3$ [amorphous]	7.1–9.4
α-$Al(OH)_3$ [gibbsite]	5.0
γ-Al_2O_3	8.0–8.5
$Fe(OH)_3$ [amorphous]	7.1–8.5
α-FeO(OH) [goethite]	3.2–6.7
γ-FeO(OH) [lepidocrocite]	5.4–7.4
α-Fe_2O_3 [hematite]	7.8
γ-Fe_2O_3 [maghemite]	6.7
MnO_2	4.0–4.5
$Mn(OH)_2$	7.0
SiO_2 [amorphous]	1.8
Organic colloids	<6.0 (usually)

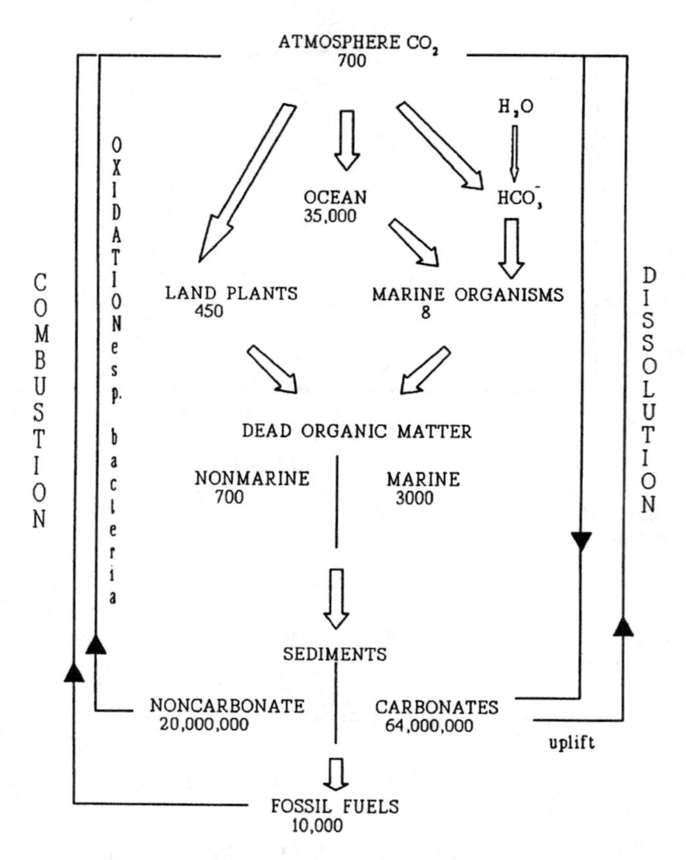

Figure 3-7. Simplified sedimentary cycle of carbon. (After Waples 1981.)

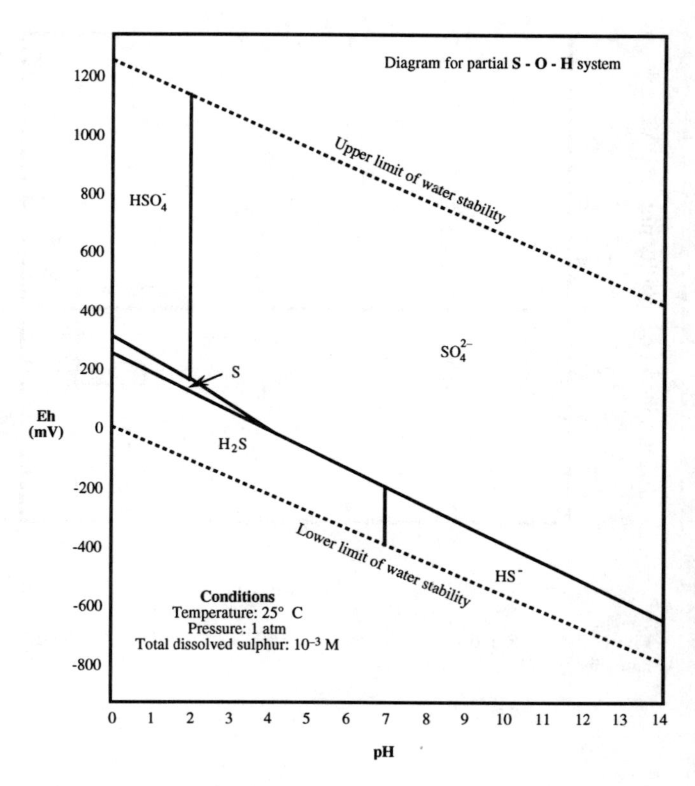

Figure 3-8. Phase diagram for common sulfur compounds at the stated conditions; fields change in size as those conditions change. (After Brookins 1988.)

Reactions Involving Sulfide and Sulfate

The sulfate (SO_4^{2-}) ion is the stable sulfur-bearing ion under oxidizing conditions in nature, and is the second most abundant anion in both seawater and river water. Because sulfate is abundant in seawater and saline lakes, it is a major constituent of evaporite minerals (see Chapter 9); sedimentary sulfates can be found in the rock record back as far as 3,450 my B.P. (e.g., Buick, Dunlop, and Groves 1981). Near some heavily industrialized areas, sulfate concentration in river water and freshwater lakes has recently risen to the point at which it is the dominant anion, largely because of the burning of sulfur-bearing fossil fuels and the smelting of sulfide ores.

Sulfides (containing S^{2-}) are the stable sulfur-bearing species in reducing sedimentary environments, where they are largely produced by microbial reduction of sulfate compounds and ions; the ratio of sulfide to sulfate in sedimentary environments is largely controlled by redox and pH conditions (e.g., Fig. 3-8). The sulfide anion is important in geochemistry because it reacts with metal ions (particularly iron) to form metal sulfides, which are largely insoluble under reducing conditions but decompose readily if conditions become oxidizing. Because many large sulfide ore deposits were originally deposited by syn-sedimentation processes, geologists developing exploration models have a particular interest in the depositional settings where metal sulfides accumulate and in the processes involved in their deposition (see further discussion in Chapter 9).

Temperature

The maximum, minimum, average temperature and its rate of change strongly influence many chemical reactions (e.g., some discussions above and see Salop 1983) or mineralogical transformations in diagenesis. Higher temperatures accelerate many chemical and biochemical reactions, and marked rises in reaction rates may result from temperature increases of only a few degrees. Temperature is also an important control on which evaporite minerals will precipitate (see Chapter 9). Overall temperature controls are climate, weather, geothermal gradient, and various exothermic and endothermic reactions within the lithosphere/hydrosphere/biosphere system.

SURFICIAL PHYSICAL PROCESSES

Physical processes acting at the earth's surface are varied in type and in scale; they include transportation and deposition by wind, water, ice, and gravity, together with accompanying abrasion of the transported particles and materials encountered in transit. Many different processes may act in a single sedimentary environment, and the same processes operate in different environments. *High-energy environments* have processes of moderate energy levels acting continuously over long time periods and their products tend to be well-sorted, well-stratified sediments (e.g., on beaches). In contrast, *high-energy processes* may be rare and episodic (e.g., floods, storms, turbidity currents, volcanic base surges) yet may have a dominant effect on the character of sequences (e.g., see Gretener, 1967; Aigner 1985; Einsele, Ricken and Seilacher 1991; Dott 1983, 1988; Hsu 1983; Tsutsui, Campbell, and Coulbourn 1987; Clifton 1988; Chough and Sohn 1990). The physical effects of major storms and other high-energy events include severe erosion as well as transport of huge quantities of sediment; subsequent deposition may result in very substantial thicknesses of sediment locally. Consequently, high-energy events not only transport and deposit very large volumes of sediment but they can rework considerable thicknesses of already deposited sediment, and in doing so they destroy original sedimentary structures and textures deposited during "normal" depositional conditions. In some sedimentary sequences, accumulated in settings that experience frequent high-energy events, it may be very difficult to determine what "normal" depositional conditions were like.

Abrasion

Abrasion is the dominant physical weathering process (see discussion in Pettijohn 1975). Subprocesses can be recognized such as grinding (by larger of smaller grains), blasting (by smaller hitting larger ones), abrasion (in the strict sense of grains rubbing together), and impact (between roughly equal-size grains). The importance of abrasion processes in reducing the size and increasing roundness of particles depends on the energy level at which abrasion occurs, the duration of abrasion, and the hardness and internal anisotropism (unequal properties in different directions) of the minerals or rock fragments involved. In general, abrasion is much less effective in destroying minerals than chemical processes; perhaps its major role is to facilitate chemical weathering by progressively reducing the size of particles, with a consequential progressive increase in the *specific area* (surface area per unit volume) and thus susceptibility to chemical attack.

Transportation and Deposition

Physical processes dictate the most obvious characteristics—structures and textures (Chapters 4 and 5)—of most sedimentary deposits. Chemical transportation in solution and deposition by precipitation account for relatively minor sedimentary accumulations, although there may be large individual deposits, such as some bedded evaporite and iron formations (Chapter 9). Rarely, processes other than currents or gravity act to transport and deposit grains (e.g., Syvitski and van Everdingen 1981 and see later discussion of biologic processes). Much research has been undertaken on the physical and gravitational processes acting in modern environments, and on their interpretation from ancient deposits. Three fundamental groups of transportational/depositional processes may be distinguished:

Passive suspension: particles that are not subject to an active vertical-lift force, or to a lateral force capable of moving them, settle because of the vertical pull of gravity through a medium that is not totally supporting them (e.g., volcanic ash or fine sediments dropped into a lake, or dead plankton falling to the seafloor).

Entrainment flow: the process by which particles are set into motion, maintained in motion, and then deposited from moving water, ice, or air (including hot air related to volcanic eruptions). Many studies of entrainment flow conditions have been made to predict their relationship to *competency*, the maximum size of particles that can be transported, and *capacity*, the total volume of sediment that can be transported.

Gravity transport: sediments move down a slope because of the downslope component of the force of gravitational acceleration that is acting on them; they accumulate when frictional retardation exceeds that force. (The term *resedimentation* has been used for these processes, but most sediments are redeposited by a variety of processes many times before finally coming to rest, and some sediments may even be transported from their source to their final site of accumulation by a single episode of gravity flow.) *Gravity flows*, comprising particles that move independently within the mass, are distinguished from other types of unit mass movement such as *slides* and *slumps*. A third category is also recognized—*rockfall*, in which particles tumble or fall directly from steep subaerial or subaqueous cliffs (such as reef fronts or fault scarps).

Distinction between these classes of processes is not always clear-cut; e.g., air or water is commonly moving when suspended sediments settle and may have some transporting effect; gravity flows may entrain some particles as a traction population; nearshore rip currents (e.g., Cook 1970) often carry both suspended and traction loads and may transport sufficient loads so that a density (sediment gravity) flow continues seaward of the rip itself.

Passive Suspension

Colloidal and clay-size grains are the most abundant sediments to settle from passive suspension, but there are exceptions (such as pebbles falling from rafts of ice or floating tree roots). Particles smaller than about 0.05–0.14 mm settle according to their size, shape, and density and the viscosity of the medium; Stokes' Law generalization for settling velocity (V_s) shows the interrelationships:

$$V_s = C_1 r^2 \quad \text{or} \quad \frac{r^2(\rho_p - \rho_w)g}{18\mu}$$

where C_1 = a constant for a given situation, r = radius of particle, ρ_p = density of particle, ρ_w = density of medium, g is the acceleration due to gravity, and μ = dynamic viscosity of medium.

Or alternatively

$$V_p = \frac{r^2(\rho_p - \rho_w)g}{7\mu}$$

where V_p (about 64% of V_s) is the settling velocity according to Wadell's sedimentation formula, which attempts to account for nonspherical grains. In the approximations, the settling velocity is proportional to the square of the particle radius.

Deposits commonly appear massive but may contain diffuse or sharp planar stratification between units if there were different episodes of sedimentation. Normal grading is present when particles of varied size and/or density were deposited at the same time. Stratification generally forms parallel to sea level (perpendicular to the gravitational field), but drapes and conforms to any irregularities present on the underlying surface. Layers may be very thin and regular (e.g., annual paired layers, known as *varves*, in lake deposits).

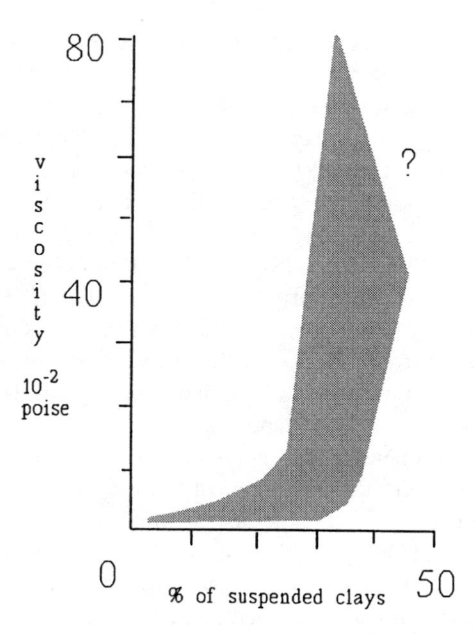

Figure 3-9. Approximate relationship between water viscosity and quantity of suspended natural clay load. (After Fritz and Moore 1988.)

Entrainment Flows

Particles entrained by water or air currents move either as a *traction load* (continuous contact with the bed of the flow), a *saltation load* (intermittent contact with the bed), and/or an (active) *suspension load* (no contact with the bed). (The distinction from passive suspension is that in deposition from entrainment flow the grains are being pushed or pulled laterally immediately before they come to rest.) The saltation and traction fraction of moving sediment comprises the *bed load*, but most techniques for measuring sediment in transport cannot differentiate between saltation and traction load or between saltation and suspension load that "accidentally" impinges on the bed (see **AS** Chapters 4 and 7). Most sediments are deposited from entrainment flows, which operate in all environments and at all depths in the ocean, and these processes can be assumed if criteria for the other mechanisms are lacking. Many papers and books treat transport and deposition by flowing water at various levels of scope and detail (e.g., Leopold, Wolmand, and Miller 1964; Middleton 1965; Allen 1985; Sternberg 1971; Moss 1972; Baker and Ritter 1975; Stanley and Swift 1976; Middleton and Southard 1978, 1984; Moss, Walker, and Hutka 1980; Dyer 1985). Aeolian processes are treated less often, but there is nonetheless a substantial literature (e.g., Folk 1971; Brookfield and Ahlbrandt 1983; Fryberger et al. 1984; Anderson and Hallet 1986; Langford 1989; Langford and Chan 1989).

Laws predicting behavior of the traction load are complex because influential variables not only include particle size, shape, and density, but also depth, width and velocity of flow, turbulence, viscosity (which varies with salinity, temperature, and quantity of suspended load, e.g., Fig. 3-9), and bed roughness; in addition, even the experimental measurement of traction load is complicated by the fact that insertion of any measuring device significantly alters transport conditions. Calculations concerning saltation load are even more difficult. The active suspension load is treated in the same way as the passive suspension load, but as particles become larger than about 0.5 mm, Stokes Law becomes invalid. Much sediment travels in the deep ocean in *nepheloid clouds*—great dispersed suspensions of clay-size sediment (e.g., Drake and Gorsline 1973; Betzer, Richardson, and Zimmerman 1974; Gorsline 1985), and both these and other suspended loads can be collected relatively easily. However, analytical problems arise in most natural systems from the fact that most suspended load travels as aggregated particles such as flocs and fecal pellets (if they didn't, much of the load would probably never settle, since the settling velocity of clay-size particles is so small as to be negated by *any* current!), and the common analytical methods disperse the samples into their constituent particles (see discussion in **AS** Chapter 5).

Flows are either *laminar* (linear flow lines) or *turbulent* (with eddies up to three times the average downstream flow velocity), depending on conditions that involve particle characteristics, bed roughness, and fluid conditions (Fig. 3-10). Most flow in nature is turbulent, and the upward component of turbulence provides the major lift force for grain movement. Fluid density is also important and variable, e.g., in water it varies with temperature, salinity, and the amount of dissolved and suspended load carried. It is involved in determining the *kinematic viscosity*, which is the ratio of dynamic viscosity (the resistance to change in shape during flow) to density. The fundamental character of natural fluids also involves their response to *shear stress* (shear force per unit area across the shearing surface such as at the bed of the flow).

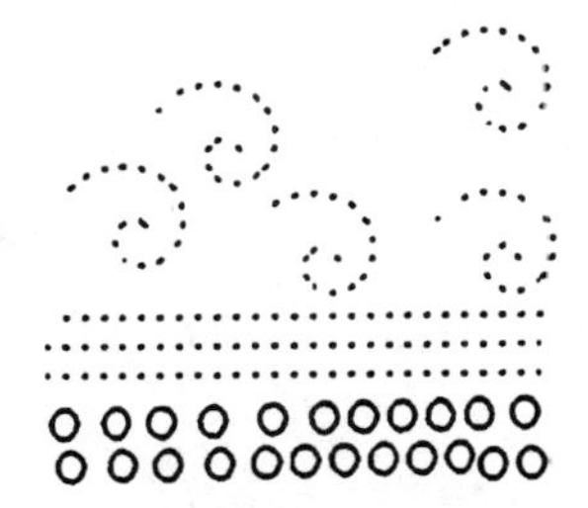

Flow lines depicting laminar flow sublayer below turbulent flow where Reynolds Number and bed roughness are low (e.g., fine sediments, low velocities)

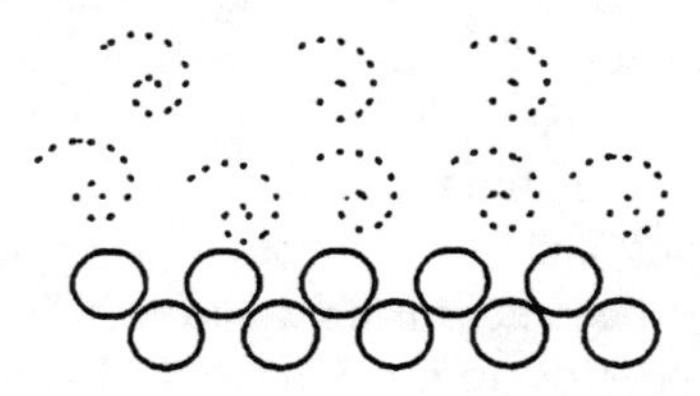

Flow lines where Reynolds Number is high or bed roughness is great (e.g., high flow velocities and/or coarse sediments)

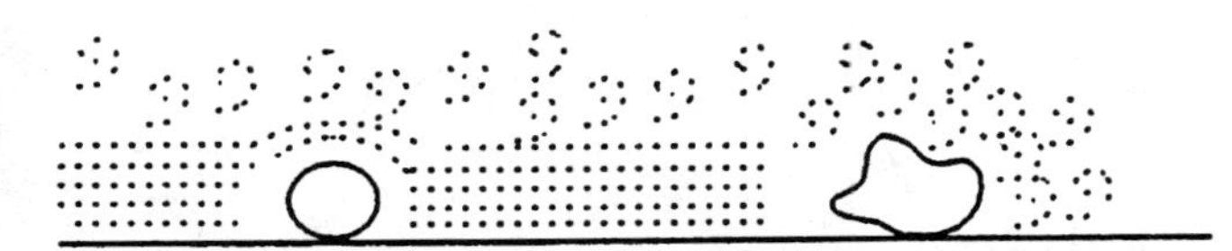

Influence of grain size on flow lines. If grains are coarse enough to project through any laminar sublayer, that sublayer is destroyed. Lift forces on grains differ. Within the laminar sublayer it results from the upward-deformed flow lines and downstream pressure shadow; where turbulence exists, turbulent eddies provide the lift. Grains may also roll.

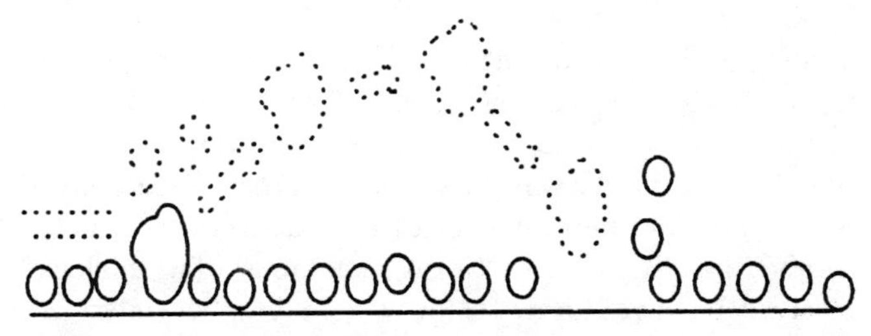

Bed movement is initiated when the first grains are lifted into saltation impact on the bed and impart extra energy on grains at the impact point, sending more into saltation.

Figure 3-10. Representation of possible flow conditions and their influence on grains at the sediment/water interface. The Reynolds number is defined and discussed in the text.

Shield's Law approximates the relationship of shear stress to initiation of grain movement:

$$\tau = f\frac{dg(\rho_s - \rho_w)V_b}{\mu}$$

where τ is the bottom shear stress; d is the particle diameter; V_b is the maximum velocity near the boundary; μ is the kinematic viscosity; ρ_p is the density of the particle and ρ_w is the density of water; and g is the acceleration due to gravity. (The relationship between shear stress and velocity of flow can be found in other graphs).

Newtonian fluids have no strength and show no change in viscosity as shear stress varies; *non-Newtonian fluids* have no strength but a variable viscosity (e.g., highly water-saturated muds have a higher viscosity and move slowly on a gentle gradient, and have a lower viscosity when they move more rapidly on a steeper gradient); *Bingham plastics* are materials that have an initial strength that must be overcome before they flow, and once they start flowing they have no strength and a constant viscosity (e.g., some sediment gravity flows).

A fundamental fluid parameter used to categorize fluid characteristics is the dimensionless *Reynolds number (R_e)*, which is a measure of inertial (causing turbulence) and viscous (resisting turbulence) forces, or drag forces acting on a particle:

$$R_e = \frac{Vd}{\mu}$$

where V is the maximum flow velocity, d is the particle diameter, and μ is the apparent kinematic viscosity. Below a critical value of the Reynolds number (dependent on conditions, but usually in the range of 500–2000), flow is *laminar* (streamlined, without eddies); above the critical value the flow is *turbulent*. Stokes' Law only applies at R_e numbers <1. Figure 3-10 depicts one aspect of the relationship of this number to flow conditions and sediment movement.

Another fundamental fluid parameter is the *Froude number* (F_r), which is the ratio between inertial and gravity forces in water:

$$F_r = \frac{V}{\sqrt{gh}}$$

where V is the average water flow velocity, g is the acceleration due to gravity (9.8 m/sec^2), and h is the water depth.

Thus, F_r depends on both the flow velocity and the water depth. Where $F_r < 1$, flow is *tranquil* (*streaming*, or *subcritical*); surface waves are moving faster than flow and can move upstream. Where $F_r > 1$, flow is *rapid* (*shooting*, or *supercritical*).

During transport of bed load, many transitions in bed-form style result from changing flow velocities and can be described mathematically using the Froude number (see Chapter 4). Although the actual transition velocities for a given depth depend on grain size, as F_r increases toward 1, ripples get larger, becoming straight crested dunes, then undulatory megaripples, and finally lunate megaripples. At $F_r = 1$, the megaripples become unstable and form the plane bed phase of the upper flow regime; the transition from lower to upper flow regime is at $F_r = 1$. At $F_r > 1$, bed forms of the upper flow regime are stable (e.g., in fast flowing rivers). In the marine environment, values of $F_r > 1$ are very rare, because even in shallow water, velocity (V) very seldom exceeds 2 m/sec; hence, small- and medium-sized ripples are the most common current-generated structures. (*Wave effects* on flow velocity are highly depth-dependent: at a depth of $\lambda/2$, where λ is the wavelength of the surface wave, there is only c. 4% of the movement that there is at the surface, and at depth λ there is only c. 0.2%.) Whereas most of us think exclusively of waves formed at the water/atmosphere interface, there are also internal waves within the water column at the interface between water masses of different density, such as at the top of salt wedges in estuaries or between cold and warm or briny and normal-salinity waters. Such waves obey the same rules, but can be initiated at substantial depths (e.g., Karl, Cacchione, and Carlson 1986). In addition, wave agitation of the substrate is not a guarantee of relatively coarse deposits; if there is abundant mud in suspension, it may be deposited above the wave base (e.g., McCave 1971).

The *Hjulstrom diagram* (Fig. 3-11) combines experimental and theoretical data to indicate approximate relationships between flow velocity for grain entrainment, sediment sizes, and fields of erosion, transport, and deposition. Although clay particles in particular will continue traveling in suspension if there is virtually any vertical lift provided by turbulence, once settled, cohesion dramatically increases the force necessary to resuspend them; this is the explanation for the graphical relationship that shows flow velocities necessary to initiate movement increasing below the size at which cohesive interparticle attraction begins (e.g., Terwindt, Breusers, and Svasec 1968; Amos and Mosher, 1985). Figure 3-12 shows settling velocities for various grain sizes in fresh water at 10°C and 30°C; at higher temperatures particles will settle slightly faster due to the reduced water viscosity, whereas at lower temperatures (or higher salinities) particles will settle a little more slowly.

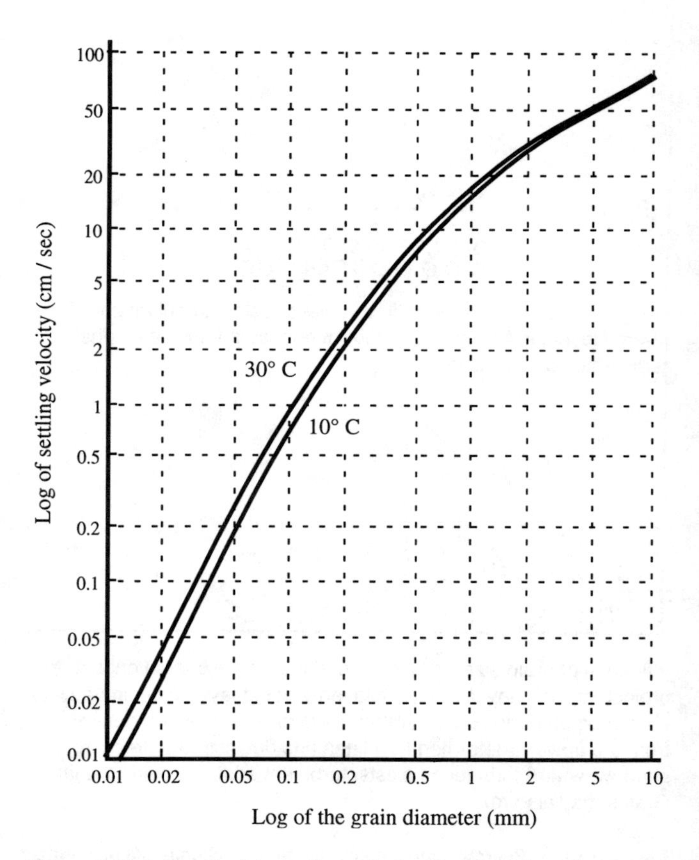

Figure 3-12. Graph of the settling velocity of quartz spheres in water.

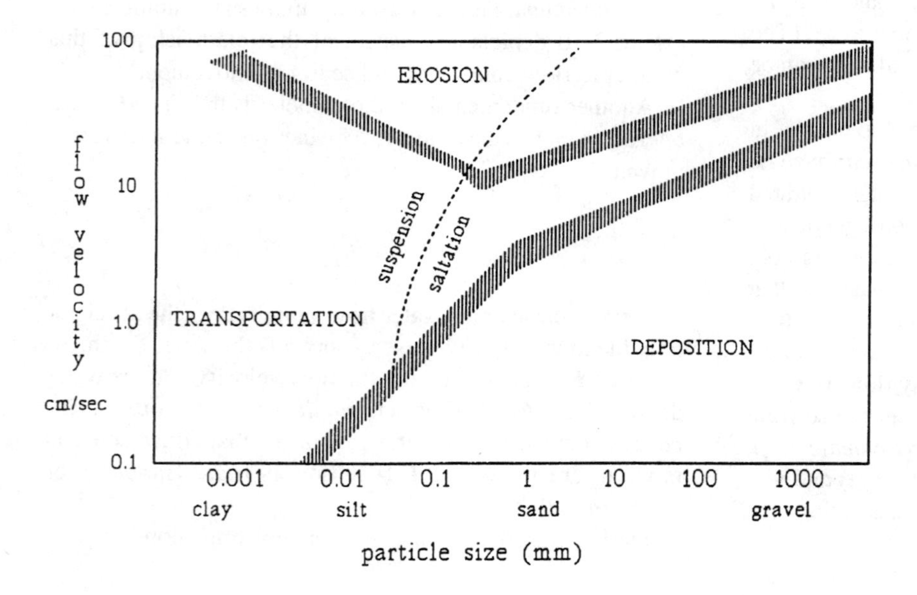

Figure 3-11. A version of the Hjulstrom diagram, showing approximate relationships between flow velocity and sediment size. Upper hachures represent the velocity of initiation for entrainment of various grain sizes; the lower represent settling velocity. (After Heezen and Hollister 1964.)

Characteristics of entrainment flows comprise the sedimentary structures: ripple marks, dune bed forms, cross-lamination, cross-bedding, and/or planar lamination (see Chapter 4). Unfortunately, these may not be present, may not be obvious (where textures and composition are homogeneous), or may have been destroyed by bioturbation after accumulation. Bottom contacts commonly are erosion surfaces, but erosion may reflect an earlier episode not connected with deposition of the sedimentation unit itself. Regular or irregular, initially subhorizontal, well–defined stratification between units is common but not diagnostic. Normal or inverse grading is possible during deposition, from waning or waxing flow velocities respectively. The main textural characteristic is some evidence of sediment sorting, and most well-sorted sediments were deposited from entrainment flows. Well–developed particle fabric is common (e.g., imbrication of pebbles, parallel alignment of the long axes of sand grains) but is not evident in all entrainment flow deposits; preferred fabrics can also develop in some gravity flows. However, the combination of well-sorted deposits with imbrication of particles is diagnostic of entrainment flows. Also generally diagnostic are the relatively rare (but economically important!) cases in which grains have been selectively sorted according to composition, as in placer deposits of heavy minerals (e.g., Komar and Wang 1984).

Exceptions to the above generalizations exist in the case of *hyperconcentrated flows*, which generally result from extreme flood conditions (e.g., Smith 1986), and *density currents*, in which the density of the water rather than the suspended load is the driving force (e.g., Harms 1974); both show transitional characteristics between entrainment and sediment gravity flows (especially debris flows, e.g., Fig. 3-13).

Gravity Transport

Sediment moving because of gravitational acceleration acting directly on the particles rather than on the surrounding medium finally comes to rest when the downslope component of that acceleration is less than the frictional resistance to motion. Gravity deposits may occur in any environment where there is a slope, an unstable accumulation of sediment, and a trigger mechanism. They are particularly common where large quantities of sediments are rapidly accumulating, as in deltas and at the base of submarine slopes (e.g., Nardin et al. 1979; Nemec 1990), but they may also occur on the continental shelf as submarine channels reflecting down-the-maximum-slope transport routes for gravity flows (these are commonly stacked vertically in the stratigraphic pile because bathymetric slope, unlike many currents, does not commonly change its direction of dip; e.g., Lewis 1982 and implication in Fig. 2-41). Because of the devastating effects such movements can have, slope stability studies, particularly by engineering geologists (e.g., Kerr, Stroud, and Drew 1971; Keefer 1984; Schiedegger 1984; Booth, Sangrey, and Fugate 1985; Rahn 1986), are widely undertaken in areas of potential failure (e.g., before construction of dams or housing developments where the prospective changes in the water table level are likely to facilitate failures). Studies of modern gravity movements in action are largely confined to a few terrestrial occurrences (e.g., Johnson 1970; Pierson 1981), but many marine, as well as nonmarine, modern and ancient deposits are known (e.g., Lewis 1976; Voigt 1978; Lewis, Laird, and Powell 1980; Saxov and Niewenhuis 1982). In the marine environment, slumps, slides, and flows can initiate on, or flow over, very gentle slopes (under 2–4°; e.g., Lewis 1971; Schwab and Less 1988) and may be of very large scale

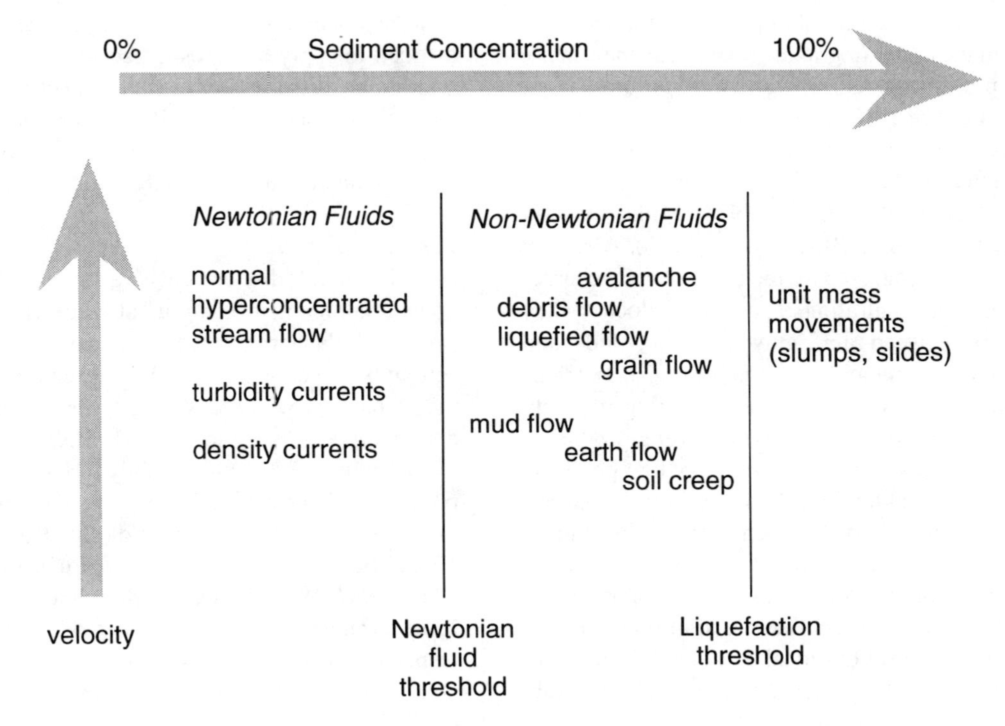

Figure 3-13. Approximate relationships between varieties of sediment/water flows. (Modified from Pierson and Costa 1984.)

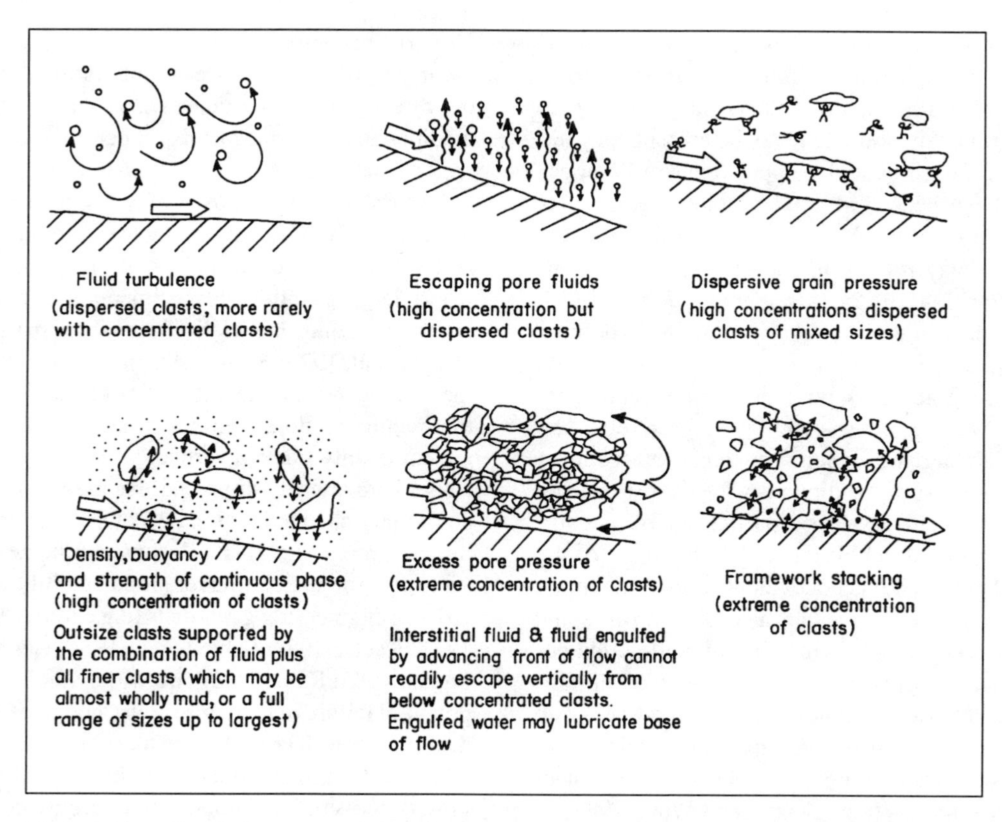

Figure 3-14. Sketch of support mechanisms in sediment gravity flows. In framework stacking, individual grains are moving, but concentration is so high that at any one time there is a vertical stack of grains in contact with each other and the substrate such that there is a rigid framework; in the next instant the stack comprises a different combination of grains. Individual mechanisms dominate in particular types of gravity flow—e.g., turbulence in turbidity currents, escaping pore fluids in liquefied flow, dispersive grain pressure in grain flows—but in fact many of the mechanisms are involved in each kind of flow, and in some types of debris flow it is not possible to attribute a dominant role to any single or even pair of mechanisms.

(e.g., Temple 1968). Because a slump or slide removes lateral support from the remaining sedimentary sequence, many automatically proliferate as *retrogressive* failures (e.g., Andressen and Bjerrum 1967; Hendry 1973).

Gravity flows and their deposits have received considerable attention in the past few decades (e.g., general principles discussed by Johnson 1970; Carter 1975; Middleton and Hampton 1973, 1976; Lowe 1982; Postma 1986). A variety of subcategories are recognized on the basis of fluid characteristics, sediment concentration, and flow velocity; they span a spectrum between unit movement of solids and hyperconcentrated entrainment flow (e.g., Fig. 3-13). Characteristics of the deposits are variable, reflecting not only these properties but also others such as the particle size range available. Transport in some flows over very low gradients may be as great as several tens of kilometres in subaerial settings and several hundreds of kilometres (to perhaps more than 1000 km) in marine settings.

Processes that oppose the force of gravity and support particles during movement (Fig. 3-14) are used to subdivide the spectrum of gravity flows. Four ideal end-members are generally recognized: turbidity current, liquefied flow, grain flow, and debris flow. However, this classification system is difficult to apply to deposits because it is genetic, based on inferred processes rather than the objective characteristics of the deposits, and because different processes may dominate in different stages in a single flow; Table 3-3 is a classification scheme that tries to emphasize the objective attributes.

Turbidity Current

Particles in turbidity currents are supported primarily by the upward component of turbulent eddies in the flowing fluid (Fig. 3-14). No large turbidity currents have been observed in action; however, small ones have been observed and experiments have provided a strong basis for understanding the natural process (e.g., Gould 1951; Middleton 1967, Normark 1989, Zeng et al. 1991). Certainly turbidity currents and their product, *turbidites*, have been very commonly interpreted for many deposits around the world (e.g., Bouma 1962; Walker 1973; Middleton and Bouma 1973; Piper and Normark 1983; Clarke et al. 1990; Porebski, Meischner, and Gorlich 1991). Most turbidites are interpreted as products of the subaqueous dispersion of incoherent sediment that failed as a slump or slide (e.g., Morgenstern 1967). Competency of turbidite flow

Table 3-3. Ideal End-Member Varieties of Sediment Gravity Flows

Characteristics of Deposit	Inferred Support Mechanism During Flow	Inferred Flow Type
Dominantly sand and silt in graded Bouma ABCD, BCD, or CD successions. Muddy matrix and generally with mudstone interbeds. A interval poorly sorted and massive.	Fluid turbulence	Turbidity current
Mainly medium and finer sands and silts, generally well sorted. Dish or sheet structures, also perhaps diffuse parallel lamination. Flat bases. Deposited near to relatively steep slopes. Absence of traction-current structures.	Escaping pore fluids	Liquified flow
Generally medium and finer sands and silts. Massive, ungraded, or with indistinct parallel lamination or dish structure. May contain sparse dispersed outsize clasts with subparallel orientation. Flat bases, though may deposit in channels. Deposited close to relatively steep slopes. Absence of traction-current structures.	Dispersive grain pressure	Grain flow
Generally showing extremely poor sorting with *large* outsize clasts, but may be well-sorted fine–very fine sands with minor (to ±20%) mud matrix. Commonly deposited in channels, but may occur as constructional mounds with flat bases. Generally not graded, but may show normal inverse, or symmetrical grading in some units. Indistinct stratification *may* be present—either subhorizontal or conforming to channel margins.	All of the above, plus density, buoyancy, and strength of the continuous phase, plus framework stacking. Laminar flow dominates during deposition and commonly during transport. Strength of continuous phase dominant in cohesive flows, other mechanisms in noncohesive flows.	Debris flow
Mainly clast-supported or without a true matrix; may be sand-matrix-supported. Largest clasts with subparallel orientation. Diffuse parallel lamination in sorted sand deposits.		noncohesive
Mainly matrix-supported, with matrix of mud. Clasts disorientated.		cohesive

appears to be varied (e.g., Komar 1985); scours are common at the base of turbidites.

When deposition begins, sediment is rapidly deposited in a normally graded *Bouma sequence* (Bouma 1962; Fig. 3-15). Relatively coarse grains accumulate in the massive poorly sorted *A interval*, which is followed by deposition of intervals *B* (parallel–laminated), *C* (cross–laminated and/or convolute–laminated), and *D* (parallel-laminated). The B, C, and D intervals are progressively finer and sorting thereby improves as coarser grains are selectively retained in lower intervals. In plan view, each interval overlaps deposits of the preceding interval in a down–current direction, and *base-cutout sequences* occur (the lower part of the sequence is left behind). The uppermost intervals of a turbidite may be eroded by subsequent flows. The Bouma sequence reflects a declining flow regime, and waning entrainment currents may produce the same sequence (e.g., Nelson 1982); distinction may be made by the facies association or by the relatively good sorting of the basal intervals in some entrainment-flow sequences.

In deep-sea environments where turbidity currents are most commonly recognized, a nonturbidite *E* interval is usually distinguished that represents fine hemipelagic muds deposited from background (mainly passive) suspension. Active suspension deposits, called *contourites* because the

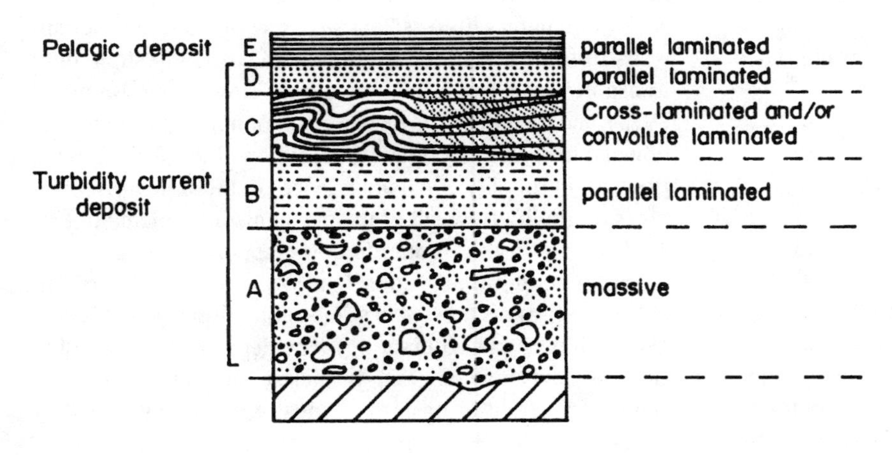

Figure 3-15. The ideal turbidite succession showing the relation of sedimentary structures (on right) and Bouma intervals (letter notation on left). Sand or silt particles may or may not persist through the uppermost D interval; if they do, they can provide a textural criterion for differentiation of the turbidite deposit from the pelagic (clay-size, with very rare exceptions), nonturbidite E interval.

suspension currents tend to flow parallel to the paleoslope, are also intercalated with turbidites at some localities (e.g., Stow and Lovell 1979). Support mechanisms other than turbulence may also be significant in flows where the concentration of sediment is very high; such deposits do not produce this ideal sequence (e.g., fluxoturbidites, see Slacza and Thompson 1981). Hence, whereas ABCD, BCD, and even CD (although with considerable faith!) vertical sequences in sedimentation units have been widely considered diagnostic of turbidites, care should be exercised in evaluating each deposit. Despite the undoubted success of the turbidite model in explaining many deposits, there is a tendency to overapply it; some of the problems pointed out by van der Lingen (1969) continue to exist and there are more turbidites inferred than actually exist.

Liquefied Flow

Liquified flow is an end-member case in which partial grain support is provided by intergranular pore fluid that is escaping from beneath settling grains (Fig. 3-14). *Fluidized flow* occurs when grains are *fully* supported by fluid moving upward and requires special and unusual situations, as in thin intervals at the top of liquefied sediments, or when compaction of subjacent sediments provides water; these flows are rare and produce few deposits (see Lowe 1976). For liquefaction to occur, the original fabric of a sediment must be disturbed by some trigger mechanism (seismic shock, wave, or current activity, even tidal fluctuations); a thixotropic transformation occurs as grains attempt to repack to a closer packing array and excess water escapes (e.g., Boswell 1948). For flows to occur, this transformation must be initiated on a slope greater than 3–4° (Lowe 1982); once moving, they can travel over low gradients, but not for great distances (<1 km, Lowe 1982), because the grains are only partially supported and are continuously settling. Liquefied flow is of primary importance *in conjunction with* other support mechanisms such as turbulence and dispersive grain pressure during transport, but may act alone in the initial and final depositional stages of other gravity flows. At low velocities the flows are probably laminar; however, if they accelerate sufficiently they may transform into turbidity currents. Deposits tend to be massive, with weakly defined water-expulsion features; most are well sorted with any original fines elutriated out.

Grain Flow

Grain flows consist of cohesionless particles in high concentrations supported during downslope movement by the upward component of impact forces as the grains hit one another (Fig. 3-14; see Bagnold 1968; Sallenger 1979). This process, termed *dispersive grain pressure* (or the *Bagnold effect*), appears to operate alone as the dominant support mechanism only with sorted sands on steep slopes; upon reaching gentle slopes, the grains freeze en masse. In the final stage of deposition, some support probably will be provided by upward movement of interstitial fluids as the grains (liquefied flow) settle (e.g., Lowe 1979).

Debris Flow

Debris flow is a class of sediment gravity flow with multiple-grain support mechanisms (Fig. 3-14); gradations to or from the other ideal end members are common. Most debris-flow deposits are gravelly and characteristically have *outsize clasts* that are very much larger than the rest of the sediment involved. Most flows occur in subaqueous environments, but studies of modern debris flows have largely focused on terrestrial flows because they are much easier to study and because they sometimes can be observed in action (e.g., Johnson 1970; Pierson 1981); terrestrial flows are the most damaging to human works (and lives). Most contain interstitial water when moving, but some have interstitial air (*sturztroms*, e.g., Hsu 1975). Most debris flows can be envisioned as having a *continuous phase* (normally water plus clays, and possibly other sediment) that behaves more like a low-viscosity plastic than a fluid such as water (e.g., Johnson 1970; Fisher 1971; Enos 1977). The continuous phase supports a *dispersed phase* of larger particles—these may range from well-sorted sands to bouldery gravels with outsized clasts up to tens (even hundreds) of meters long. In ideal or Newtonian fluids, viscosity is constant and independent of the shear stress, whereas in low-viscosity plastics, while viscosity may be largely independent of shear stress, there is a finite yield strength that must be overcome before flow begins (and which, when attained as the flow decelerates, causes the flow to freeze en masse). As defined by many, the yield strength of a debris flow is provided by the cohesion of the continuous phase; this mechanism is typified by high-viscosity mudflows, and their deposits have an abundant mud matrix (mud-supported gravels). The typical mudflow or pebbly mudstone deposits are good examples of these kinds of debris flows (e.g., Crowell 1957).

In contrast to these *cohesive debris flows*, other *cohesionless debris flows* (Table 3-3; e.g., Lewis 1976, 1980; Lewis, Laird, and Powell 1980; Nemec and Steel 1984) have a yield strength dictated by frictional forces between the grains. Cohesionless debris flows may not have a continuous phase as such (although as little as 2 wt % of clay in marine water appears to form a continuous phase with sufficient strength to transport fines and sands as debris flows, see Hampton 1975), and their deposits have a small or negligible clay fraction and may show a complete gradation of grain sizes in the flow, all in roughly equal abundance. In these low-viscosity types of gravity flow, all of the mechanisms depicted in Fig. 3-14 act in varying proportions to support the largest particles. In addition, a traction load equivalent to that of entrainment flows is present in some debris flows (e.g., Pierson 1981 and personal observation). The deposits of these cohesionless flows may be grain- or matrix (sand or fine gravel)-supported, sorted or unsorted, sands or gravels; unless they contain outsize clasts (the most distinctive characteristic of

Table 3-4. Common Diagenetic Processes

Siliciclastics	Carbonates
Cementation Precipitation of new minerals around and between grains, often as grain overgrowths	
Burrowing Activities	
Pedogenesis chemical and biochemical modifications	Boring Activities on particles or hard-ground substrates
Compaction dramatic thinning of clay- or organic-rich sediments (e.g., porosity decrease in mudstone volume from 60–80% to 10–20%) intergrain distortion of micas, fine rock fragments, perigenic clasts	Inversion replacement of mineral by another of the same chemical composition but different crystal form (e.g., Mg-calcite to calcite)
Pressure solution[a] can produce stylolite seams as well as intergranular stylolites grain overgrowths compaction	Compaction
Intrastratal Solution[a] partial grain corrosion complete grain dissolution	Pressure-Solution Stylolites[a]
Alteration, Replacement and Recrystallization mineralogical changes, e.g., feldspars alter to clays, leucoxene replaces ilmenite products may pseudomorph original grains	Dissolution[a] selective, e.g., of finest shells, or Mg-calcite/aragonite nonselective, e.g., cavernous
	Recrystallization aggradational degradational (e.g., micritization)
	Dolomitization
	Dedolomitization

[a]Involve compaction.

debris flows), they may differ from some entrainment-flow deposits only by the absence of entrainment structures and the presence of small quantities of mud that are apparent only after laboratory analysis (e.g., the fine-grained debris flows with well-sorted sand grains described by Hampton 1975, Lewis 1976). Whereas laminar flow may pertain during deposition from, and probably during much of the transportation stage of, high-viscosity cohesive debris flows, it pertains less commonly in low-viscosity cohesionless debris flows, many of which contain rafts of locally derived sediment that have been contorted by the torque of eddies in the flow. Turbulent low-viscosity debris flows would be called *high-density turbidity currents* by many (e.g., Lowe 1982); laminar low-viscosity debris flows may be called *density-modified grain flows* (e.g., Lowe 1982). In terrestrial settings, deposits of these types of flow may be difficult to distinguish from deposits of hyperconcentrated stream flows (e.g., Pierson and Costa 1984, and Fig. 3-13).

Outsize clasts similar to those carried in debris flows may occur in deposits from passive suspension settling of pebbles *rafted* into the depositional environment incorporated in ice or attached to floating plant debris (rarely, there may also be local deposits of animal gizzard stones), but these are generally much fewer than in any debris-flow deposit. In low-viscosity, noncohesive debris-flow deposits there is a preferred subparallel orientation of the clasts (cross-sections subhorizontal or poorly imbricated); in high-viscosity, cohesive-flow deposits the clast fabric is random. Although they can occur in almost any environment, most debris-flow deposits are associated with other kinds of finer sediment gravity flow in submarine fans, and with other gravelly deposits in alluvial and coastal settings (e.g., Nemec and Steel 1984; 1988). There appears to be a predominance of the noncohesive deposits in shoreline, shelf, and upper fan paleoenvironments (generated from sorted sediments) and of the cohesive deposits in mid- and lower-fan settings (generated on muddy slope settings); both occur in terrestrial environments, depending on the character of the source sediments.

DIAGENESIS

Diagenesis is the general term for all processes that affect sediments after their final deposition (e.g., Table 3-4; Packham and Crook 1960; see de Segonzac 1968 for an exhaustive history of the concept). Processes and consequences are varied and can strongly influence the characteristics of rocks and their storage potential for economic accumulations. Boundaries between diagenetic, hydrothermal, and metamorphic processes and effects are gradational and difficult to define strictly. At the other end of the spectrum, boundaries between diagenesis, pedogenesis, and weathering are equally difficult to define; if the sediment is not subsequently transported, both initial weathering and pedogenesis can be properly considered to be part of diagenesis.

Interpretation of diagenetic history is commonly difficult; generally a suite of samples from a given unit or succession is required to determine the *paragenetic sequence* of modifications. Because it is the original characteristics of the sediment that are used to infer the depositional and predepositional history, diagenetic effects must be identified and "subtracted" from the present rock character to determine them. Larsen and Chilingar (1978), Scholle and Schluger (1979), Parker and Sellwood (1983), McDonald and Surdam (1984), and Marshall (1987) provide a variety of papers that indicate the scope and importance of diagenesis. Figure 3-16 shows relationships between various important diagenetic effects. Discussion in this section attempts a general overview; further discussion is provided for detrital sediments in Chapter 6, for clays in Chapter 7, for carbonates in Chapter 8, and for chemical sediments in Chapter 9.

Physical diagenesis results primarily from increasing pressures as sediments are progressively buried (sometimes resulting in overpressured sediment when contained fluids cannot escape; e.g., Bredehoeft and Hanshaw 1968; Swarbrick 1968; Plumley 1980). Chemical diagenesis, such as dissolution and alteration, result from chemical interaction between the minerals and aqueous solutions that fill the interstitial pores (e.g., Runnels 1969). However, most workers attempt to separate these processes from *epidiagenetic* or *epigenetic* processes (e.g., Perel'man 1967), which affect sedimentary rocks after uplift brings them into the influence of circulating fresh groundwaters.

Early diagenesis involves shallow–burial, prelithification, geologically rapid processes and is characterized by biological activity (such as bacterial–induced effects), oxidation, reduction, and dewatering accompanying initial compaction by physical rearrangement of particles. If oxygen is present in the system, oxidation reactions dominate initially, and aerobic bacteria break down organic matter to produce carbon dioxide. Where and when oxygen is depleted, anaerobic bacteria break down sulfates and produce hydrogen sulfide and sometimes calcite; the active H_2S can lead to metal sulfide precipitation (e.g., pyrite). Most of these earliest diagenetic reactions involve interactions between the sediment and pore waters, which are usually compositionally quite different from the water above the sediment/water interface (in particular, the salinity of pore water brines can be much higher as a consequence of the lack of free circulation and higher brine density). These early diagenetic reactions taking place at, or very near, the sediment/water interface are termed *syndiagenetic* (or syngenetic). They are particularly important in the formation of many sedimentary ore deposits (e.g., Maynard 1983). Syndiagenetic processes include the formation of authigenic minerals and mineral transformations (dissolution, displacive crystal growth, alteration, oxidation, reduction, adsorption, desorption, and ion exchange) that are induced as the sediment passes through the near–surface, high chemical, physical, and biological gradients with continuing deposition.

Late diagenesis involves deeper burial, commonly postlithification and generally longer-term (up to 10^8 yr) processes that include compaction, most cementation, alteration and creation of silicate minerals, and conversion of organic compounds to hydrocarbons (maturation). Negative Eh and alkaline pH conditions prevail, but solution chemistry may vary extensively and complexly; temperature and pressure increases are influential factors (e.g., various papers in Scholle and Schluger 1979). Clay mineral changes are commonly substantial; their late-stage dewatering promotes further compaction, and provides both a driving mechanism for petroleum migration and a reaction medium for many mineralogical transformations (e.g., Boles and Franks 1979; Lee, Ahn, and Peacor 1985; Velde and Nicot 1985).

Diagenetic effects may be negligible or extensive, local or widespread, essentially chemical or dominantly physical. They may be obvious in hand specimens or thin sections (e.g., Fig. 3-17), they may be apparent only after analysis with special equipment (such as cathodoluminescence, scanning electron microscopy, x-ray diffraction, x-ray fluorescence, or electron probe; see **AS**), or study of a rock sequence may be necessary to determine them. The rank of any coaly material present in the sediments is commonly used as an index to the extent of diagenesis. Rank (measured from moisture content, volatile–matter content, carbon and hydrogen content, and/or optical reflectivity) reflects essentially the degree and duration of temperature increases; however, both coal diagenesis and the relationship between it and other diagenetic modifications to sediments is complex and not fully known. In particular, rank will seldom indicate the extent of diagenetic reactions caused by interactions between the sediment and pore waters, in which the reactions are controlled by a progressive change in pore-water composition (e.g., due to compositional evolution of trapped brines along a hydraulic gradient) or by addition of compositionally different brines (e.g., hydrothermal solutions).

Compositional Changes

Although diagenesis does not generally destroy the dominant detrital components of a sediment, grains of chemically unstable minerals may be completely dissolved or altered and replaced to such an extent that their original character is obscured (e.g., Milliken, McBride, and Land 1989). When minerals are completely replaced, the diagenetically formed replacement mineral often pseudomorphs the original grains, and it may be possible to determine the identity of the minerals that were originally present. However, fossil fragments may be completely dissolved; feldspar or volcanic rock fragments may be so altered to clay minerals that they become difficult to distinguish from clayey sedimentary rock fragments or clayey matrix; ilmenite and other detrital titaniferous minerals may be altered to leucoxene, which precipitates after some movement of the components in solution; and iron-bearing minerals may be oxidized in place or may be

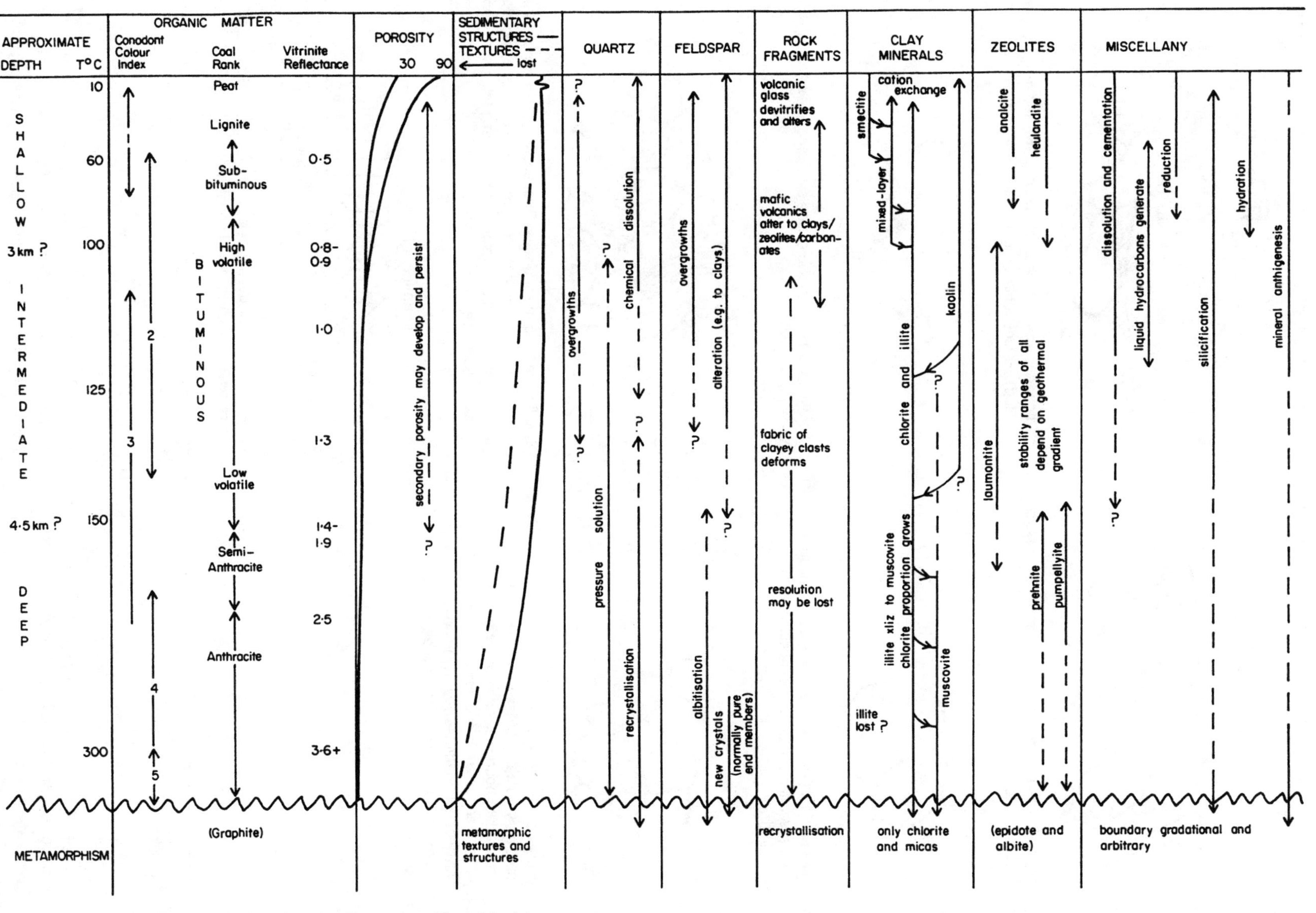

Figure 3-16. Generalized relationships between diagenesis and burial depth/temperature.

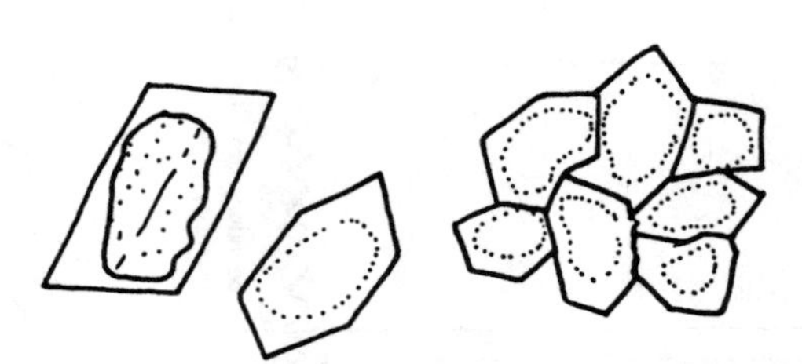

Euhedral overgrowths on rounded detrital grains; rims of quartz grains outlined by dust (e.g., iron oxides, clays).

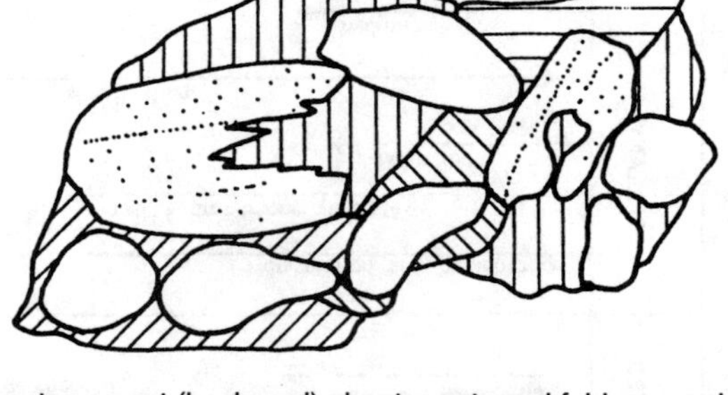

Carbonate cement (hachured) about quartz and feldspar grains (mottled). Patchy replacement of several feldspars.

Concavo-convex contact between two grains subjected to a low degree of pressure-solution. At time of deposition, point contacts occur generally.

Microstylolite between two grains subjected to extensive pressure solution.

Detrital quartz grains without dust rims have developed overgrowths (and may be partly pressolved) to produce this texture. The concordant boundaries and occasional thin, irregular projections demonstrate the diagenetic modifications—such concordancy of boundaries is virtually impossible from original packing and the projections are unlikely to have survived abrasion.

Stylolite crossing many grains. Dust and insoluble minerals will ultimately be concentrated along the stylolitic surface and block further movement of solutions, thereby halting the dissolution of grains on both sides.

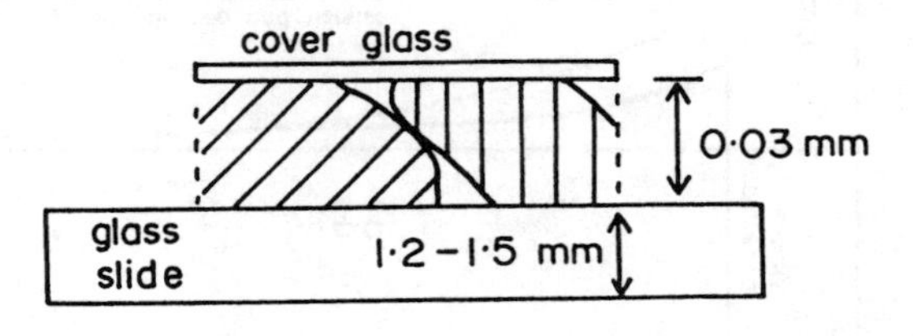

Note: Watch out for overlap relationships in thin sections, which can obscure true character of contacts.

Figure 3-17. Examples of some diagenetic effects.

reduced and the iron translocated. Silica is particularly prone to remobilization in diagenesis and various phases precipitate (e.g., Williams and Crerar 1985).

Solutions of differing chemistry migrate through the sediments at many stages after burial: hydraulic gradients result from compaction of muddy sediments deposited as a lateral facies of the unit being studied, from folding or faulting, from differential heating (e.g., due to subjacent igneous intrusion), and from other causes. In cases in which units are isolated by impervious lithologies soon after burial and fluid circulation is prevented, diagenetic modifications may be insignificant; impermeable sediments or early concretions may display the least affected primary characteristics.

New (*authigenic*) minerals may be formed during diagen-

Table 3-5. Geochemical Classification of Stages in Diagenesis Indicated by Authigenic Mineralogy

Geochemical Environment			Characteristic Authigenic Minerals
Oxic—dissolved oxygen present			Hematite, goethite, MnO_2 minerals; no organic matter
Anoxic	Sulfidic—dissolved sulfide (H_2S and HS) present		Pyrite, marcasite, rhodochrosite; abundant organic matter
	Nonsulfidic	Postoxic—weakly reducing, no oxygen; sulfates not reduced	Glauconite and other Fe^{+2} - Fe^{+3} silicates; siderite, vivianite, rhodochrosite; no sulfide minerals; minor organic matter
		Methanic—strongly reducing; no oxygen; sulfates reduced with consequent methane formation	Siderite, vivianite, rhodochrosite; earlier-formed sulfide mineral; organic matter

The general early diagenetic sequence is:

Oxic ⟶ Postoxic ⟶ Sulfidic ⟶ Methanic

Source: After Berner (1981).
Note: The scheme is theoretically independent of changes in salinity and pH. It assumes that both Fe and Mn are present in the initial sediment. Microenvironments (e.g., within shells) may differ from the general environment in geochemical conditions. Interpretation of the paragenetic sequence is usually difficult because changes in diagenetic conditions can occur independently of the general trend.

esis not only by replacement of primary components but also by direct precipitation from interstitial fluids, containing ions derived from dissolution of other minerals or inherited from the initial pore waters (e.g., Teodorovich 1961). Precipitated authigenic minerals tend to form euhedral crystals, but interference caused by adjacent clastic (*allogenic*) minerals or the growth of other authigenic components may prevent the development of ideal crystal forms. The authigenic components may form as discrete, isolated new crystals or as syntaxial *overgrowths* (formed around the edges of primary "seed" crystals and commonly showing the same crystallographic/optical orientation as the parent, as on quartz, alkali feldspars or the single calcite crystals of echinoderm fragments). Or they may partly to completely fill the original interstitial pore space as cements (carbonate, silica, iron oxide, pyrite). Some components (such as glauconite) may form on the seafloor and may be transported locally before final burial; it is useful to use the term *perigenic* (rather than allogenic or authigenic) when such grains can be identified.

A practical classification scheme that can be used as an initial guide for subdividing stages of diagenesis in detrital sediments has been suggested by Berner (1981; Table 3-5). It is based on the presence of various authigenic minerals that originate under particular geochemical conditions. Because pH and salinity conditions can rarely be adequately inferred from the final diagenetic mineral assemblage, the scheme is based on the presence of sulfur compounds, which are strongly influenced by the presence/absence of oxygen, and which strongly influence authigenic mineralogy. Depositional, diagenetic, and epigenetic conditions are normally superimposed, and products of an earlier stage may survive through a later stage; hence interpretation of the postdepositional history can be difficult. Interpretation is even more complicated because chemical/biochemical changes can occur in the environment during deposition or diagenesis, with consequential complications on the resultant suite of authigenic minerals.

Textural Changes

Diagenetic modifications of primary sediment texture may take place either as a result of chemical modifications (such as dissolution or replacement), or as a result of physical compaction, or both. The precipitation of authigenic mineral overgrowths may obscure the original roundness and size of grains. *Pressure solution* occurs when two grains are touching and differential pressure causes one or both grains to dissolve preferentially at the points of greatest pressure, so that the grains interpenetrate and interlock (overgrowth interlocking commonly results in a similar texture). When pressure solution occurs along an irregular surface marking the escape route of solutions carrying the dissolved constituents, stylolites form; insoluble materials tend to be concentrated along them. The net result of these processes may involve substantial modification of original grain roundness and even original grain size. Recrystallization (mainly to larger, but occasionally to smaller, crystals) is common in carbonate sediments and may also obliterate original texture. Dramatic reorganization of sediment fabric may occur from physical or biological diagenetic processes (e.g., burrowing or hydroplastic to quasi-fluid deformation and dewatering structures, see Chapter 4). Changes in porosity and permeability resulting from all these modifications are particularly important to considerations of migration and entrapment of petroleum and other ore solutions; diagenetic changes can destroy initial porosity, but secondary porosity may be developed (Fig. 3-18; e.g., Rittenhouse 1971; papers in Scholle and Schluger 1979).

BIOLOGICAL PROCESSES

Biological processes have many diverse chemical and physical effects on sedimentation (e.g., Table 3-6). Both kinds of effects may result from the same organic activity, or the effects may result indirectly from organic processes. For ex-

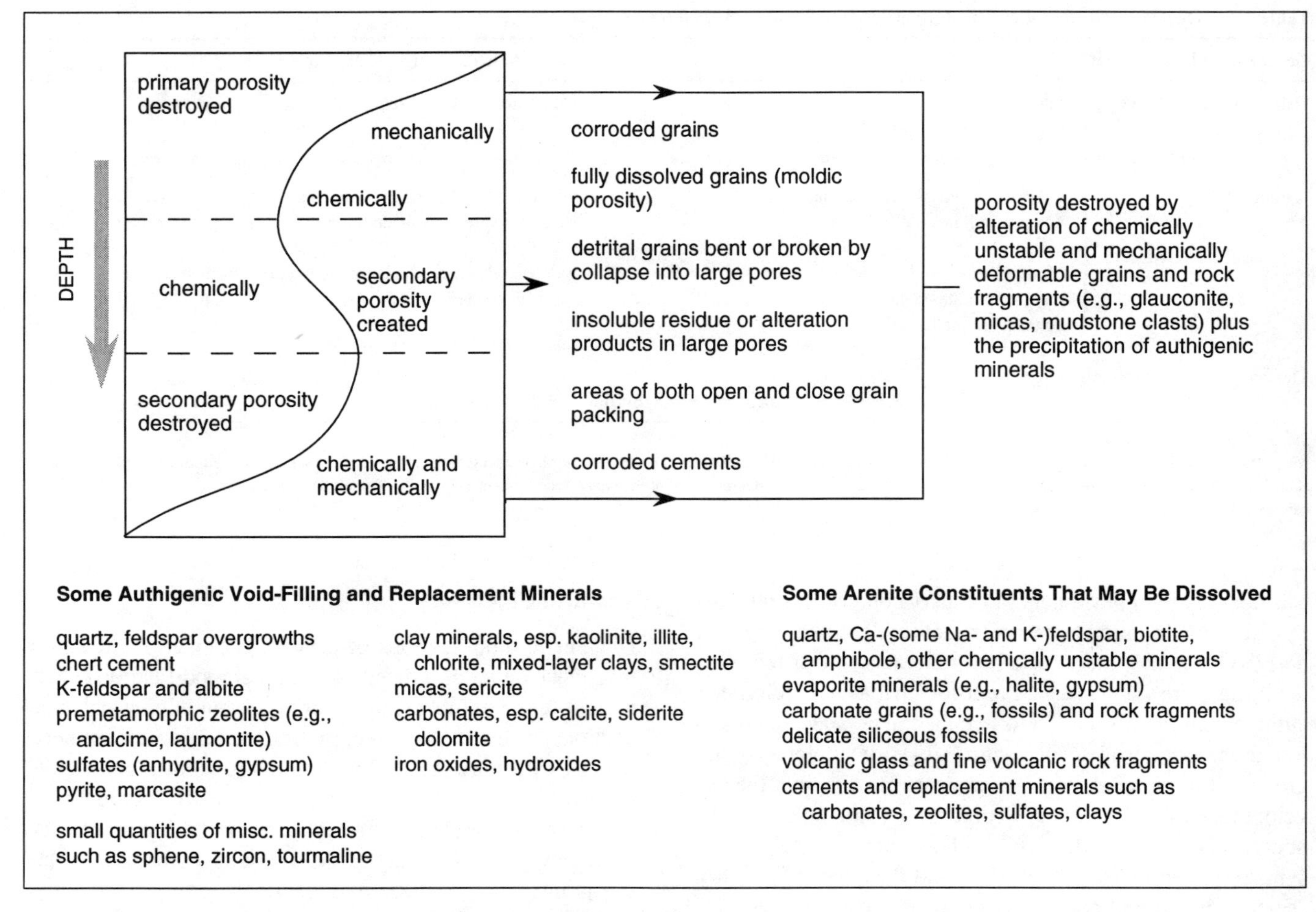

Figure 3-18. Factors related to diagenetic changes in porosity of detrital sediments. (Modified from Hayes in Scholle and Schluger 1979.)

ample, plant roots may mechanically fracture rocks (permitting increased water entry) and biochemically break down minerals either directly for nutrients (e.g., potassium) or indirectly as a by-product of their decomposition (e.g., humic acid formation); plants may also bind partly altered minerals, thereby retaining them in the weathering zone for further operations by physical, chemical, and biological processes. Salient direct and indirect processes and effects of biota are reviewed below; most of the relevant literature has been cited or is listed in Chapter 2, and other aspects are discussed in Chapter 4 (biogenic structures that may be created by organic activity) and in Chapter 8 (influence of biota on carbonate sediments).

Direct Effects

The most evident effects of biological processes on sedimentation are the precipitated carbonate endo- or exoskeletons that make up lime sediments and limestones (see Chapter 8), as well as the ubiquitous shells or shell fragments in detrital marine sediments. These comprise not only large and obvious local masses such as reefs but also the more widespread and much larger volumes of lime mud, for which an organic origin is only evident under the microscope: calcareous tests of microorganisms, both plants such as algae and animals such as foraminifera, and fine sediment formed by the abrasion of biogenic carbonates originally formed by larger organisms. Whereas almost all modern calcareous sediments are biogenic, the contribution from inorganic carbonate deposition has been higher in the geologic past, precipitated in giant evaporite basins (e.g., Chapter 9) and prior to the evolution of carbonate-secreting biota. Limestone deposits provide by far the major carbon reservoir in the modern global carbon cycle; thus, they have a significant influence on the carbon dioxide content of the atmosphere and hydrosphere. They have played, and will in the future play, a vital role in controlling human habitat; e.g., in relation to the greenhouse effect, note that carbonate rocks (made of organic hardparts) store over 90,000 times more CO_2 than the global atmosphere, over 6,000 times more than all known fossil fuels, and over 140,000 times more than all the world's land plants (cf. Fig. 3-7). And it is the biological deposits of bituminous and carbonaceous matter that are progenitor materials for fossil fuels (coal, oil, and gas).

Organisms with siliceous shells or skeletal parts (e.g., diatoms, sponges, and radiolaria) are widespread in sedimentary

Table 3-6. Some Roles Played by Organisms in Sedimentation

Organisms	Role in Sedimentation
Blue-green algae (cyanobacteria)	Trap fine sediment in mats and as stromatolites
Green algae, sponges, echinoids, barnacles, crinoids	Fragment on death to produce sand-size clasts; all fragments of echinodermata comprise single calcite crystals
Planktonic foraminifera, coccoliths (post-Jurassic), benthic foraminifera, red and green algae	Form major portion of most lime muds, together with fine abrasion products of other biota
Encrusting worms, foraminifera, bryozoa, coralline algae, stromatoporoids	Encrust grains and other substrates; may cement sediments into substantial masses (e.g., modern reefs); may break into gravel or sand clasts
Bivalves, brachiopods, cephalopods, gastropods, benthic foraminifera, bryozoa, trilobites	Whole or broken, form major component of most carbonate sands and gravels
Corals, oysters, archaeocyathids, stromatoporoids, rudistid bivalves, bryozoa	Framework organism of bioherms; fragment into gravel-size clasts
Worms, crustaceans, bivalves and representatives of many other groups	Sediment ingestion and excretion modify texture and composition; burrowing modifies primary structures

deposits and also account for large relatively pure siliceous accumulations (e.g., chert, diatomite, spicularite, and radiolarian oozes). These deposits are by far the most important factor in the geochemical cycle of silica (Fig. 3-19), although the situation was rather different prior to the evolution of silica secretors (e.g., Cloud 1973 and papers in Windley 1976). Except for sediments in which the silica has been extensively remobilized in diagenesis, biogenic silica accumulations are the most concentrated in freshwater lakes (diatomites) and in marine sediments deposited in deep water below the *carbonate compensation depth*, where cold waters hold sufficient CO_2 so that all carbonate sediment supplied is dissolved by the bicarbonate reaction discussed above.

Organisms have a marked physical impact on many rocks and sediments by a miscellany of effects grouped under the term *bioerosion*. Herbivores such as gastropods and limpets repeatedly scrape epibionts from their localized rock "farms"—the rasping action wears away the rock. Lichens (and other plants) exude acids and complexing agents to decompose minerals for nutrients, but also as they contract in dry times they passively erode the rock by pulling away the weakened crust to which they are attached. Sediment eaters (e.g., the coral-consuming parrotfish and holothurians) and borers (e.g., some sponges, bivalves, and even some algae and fungi) erode rocks and abrade sediment. Burrowing organisms (e.g., crustaceans, worms, some bivalves) rework

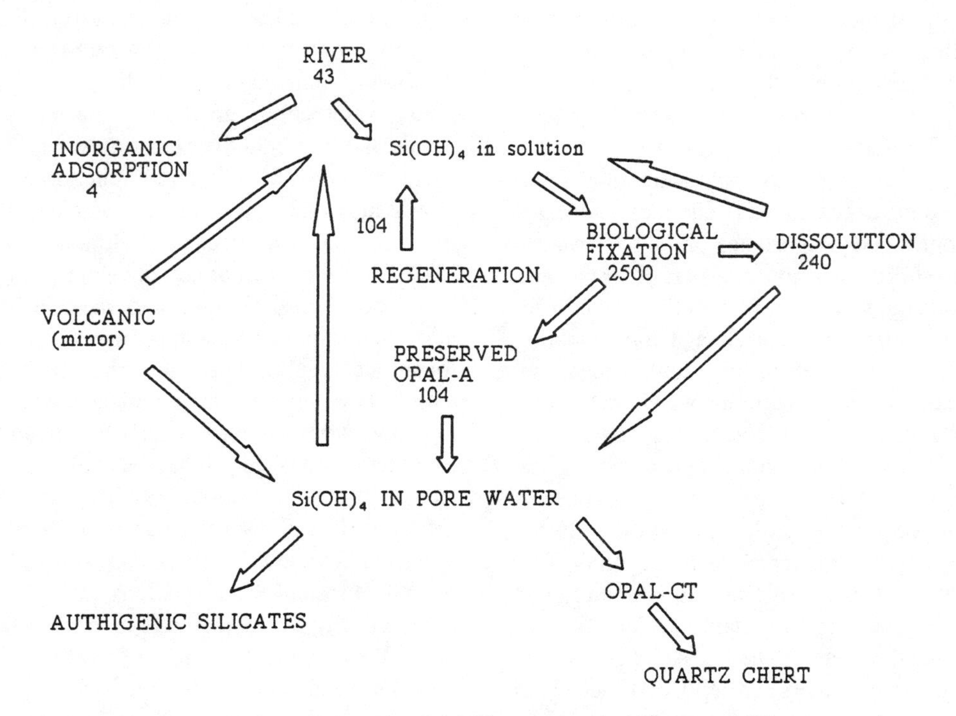

Figure 3-19. Simplified sedimentary cycle of silica; numbers in 10^{13} grams of SiO_2 transported or fixed per year. (Modified from Calvert 1974.)

and mix vast quantities of sediment because of their constant activity over time; in places they may concentrate either finer or coarser sediment at the surface than would accumulate under the inorganic physical processes acting in a particular environment; they also aerate substrates that would otherwise be anoxic, with consequential chemical ramifications. Fecal and feeding pellets agglutinated by invertebrates account for much sedimentation of clay particles, and fecal ejection can also result in the suspension and consequent transportation of clays (see discussion and references in Chapter 5).

In contrast to bioerosion, some biota (mainly plants) also act to stabilize sediment accumulations. Examples are algae and cyanobacteria, grass (terrestrial and marine), and tree roots (terrestrial and mangroves in the marginal marine setting); processes include both direct *binding* of the sediment itself, and indirect *baffling* of the currents that might otherwise have sufficient velocity and turbulence to erode them (e.g., Hine et al. 1987; Ward, Boynton, and Kemp 1984; and see references in Chapter 2). Algae and cyanobacteria can provide a continuous protective cover over the sediment (causing even sands to act as a coherent mass and reducing the surficial friction coefficient). Grasses form a discontinuous cover, but their roots commonly reach much deeper than the surface blades (especially in marine grasses), and as a current strong enough to cause erosion flows over the grasses, the blades flatten over the sediment and protect it. Mangroves and similar plants are not as effective at binding sediment as grasses and algae because they do not provide a dense vegetation cover near the sediment/water interface, but they form excellent baffles (particularly species with well-developed pneumatophore or stilt root systems) that reduce the flow velocity of passing currents. The reduction in flow velocity causes both an increase in the rate of sediment deposition at the site and a reduction in the vulnerability of existing sediment at the site to scouring and erosion (e.g., Bird 1971; Stephens 1962; references in Chapter 2).

In addition to the effects of the larger biota mentioned, microbes have a marked influence on many chemical processes in natural environments. The importance of the bacterial reduction of sulfate to sulfide was noted earlier; other bacteria control most of the nitrogen cycle, oxidize sulfides, aid in decomposing organic matter, and can change the oxidation state of transition elements such as iron and manganese. There are very few sedimentary environments where bacteria have no effect on chemical processes. Bacterial content of a sediment is usually related to the amount of decaying organic matter present. In general, the organic matter content of sediments increases with decreasing grain size, particularly because decomposing organic matter is adsorbed onto clays and colloids. However, even in sediments in which organic matter is sparse, it may be very important in microenvironments (e.g., within fecal pellets or shells). Aerobic bacteria generally attain maximum numbers at a depth of 40–60 cm in well-oxygenated sediment and are rapidly replaced by anaerobic types below this depth; in many muddy sediments, aerobic bacteria live only in a very thin surficial layer; in euxinic (anoxic) conditions they are not present at all.

Indirect Effects

Biological activity commonly has indirect as well as direct effects on sedimentation. One example is the effect that plant respiration has on the solubility of calcium carbonate through its influence on atmospheric CO_2 and on aqueous HCO_3^- (see above), and consequently on the redox status of surficial environments and on the pH of the oceans. Particularly Precambrian (but also some Phanerozoic) *stromatolites* probably formed because cyanobacteria, by removing CO_2 from the water, initiated precipitation of $CaCO_3$ that buried the community; a new community colonized the substrate so formed, and the cycle repeated many times to produce these finely laminated lime mudstone structures (e.g., Kempe and Kazmierczak, in Ittekkot et al. 1990). Alternatively, or in addition, the formation of these structures involved trapping and binding of muds by the sticky mucilaginous sheaths of the organisms. These indirect biological processes are some of the earliest evidence of life on Earth (e.g., Buick 1984; Walter and Hofmann 1983; references in Chapters 2 and 4).

Perhaps the most obvious indirect influence of biota is when the development of a major biogenic structure (e.g., a coral reef or shell bed) forms a barrier that reduces the energy experienced by sedimentary environments on its lee side, which consequently markedly affects the sedimentary structures, textures, and compositions of sediments that accumulate in these lee-side environments. Other effects of biological processes are the transport of gravels by floating vegetation (*rafting*) and as gizzard stones (large reptiles, birds, and some mammals); where large grains occur in very quiet water environments that normally accumulate only muds, the only other explanation is supply from ice rafting, with consequential climatic implications.

Living organisms influence chemical processes by removing potential reactants from solution—not only $CaCO_3$ and SiO_2 for skeletons, but cell development also consumes carbon, phosphorus, nitrogen, and trace elements. Biota alter the chemical form of potential reactants (e.g., bacterial reduction of sulfate to sulfide). The great influence that biological processes have on surficial chemical processes is most clearly demonstrated by comparisons between modern sediments and sediments deposited prior to the evolution of major biochemical processes (e.g., Cloud 1973, 1974; Holland 1978, 1984; McConchie 1987; several papers in Windley 1976 and Holland and Schidlowski 1982). For example, chert bands, found in many Precambrian formations, must have been precipitated inorganically; such precipitation of silica became improbable after the evolution of silica-secreting organisms such as radiolaria and diatoms (cf. Fig. 3-19).

After death, decomposing biotic softparts continue to influence the chemistry of the diagenetic environment. For example, plant remains commonly produce humic and fulvic acids that are very effective chelating agents with a high ion-exchange capacity. Plant-derived chelating agents provide an efficient means of transporting and trapping metal ions; plant debris, which is transported by physical processes, carries to the site of deposition any elements bioaccumulated during its growth. Gasses generated during decomposition can fill small shells (e.g., of foraminifera) and cause them to float for substantial distances. Bioturbation and the production of burrowing, grazing, and other traces by a wide variety of organisms (see discussion in Chapter 4) constitutes a biological process that has both direct and indirect effects.

VOLCANIC PROCESSES

Many volcanic processes (e.g., Table 3-7) cause the deposition of volcanogenic material in what would normally be regarded as subaerial and subaqueous sedimentary settings (e.g., Fisher and Smith 1991). Where this volcanogenic debris is a minor constituent of a sediment, its presence may or may not be noted (much volcanic material alters readily to indistinguishable clayey material) and volcanic processes are not commonly invoked. However, where volcanogenic debris dominates or is the sole constituent of a sediment, more specific volcanogenic terms are used to describe the resulting deposit (e.g., Table 3-8; Fisher and Schmincke 1984). Some fine-grained volcanogenic material is transported as airborne ash and may be deposited considerable distances from its source from either passive or active suspension; distinction between volcanic and eolian processes may sometimes be made on the basis of texture (e.g., Smith and Kalzman 1991). Pyroclastic flows may initiate in subaerial environments and continue to flow subaqueously when they enter lakes or the ocean (e.g., Cole and DeCelles 1991). Other fine- and coarse-grained volcanogenic material is transported as mass flow deposits, the movement of which is usually aided by the lift imparted by escaping steam and other gasses (e.g., Fig. 3-20; Smith 1986; Chough and Sohn 1990); some criteria for distinguishing between these pyroclastic flows and epiclastic flows are presented in Table 3–9. Irrespective of its mode of transport and deposition, a large proportion of any fine volcaniclastic material that has not fused into large solid masses (e.g., lava flows) is highly likely to be reworked by physical sedimentary processes and diluted by mixing with

Table 3-7. Common Types of Volcanic Eruption and Their Products

Common Types of Eruption

Hawaiian—highly fluid basaltic lavas of low (or easily lost) gas content; relatively little pyroclastic debris.

Strombolian—more pyroclastic debris due to explosive activity of relatively gas-rich basaltic-intermediate magma.

Plinian—mainly flows, with some pyroclastic debris due to explosive activity of gas-rich intermediate-acidic magma.

Plinian Deposits	*Strombolian Deposts*
Felsic composition (rhyolite, trachyte, andesite, phonolite, dacite), sometimes becoming more mafic with time; no lava flows	Basaltic/andesitic composition, lava flows common
Massive to thick-bedded sheets, common grading	Massive beds in spatter and scoria cones, with some alignment of deformed pyroclasts and clast-size concentrations
Angular, highly vesicular, juvenile clasts, commonly well sorted	Vesicular juvenile and cognate bombs to glassy clasts, commonly well sorted
Clast size decreases lognormally away from source	Fine ash rare, wide range clast size
Commonly associated with welded or unwelded pyroclastic-flow deposits (ignimbrites) comprising more than 50% ash; some may resemble lava flows	Welding (agglutination) common in some

Source: After Fisher and Schmincke (1984).

Table 3-8. Pyroclastic Deposit Terminology

Pyroclastic Deposits		*Epiclastic Deposits* (volcanic and nonvolcanic)
Agglomerate or pyroclastic breccia	Intermediate deposits, with 25–75% pyroclasts,	Conglomerate or breccia
Lapillistone or lapilli-tuff	may be grouped as *tuffites*, and are described	Sandstone
Ash or tuff	as tuffaceous (epiclastic name)	Mudstone or siltstone and claystone
coarse—sand size		
fine—mud size		

Pyroclasts—fragments produced by many processes connected with volcanic eruptions.
Hydroclasts—pyroclasts formed from steam explosions and where magma or lava contacts water these may be:
Juvenile—freshly derived from magma; Cognate—fragments from previous eruptions of the same volcano; or
Accidental—derived from the subvolcanic basement (any kind)

Blocks—angular pyroclasts larger than 64 mm; Bombs—juvenile pyroclasts larger than 64 mm and partly molten when ejected, thus shaped in flight (and on impact if still plastic)

Lapilli—2–64 mm pyroclasts, angular or rounded, including special case of *accretionary lapilli* made of ash aggregates

Ash—pyroclasts finer than 2 mm or a deposit with more than 75% of these particles; may be *vitric* (glassy), *crystal,* or *lithic* particles

Tuff—consolidated equivalent of ash and lapilli

Tephra—unconsolidated pyroclastic materials, regardless of particle size

Epiclasts—fragments eroded by nonvolcanic processes from older volcanic rocks (do not include particles derived from unconsolidated pyroclastic debris); generally include relatively few glass shards or pumice clasts

Volcaniclasts—all fragmental volcanic materials produced by any process, transported by any process, and deposited anywhere

Source: Data from discussions in Fisher and Schmincke (1984).

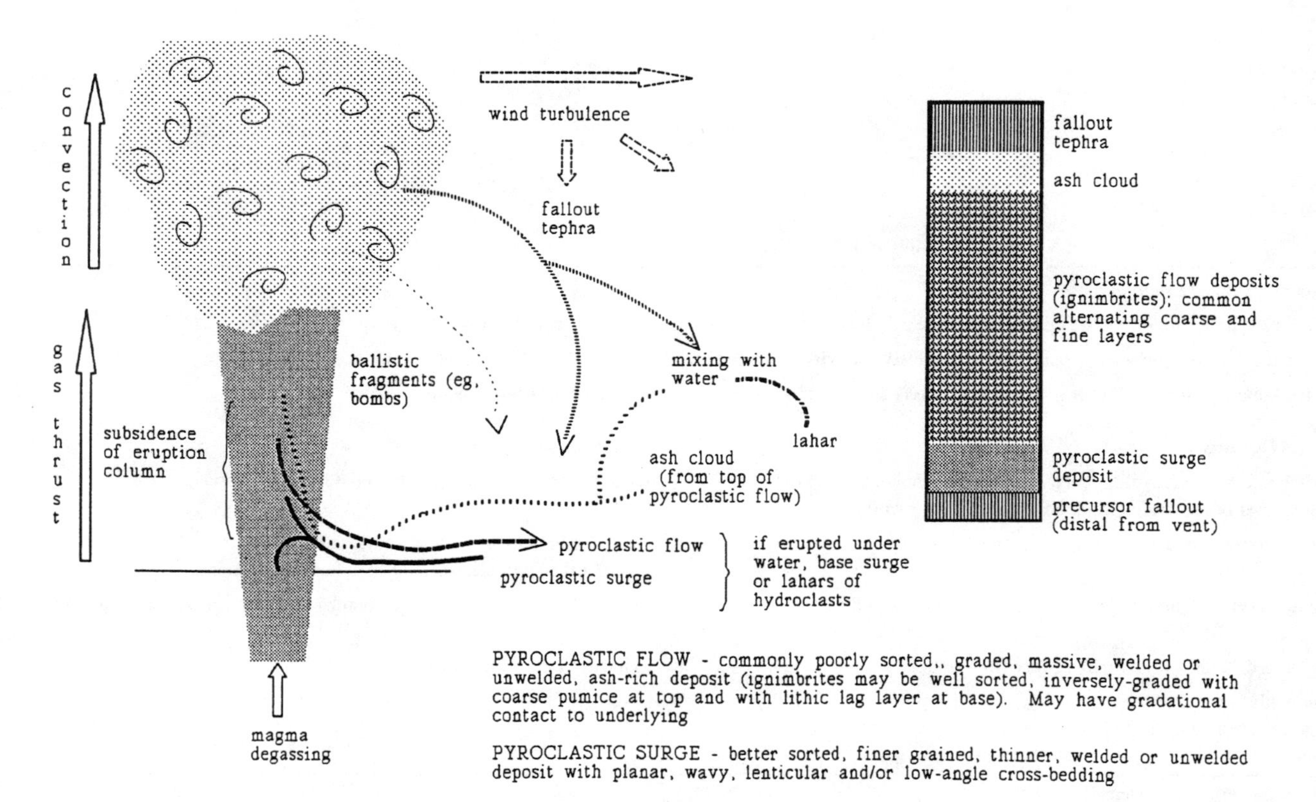

Figure 3-20. Schematic representation of some Plinian volcanic processes and products. (After Fisher and Schmincke 1984.)

Table 3-9. Some Criteria for Distinguishing between Pyroclastic and Epiclastic Flows

Properties	Pyroclastic Deposits	Epiclastic Deposits
Variability	High, depending on distance from source, kind of eruption, changes in style during eruption	High, depending on flow regime, grain sizes available, agent of deposition (wind or water)
Plan-view geometry	Circular or fan-shaped distribution relative to source (air-fall deposit geometry influenced by prevailing wind); units show wedge-shape cross-section	Sheet or larger-scale fan and wedge-shaped geometries (exceptions: small alluvial fans and deltas)
	Thick units rarely extend over wide areas (1000s km); rare exceptions with large pyroclastic flows	Thick units may be very extensive laterally
Relation to paleotopography	Distal airfall deposits have parallel beds over gentle topographic irregularities; proximal flows and surges may fill valleys but also may overflow	Loess deposits may mimic bedding geometry of air-fall pyroclastics; subaqueous beds are thicker in lows and thinner on highs, or confined to lows
Contacts	Generally sharp, with planar bedding planes between eruptive units, but lower contacts commonly nonerosive	Sharp, planar to irregular between thinner sedimentation units (exceptions); lower contacts commonly erosive
Structures	Internal bedding generally poorly defined (exceptions, particularly in surge deposits); alternating grain-size layers common in air-fall deposits	Internal bedding commonly well-defined; alternating grain-size layers not common within sedimentation units
	Ripples, cross-lamination and cross-bedding rare; main exception as medium to large scale, poorly defined single internal cross-bed (including antidune) sets within surge deposits	Ripples, cross-laminations, cross-bedding common at all scales; antidune bedding very rare
	Inverse and/or normal grading; Bouma sequences rare	Inverse grading rare; normal grading and Bouma successions common in turbidite sequences
	Bioturbation generally rare	Bioturbation common
Maximum grain size	Maximum grain size decreases with distance from source in single layers (exceptions with pumice)	Maximum grain size decreases downcurrent, but in sediment gravity-flow sequences, coarser sediments may bypass finer proximal deposits
Floating outsize clasts	Common (exception: distal air-fall tuffs)	Rare (exception: debris flows and ice- or tree-rafted clasts)
Particle shape	Large lithic clasts highly irregular (to regular aerodynamic shapes of bombs)	No highly irregular or aerodynamic volcanic lithic clasts
Sorting	Using Inman measures of central two-thirds of distribution only—poor in proximal pyroclastic flow (and laharic) deposits, but up to moderate or good in some surge deposits; moderate to good in distal deposits, particularly air-fall tuffs	Using more precise Folk and Ward or method of moments analysis because grains less weathered or likely to break up in analysis—poor to well sorted; poorest sorting (tills, mudflows) generally worse than in pyroclastic deposits (exception: lahars)
Composition	Dominantly juvenile and cognate particles; vesiculated lithic clasts (e.g., scoria, pumice), crystals of mafic minerals (e.g., hornblende, augite), vitric particles (highly irregular and angular glass shards); all angular and showing same degree of weathering (postdepositional); fragments of pillows with palagonitic rims may be common	Virtually no juvenile particles (especially vitric and crystal fragments); juvenile lithics and cognate particles generally subrounded-rounded with notably weathered rims; pumice can float large distances, but is generally sparse and dispersed; no palagonitic fragments except where redeposition has been almost immediately after eruption
Compositional trends	Vertical changes may be sharp and distinctive; lateral trends may be evident in outcrop	Vertical changes very gradual (even over most unconformities); lateral trends not visible in outcrop
Organic matter	Rare (charred wood or thin layers of fossils may occur within or between eruptive units)	Common and widely distributed
Associations	Association with lava flows or domes, agglomerates composed almost exclusively of juvenile or cognate pyroclasts	Reworking and redeposition may occur in close proximity to primary pyroclastic deposits, but in other associations, such as with limestones, thick mudstones are much more common

Note: In comparing airfall volcanic and epiclastic deposits, the major problems lie in distinguishing between gravity flow deposits generated from volcanic versus water activity.

other detrital sedimentary grains. Where such reworking takes place, the resulting deposits usually have typically sedimentary characteristics and are classified using standard sediment description nomenclature (see Chapters 4, 5, and 6).

SELECTED BIBLIOGRAPHY

Chemical Processes

Anderson, M., and A. Rubin (eds.), 1981, *Adsorption of Inorganics at the Solid–Liquid Interface.* Ann Arbor Science Publications, Ann Arbor, Mich., 357p.

Berner, R. A., 1971, *Principles of Chemical Sedimentology.* McGraw-Hill, New York, 240p.

Brookins, D. G., 1988, *Eh–pH Diagrams for Geochemistry.* Springer-Verlag, New York, 176p.

Dean, W. E., and B. C. Schreiber, 1978, *Marine Evaporites.* Short Course 4, Society for Sedimentary Geology, Tulsa, Okla., 188p.

Drever, J. I., 1988, *The Geochemistry of Natural Waters,* 2d ed. Prentice-Hall, Englewood Cliffs, N.J., 437p.

Förstner, U., and G. T. W. Wittmann, 1981, *Metal Pollution in the Aquatic Environment.* Springer-Verlag, New York.

Garrels, R. M., and C. L. Christ, 1965, *Solution, Minerals and Equilibria.* Harper & Row, New York, 450p.

Gibbs, R. J., D. M. Tshudy, L. Konwar, and J. M. Martin, 1989, Coagulation and transport of sediments in the Gironde Estuary. *Sedimentology* 36:987–1000.

Grim, R. E., 1968, *Clay Mineralogy.* McGraw-Hill, New York, 596p.

Harder, H., and W. Flemhig, 1970, Quarzsynthese bei tiefen temperaturen. *Geochimica et Cosmochimica Acta* 34:295–305.

Holland, H. D., 1972, The geological history of sea water—an attempt to solve the problem. *Geochimica et Cosmochimica Acta* 36:637–651.

Holland, H. D., 1978, *The Chemistry of the Atmosphere and Oceans.* Wiley-Interscience, New York, 351p.

Holland, H. D., 1984, *The Chemical Evolution of the Atmosphere and Oceans.* Princeton University Press, Princeton, N.J., 582p.

Iler, R. K., 1979, *The Chemistry of Silica.* Wiley-Interscience, New York, 866p.

Kabata-Pendias, A., and H. Pendias, 1992, *Trace Elements in Soils and Plants,* 2d ed. CRC Press, Boca Raton, Fla., 365p.

Krauskopf, K. B., 1979, *Introduction to Geochemistry,* 2d ed. McGraw-Hill, New York, 617p.

Maynard, J. B., 1983, *Geochemistry of Sedimentary Ore Deposits.* Springer-Verlag, New York, 305p.

Mueller, R. F., and S. K. Saxena, 1977, *Chemical Petrology.* Springer-Verlag, New York.

Rosler, H. J., and H. Lange, 1972, *Geochemical Tables.* Elsevier, Amsterdam.

Salop, J. L., 1983, *Geological Evolution of the Earth During the Precambrian.* Springer-Verlag, Berlin, 459p.

Santschi, P., P. Höhener, G. Benoit, and M. Buchholtz-ten Brink, 1990, Chemical processes at the sediment-water interface. *Marine Chemistry* 30:269–315.

Stevens, R. E., and M. K. Carron, 1948, Simple field test for distinguishing minerals by abrasion pH. *American Mineralogist* 33:31–50.

Stumm, W., and J. J. Morgan, 1981, *Aquatic Chemistry,* 2d ed. Wiley-Interscience, New York, 780p.

Walker, J. C. G., 1983, Possible limits on the composition of the Archaean Ocean. *Nature* 302:518–20.

Yariv, S., and H. Cross, 1979, *Geochemistry of Colloid Systems for Earth Scientists.* Springer-Verlag, Berlin.

Physical Processes

Aigner, T., 1985, *Storm Depositional Systems. Lecture Notes in Earth Sciences 3.* Springer-Verlag, New York, 174p.

Allen, J. R. L., 1985, *Principles of Physical Sedimentology.* George Allen & Unwin, London, 272p.

Amos, C. L., and D. C. Mosher, 1985, Erosion and deposition of fine-grained sediments from the Bay of Fundy. *Sedimentology* 32:815–32.

Anderson, R. S., and B. Hallet, 1986, Sediment transport by wind: Toward a general model. *Geological Association of America Bulletin* 97:523–35.

Bagnold, R. A., 1968, Deposition in the process of hydraulic transport. *Sedimentology* 10:45–56.

Baker, V. R., and D. F. Ritter, 1975, Competence of rivers to transport coarse bedload material. *Geological Society of America Bulletin* 86:975–8.

Betzer, P. R., P. L. Richardson, and H. B. Zimmerman, 1974, Bottom currents, nepheloid layers and sedimentary features under the Gulf Stream near Cape Hatteras. *Marine Geology* 16:21–9.

Brookfield, M. E., and T. S. Ahlbrandt, 1983, *Eolian Sediments and Processes.* Developments in Sedimentology 38. Elsevier Scientific Publications, New York, 660p.

Clifton, H. E. (ed.), 1988, *Sedimentologic Consequences of Convulsive Geologic Events.* Geological Society of America Special Paper 229, New York, 157p.

Cook, D. O., 1970, The occurrence and geologic work of rip currents off southern California. *Marine Geology* 9:173–86.

Dott, R. H., Jr., 1983, Episodic sedimentation—how normal is average? How rare is rare? Does it matter? *Journal of Sedimentary Petrology* 51:5–23.

Drake, D. E., and D. S. Gorsline, 1973, Distribution and transport of suspended particulate matter in Hueneme, Redondo, Newport and La Jolla Submarine Canyons, California. *Geological Society of America Bulletin* 84:3949–68.

Dyer, K. R., 1985, *Coastal and Estuarine Sediment Dynamics.* Wiley-Interscience, New York, 342p.

Einsele, G., W. Ricken, and A. Seilacher (eds.), 1991, *Cyclic and Event Stratification.* Springer-Verlag, Berlin, 1040p.

Fangers, J. C., and D. A. V. Stow, 1993, Bottom-current-controlled sedimentation—A synthesis of the contourite problem. *Sedimentary Geology* 76:177–185.

Folk, R. L., 1971, Longitudinal dunes of the northwestern edge of the Simpson Desert, Northern Territory, Australia. 1. Geomorphology and grain size relationships. *Sedimentology* 16:5–54.

Fryberger, S. G., A. M. Al-Sari, T. J. Clisham, S. A. R Rizvi, and K. G. Al-Hinai, 1984, Wind sedimentation in the Jafurah sand sea, Saudi Arabia. *Sedimentology* 31:413–32.

Gorsline, D. S., 1985, Some thoughts on fine-grained sediment transport and deposition. *Sedimentary Geology* 41:113–30.

Gretener, P. E., 1967, Significance of the rare event in geology. *American Association of Petroleum Geology Bulletin* 51:2197–206.

Heezen, B., and C. Hollister, 1964, Deep-sea current evidence from abyssal sediments. *Marine Geology* 1:141–74.

Hsu, K. J., 1983, Actualistic catastrophism; address of the retiring president of the International Association of Sedimentologists. *Sedimentology* 30:3–9.

Karl, H. A., D. A. Cacchione, and P. R. Carlson, 1986, Internal-wave currents as a mechanism to account for large sand waves in

Navarinsky Canyon head, Bering Sea. *Journal of Sedimentary Petrology* 56:706–14.

Komar, P. D., and C. Wang, 1984, Processes of selective grain transport and the formation of placers on beaches. *Journal of Geology* 92:637–55.

Langford, R. P., 1989, Fluvial-aeolian interactions: Part I, modern systems. *Sedimentology* 36:1023–36.

Langford, R. P., and M. A. Chan, 1989, Fluvial-aeolian interactions: Part II, ancient systems. *Sedimentology* 36:1037–51.

Leopold, L. B., M. G. Wolman, and J. P. Miller, 1964, *Fluvial Processes in Geomorphology.* W. H. Freeman & Co., San Francisco.

McCave, I. N., 1971, Wave effectiveness at the sea bed and its relationship to bed-forms and deposition of mud. *Journal of Sedimentary Petrology* 41:89–96.

Middleton, G. V., (ed.), 1965, *Primary Sedimentary Structures and Their Hydrodynamic Interpretation.* Society for Sedimentary Geology Special Publication 12, Tulsa, Okla., 265p.

Middleton, G. V., 1976, Hydraulic interpretation of sand size distributions. *Journal of Geology* 84:405–26.

Middleton, G. V., and J. B. Southard, 1978/1984, *Mechanics of Sediment Movement,* 1st & 2d eds. Short Course 3, Society for Sedimentary Geology, Tulsa, Okla.

Moss, A. J., 1972, Bed-load sediments. *Sedimentology* 18:159–219.

Moss, A. J., P. H. Walker, and J. Hutka, 1980, Movement of loose, sandy detritus by shallow water flows: An experimental study. *Sedimentary Geology* 25:43–66.

Nairn, A. E. M., 1965, Uniformitarianism and environment. *Palaeogeography, Palaeoclimatology, Palaeoecology* 1:5–11.

Pettijohn, F. J., 1975, *Sedimentary Rocks,* 3d ed. Harper & Row, New York, 628p.

Pierson, T. C., and J. E. Costa, 1984, A rheological classification of subaerial sediment-water flows (abstract). *Geological Society of America Abstracts with Programs* 16:623.

Rust, B., and G. C. Nanson, 1989, Bedload transport of mud as pedogenetic aggregates in modern and ancient rivers. *Sedimentology* 36:291–306.

Stanley, D. J., and D. J. P. Swift (eds.), 1976, *Marine Sediment Transport and Environmental Management.* John Wiley & Sons, New York, 602p.

Sternberg, R. W., 1971, Measurements of incipient motion of sediment particles in the marine environment. *Marine Geology* 10:113–20.

Stow, D. A. V., and J. P. B. Lovell, 1979, Contourites: Their recognition in modern and ancient sediments. *Earth-Science Reviews* 14:251–91.

Terwindt, J. H. J., H. N. C. Breusers, and J. N. Svasek, 1968, Experimental investigation on the erosion-sensitivity of a sand–clay lamination. *Sedimentology* 11:105–14.

Tsutsui, B., J. F. Campbell, and W. T. Coulbourn, 1987, Storm-generated, episodic sediment movements off Kahe Point, Oahu, Hawaii. *Marine Geology* 76:281–99.

Sediment Gravity Transport

Andressen, A., and L. Bjerrum, 1967, Slides in subaqueous slopes in loose sand and silt. In A. F. Richards (ed.), *Marine Geotechnique,* University of Illinois Press, Urbana, pp. 221–39.

Booth, J. S., D. A. Sangrey, and J. K. Fugate, 1985, A nomogram for interpreting slope stability of fine-grained deposits in modern and ancient marine environments. *Journal of Sedimentary Petrology* 55:29–36.

Boswell, P. G. H., 1948, The thixotropy of certain sedimentary rocks. *Science Progress* 36:412–22.

Bouma, A. H., 1962, *Sedimentology of Some Flysch Deposits.* Elsevier, Amsterdam, 168p.

Carter, R. M., 1975, A discussion and classification of subaqueous mass-transport with particular application to grain flow, slurry flow and fluxoturbidites. *Earth-Science Reviews* 11:145–77.

Clarke, J. E. H., A. N. Shor, D. J. W. Piper, and L. A. Mayer, 1990, Large-scale current-induced erosion and deposition in the path of the 1929 Grand Banks turbidity current. *Sedimentology* 37:613–29.

Crowell, J. C., 1957, The origin of pebbly mudstones. *Geological Society of America Bulletin* 68:993–1009.

Enos, P., 1977, Flow regimes in debris flows. *Sedimentology* 24:133–42.

Fisher, R. V., 1971, Features of coarse-grained, high-concentration fluids and their deposits. *Journal of Sedimentary Petrology* 41:916–27.

Gould, H. R., 1951, Some quantitative aspects of Lake Mead turbidity currents. In *Turbidity Currents and the Transportation of Coarse Sediments to Deep Water.* Society of Sedimentary Geology Special Publication 2, Tulsa, Okla., pp. 34–52.

Hampton, M. A., 1975, Competence of fine-grained debris flows. *Journal of Sedimentary Petrology* 45:834–44.

Harms, J. C., 1974, Bushy Canyon Formation, Texas: A deep-water density current deposit. *Geological Society of America Bulletin* 85:1763–84.

Hendry, H. E., 1973, Sedimentation of deep water conglomerates in lower Ordovician rocks of Quebec—composite bedding produced by progressive liquefaction of sediment. *Journal of Sedimentary Petrology* 43:125–36.

Hsu, K. J., 1975, Catastrophic debris streams (*sturzstroms*) generated by rockfalls. *Geological Society of America Bulletin* 86:129–40.

Johnson, A. M., 1970, *Physical Processes in Geology.* Freeman Cooper, San Francisco, 577p.

Keefer, D. K., 1984, Landslides caused by earthquakes. *Geological Society of America Bulletin* 95:406–21.

Kerr, P. F., R. A. Stroud, and I. M. Drew, 1971, Clay mobility in landslides, Ventura, California. *American Association of Petroleum Geologists Bulletin* 55:267–91.

Komar, P. D., 1985, The hydraulic interpretation of turbidites from their grain sizes and sedimentary structures. *Sedimentology* 32:395–408.

Kuenen, Ph. H., 1967, Emplacement of flysch-type sand beds. *Sedimentology* 9:203–43.

Lewis, D. W., 1976, Subaqueous debris flows of early Pleistocene age at Motunau, North Canterbury, New Zealand. *New Zealand Journal of Geology and Geophysics* 19:535–67.

Lewis, D. W., 1980, Storm-generated graded beds and debris flow deposits with *Ophiomorpha* in a shallow offshore Oligocene sequence at Nelson, South Island, New Zealand. *New Zealand Journal of Geology and Geophysics* 23:353–69.

Lewis, D. W., 1982, Channels across continental shelves: Corequisites of canyon-fan systems and potential petroleum conduits. *New Zealand Journal of Geology and Geophysics* 25:209–55.

Lewis, D. W., M. G. Laird, and R. D. Powell, 1980, Debris flow deposits of early Miocene age, Deadman Stream, Marlborough, New Zealand. *New Zealand Journal of Geology and Geophysics* 27:83–118.

Lewis, K. B., 1971, Slumping on a continental slope inclined at 1°–4°. *Sedimentology* 16:97–110.

Lowe, D. R., 1976, Subaqueous liquified and fluidized sediment flows and their deposits. *Sedimentology* 23:285–308.

Lowe, D. R., 1979, Sediment gravity flows: Their classification and some problems of its application to natural flows and deposits. In L. J. Doyle and O. H. Pilkey (eds.), *Geology of Continental Slopes*. Society of Sedimentary Geology Special Publication 27, Tulsa, Okla., pp. 75–82.

Lowe, D. R., 1982, Sediment gravity flows: II. Depositional models with special reference to the deposits of high-density turbidity currents. *Journal of Sedimentary Petrology* 52:279–97.

Middleton, G. V., 1967, Experiments on density and turbidity currents, III: Deposition of sediments. *Canadian Journal of Earth Science* 4:475–506.

Middleton, G. V., and A. H. Bouma (eds.), 1973, *Turbidites and Deep-Water Sedimentation.* Short Course Lecture Notes, Society for Sedimentary Geology, Pacific Section, Los Angeles, Calif., 157p.

Middleton, G. V., and M. A. Hampton, 1973, Sediment gravity flows—mechanics of flow and deposition. In. G. V. Middleton and A. H. Bouna (eds.), *Turbidites and Deep-Water Sedimentation*. Short Course Lecture Notes, Society for Sedimentary Geology, Pacific Section, Los Angeles, Calif., pp. 1–38.

Middleton, G. V., and M. A. Hampton, 1976, Subaqueous sediment transport and deposition by sediment gravity flows. In D. J. Stanley and D. J. P. Swift (eds.), *Marine Sediment Transport and Environmental Management.* Wiley, New York, pp. 197–218.

Morgenstern, N., 1967, Submarine slumping and the initiation of turbidity currents. In A. F. Richards (ed.), *Marine Geotechnique.*University of Illinois Press, Urbana, pp. 189–220.

Nardin, T. R., F. J. Hein, D. S. Gorsline, and B. D. Edwards, 1979, A review of mass-movement processes, sediment and acoustic characteristics, and contrasts in slope and base-of-slope systems versus canyon-cum-basin floor systems. In L. J. Doyle and O. H. Pilkey, Jr. (eds.), *Geology of Continental Slopes.* Society of Sedimentary Geology Special Publication 27, Tulsa, Okla., pp. 61–73.

Nelson, C. H., 1982, Modern shallow-water graded sand layers from storm surges, Bering Shelf: A mimic of Bouma sequences and turbidite systems. *Journal of Sedimentary Petrology* 52:537–45.

Nemec, W., 1990, Aspects of sediment movement on steep delta slopes. In A. Colella and D. B. Prior (eds.), *Coarse-Grained Deltas.* International Association of Sedimentologists Special Publication 10, Blackwell Scientific Publications, Oxford, pp. 29–73.

Nemec, W., and R. J. Steel, 1984, Alluvial and coastal conglomerates: Their significant features and some comments on gravelly mass-flow deposits. In E. H. Koster and R. J. Steel (eds.), *Sedimentology of Gravels and Conglomerates.* Canadian Society of Petroleum Geologists Memoir 10, Calgary, Canada, pp. 1–31.

Nemec, W., and R. J. Steel, (eds.), 1988, *Fan Deltas: Sedimentology and Tectonic Settings.* Blackie & Son, Ltd., Glasgow, 464p.

Normark, W. R., 1989, Observed parameters for turbidity-current flow in channels, Reserve Fan, Lake Superior. *Journal of Sedimentary Petrology* 59:423–31.

Pierson, T. C., 1981, Dominant particle support mechanisms in debris flows at Mt. Thomas, New Zealand, and implications for flow mobility. *Sedimentology* 28:49–60.

Piper, D. J. W., and W. R. Normark, 1983, Turbidite depositional patterns and flow characteristics, Navy Submarine Fan, California borderlands. *Sedimentology* 30:681–94.

Porebski, S. J., D. Meischner, and K. Gorlich, 1991, Quaternary mud turbidites from the South Shetland Trench (West Antarctica): Recognition and implications for turbidite facies modelling. *Sedimentology* 38:691–716.

Postma, G., 1986, Classification for sediment gravity flow deposits based on flow conditions during sedimentation. *Geology* 14:291–4.

Rahn, P. H., 1986, *Engineering Geology: An Environmental Approach.* Elsevier, Amsterdam.

Sallenger, A. H., Jr., 1979, Inverse grading and hydraulic equivalence in grain-flow deposits. *Journal of Sedimentary Petrology* 49:553–62.

Saxov, S., and J. K. Nieuwenhuis (eds.), 1982, *Marine Slides and Other Mass Movements.* Plenum Press, New York, 353p.

Scheidegger, A. E., 1984, A review of recent work on mass movements on slopes and on rock falls. *Earth-Science Reviews* 21:225–49.

Schwab, W. C., and H. J. Less, 1988, Causes of two slope-failure types in continental-shelf sediment, northeastern Gulf of Alaska. *Journal of Sedimentary Petrology* 58:1–11.

Slaczka, A., and S. Thompson, III, 1981, A revision of the fluxoturbidite concept based on type examples in the Polish Carpathian flysch. Annales Societatis Geologorum Poloniae 51:3–44.

Temple, P. G., 1968, Mechanism of large-scale gravity sliding in the Greek Peloponnesos. *Geological Society of America Bulletin* 79:687–700.

van der Lingen, G. J., 1969, The turbidite problem. *New Zealand Journal of Geology and Geophysics* 12:7–50.

Voigt, I. B. (ed.), 1978, *Rock Slides and Avalanches,* 2 vols. Elsevier, Amsterdam.

Walker, R. G., 1973, Mopping-up the turbidite mess. In R. N. Ginsburg (ed.), *Evolving Concepts in Sedimentology.* Johns Hopkins University Press, Baltimore, pp. 1–37.

Zeng, J., D. R. Lowe, D. B. Prior, W. J. Wiseman, Jr., and B. D. Bornhold, 1991, Flow properties of turbidity currents in Bute Inlet, British Columbia. *Sedimentology* 38:975–96.

Diagenesis

Boles, J. R., and S. G. Franks, 1979, Clay diagenesis in Wilcox sandstones of southwest Texas: Implications of smectite diagenesis on sandstone cementation. *Journal of Sedimentary Petrology* 49:55–70.

Bredehoeft, J. D., and B. B. Hanshaw, 1968, On the maintenance of anomalous fluid pressures: I, Tidal sedimentary sequences; II, Source layer at depth. *Geological Society of America Bulletin* 79:1097–106; 1107–22.

de Segonzac, G. D., 1968, The birth and development of the concept of diagenesis (1866–1966). *Earth-Science Reviews* 4:153–207.

Faas, R. W., and C. A. Nittrouer, 1976, Post-depositional facies development in the fine-grained sediments of the Wilkinson Basin, Gulf of Maine. *Journal of Sedimentary Petrology* 46:337–44.

Hayes, J. B., 1979, Sandstone diagenesis—the Hole truth. In P. A. Scholle and P. R. Schluger, (eds.), *Aspects of Diagenesis.* Society for Sedimentary Geology Special Publication 26, Tulsa, Okla., pp. 127–40.

Larsen, G., and G. V. Chilingar (eds.), 1978, *Diagenesis in Sediments and Sedimentary Rocks.* Developments in Sedimentology 25A, Elsevier, Amsterdam, 579p.

Lee, J. H., J. H. Ahn, and D. R. Peacor, 1985, Textures in layered silicates: Progressive changes through diagenesis and low-temperature metamorphism. *Journal of Sedimentary Petrology* 55:532–40.

McCall, P. L., and M. J. S. Tevesz, 1982, *Animal–Sediment Relations: The Biogenic Alteration of Sediments.* Plenum Press, New York, 336p.

McDonald, D. A., and R. C. Surdam (eds.), 1984, *Clastic Diagenesis.* American Association of Petroleum Geologists Memoir 37, Tulsa, Okla., 434p.

Marshall, J. D. (ed.), 1987, *Diagenesis of Sedimentary Sequences.* Geological Society Special Publication 36, Blackwell Scientific Publications, Oxford, 360p.

Maynard, J. B., 1983, *Geochemistry of Sedimentary Ore Deposits.* Springer-Verlag, New York, 305p.

Milliken, K. L., E. F. McBride, and L. S. Land, 1989, Numerical assessment of dissolution versus replacement in the subsurface destruction of detrital feldspars, Oligocene Frio Formation, South Texas. *Journal of Sedimentary Petrology* 59:740–57.

Packham, G. H., and K. A. W. Crook, 1960, The principle of diagenetic facies and some of its implications. *Journal of Geology* 68:392–407.

Parker, A., and B. W. Sellwood (eds.), 1983, *Sediment Diagenesis.* D. Reidel Publishing Co., Dordrecht, Holland 427p.

Perel'man, A. I., 1967, *Geochemistry of Epigenesis.* Plenum Press, New York, 266p.

Plumley, W. J., 1980, Abnormally high fluid pressure: Survey of some basic principles. *American Association of Petroleum Geologists Bulletin* 64:414–30.

Rittenhouse, G., 1971, Pore-space reduction by solution and cementation. *American Association of Petroleum Geologists Bulletin* 55:80–91.

Runnells, D. D., 1969, Diagenesis, chemical sediments and the mixing of natural waters. *Journal of Sedimentary Petrology* 39:1188–201.

Scholle, P. A., and P. R. Schluger (eds.), 1979, *Aspects of Diagenesis.* Society of Sedimentary Geology Special Publication 26, Tulsa, Okla., 443p.

Siever, R., 1979, Plate-tectonic controls on diagenesis. *Journal of Geology* 87:127–55.

Suttner, L. J., and P. K. Dutta, 1986, Alluvial sandstone composition and paleoclimate, I. Framework mineralogy. *Journal of Sedimentary Petrology* 56:329–45.

Swarbrick, E. E., 1968, Physical diagenesis: Intrusive sediment and connate water. *Sedimentary Geology* 2:161–75.

Teodorovich, G. I., 1961, *Authigenic Minerals in Sedimentary Rocks.* Consultants Bureau, New York, 120p.

Velde, B., and E. Nicot, 1985, Diagenetic clay mineral composition as a function of pressure, temperature, and chemical activity. *Journal of Sedimentary Petrology* 55:541–7.

Williams, L. A., and D. A. Crerar, 1985, Silica diagenesis: I, Solubility controls; II, General mechanisms. *Journal of Sedimentary Petrology* 55:301–11; 312–21.

Biological Processes

Aiken, G. R., D. M. McKnight, R. L. Wershaw, and P. MacCarthy, 1985, *Humic Substances in Soil, Sediment and Water.* Wiley-Interscience, New York, 692p.

Almasi, M. N., A. Al-Zamel, D. J. Shearman, and A. Reda, 1987, Effects of natural and artificial Thalassia on rates of sedimentation. *Journal of Sedimentary Petrology* 57:901–6.

Bird, E. C. F., 1971, Mangroves as land builders. *Victoria Naturalist* 88:189–97.

Buick, R., 1984, Carbonaceous filaments from North Pole, Western Australia: Are there fossil bacteria in Archaean stromatolites? *Precambrian Research* 24:157–72.

Buick, R., J. S. R. Dunlop, and D. I. Groves, 1981, Stromatolite recognition in ancient rocks; an appraisal of irregularly laminated structures in an Early Archaean chert–barite unit from North Pole, Western Australia. *Alcheringa* 5:161–81.

Calvert, S. E., 1974, Deposition and diagenesis of silica in marine sediments. In K. J. Hsu and H. C. Jenkyns, *Pelagic Sediments on Land and Under the Sea.* International Association of Sedimentologists, Special Publication 1, Blackwell Scientific Publications, Oxford, pp. 273–300.

Cloud, P. E., 1973, Paleoecological significances of the banded iron-formation. *Economic Geology* 68:1135–44.

Cloud, P. E., 1974, Evolution of ecosystems. *American Scientist* 62:54–66.

Degens, E. T., 1989, *Perspectives on Biogeochemistry.* Springer-Verlag, Berlin.

Dillon, W. P., and H. B. Zimmerman, 1970, Erosion by biological activity in two New England submarine canyons. *Journal of Sedimentary Petrology* 40:542–7.

Frankel, L., and D. J. Mead, 1973, Mucilaginous matrix of some estuarine sands in Connecticut. *Journal of Sedimentary Petrology* 43:1090–5.

Frydl, P., and C. W. Stearn, 1978, Rate of bioerosion by parrotfish in Barbados reef environment. *Journal of Sedimentary Petrology* 48:1149–58.

Gilbert, R., 1984, The movement of gravel by the alga *Fucus vesiculosis* (L.) on an arctic intertidal flat. *Journal of Sedimentary Petrology* 54:463–8.

Hine, A. C., M. W. Evans, R. A. Davis, and D. F. Belknap, 1987, Depositional response to seagrass mortality along a low-energy, barrier-island coast: West-Central Florida. *Journal of Sedimentary Petrology* 57:431–9.

Holland, H. D., and M. Schidlowski (eds.), 1982, *Mineral Deposits and the Evolution of the Biosphere.* Springer-Verlag, New York.

Ittekkot, V., S. Kempe, W. Michaelis, and A. Spitzy, 1990, *Facets of Modern Biogeochemistry.* Springer-Verlag, Berlin, 433p.

Lasserre, P., and J.-M. Martin, 1986, *Biogeochemical Processes at the Land–Sea Boundary.* Elsevier Oceanography Series 43, Elsevier, Amsterdam, 214p.

McCave, I. N., 1988, Biological pumping upwards of the coarse fraction of deep-sea sediments. *Journal of Sedimentary Petrology* 58:148–58.

McConchie, D. M., 1987, The geology and geochemistry of the Joffre and Whaleback Shale members of the Brockman Iron Formation, Western Australia. In P. Appel and G. LaBerge (eds.), *Precambrian Iron Formations.* Theophrastus Publications, Athens, pp. 541–601.

McConchie, D. M., and L. M. Lawrance, 1991, The origin of high cadmium loads in some bivalve molluscs from Shark Bay, Western Australia: A new mechanism for cadmium uptake by filter feeding organisms. *Archives of Environmental Contamination and Toxicology* 21:303–10.

Pryor, W. A., 1975, Biogenic sedimentation and alteration of argillaceous sediments in shallow marine environments. *Geological Society of America Bulletin* 86:1244–54.

Raymond, P. E., and H. C. Stetson, 1931, A new factor in the transportation and distribution of marine sediments. *Science* 73:105–6.

Redfield, A. C., 1958, The biological control of chemical factors in the environment. *American Scientist* 46:205–21.

Scoffin, T. P., 1970, The trapping and binding of subtidal carbonate sediments by marine vegetation in Bimini Lagoon, Bahamas. *Journal of Sedimentary Petrology* 40:249–73.

Stearley, R. F., and A. A. Ekdale, 1989, Modern marine bioerosion by macroinvertebrates, northern Gulf of California. *Palaios* 4:453–67.

Stephens, W. M., 1962, Trees that make land. *Sea Frontiers* 8: 219–30.

Swinbanks, D. B., 1981, Sediment reworking and the biogenic formation of clay laminae by *Arenicola pacifica. Journal of Sedimentary Petrology* 51:1137–45.

Syvitski, J. P. M., and D. A. van Everdingen, 1981, A reevaluation of the geologic phenomenon of sand flotation: A field and experimental approach. *Journal of Sedimentary Petrology* 51:1315–22.

Tudhope, A. W., and M. J. Risk, 1985, Rate of dissolution of carbonate sediments by microboring organisms, Davies Reef, Australia. *Sedimentology* 32:440–7.

Walter, M. R., and H. J. Hofmann, 1983, The palaeontology and palaeoecology of Precambrian iron-formations. In A. F. Trendall and R. C. Morris (eds.), *Iron-Formation Facts and Problems.* Elsevier, Amsterdam, pp. 373–400.

Waples, D., 1981, *Organic Geochemistry for Exploration Geologists.* Burgess, CEPCO Division, Minneapolis, Minn., 151p.

Ward, L. G., W. R. Boynton, and W. M. Kemp, 1984, The influence of waves and seagrass communities on suspended particulates in an estuarine embayment. *Marine Geology* 59:85–103.

Windley, B. F. (ed.), 1976, *The Early History of Earth.* Wiley & Sons, London.

Volcanic Processes

Cole, R. B., and P. G. DeCelles, 1991, Subaerial to submarine transitions in early Miocene pyroclastic flow deposits, southern San Joaquin Basin, California. *Geological Society of America Bulletin* 103:221–35.

Chough, S. K., and Y. K. Sohn, 1990, Depositional mechanics and sequences of base surges, Songaksan tuff ring, Cheju Island, Korea. *Sedimentology* 37:1115–35.

Fisher, R. V., and H.-U. Schmincke, 1984, *Pyroclastic Rocks.* Springer-Verlag, New York, 472p.

Fisher, R. V., and G. A. Smith (eds.), 1991, *Sedimentation in Volcanic Settings.* Society for Sedimentary Geology Special Publication 45, Tulsa, Okla., 257p.

Smith, G. A., 1986, Coarse-grained nonmarine volcaniclastic sediment: Terminology and depositional process. *Geological Association of America Bulletin* 97:1–10.

Smith, G. A., and D. Kalzman, 1991, Discrimination of eolian and pyroclastic-surge process in the generation of cross-bedded tuffs. Jemez Mountains volcanic field, New Mexico. *Geology* 19:465–8.

4
Sedimentary Structures

A wide variety of structures are formed by physical, chemical, and biological processes during or shortly after deposition; Fig. 4-1 categorizes many by their mode of origin. These structures are the best indicators in any deposit of processes operating in the depositional environment because they formed where they are presently found. Hence, they provide powerful tools for assessing the influence of processes that could not be observed because the processes operate infrequently and/or in remote sites (e.g., current patterns and sediment movement during a major storm; concretions forming below the sediment-water interface), or because they operate at a rate unsuited to direct observation (either too rapidly or too slowly). Although there are instrumental techniques for assessing sediment dynamics and other processes in modern environments (e.g., see **AS** Chapters 4 and 6), a skilled observer can make reliable assessments of the nature and extent of sediment movement, and of the processes controlling sediment movement, solely on the basis of preserved sedimentary structures. Use of sedimentary structures in this way requires an extensive knowledge of the types of structures formed by particular processes acting on different sediments in various environments. When dealing with sedimentary rocks, interpretation of the processes controlling deposition is a fundamental step in paleoenvironmental analysis.

Most sedimentary structures are best studied in the field, where it is essential to describe them fully, with sketches or photographs so that a later return to the exposure for details is not necessary. In general, only small-scale structures can be sampled from both friable and indurated sediments, but larger-scale peels may be collected from subaerial loose sediment exposures (see **AS** Chapters 4 and 5 for reviews of sampling and peel techniques). In some sediments, the apparent absence of sedimentary structures is due to the inability of the eye to identify them (e.g., in the case of homogeneous textures and composition); such structures may be revealed by laboratory analysis using dyes or X-ray radiography (see **AS** Chapter 6). When structures are truly missing, the information provided by their absence can be almost as useful as when they are present: either bioturbation has homogenized the sediment (indicating the presence of burrowers), or sedimentation was exceedingly rapid and from a turbulent mechanism (coarse sediments) or exceedingly slow and continuous (very fine sediment settling from passive suspension). It must always be remembered that those structures which *are* found in a particular sediment are in most cases only part of the original array of structures that may have existed at an earlier time: only some are preserved, and those that are may not be the best guides to the average conditions of sedimentation (e.g., Allen 1967). When multiple structures are present, studies are facilitated by dividing them into categories of scale and type, as in architectural analysis of sedimentary environments (Fig. 4-2; e.g., see DeCelles et al. 1991).

Published literature on sedimentary structures is extensive; some of the best general treatments are given in Coneybeare and Crook 1968; Potter and Pettijohn 1977; Pettijohn and Potter 1964; Reineck and Singh 1980; Collinson and Thompson 1985; Jones and Preston 1987 (see also Allen 1982 for a general treatment focusing on processes). Interpretations from sedimentary structures range from formative process through way-up determination in deformed rock sequences (e.g., Shrock 1948), to determining paleocurrent directions and/or paleoslopes (e.g., Bailey 1967; Klein 1967), to indicators of the presence and activities of organisms at or shortly after deposition, and to environmental recognition (e.g., Shawa 1974). Although the focus of most published discussions is on structures in detrital sediments, the same range exists in carbonate sediments (e.g., Schwartz 1975).

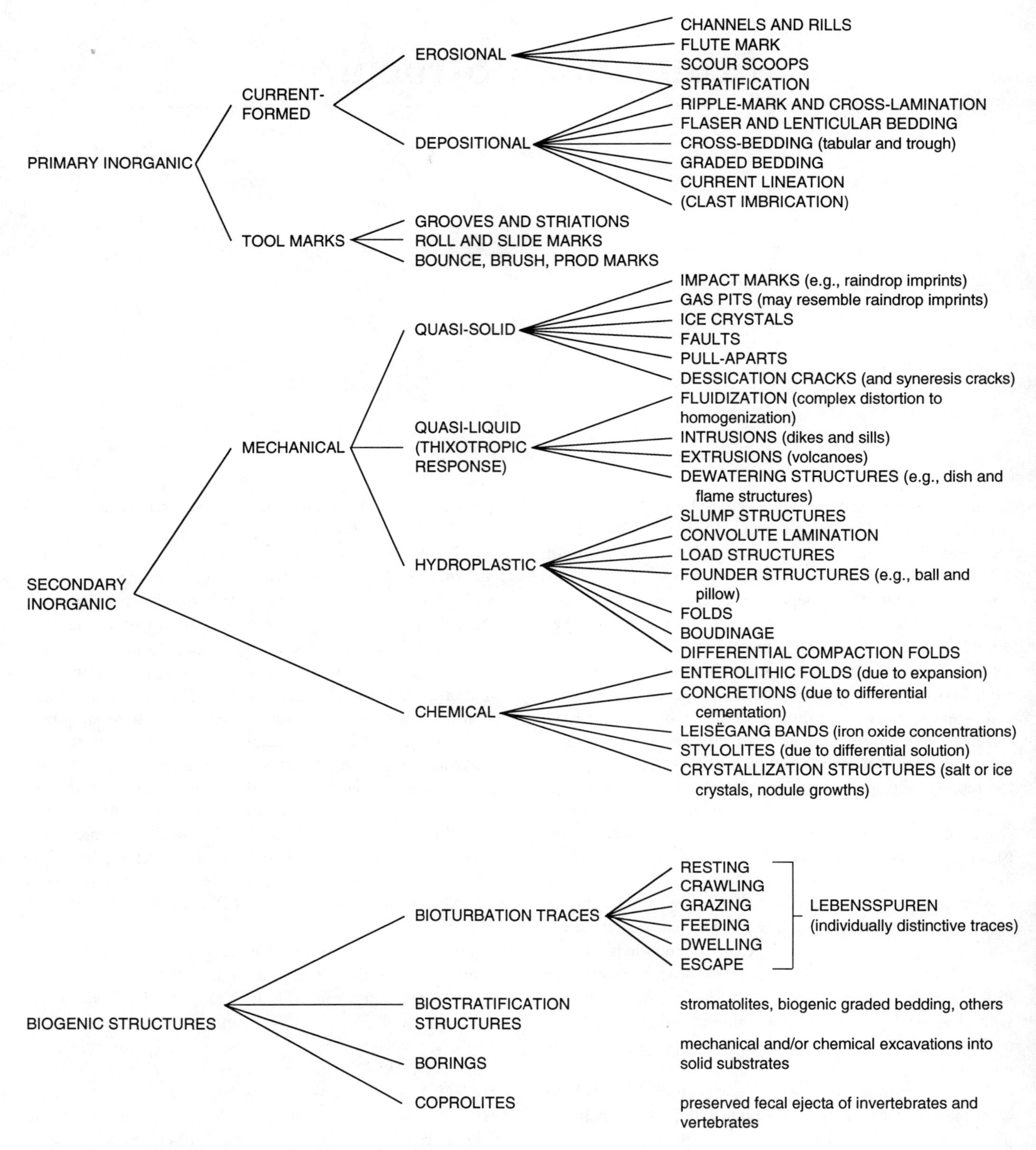

Figure 4-1. A genetic classification of sedimentary structures.

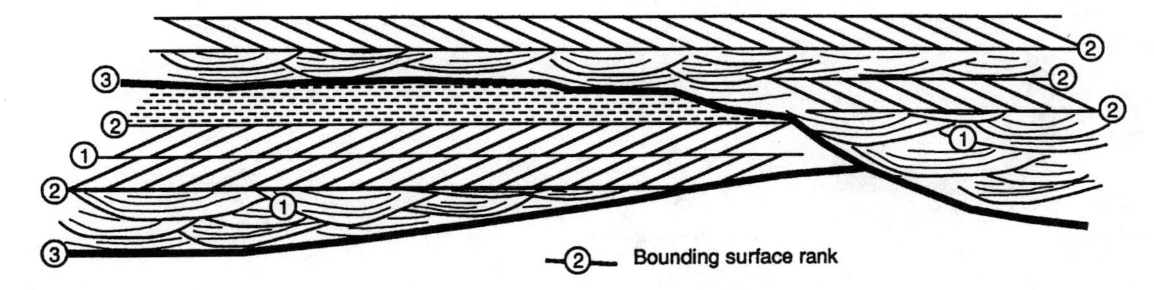

Figure 4-2. Example of a step in architectural analysis. Surfaces within the sediment are grouped according to their character, then ranked in order of their importance to the history of the deposit (e.g., unconformities would be the most important, and there could be different rankings of unconformities if more than one type is present). A major criterion for ranking is *truncation*, i.e., if one surface truncates another, it is the higher rank (higher number). Although surfaces are theoretically ranked on purely objective criteria (such as truncation), personal experience (bias) or inferred means of origin may be involved, and different interpretations of the history of a complex series of architectural units are possible. Current practice by most workers is to rank bounding surfaces from minor to major on an ascending scale (1–*n*); numbering appears to begin at the boundary of sedimentation units (sediments deposited under essentially constant physical conditions, such as cross-bedded sands from a migrating dune bed form), but individual bias in selection is inescapable. In the example, 1 is given to bounding surfaces for cross-bed cosets and 3 to major erosional surfaces that truncate all lesser surfaces.

INORGANIC STRUCTURES

Primary Structures

Primary sedimentary structures inform about processes acting in the paleoenvironment. Some primary mechanical structures provide information on the *direction* in which the formative water or wind currents were flowing (e.g., Wells 1988); others may provide an *orientation* but not a sense of direction (i.e., the currents flowed in one or the other direction along a line; Table 4-1 lists common structures in each category). Both types are used to interpret paleocurrent patterns (Fig. 4-3; see **AS** Chapter 6 for treatment of data). In ancient sedimentary sequences, they provide information on paleogeography as well as the nature and energy levels of depositional processes. Some primary (and secondary) structures can be used to distinguish stratigraphic sequence—i.e., the original younging direction or way up (Table 4-2). Such structures must be sought in highly deformed rock sequences where strata may have been overturned (for a comprehensive discussion, see Shrock 1948). Because they are related to processes often associated with particular depositional settings, primary (and to a lesser extent secondary) sedimentary

Table 4-1. Some Paleocurrent Indicators

Directional Structures	Orientational Structures
Asymmetric ripples perpendicular to crest toward dip direction of steep face	Symmetric ripples perpendicular to crest
Cross-bed or lamination azimuth of steepest dip of foreset (fails with point-bar surfaces, and may reflect local vs. regional current direction; must be down the axis of troughs)	Channels and scours axial orientation
Flute marks long axis, away from the direction of the bulbous head	Current and parting lineation along the lineation
Gravel imbrication direction in which clasts overlap each other	Groove casts, striations, roll and slide marks along the lineation
Trends of grain size fining downcurrent	Bounce, brush, prod marks along line of similar prod marks; rarely can give direction if the shape of the tool is known or if sediment shows slight impact deformation

Standard axes used for reference with sedimentary structures:

c = vertical to principal bedding
a = paleocurent azimuth, at right angles to *c*
b = perpendicular to the *ac* plane

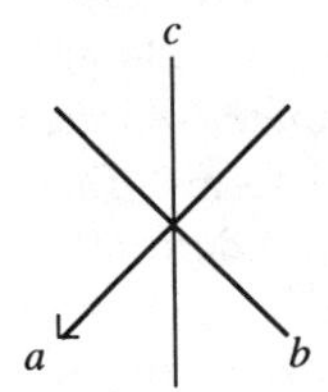

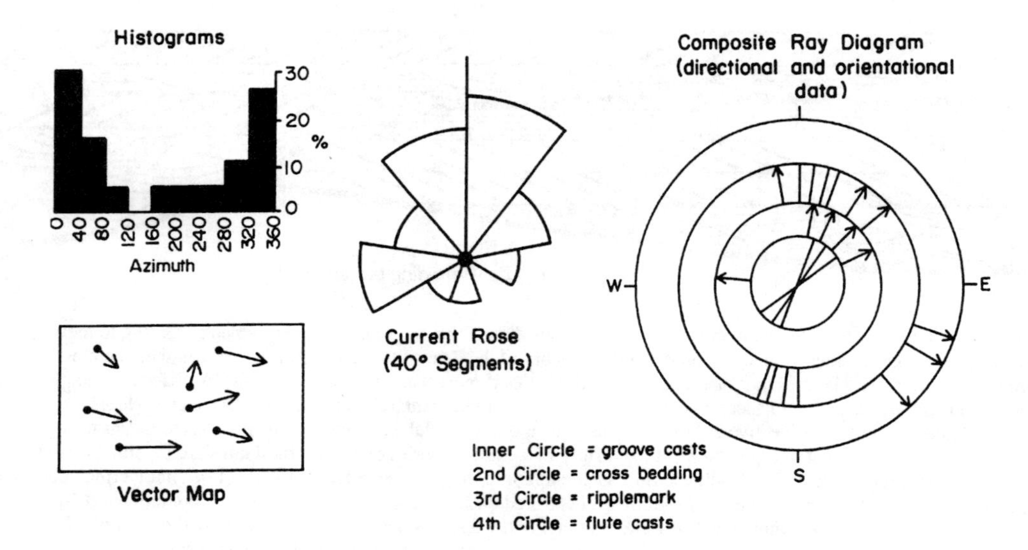

Figure 4-3. Examples of diagrams used to represent directional and orientational data. The ordinary histogram is rarely used because it can obscure the pattern of azimuthal (360°) data (e.g., the example shown is unimodal but appears bimodal); current roses are used instead (but see **AS** Chapter 6 for the correct way in which to compose them). The composite ray diagram permits different classes of structure to be plotted in different circles: those providing a direction are shown by arrows, those showing orientation by lines in both azimuthal positions.

structures can be useful paleoenvironmental indicators (e.g., Table 4-3); associations are more indicative than the occurrence of any one structure, and exceptions can occur in virtually every environment.

General

Stratification, or bedding, is the most common primary inorganic structure. Two adjacent bedding planes (two-dimensional surfaces of varied irregularity) generally delimit a *sedimentation unit*—a volume of sediment deposited under essentially a single set of processes, although part of that unit may have been eroded below the uppermost bedding plane. A standard descriptive system should be used for the scale of beds (e.g., Fig. 4-4a). A sedimentation unit may have smaller-scale, internal bedding such as a set of cross-strata deposited by ripple or dune bed forms migrating under constant current conditions during deposition of the unit (e.g., Fig. 4-4b). Relatively rarely, when physical processes have changed substantially but gradually during deposition, there are gradational contacts between sedimentation units and an arbitrary bedding plane must be defined.

Bedding surfaces originate during deposition, either from erosion or as a result of times of nondeposition; occasionally in nondetrital sediments, they form as a result of chemical diagenetic processes (secondary bedding). Bedding irregularities reflect draping of younger sediments over erosional features and depositional bed forms of the older sediments, or postdepositional (secondary) deformation. It is generally difficult to determine the time taken to form a principal bedding plane: most are diastems, reflecting relatively minor and short-term fluctuations in environmental conditions (e.g., flood events, storms, seasonal changes, normal lateral migration of current routes), but others are erosional or nondepositional *unconformities* of substantial temporal span (e.g., Table 4-4).

Whereas principal stratification surfaces are generally considered to be parallel to sea level at the time of deposition, and thus any present differences in attitude are used to infer tectonic deformation, this generalization works only on the outcrop scale and can be false even then! Almost all sediments are deposited on dipping surfaces; even if the dip is only 3°, on an expanded cross-section, particularly with a vertical exaggeration that is commonly applied, the primary dip can appear substantial over a distance of a few kilometres. Primary dips may be much greater than a few degrees—e.g., where sediments are deposited over an irregular bathymetry or at the front of lake deltas and in eolian dunes (orogenies have been mistakenly inferred when delta foresets have been interpreted to result from tectonic deformation). In addition, secondary dips may eventuate from nontectonic factors such as differential compaction (e.g., over a buried reef).

The origin of a bedding plane can be visualized as the preservation of a *profile of equilibrium*, a dynamic surface along which there was a balance between supply and removal of sediment (Fig. 4-5). Under a given set of conditions (flow velocity, rate of supply of texturally similar sediments, etc.), a surface exists above which sediment continuously moves past any reference position without net deposition or erosion. The sediment may move as a train of transitory structures (such as ripple bed forms), but these structures will be preserved only if a rapid change of condi-

Table 4-2. Criteria for Determining the Younging Direction in Sedimentary Sequences

Geopetal structures—such as half-filled shells (sediment on the bottom, void space or void-filling cement above) that indicate the position of the gravity field during deposition (sediment falls as far as it can go!) and are the best way-up indicators.

Truncation—of any other structures such as bedding planes or ripples, or of any other primary feature. A wide variety of features are included under this criterion—any truncation of beds (as in many kinds of scour structures and tool marks), of particles (such as shell fragments), of tectonic structures (such as faults), or of igneous masses (such as dikes). *Problems*: (1) stylolites developed by dissolution after deposition may truncate fossils, other grains, and even structures; the original way-up cannot usually be distinguished from stylolites; (2) tectonically induced faults, commonly subparallel to bedding planes, may truncate sedimentary structures in any sense.

Tangential lower contact of cross-beds or cross-laminae—because finer grains are carried farther by inertia than coarser grains, foreset beds on the lee side of ripples and dunes are concave-up and grade downslope to a progressively lower angle of repose. The constant downcurrent movement of the bed forms causes the erosive stoss side to truncate the lee-side layers; thus, there are usually both criteria in the same sediment units. *Problem*: in some settings where constant winnowing has removed essentially all the fines, as in some tidal compartments and eolian settings, dune migration is by avalanching of grains down the foreset face; consequent cross-bedding may have coarser grains at the toe of the slope and abut the underlying beds abruptly, giving the appearance in cross-section of truncation at the base. Cross-checking with other way-up criteria and the absence of fines in the beds will warn of this potential exception to the rule.

Shape of symmetrical ripple mark—crests are sharper and narrower than troughs.

Mud cracks—wedge downward and commonly filled with sandy sediment from above.

Load structures—where coarser or denser overlying sediments have sunk differentially into underlying finer or less dense, water-saturated deposits (e.g., along original depressions such as ripple troughs, flute scours).

Flame structures—where water-saturated sediments have been squeezed upward into overlying, commonly denser or coarser, sediments because of their differential loading or fracturing.

Differential compaction—of sediments (particularly muds) over resistant features such as large particles (e.g., shells, rafted pebbles), irregular relief, or reefs.

Pinch-outs—of sedimentary units (laminae to full formations) onto a feature with some relief (e.g., a clast, a reef, a cliff, a hill); bedding planes will normally terminate against the feature in contrast to the situation with differential compaction.

Impact structures—such as rain/hail imprints, or where large particles rafted by ice or plant roots have landed on the substrate. Laminations below the particle may be disrupted (compaction may attenuate but does not break overlying layers), subsequent sedimentation results in termination of laminations against the edges of the particles (laminations drape around them with compaction), or there may be some asymmetry to the structures (generally absent with compaction). Displacive diagenetic growth of crystals such as gypsum can locally mimic these features.

Sole marks—original depressions in fine sediment filled by usually coarser sediment and thus now preserved as castings or moldings on the base of the more resistant lithologies. Specific structures include flute and tool marks, raindrop imprints, organism tracks.

Normally graded bedding—widely used in deformed sequences, this criterion is much less reliable than many others because reverse grading can occur commonly in some sequences deposited under entrainment flow.

Biogenic structures—such as tracks where the underside can be determined, burrows filled with sediment from above, characteristic form (e.g., escape structures or conical-upward shape of *Zoopycos*), or internal structure (concave-upward shape of meniscate internal structures reflecting the edges of old U-shaped burrows). U-shaped burrows always open upward; rootlet structures taper downward.

Presence of encrusting organisms—on the top of hardgrounds or rockgrounds.

Tracing of beds laterally—puts beds above and below into context.

Recognition of known sequences—between areas when the way up is known in one of them.

Note: See Shrock (1948) for comprehensive coverage of this theme.

tions (e.g., increased sediment supply or decreased flow velocity) permits net accumulation. Erosion eliminates many old profiles of equilibrium in the process of creating new ones. Discussions on the origin of horizontal stratification are surprisingly few in the literature, since it is perhaps the most characteristic feature of sedimentary deposits in general (but, see Southard and Boguchwal 1973; Bridge 1978; Lindholm 1981; Cheel 1990).

A profile of equilibrium may also reflect a passive attainment of balance with gravitational forces controlling deposition at the *angle of repose* (Fig. 4-6). The maximum angle of repose for dry angular sands is approximately 35°, but post-depositional syn-sedimentary deformation may oversteepen or reduce the angle; the maximum angle of repose for most sediment deposited under water is much lower.

When the dynamic profile of equilibrium is ascending, such that there is net deposition, the character of most stratification reflects bed forms whose geometry depends on entrainment-flow conditions (see Chapter 3). The relationship between flow conditions, sediment character and transport, and bed form is contained in the flow-regime concept (Fig. 4-7 and see Middleton 1965, 1977; Harms et al. 1975; Middleton and Southard 1984; Southard and Boguchwal 1990). Bed forms can be very large (e.g., 15 m in the fluvial system described by Jones and Rust 1983; see also McCabe 1977).

(Text continues on page 98.)

Table 4-3. Generalized Occurrence Pattern of Some Structures in Various Sedimentary Environments

Structure	Environment							
	Eolian	Glacial Outwash	Fluviatile	Lacustrine	Beach	Tidal Flat	Neritic	Bathyal
Primary								
Small symmetrical ripples	–	–	r	P	P	P	P	–
Small straight asymm. ripples	P	r	P	P	P	P	P	P
Small linguoid asymm. ripples	–	–	P	–	r	P	–	–
Anastomosing asymm. ripples	r	–	P	r	r	P	P	–
Ladder-back ripples	–	–	–	–	P	P	–	–
Flat-topped ripples	–	–	–	–	P	P	–	–
Symmetrical megaripples	–	–	–	–	–	–	P	–
Asymmetrical megaripples	–	–	P	–	P	r	P	–
Small-scale cross-strat.	r	–	P	P	r	P	r	P
Medium-scale cross-strat.	P	–	P	r	P	P	–	r
Large-scale cross-strat.	P	–	P	–	P	–	–	–
Very large-scale cross-strat.	P	–	–	–	P	–	–	–
Hummocky cross-strat.	–	–	–	–	–	–	P	–
Flaser and lenticular bedding	–	–	–	r	r	P	P	r
Clast imbrication	–	–	P	–	P	–	–	r
Current lineation	r	–	r	–	P	P	P	P
Graded bedding (normal)	–	P	P	r	–	–	r	P
Graded bedding (reverse)	–	P	P	–	–	–	–	r
Channels and scours	–	P	P	r	P	P	P	P
Tool marks	–	–	r	–	r	P	–	P
Flute marks	–	–	r	r	–	–	r	P
Rill marks	r	r	r	–	P	P	–	–
Salt crystal imprints	–	–	–	P	–	P	–	–
Secondary								
Mud and sand volcanoes	–	r	P	P	–	–	P	P
Clastic dykes	–	P	P	–	P	–	P	P
Flame structures	–	–	P	P	–	–	–	P
Slump structures	r	P	r	P	–	r	r	P
Ball and pillow structures	–	–	P	–	–	–	P	–
Load casts	–	P	P	P	–	P	P	P
Convolute lamination	–	r	–	P	–	–	r	P
Gas pits	–	–	–	r	P	r	–	–
Desiccation cracks	P	–	P	P	–	P	–	–
Rain and hail imprints	P	r	P	P	P	P	–	–
Biogenic								
Plant root casts	–	–	P	r	–	P	–	–
Dwelling traces	–	–	–	–	P	P	P	–
Feeding traces	–	–	–	–	r	P	P	r
Crawling traces	r	–	r	P	r	P	P	r
Resting traces	–	–	–	–	–	r	P	–
Coprolites	–	–	–	r	–	P	P	–
Stromatolites	–	–	–	r	–	P	r	–

Note: Exceptions exist!
P = often present, r = rarely present, – = absent.

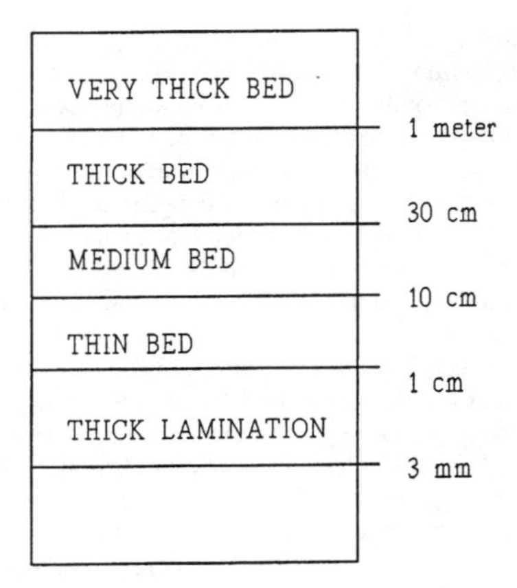

Figure 4-4a. Standard terminology for bedding thickness. The term cross is added before *bedded* or *laminated* for internal cross-stratification. Also noted in the field are (1) bed regularity or irregularity (scale and shape); (2) lateral persistence of beds; (3) textures, composition, and other internal structures of principal beds; (4) strike and dip of beds; and (5) nature of top and bottom contacts. (After Ingram 1954.)

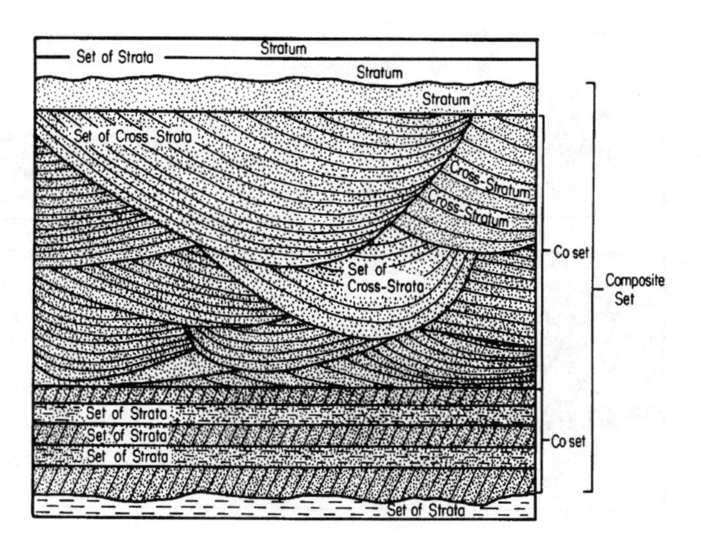

Figure 4-4b. Terms for primary and internal stratification. (After McKee and Weir 1953.)

Table 4-4. Some Criteria for Recognizing Unconformities

Contact on igneous or metamorphic rocks—a *nonconformity* or *lithological unconformity*; however, in some cases the contact may represent an intrusive contact (whereupon there should be evidence of contact metamorphism).

Regional truncation of older sedimentary units—an *angular unconformity*. Local truncation may reflect slumping or the junction between topset and foreset beds of a delta. Keep in mind that the original extent of units may be limited.

Truncation of secondary sedimentary or tectonic structures—an erosion surface cutting across, for example, contorted beds, folds, or faults. Even in these cases, however, a diastem may be represented if the secondary deformation was syn-sedimentary—i.e., developed in unlithified sediment shortly after deposition in the same environment, such as when slumping occurs in response to excessive loading or seismic shock.

Truncation of grains—implying that the underlying sediment was lithified at the time. Rare exceptions may exist with carbonate sediments, and common exceptions exist with evaporites.

Presence of a paleosol—old soil features may comprise organic accumulations, iron oxide stains, caliche horizons, various soil textures, detectable zonation, or other characteristics that may or may not be readily identifiable.

Age differences—a *paraconformity* or *parallel unconformity* discovered by breaks in the evolutionary lineage of organic remains, or by radiometric age determinations on either side of the surface.

Depressions or burrows filled with sediment that is not present in overlying beds—implying original sedimentation of another unit prior to that which now rests on the underlying unit.

Presence of encrusting organisms—implying that the underlying sediment was lithified. There may be rare exceptions where submarine cementation of seafloor carbonates has been rapid.

Presence of karstic features—sinkholes and other dissolution features should be filled with bedded new sediment (vs. collapse breccias or insoluble residue accumulations, for example). On a local scale, problems arise in distinguishing solution features that have developed at depth from those that developed along the original sedimentary substrate; however, both situations represent different types of unconformity.

Presence of a conglomerate or breccia comprising rock fragments of underlying units—implying lithification of those units prior to the derivation of the clasts. A problem may lie in distinguishing the original state of sediment consolidation.

Presence of a conglomerate or breccia comprising rock fragments derived from distant sources—with an exception in the case of ice-rafted clasts, which generally deposit only local lenses.

Concentration of materials representative of slow sedimentation—such as glauconite, phosphates, manganese. However, the concentration may reflect later reworking of originally scattered grains, in which case the unconformity may not immediately underlie them.

Superposition of sedimentary units that could not have been deposited in laterally adjacent facies (an application of Walther's Law)—Care must be taken with the application of this criterion to eliminate alternative explanations but, e.g., an unconformity would be expected at a contact between nonmarine coal measures and open shelf muds.

Note: Many other features *may* be associated with, or *suggestive* of, unconformities. They must be considered in association with each other and with all other possible explanations for their presence. Because unconformities are distinguished from diastems on the basis of the time span represented, distinction between them is commonly difficult and there are boundary problems.

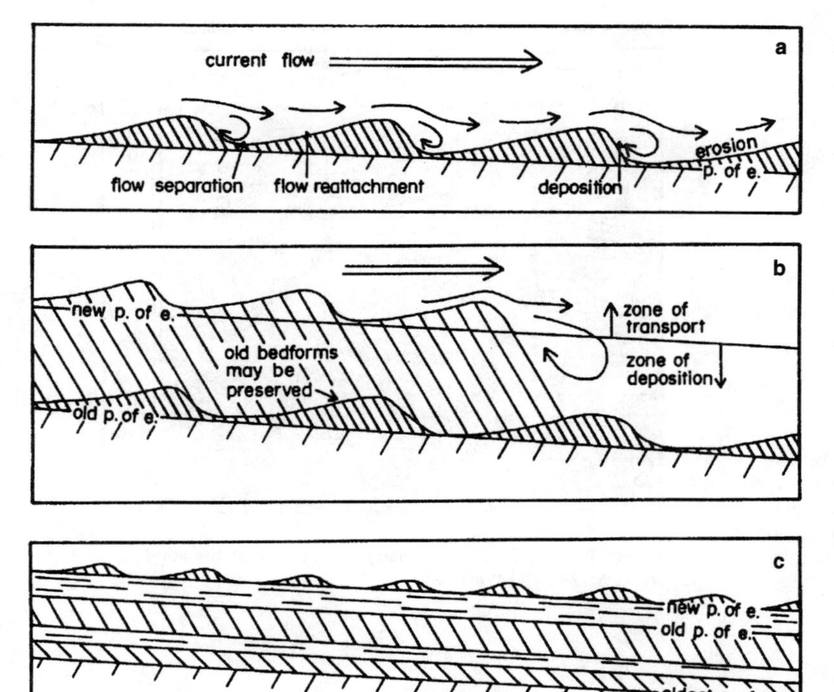

Figure 4-5. Accretionary stratification and cross-stratification. *a: Constant flow:* bed forms migrate along a profile of equilibrium (p. of e.) with no net deposition or erosion. Because of expanding flow, and flow separation over, the crests of ripples and dunes, grains on the lee side effectively accumulate under conditions of zero velocity; grains avalanche down the lee face and accumulate at the angle of repose. *b:* Relatively large-scale and/or rapid upward movement of the p. of e. (consequential to a rise in base level, a decrease in flow velocity, an increase in load, or other causes). Net accumulation of foresets occurs at the angle of repose, and a cross-laminated or cross-bedded unit (depending on the scale of the bedform) is generated between the old and the new profiles of equilibrium. Sediment passes continuously over the new p. of e. *c.* Slow and/or small-scale upward movement of the p. of e. A thin layer of sediment accumulates, so thin that it commonly has no obvious internal structure. Each p. of e. forms a stratification plane and units may be preserved with or without obvious cross-stratification. Downward shifts of the p. of e. (due to such causes as lowering of base level, increase in velocity, decrease in load): erosion removes all sediment above the new p. of e., which may itself be preserved as a major bedding plane if the p. of e. subsequently moves upward. The dynamic p. of e. concept is useful as a guide to understanding the formation of stratification in general, and in particular of widespread subhorizontal bedding planes. Whereas erosion and deposition are rarely balanced for any lengthy period, there is a tendency for balance under any set of conditions. (After Jopling 1966.)

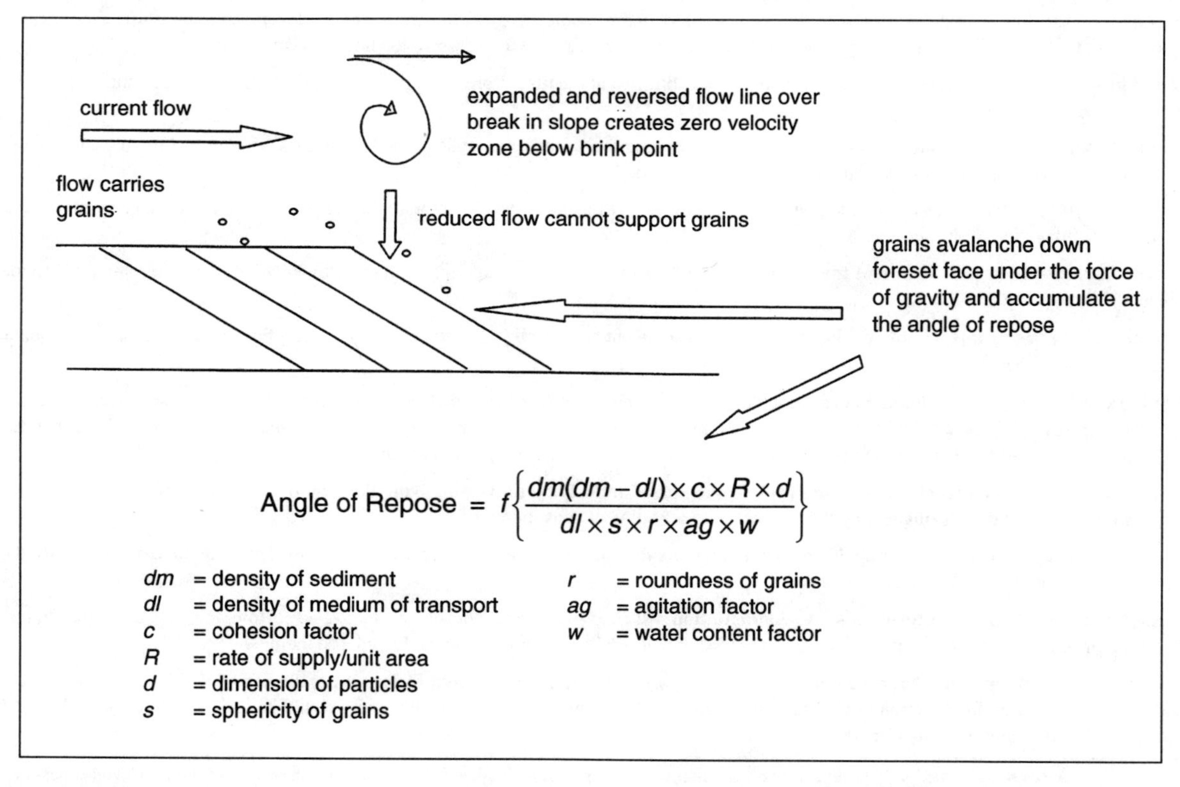

Figure 4-6. Schematic representation of the controlling factors for the natural *angle of repose* on which sediments will accumulate. Upper diagram depicts the common situation at a change in slope where a current is transporting bed load.

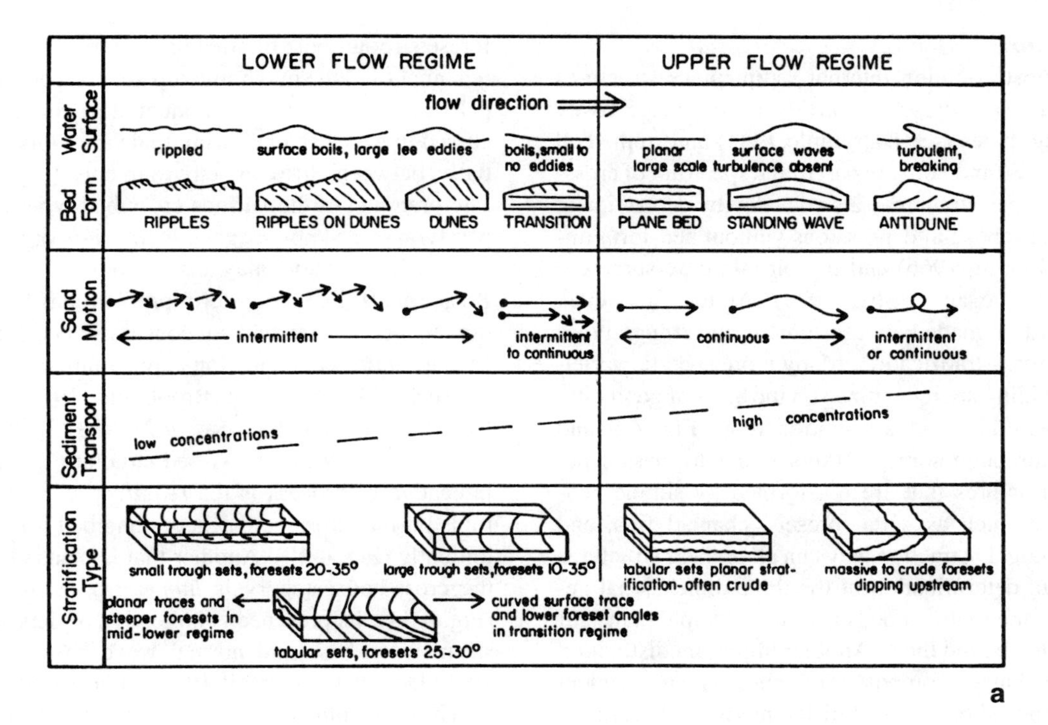

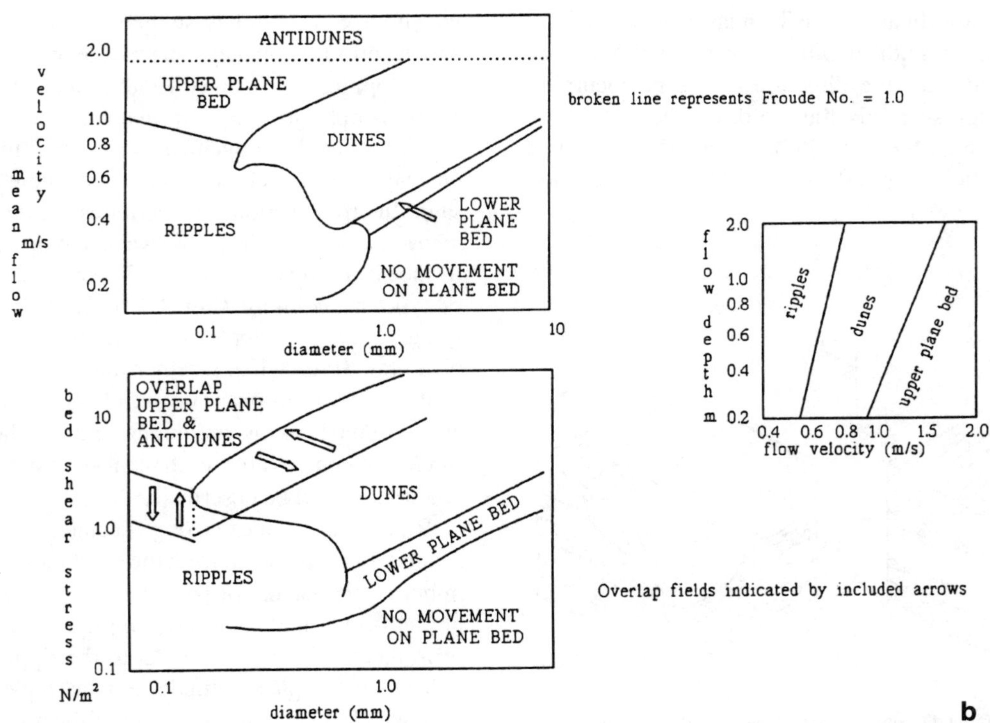

Figure 4-7. Form-drag bed forms and internal stratification related to flow velocities and textures. *a:* External and internal structures and transport of sand in relation to the flow regime. Derived for flow in alluvial channels, but widely applied as a guide for interpreting all subaqueous current structures. (After Harms and Fahnestock, in Middleton 1965.) *b:* Approximate fields for form-drag bed forms at 10°C; mean flow velocity versus bed load and versus flow depth, and relation of bed shear stress to grain size. (After Southard and Boguchwal 1990; see reference for details.)

Specific Structures

Perhaps the most common internal sedimentary structures are *cross-bedding* and *cross-lamination*, formed respectively from migrating dunes (see Dalrymple 1984) and ripple bed forms (see below), and both are part of the spectrum of cross-stratification. Cross-strata may also develop by deposition at a steep angle of repose in depressions without bed-form migration (e.g., Jopling 1966) and in volcanic base-surge deposits (e.g., Crowe and Fisher 1973). At the first level, distinction must be made between *tabular* and *trough* types (Fig. 4-8) of cross-stratification. Many more specific varieties of cross-bedding are recognized on the basis of geometry, grouping, magnitude, and association (e.g., Fig. 4-9 and Allen 1963), although some workers prefer to class separately those structures that are not formed by simple bed-form migration, such as delta foresets, channel fills, and point-bar bedding. Distinction between all of these structures requires careful determination of the three-dimensional geometry of both the inclined beds and the principal planes of stratification that bound them. Another important distinction is between sets that are composed of homogeneous sediment and those composed of layers of different kinds of sediment (i.e., heterolithic stratification, see Thomas et al. 1987).

Generally, the azimuth of dip of the internal layers of cross-bed sets indicates the direction of the paleocurrent. However, in trough sets only the dip down the axis of the trough is correct (e.g., Meckel 1967), and in other cross-bed sets the dip direction may reflect other controls: in epsilon foresets (point bars of streams) and zeta cross-stratification (channel fills) the maximum dip may be perpendicular to the paleocurrent direction. Treatment of paleocurrent data is discussed in **AS** Chapter 6; azimuthal variability varies substantially between different environments (e.g., Miall 1974). Interpretation of the various cross-bed types commonly depends on some knowledge of their origin or setting, and lack of this information may cause confusion (e.g., Dott and Roshardt 1972). A long-term problem has been the distinction between eolian and aqueous cross-bed sets; differences can be subtle and distinction may require very careful analysis (e.g., McKee 1966; Brookfield 1977; Hunter 1977; Kocurek and Dott 1981; Kocurek 1988).

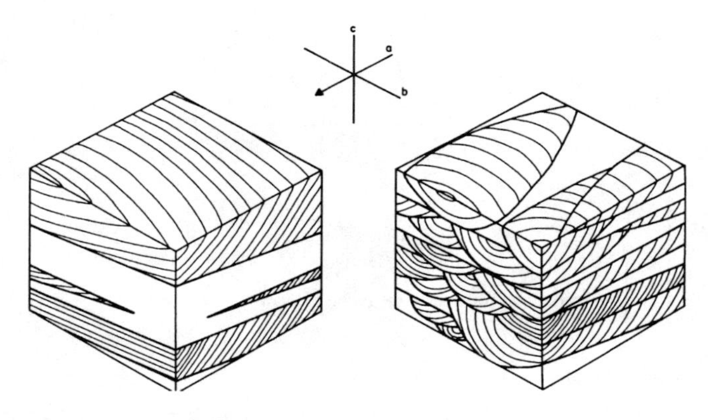

Figure 4-8. Tabular and trough cross-bedding. Tabular sets have straight planar boundary surfaces, which may be either erosional or nonerosional. Trough sets have curved erosional boundary surfaces, and each individual cross-bed is basically trough-shaped. The three-dimensional geometry of the sets must be known before the terminology can be applied. As with cross-lamination from asymmetrical ripples, cross-bedding can be used to infer the direction of flow (toward the azimuth of maximum dip) and the younging direction (truncated tops, tangential bases unless a fine tail of grain sizes is not available). To obtain directional data in trough sets, dips must be measured in the axial plane of the troughs; otherwise there will be spurious variations in the directional data. Beware of cross-sections of point-bar or channel-filling sequences where apparent maximum dips may be at right angles to current flow.

A special type of cross-bed structure that is particularly indicative of process is the *reactivation surface* that interrupts normal cross-beds where dune bed forms move intermittently (Fig. 4-10). Another that is placed by many under the cross-bed category is *hummocky (cross-)stratification* (Fig. 4-11); its presence is indicative of deposition between storm wave base and normal wave base (e.g., Bourgeois 1980; Dott and Bourgeois 1982; Duke 1985).

The most common bed forms seen in plan view are *ripples*, which have been widely studied because of their variability and the information they can provide on depositional process (e.g., Jopling and Walker 1968; Allen 1968, 1977; Harms, 1969). Ripples form only from cohesionless sediment where free intergranular movement is possible; they will not form on clay or clay-rich sediments, or where algal mats bind sand. In cross-section, two varieties are distinguished: *symmetric*, formed entirely by wave action, and *asymmetric*, formed by currents and waves. Several varieties of asymmetric types are common (Fig. 4-12). Another method of classifying varieties of ripples is by the pattern of their crests in plan view (Fig. 4-13), whether the crestal patterns for adjacent ripples are in-phase or out-of-phase, whether the crests are continuous or discontinuous, and whether or not they bifurcate. When possible, both plan-view and cross-section geometries should be described.

Ripple-mark indexes (e.g., Tanner 1967; Reineck and Singh 1975) calculated from the combined characteristics are important indicators of paleoenvironment and paleogeography. Two particularly useful ripple indices are the vertical form index RI, which is equal to the ripple wavelength divided by the ripple amplitude, and the ripple symmetry index RSI, which is equal to the length of the horizontal projection of the stoss side divided by the length of the horizontal projection of the lee side (Fig. 4-12). Generally, wave-generated ripples have an RI ≤ 4 and an RSI ≤ 2.5, whereas current-generated ripples have an RI ≥ 15 and an RSI ≥ 3. Wave-generated ripples are also commonly straight crested and frequently bifurcate, whereas current ripples seldom bifurcate, although crests may terminate and be replaced by other crests, giving a false impression of bifurcation. Ripple amplitude tends to increase with increasing current velocity and

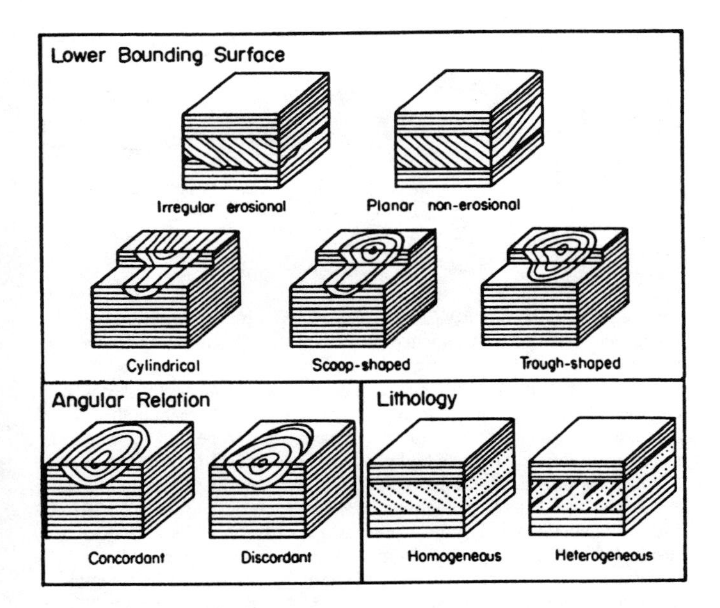

Figure 4-9. Detailed descriptive terminology for varied cross-stratification geometries. Few of these are commonly used by most workers, but each geometry represents a different set of processes, and careful examination of cross-bed sets can provide extra information. (After Allen 1963; see reference for details and implications.)

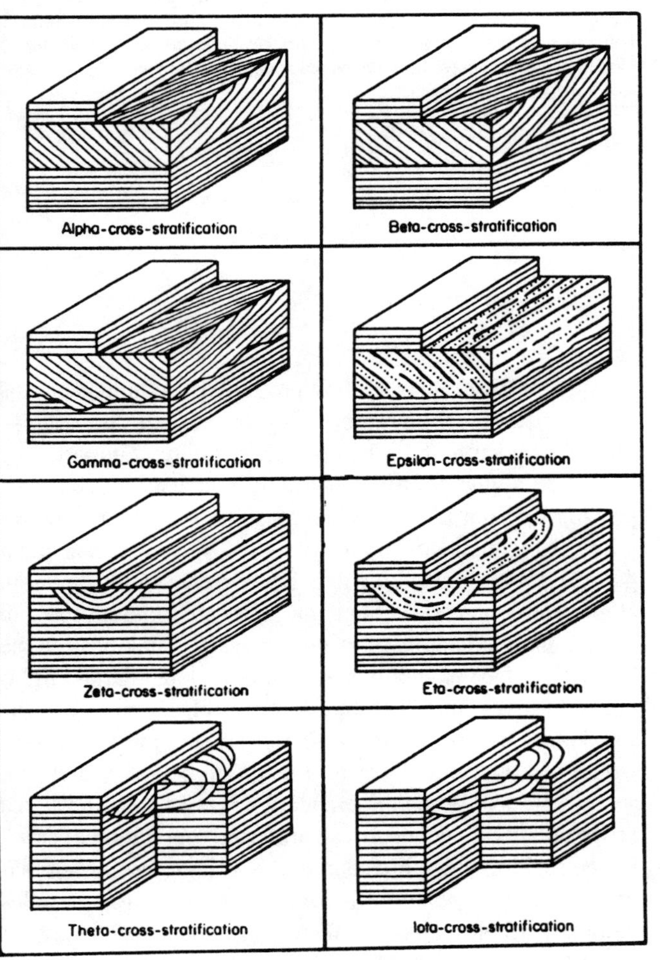

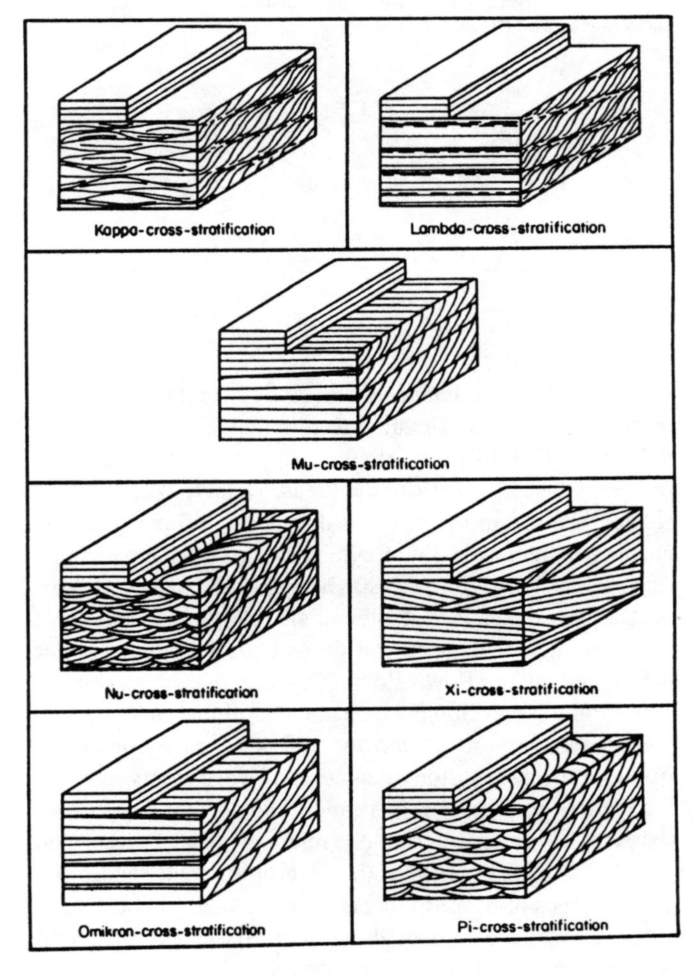

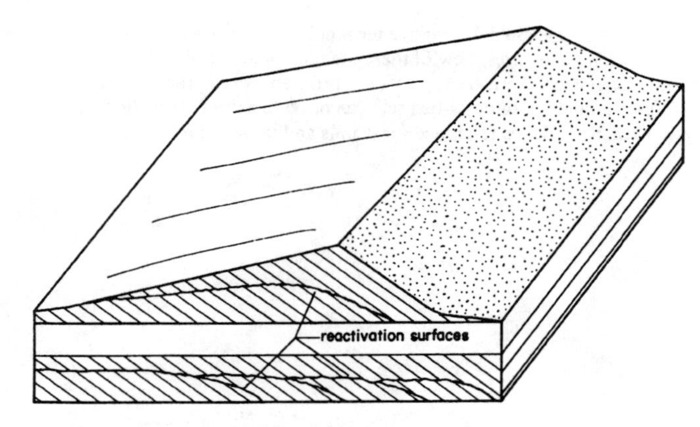

Figure 4-10. *Reactivation surfaces* reflect interruptions in the advance of sand waves (dune bed forms). Erosion of the top of the sand waves accompanies a change in flow strength or direction, such as during low river stages or tidal cycles. When flow conditions return to those conducive to sand wave migration, the eroded sand waves are reactivated and build up to their previous shape before migrating downcurrent. Cross-beds dip at the same angle of repose above and below the erosion (reactivation) surfaces if sediment texture remains constant; hence the surfaces may be difficult to find.

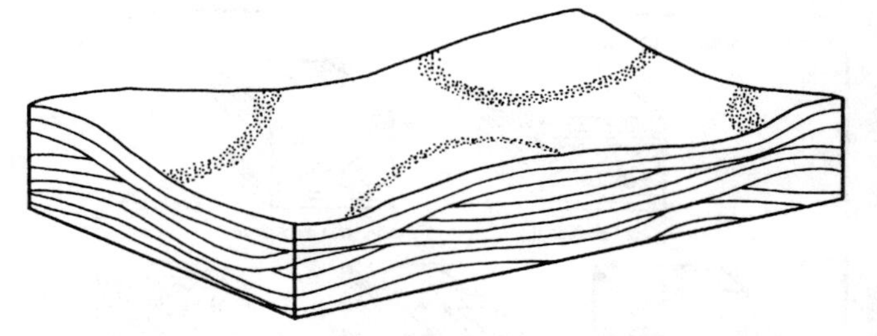

Figure 4-11. *Hummocky (cross-) stratification* is seen in sediment sequences up to c. 1 m thick wherein internal stratification is smoothly wavy with low-angle (<15°) truncation of some of the positive wave forms. The sequences commonly extend for hundreds of metres laterally, and for field recognition it is generally necessary that there be both continuous vertical exposure for tens of metres and some cross-sections at right angles showing that the hummock surfaces extend in three dimensions. Internal stratification is diffuse and parallel to the hummocks and swales of the initial depositional surface; each surface was a transient bed form generated during the waning stages of storm action as sediment suspended by the storm settles from active suspension (the irregular surfaces reflect the complex pattern of storm wave action). Wavelengths of the hummocks or swales are commonly between 1 and 5 m, and amplitudes are about 40 cm or less. Amalgamated swales with few or no hummocks characterize swaley cross-stratification; sequences with this structure appear to have been deposited at slightly shallower depths where storm wave action was more intense.

ripple crest form changes from straight at low velocities through sinuous and linguoid to rhombohedral at high flow velocities (linguoid megaripples are uncommon and when present generally comprise the lunate variety). Variations in ripple type occur over short distances, reflecting subtle changes in the lower flow-regime processes that create them (e.g., Reineck and Singh 1980). Megaripples, sand waves, or subaqueous dunes (see Ashley et al. 1990) commonly have small ripples on their stoss side and require greater water depths and greater flow velocities to form than small ripples.

Asymmetrical ripples indicate the direction of paleocurrent and sediment movement, whereas symmetrical ripples indicate only the orientation of wave fronts. Internal lamination in asymmetric ripples and the cross-sectional shape of symmetrical ripples (relatively sharp crests and rounded troughs) can give the way-up of rock sequences. Special types of ripple mark can be particularly indicative of process and depositional setting, e.g., flaser and lenticular bedding (Fig. 4-14 and see Reineck and Wunderlich 1968), flat-topped and ladder-back ripples (which have been widely used as indicators for intertidal deposition in areas with a high tidal range, although they have been reported in subtidal conditions by Reddering 1987 and rarely in fluvial settings). Wind ripples (e.g., Sharp 1963) are as common as water ripples in modern environments, and distinguishing between them in deposits is as difficult as distinguishing between eolian and aqueous cross-beds.

Current lineation is a fabric property of sediments, but is usually discussed with sedimentary structures because it provides a paleocurrent orientation. This feature is imparted by the subtle alignment of grains during deposition from suspension or from the plane-bed stage of the upper flow regime. Alignment of large particles (e.g., shells, plant particles) is obvious to the eye. However, in sands the grain alignment is barely visible even under the microscope and the common indication of its presence is a streakiness on the bedding surface (sometimes visible only under certain lighting and usually more difficult to see the closer you get). The streakiness will be consistent in one orientation on an individual bed, but may differ in orientation between beds that are superposed. When the orientation is consistent through a series of beds at one locality, a preferred orientation of small joints, each of which breaks through a single bed or lamination, produces *parting lineation.* When large particles (pebbles, cobbles, and mud flakes) have been moved by the currents, clast *imbrication* (a fabric of the sediment) may develop; some regard it as a form of current lineation.

Graded bedding is also an internal fabric in sedimentation units, but is discussed together with sedimentary structures in most books. *Normal* grading occurs when grain-size gradually fines upward in a sedimentation unit. It can originate in most environments either from settling from passive suspension, from a gradual waning in the competency of an entrainment flow (sorting tends to be good in horizontal planes, but

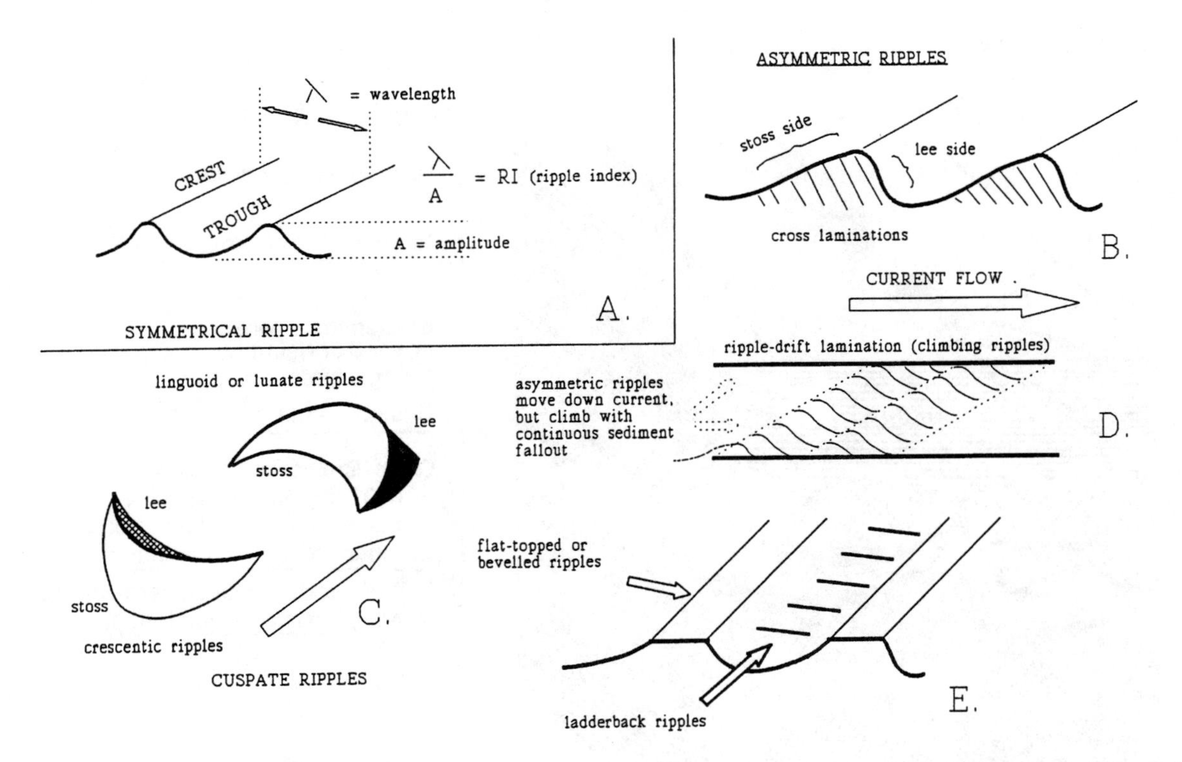

Figure 4-12. Some varieties of ripple marks. Ripple bed forms are created in the lower flow regime from the frictional drag on cohesionless sediment by eolian or aqueous currents. The bed forms are constantly moving over a dynamic profile of equilibrium and can be preserved only when there is continuous fallout of sediment (the dynamic profile is moving upward) or there is a rapid loss of flow (in which case the ripples may be buried below much finer sediment). *A: symmetrical ripple mark* showing the relation of wavelength to amplitude that can be used with any kind of ripple mark. Most symmetrical ripples have wavelengths of 0.9–200 cm, amplitudes of 0.3–22.5 cm, and ripple indices of 4–13 (most are 6–7). Most form as a result of the interaction between waves and the sediment surface, but waves occur in waters of virtually all environments and depths, hence there is no environmental restriction. Because the crest is sharper than the trough (unlike the sketch!), this structure can be used as a way-up indicator. *B: asymmetric ripples* show a wide range of geometries, reflecting a wide range of formative current (and some wave-drag) conditions. Wavelengths are usually greater than 5 cm (mainly in the 8–15 cm range). Internal foreset laminations, formed from the avalanching of grains down the lee face of the bed forms, provide information on the current direction; whereas the overall bed-form geometry cannot be used as a way-up indicator, truncations at the top and tangential lower contacts at the base of these cross-laminations are excellent way-up indicators. *C: cuspate ripples* include a variety of short-crested asymmetric ripples deposited under more rapid and complex flow conditions than ripples with continuous crests. If the steep leeward face is on the concave side of the crest, they are crescentic ripples; if on the convex side, linguoid or lunate. *D: ripple-drift lamination* formed by climbing ripples. Continuous sedimentation while asymmetric ripples are migrating downcurrent results is a continuous upward movement of the dynamic profile of equilibrium and the construction of a set of foreset laminations that "climb" between bedding planes marking previous and later profiles of equilibrium. *E: flat-topped ripples* are indicative of intermittent times of water depths less than the ripple amplitude, when erosion bevels the originally rounded crests. *Ladderback ripples* commonly form in the troughs because of late-stage water flow confined to the troughs; ladderback ripples reflect flow at right angles to the dominant current flow that creates the main ripples.

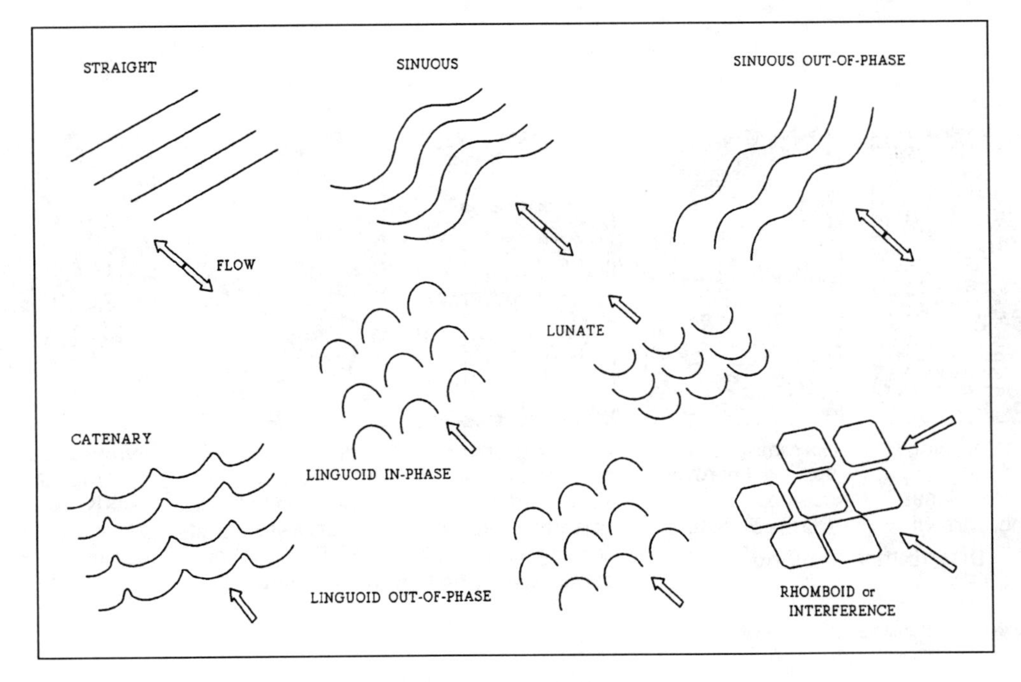

Figure 4-13. Plan-view representation of ripple crest terminology.

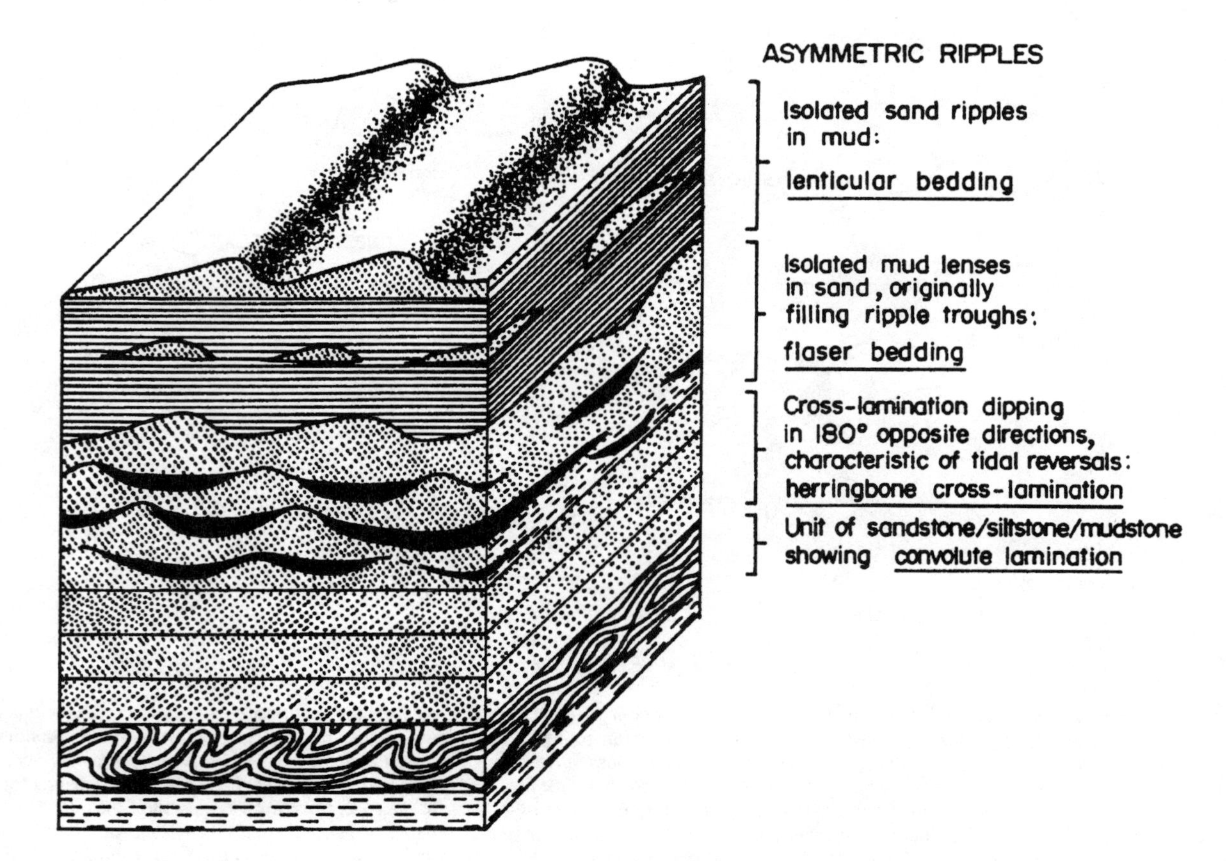

Figure 4-14. Block diagrammatic representation of some varieties of small-scale internal sedimentary structures. *Flaser bedding* is a structure in which mud once draping the entire bed surface (after a fall in flow velocity) erodes from the ripple crests as flow velocity is reestablished; thin mud lenses remain in the troughs of the original ripples. *Lenticular* or *linsen bedding* is the result of individual ripple forms isolating within mud; they reflect not only times of increased flow velocities in environments that normally accumulate mud but also situations in which silt or sand is not sufficiently abundant to form a continuous sheet of moving grains. A more explicit term for the latter structures are *starved ripples. Herringbone cross-lamination* or *cross-bedding* results when current flows sufficient for ripple or dune migration reverse direction in situations where the profile of equilibrium is rising and sediment can accumulate (e.g., in some tidal settings or on the margins of some river channels).

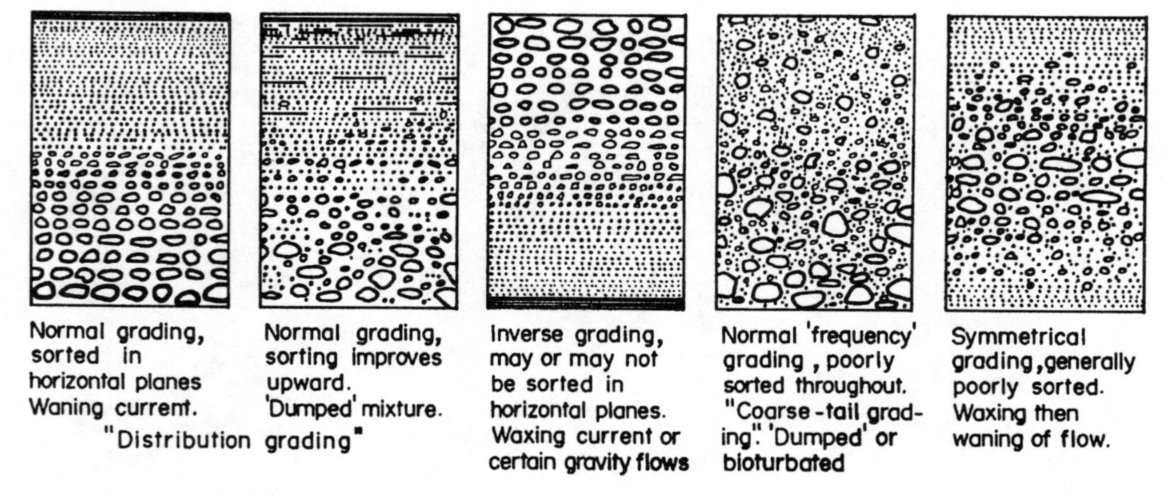

Figure 4-15. Some varieties of graded bedding.

there are exceptions with unstable flows), or from a dumping process—a sudden drop in velocity and consequently capacity and competency of the transporting agent, as in the case of storm or turbidity current events (frequency or coarse-tail grading tends to result, with poor sorting improving upward as the coarsest grains drop out faster). *Inverse* or *reverse* grading occurs when the grain size increases upward, generally because of an increase in transport competence at the same time as the profile of equilibrium moves upward, as in the case of a waxing current supplied with more sediment than it can carry (e.g., slumping of river banks during a flood phase). Figure 4-15 illustrates a variety of different types of grading that have been recognized in various settings (see also Bishop and Force 1969; Nelson 1982). Much more subtle grading may be revealed in some units by an upward decrease in the abundance of heavy minerals. When turbidity currents appear to be the cause of deposition, variation in the character of the vertical sequence of structures may be indicative of relative depositional position (e.g., Walker 1967).

Tool marks are the structures caused by particles (e.g., shells, grains, pieces of wood, dead bodies) carried as a traction or saltation load across a substrate. As such they include striations and *chattermarks*, continuous and discontinuous (respectively) scratchings formed on lithified substrates most commonly by clasts carried in glacial ice, but also by boulders in gravitational slides and flows. The equivalents in soft sediment are *grooves*, which are distinguished by their continuity from *roll, slide, bounce, brush,* and *prod marks*. They too can be formed by ice (e.g., Thomas and Connell 1985), but most are formed by clasts carried in aqueous currents. Tool marks indicate the orientation of movement of the agent of transport, but rarely the direction of movement.

Tool marks most commonly are found in the rock record as *sole marks*—a category of sedimentary structure that includes other primary as well as some secondary sedimentary structures on bedding planes. These are casts (American usage) or molds (British usage) on the base of more resistant lithologic units that were deposited after the structures formed. Common primary structures preserved in this way are *flutes* (concave-up, teardrop-shaped scours made by the brief impact, then dispersal, of turbulent current eddies—the deep bulbous end is upcurrent). Secondary structures preserved as sole marks include rain and gas pits, ice crystals, mud cracks.

Current-formed erosional structures, such as shallow rills and discontinuous scour marks (such as flutes) as well as the more continuous channel structures, are also primary sedimentary structures. They reflect the selective removal of sediment rather than addition of new material. Some scours form behind or around obstacles that have often subsequently moved away (e.g., pebbles or shells on a sandy beach). Although the geometry of erosional structures commonly only provides the orientation of the causal current, characteristics of the fill (e.g., cross-stratification or longitudinal-size grading) commonly provide information on its direction.

Block diagrams (e.g., Fig. 4-14) are commonly used to represent the three-dimensional geometry and association of primary sedimentary structures; vertical associations are also illustrated on graphic logs (discussed in **AS** Chapters 3 and 10).

Secondary Structures

Secondary sedimentary structures indicate physical or chemical conditions in the diagenetic environment. Insofar as physical remobilization requires a trigger cause (such as excessive loading and/or seismic shock), the presence or abundance of secondary mechanical structures can combine with other evidence to indicate a general depositional setting (e.g., delta front environments, because of generally rapid sedimentation creating unstable accumulations, are particularly prone to abundant failures). Careful study of chemical sedimentary structures such as concretions may reveal a complex history of both early and late diagenetic modification (e.g., Kennedy et al. 1977; Boles, Landis, and Dale 1985). In some sequences in which diagenetic processes have destroyed syn-depositional minerals (e.g., various evaporite minerals or ice crystals) that could provide important clues to depositional conditions, chemically formed structures (such as crystal casts) may provide the only remaining evidence of the original presence of those minerals.

Physical Structures

A variety of structures are formed in the sediment after deposition by mechanical processes such as quasi-solid failure (unconsolidated sediment that behaves as a brittle substance), hydroplastic deformation (sediment that behaves as a high-viscosity plastic as a result of interstitial water reducing intergrain friction), or quasi-liquid flowage (resulting from excess pore fluid pressures); these states of rheologic behavior are discussed in detail by Elliott (1965). Causes of deformation are most commonly excessive sediment loading (e.g., sudden deposition of thick coarse sediments), sudden unloading (e.g., initial slumping removes lateral support for sediment behind the scarp and a series of retrogressive slumps develop at the same locality), slope instabilities (e.g., oversteepening by sediment accumulations on the edge of a depression), development of excess pore pressures (e.g., by seismic shock leading to fabric rearrangements), and combinations of these factors; tectonic forces may also play a dominant role as they do in deforming lithified sediments.

Quasi-solid faults can be distinguished from postlithification tectonic faults by evidence of soft-sediment deformation next to the fault plane, such as a smearing-out of laminations (e.g., Thomson 1973). Commonly, they are scarps marking the locus of slumps (eg, Laird 1968). *Sedimentary folds* may be just as complex and almost as large-scale as postlithification tectonic ones (e.g., Gregory 1968); internal soft-sediment deformation or inconsistency of the fold pattern may indicate the difference. Small-scale *intraformational folds* may be confined entirely to individual layers that are centimetres up to c. 1 m thick, above and below which beds and their internal structures show no deformation (e.g., *convolute lamination*, as in part of Fig. 4-14); presumably the water content of that layer was particularly high, hence the internal frictional resistance to an imposed shear was low (e.g. Sanders, 1960). Folding, together with small-scale features like *raindrop* (or *hail*) *imprints* and/or gas pits (e.g., Moussa 1974), and *pull-aparts* (where sedimentary layers subjected to slow downslope creep have broken apart) reflect sediment behavior that is transitional to hydroplastic. Shrinkage cracks (*desiccation* or *syneresis* cracks, e.g., Plummer and Gostin 1981) can also be regarded as quasi-solid response structures. *Slump structures* away from the scarp include complex contortions that are in fact folds, but show much more small-scale and incompetent irregularities than in most tectonic folds; paleoslope direction may or may not be interpretable from them (e.g., Morris 1971; Lajoie 1972; Woodcock 1979). *Load* and *founder* structures are closely related: both represent denser, commonly coarser, sediment sinking into less dense underlying beds, but the latter term is conveniently applied when the denser sediment is in a thin bed that actually disarticulates and founders into the underlying sediment as a series of discrete masses (generally with deformed internal laminations that may bend to form near circular ball and pillow forms; e.g., McKee and Goldberg 1969, Mills 1983).

Boudinage refers to the development of a regular pinching and swelling structure (similar to a string of sausages) in relatively competent beds in response to the uniaxial stress applied by burial pressure; tectonic boudinage is more commonly described in rock successions (e.g., Ramberg 1955). An unusual form of boudinage is produced by differential loading and lateral plastic creep of gelatinous silica precursors of chert bands in banded iron formations (McConchie 1987). The resulting "chert pods" (Trendall and Blockley 1970) range from a series of pinch and swell structures in chert-rich bands to disconnected lenses, some of which are rotated relative to bedding ("cross-pods"); almost all retain the internal fine-banded substructure of the original layers. Similar lateral plastic creep processes may also account for the origin of the roughly circular *macules* in banded iron formations (described and named by Trendall and Blockley 1970).

Quasi-fluid structures generally result from thixotropic transformations brought about when watery sediments repack to a closer grain fabric and the water cannot escape fast enough (e.g., Selley and Shearman 1962). *Dewatering structures*, such as diffuse parallel, subvertical sheet structures (Laird, 1970) or dish and pillar structures (e.g., Lowe and LoPiccolo 1974; Lowe 1975) may develop during sedimentation from rapidly depositing gravity flows, but the effects of dewatering usually are more obscure, such as the elutriation of fines from the deeper levels to the surface of the deposit, whence they may be subsequently removed by other currents in the environment. Dewatering may be delayed until a late burial stage, particularly where permeable sediment lenses have been surrounded by impermeable deposits such as muds. In these cases, compaction and cementation will be prevented by the inability of the interstitial water to escape, and the fluid bears a growing proportion of the burial load. Such "overpressured" muds and sands finally burst out along any line of weakness (late-stage joints, drillholes), and the unconsolidated sediment is swept along with the water to form sedimentary *dikes* and *sills* of various thicknesses, and if and when they reach the surface, *mud* and *sand volcanoes* (e.g., Swarbrick 1968).

In the oil industry, units of overpressured sands or muds (often encountered in deltaic sequences) present a significant drilling hazard and can cause "blowouts" when encountered unexpectedly. Sedimentary dikes are probably more common than igneous ones and can show complex internal structures and contain xenoliths of country rock plucked from the sides of their route (e.g., Diller 1890; Peterson 1968; Lewis, Smale, and van der Lingen 1979). Some mud diapirs, resulting from upward flowage of less dense muds where they are overlain by thick accumulations of denser sediments (as where delta front sediments prograde over prodelta deposits), are of very large scale, covering hundreds of square kilometres and producing islands (e.g., Freeman in Braunstein and O'Brien 1968). Some dikes are filled with sediment from above rather than from below, and in some situations can have complex origins involving filling from both above and below (e.g., Lewis 1973).

Salt diapirs are also a form of secondary sedimentary structure (e.g., Braunstein and O'Brien 1968), although they form long after the sediment has been deposited. As with other types of sedimentary intrusion, they form as a result of differential loading (most sediment is denser than massive evaporites), but the evaporite units need not have a high water content because flow of the salts along planes of weakness is achieved by easy and rapid recrystallization of the evaporite minerals. Once localized, the upwelling low-density salts are continuously fed by lateral flowage from the source beds and develop into large-scale intrusions (some apparently even continue moving after the connection to the source beds is broken, like hot-air balloons). Deformation of sedimentary rocks displaced by the mobile salts can be extremely complex and both the fold, fault, and piercement structures themselves provide a variety of petroleum traps. Because salt domes (the swelling of the earth's surface over the rising diapirs) may affect tens to hundreds of km^2 and persist over geologically long periods of time, they may play a significant role in sedimentation (e.g., stratigraphic units may pinch out against the rising highs).

Chemical Structures

Structures formed predominantly as a result of chemical processes after burial include a variety formed as a result of dissolution, precipitation of cements, crystal growth, or expansion and contraction accompanying hydration, dehydration, or recrystallization. *Enterolithic folds* result from hydration of anhydrite to gypsum; the addition of the H_2O molecules results in an increase in volume of the solid by about 38%, and if burial pressure is great enough, expansion is largely lateral with the development of intense small-scale folding confined to thin layers of evaporite sulfates. *Concretions* form by migration of ions down geochemical gradients toward loci of precipitation (commonly plant, shell, or bone fossils), where they are removed by crystal growth in the water-filled pore spaces of sediments at any time from immediately after deposition to deep-burial, postlithification stages. Many varieties of concretions have been distinguished on the basis of composition, shape, time, or nature of origin (e.g., Raiswell 1971; Boles, Landis, and Dale 1985); Strakhov (1969) provides a good discussion in the final chapter of his book. Some concretions form so early that they are exposed at the sediment/water interface by local erosion of the still-unindurated surrounding sediment; the results can be very complex (e.g., Kennedy et al. 1977).

Liesegang bands represent a special type of concretionary development, in which iron ions (Fe^{+2}) appear to have oxidized to Fe^{+3} hydroxides along highly irregular diffusion fronts (e.g., Carl and Amstutz 1958); they commonly are mistaken for bedding, but careful tracing of their shape will show that they frequently cross true bedding or other primary features of a deposit. In some cases, they emphasize obscure original structures, e.g., some cross-bedding in which foreset beds differ slightly in texture.

Stylolites, irregular dissolution seams on a scale of up to a few millimetres of relief (and *microstylolites*, of finer scale), reflect pressure-assisted differential dissolution (e.g., Fig. 3-17; McClay 1977). These structures are particularly common in easily dissolved carbonate and evaporite sediments (e.g., Park and Schot 1968), but may also be present (less obviously) in pure detrital sediments. Estimates have been made of up to 50% dissolution of the original volume of some carbonates. Stylolites are generally identified by the insoluble particles concentrated along the seams and ultimately clogging the escape routes of dissolving constituents. They may form pseudo-bedding in what appear to have been homogeneous carbonates (e.g., Simpson 1985).

Other chemically generated structures, commonly observed in sediments deposited in arid zones (particularly in lacustrine and tidal flat settings), are produced by the displacive diagenetic growth of crystals (commonly gypsum) within unconsolidated sediment (e.g., Arakel and McConchie, 1982). The diagenetically formed crystals may cut across primary sedimentary laminae, but more commonly they deform the laminae; the resultant structure sometimes mimics deposition of sediment as a drape over crystals initially present on the sediment surface. In addition, just as appropriate conditions may cause crystal growth within the sediment, a change in conditions may lead to their dissolution or replacement; when dissolution has removed crystals from a sediment that is sufficiently lithified to preserve the void, casts of the original crystals may remain (if interfacial angles are adequately preserved, the identity of the original crystal can be determined).

BIOGENIC STRUCTURES

Origin and Classification

Trace fossils constitute a field of study (ichnology) that combines the disciplines of ecology and sedimentology (for a good introduction, see Frey 1973 or the first chapter in Frey 1975). The definition given by Frey (1973)—"evidence of activity by an organism, fossil or recent, other than the production of body parts"—embraces a wide variety of fundamentally different structures, such as bioturbation structures (disruption of physical structures or sediment fabrics), biostratification structures (e.g., stromatolites; Hofmann 1973), bioerosion structures (e.g., borings; see Carriker, Smith, and Wilce 1969; Budd and Perkins 1980), and even coprolites (fossilized excreta, see Hantzschel et al. 1969). In modern sediments, many of these structures are used by ecologists to determine the behavior (e.g., feeding operations) of organisms operating below the sediment surface. In ancient sediments, they also indicate the presence of otherwise unpreservable soft-bodied organisms, although the species of organism that created the trace is usually unknown. Albeit rarely, they reveal the soft-part morphology of extinct organisms or the way in which unpreserved appendages operated.

Burrow (bioturbation) structures are the most widely used biogenic structures for sedimentological interpretations. The most important first step in describing these structures is to discover their three-dimensional geometry. In modern sediments, even if there is sufficient cohesion for observations by cutting serial sections into a trench face, all that can usually be observed is the latest, open burrow—which is usually only one part of a series of connected burrowing operations that results in a *spreite*, a composite trace generally better defined in ancient sediments (Fig. 4-16). Two examples of descriptive classifications are shown in Fig. 4-17 (see Simpson in Frey 1975 for discussion of the original version of Fig. 4-17b). Specific terms also exist for certain individual structural elements within a trace (e.g., Fig. 4-17c).

Once the overall morphology of the trace is known, an individually distinctive trace may be assigned to a form genus and form species, following as closely as feasible the system for body fossils of the International Commission on Zoological Nomenclature (see Hantzschel 1975 for the most comprehensive current listing of names). For example, *Ophiomorpha nodosa* identifies a characteristic knobbly cylindrical burrow form (e.g., Frey, Howard, and Pryor 1978). The name of the creator organism is not used (in modern deposits the maker of this particular trace is the shrimp *Calianassa*), because the creator is generally not known, because the same organism can create distinctively different

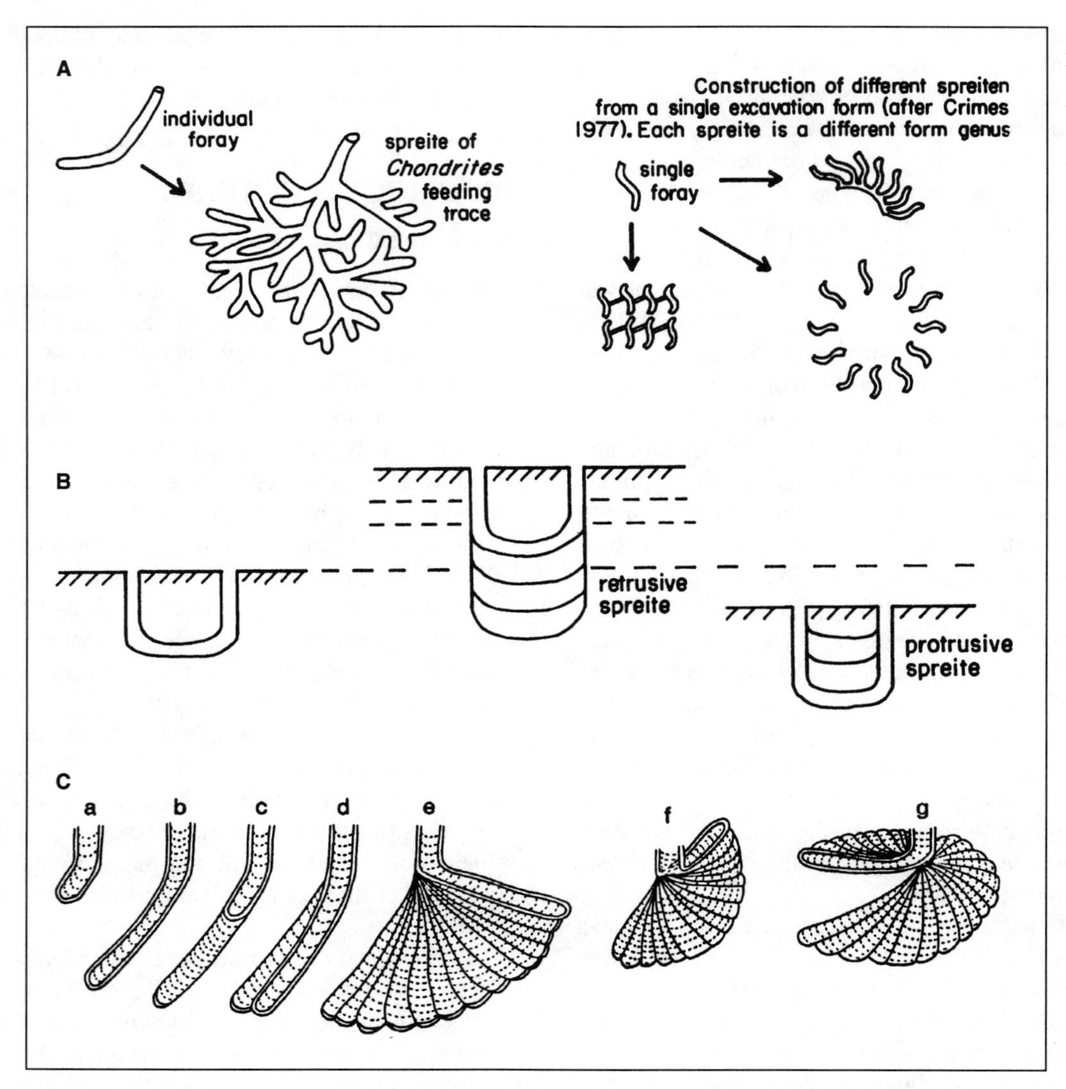

Figure 4-16. Examples of trace fossil characteristics. *A: Spreiten* are the composite traces resulting from multiple activities of the creator organism, e.g., numerous feeding forays, each overlapping and connected to a central burrow. The final structure may differ very substantially from that of an individual foray. *B: Diplocraterion* is a form genus comprising a vertical spreite made up of a U-tube, the base of which has been progressively shifted by the organism either in the search for food in the sediment or to maintain a given distance from the sediment/water interface. Multiple species of *Diplocraterion* can be distinguished on the basis of scale and other morphological characteristics, can be found in the same sedimentary sequence, and are probably formed by a variety of different organisms (e.g., Ekdale and Lewis 1991*a*). In all U-tube traces (inclined, vertical, horizontal), if the base of the tube has progressively moved away from the sediment water interface, it is termed a *protrusive* spreite; if it has moved toward the interface, it is termed *retrusive*. *C:* Stages in the construction of a *Zoophycos* spreite, probably the most complex of the feeding traces. As in *Diplocraterion*, a U-tube is constructed for respiration; repeated passes are made in the search for food within the sediment via a systematic pattern in which each pass is marginal to the previous one and the overall spreite follows a helix to avoid reworking sediment. (*C:* After Ekdale and Lewis 1991*b*.)

Character	*Name*
Trace occurs on top surface of stratum	epichnia or epirelief structure—concave or groove, convex or ridge
Trace occurs on bottom of stratum	hypichnia or hyporelief—convex ridge or concave groove
Trace is within stratum and infilled with similar sediment	endichnia } full relief structures
Trace is within stratum and filled with different sediment to that surrounding it	exichnia } full relief structures

A

INCREASING LEVEL OF REFINEMENT ⟹

Track of bedding plane ("prods" or "scratches")		all alike; different	clustered scratches; rows of prods	FORM GENUS NAME → FORM SPECIES NAME
Trail on bedding plane	freely winding; patterned or windings in	simple; bilobed; or?	transverse ornament; no ornament	
Radially symmetrical in a horizontal plane	without axial vertical structure; with axial vertical structure	5-rayed; multirayed; circular outline	rays with grooves; club-shaped rays; conical depression; radial branches	
Tunnels and shafts	uniform diameter; variable diameter	vertical; horizontal; U-shaped; regular branching; irregular network	isolated; multiple; winding	
Forms with a spreite		U-shaped; spiral; branched	vertical plane; horizontal plane; inclined plane	
Pouch-shaped		smooth surface; transverse ornament		
Miscellaneous		e.g., net pattern		

B

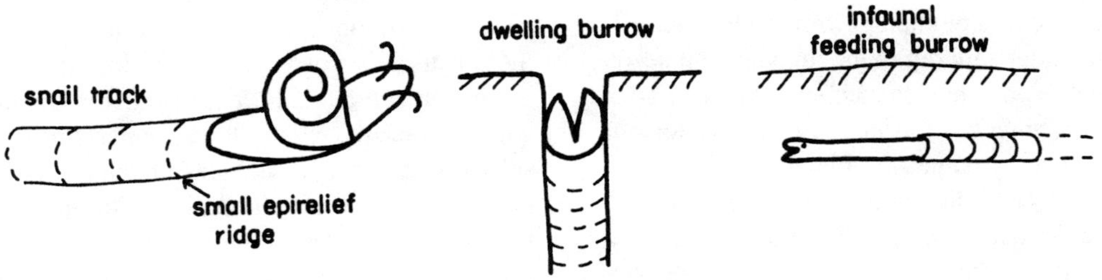

snail track
small epirelief ridge
dwelling burrow
infaunal feeding burrow

C

Figure 4-17. Examples of descriptive classification systems for trace fossils. *A:* Toponomic (mode of preservation) classification. *B:* Example of morphologic classification scheme for trace fossils; the successive refinement levels are analogous to the sequence from phylum to species in naming modern organisms. (Simplified from Simpson in Frey 1975.) Individually distinctive forms are given Linnean genus and species names as for body fossils; only the presence of *-ichnus* (if present!) at the end of some names is a key to their distinction. Before any trace is assigned a form genus or species name, its three-dimensional geometry must be known! *C: Meniscus* structures are sets of curved lamellae or small ridges between the principal boundaries of a trace. They represent intermediate resting positions of the moving progenitor organism and are generally concave in the direction of movement. Passive later fill of burrows may also show internal structures, but they are generally irregular and more variable in character. (*A:* After Martinsson 1965 as depicted in Frey 1975, table 3.5.)

traces, and because different organisms can create similar traces (e.g., several genera and species of arthropods are known to dig similar burrows today). Hence, an *Ophiomorpha* trace may connect to a trace named *Thalassinoides* (a horizontal burrow generally without a knobbly exterior), but it is only when the connection is found that one can be sure the same organism created the distinctively different trace forms (commonly one or the other trace is absent or no connections are seen). Unlike the link between organisms and body fossils, variation in trace morphology must be permitted within the same form genus or species because each creator may have its own behavioral eccentricities; conversely, the same trace species can belong to several different form genera.

Burrows and *spreiten* reflect animal behavior patterns. Although the creator organism can seldom be identified, deduction of its behavior and the stimuli that led to a particular behavior can provide useful information about its living environment and/or sedimentation processes therein. For example, animals operating below the sediment-water interface, where the conditions are normally anoxic, generally must maintain a connecting shaft to overlying oxygenated waters for respiration. Many organisms, even under aerobic substrate conditions, maintain a continuously open tube (basically U-shaped) with two connections to the water to aid circulation and to eject fecal matter.

In traces such as *Diplocraterion* (Fig. 4-16b, and see Cornish 1986), the U-tube is characteristic of the trace, but in traces such as *Ophiomorpha*, the basic U-tube geometry cannot be determined in typical exposures and became apparent only from studies of the modern burrow systems. Only one vertical shaft can be seen in most *Zoophycos* traces, but a complex open U-tube must have been maintained by the deep-burrowing creator because the bulk of the *spreite* is deep in the sediment and the organism had to breathe (see Fig. 4-16c and Ekdale and Lewis 1991b). The kind of behavior that produced each spreite is a common target of investigations and an ethological (function) classification has been specifically designed for traces (Fig. 4-18); in some investigations, workers class traces entirely on the basis of their ethology and do not attempt form-genus identification. However, an understanding of the fundamental needs of organisms, as well as of physicochemical characteristics of sedimentary environments, is necessary for ethological interpretation; even then, boundary problems remain.

A useful terminology that distinguishes between suites of burrows made at different times has been introduced by Bromley (e.g., Bromley in Frey 1975). When a bedding surface represents a time of substantial nondeposition or erosion and nondeposition (something between a diastem and an unconformity), it can be termed an *omission* surface. Preomission, omission, and postomission trace assemblages can be distinguished on the basis of their relationship to the surface and each other. Recognition of these episodes of bioturbation considerably increases the resolution of interpretations of the history of sedimentation.

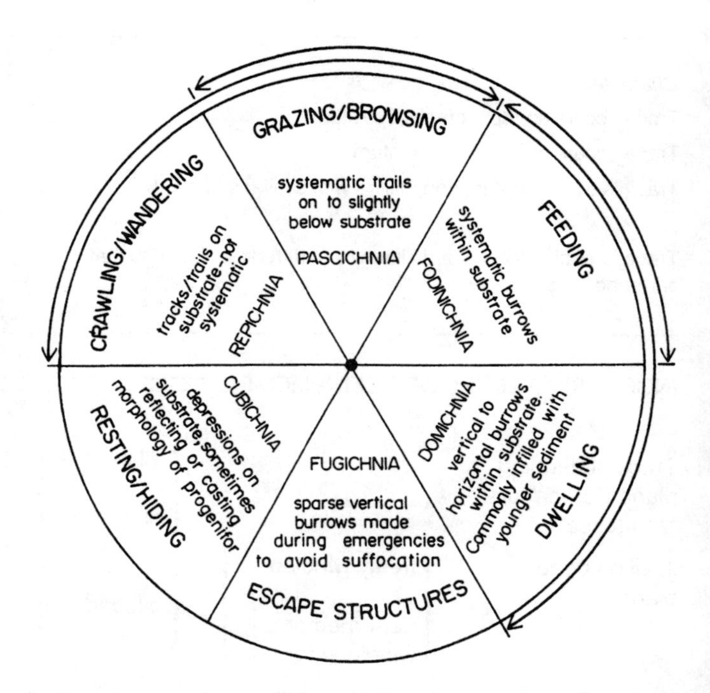

Figure 4-18. Ethological (functional) classification of trace fossils. External arrows show possible overlap of functional classes. Application of the terminology requires knowledge or inference of the behavior of the creator organism, and is complicated by the fact that organisms may produce very similar traces when performing different functions; e.g., they may use the same burrow both as a dwelling and as a site from which to feed. (Based on the system of A. Seilacher, see Frey 1973 for a full discussion of the classes.)

Interpretation

Organisms can have many effects on the physical and chemical properties of the depositional and early diagenetic environment (e.g., McCall and Tevesz 1982). For example, the mere presence of benthic traces indicates oxygenated bottom waters, since organisms must respire. In general, vertical burrows dominate where high-energy conditions keep the surficial sediment mobile (such as on beaches and in barfront settings), whereas horizontal burrows dominate where energy conditions are low and nutrients are abundant within the substrate. Suspension feeders dominate where substrates are firm and there is little suspended sediment in the water, whereas detritivores (sediment-eaters) dominate where the substrate is soft and includes abundant nutrients. Intensely bioturbated intervals alternating with intervals showing fine inorganic structures commonly indicate episodic sedimentation (periodic deposition of intervals too thick for the burrowers to penetrate).

Parallel to the concept of homeomorphy in zoology and paleontology, similar traces are created by different organisms if the stimuli are similar. Consequently, particular assemblages of traces characterize similar environments throughout the history of life. Characteristic assemblages of traces have been recognized in particular environments, and specific ichnofacies have been named on the basis of the most common member of each assemblage (e.g., Fig. 4-19, and see discussions in Seilacher 1964; Crimes and Harper 1971; Frey 1975; Ekdale 1978, 1988; Basan 1978; Miller, Ekdale, and Picard 1984; Bromley 1991). The nature of the assemblage of traces can also indicate elements of post-depositional history, as in the development of unconformities (e.g., Fig. 4-20 and see Lewis and Ekdale 1992). The greatest interpretive power of traces is achieved when the assemblages and implications of the various types are integrated with information from the associated array of body fossils and the physical and chemical sedimentary structures: concurrent changes will indicate precisely the causative changes in environmental conditions. In addition, insofar as burrowing organisms have inhabited most marine environments throughout the history of life, the *absence* of burrows in marine strata is almost as informative as their presence; inimical conditions may have been absence of oxygen in bottom waters of the basin, hypersalinity, excessive mobility of sediments (preventing preservation, as on high-energy beaches), excessive suspended load in the bottom waters, or highly episodic sedimentation (reworking and/or depositing greater thicknesses of sediment than can be subsequently burrowed). Additional evidence for one of these explanations should be found in the structures, textures, or composition of the sediment.

When interpreting burrow assemblages, an important principle is that the youngest trace (the one that cuts the others) generally reflects burrowing by the organism that operated at the greatest depth within the sediment because the effects of each deeper operator will be progressively superposed on the effects of shallower forms as sedimentation continues (e.g., Ward and Lewis 1975; Bromley and Ekdale 1986). Exceptions exist when penecontemporaneous erosion lowers the sediment-water interface, and shallower forms operate within originally deeper layers. The resultant *ichnofabrics* may be complex, and there is a need for a classification of the various stages in complexity (e.g., Droser and Bottjer 1986).

Other forms of biogenic structure that provide paleoenvironmental information include stromatolites and algal mats, and coprolites (fecal pellets produced by both vertebrates and invertebrates). Stromatolites and algal mats are largely restricted today to shallow water, tidal flat, and some lacustrine and sabkha settings where the influx of clastic sediment is low and herbivores are rare; physical structures and deposit geometry can usually be used to distinguish between these settings (e.g., Logan, Rezak, and Ginsburg 1964; Logan et al. 1970; Burne and Moore 1987, 1988; and papers in Monty 1981). Many varieties of stromatolites (e.g., laminar, columnar, domal, branching, etc.), microbialites, and algal mats have been described in the literature, and it is becoming increasingly possible to use varieties of these structures as indicators of specific environmental settings and physicochemical conditions. Stromatolites and related algal structures are particularly interesting because they have been identified in sediments as old as 3450 my B.P. (Walter,

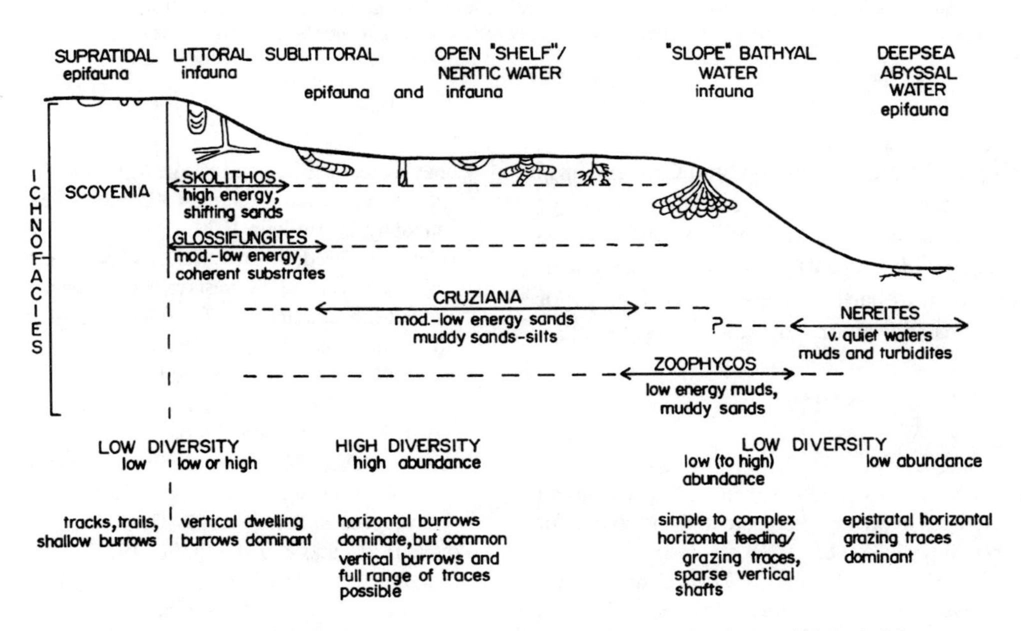

Figure 4-19. Idealized environmental zonation of trace fossils. Physical, chemical, and biological factors establish ecological habitats that may be similar at different water depths or distances from shorelines (e.g., see Ekdale 1988). Thus, this generalized ichnofacies model can only be appropriately used when *trends* in ichnofacies are determined in the sedimentary association. It is the assemblage of traces that determine the name of the ichnofacies: the name giver for a particular ichnofacies may not in fact be present in that facies, or it may be present in another ichnofacies. Subdivisions of each ichnofacies may be recognized in any particular study; the nonmarine *Scoyenia* ichnofacies is particularly coarse because nonmarine traces tend to be low in diversity in any particular sediment, but there are many types of nonmarine traces.

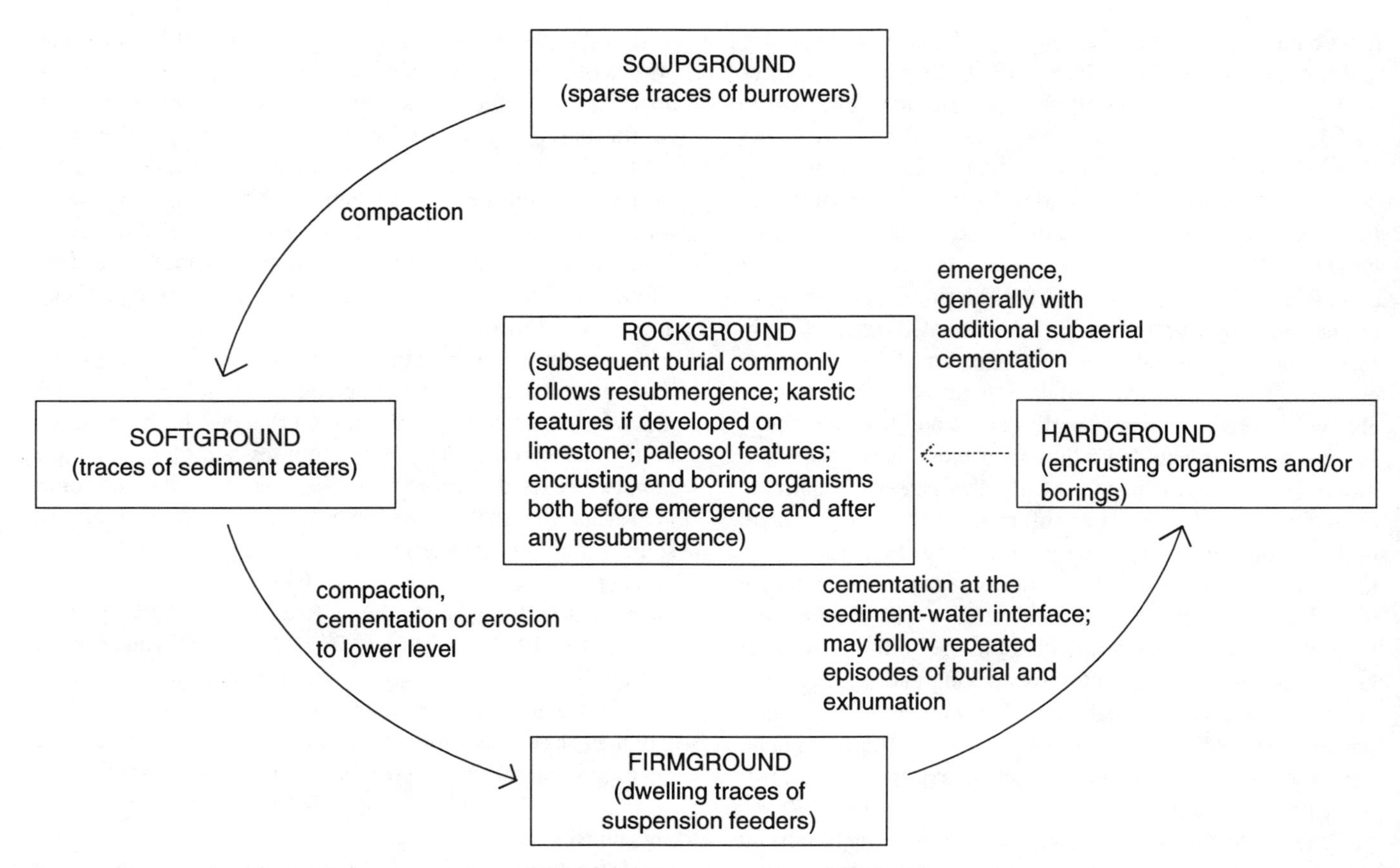

Figure 4-20. A classification of substrates. Soupground sediment is highly water-saturated and not compacted; burrows formed are indistinct and deformed by subsequent compaction. Softground burrow fillings differ from surrounding sediment only slightly, due to the effects of ingestion, concentration of organic matter in fecal packets (e.g., pellets), or the oxidizing or reducing effects that the organism's activities have produced. Firmground dwelling burrows and hardground or rockground borings are filled by sediment that was deposited on the substrate surface at a later date; erosion may have removed that sediment locally except for that which is trapped in these burrows. Interpretation in the rock record is complicated by the fact that these different substrates may develop consecutively on the same surface (e.g., as cementation consolidates the sediment), or may reflect erosion through a vertical stack of the different types (compaction below the sediment/water interface commonly results in firmgrounds below softgrounds below soupgrounds).

Buick, and Dunlop 1980; Buick, Dunlop, and Groves 1981) and appear to have formed in the same manner as those of today. Coprolites, produced by both vertebrate and invertebrate fauna, come in a wide range of sizes and forms, some of which can be attributed to particular organisms (e.g., Pryor 1975) and thus to particular environmental settings.

SELECTED BIBLIOGRAPHY

General

Allen, J. R. L., 1967, Notes on some fundamentals of paleocurrent analysis with reference to preservation potential and sources of variance. *Sedimentology* 9:75–88.

Allen, J. R. L., 1982, *Sedimentary Structures: Their Character and Physical Basis.* Developments in Sedimentology 30A, Elsevier, Amsterdam, 593p.

Allen, J. R. L., 1985, *Principles of Physical Sedimentology.* George Allen & Unwin, London, 272p.

Bailey, R. J., 1967, Paleocurrents and paleoslopes. *Journal of Sedimentary Petrology* 37:1252–5.

Collinson, J. D., and D. B. Thompson, 1985, *Sedimentary Structures.* George Allen & Unwin, London, 194p.

Coneybeare, C. E. B., and K. A. W. Crook, 1968, *Manual of Sedimentary Structures.* Bureau of Mineral Resources, Geology and Geophysics Bulletin 102, 327p.

DeCelles, P. G., M. B. Gray, K. D. Ridgway, R. B. Cole, D. A. Pivnik, N. Pequera, and P. Srivastava, 1991, Controls on synorogenic alluvial-fan architecture, Beartooth Conglomerate (Palaeocene), Wyoming and Montana. *Sedimentology* 38:567–90.

Dzulynski, S., and E. K. Walton, 1963, *Sedimentary Features of Flysch and Greywacke.* Developments in Sedimentology 7, Elsevier, Amsterdam, 274p.

Klein, G. deV., 1967, Paleocurrent analysis in relation to modern marine dispersal patterns. *American Association of Petroleum Geologists Bulletin* 51:366–82.

Potter, P. E., and F. J. Pettijohn, 1977, *Paleocurrents and Basin Analysis.* Springer-Verlag, New York, 425p.

Reineck, H.-E., and I. B. Singh, 1980, *Depositional Sedimentary Environments,* 2d ed. Springer-Verlag, New York, 439p.

Schwartz, H.-H., 1975, Sedimentary structures and facies analysis of shallow marine carbonates. *Contributions to Sedimentology* 3:1–100.

Shawa, M. S. (ed.), 1974, *Use of Sedimentary Structures for Recognition of Clastic Environments.* Canadian Society Petroleum Geologists, Calgary, Canada, 66p.

Shrock, R. R., 1948, *Sequence in Layered Rocks.* McGraw-Hill, New York, 507p.

Primary Sedimentary Structures

Allen, J. R. L., 1963, The classification of cross-stratified units, with notes on their origin. *Sedimentology* 2:93–114.

Allen, J. R. L., 1968, *Current Ripples*. Elsevier, Amsterdam, 433p.

Allen, J. R. L., 1977, The plan shape of current ripples in relation to flow conditions. *Sedimentology* 24:53–62.

Ashley, G. M. (chairperson), and others, 1990, Classification of large-scale subaqueous bedforms: A new look at an old problem. *Journal of Sedimentary Petrology* 60:160–72.

Bishop, D. G., and E. R. Force, 1969, The reliability of graded bedding as an indicator of the order of superposition. *Journal of Geology* 77:346–52.

Bourgeois, J., 1980, A transgressive shelf sequence exhibiting hummocky stratification—the Cape Sebastian Sandstone (Upper Cretaceous), southwestern Oregon. *Journal of Sedimentary Petrology* 50:681–702.

Bridge, J. S., 1978, Origin of horizontal lamination under turbulent boundary layers. *Sedimentary Geology* 20:1–16.

Brookfield, M. E., 1977, The origin of bounding surfaces in ancient aeolian sandstones. *Sedimentology* 24:303–32.

Cheel, R. J., 1990, Horizontal lamination and the sequence of bed phases and stratification under upper-flow-regime conditions. *Sedimentology* 37:517–29.

Clifton, H. E., 1976, Wave-formed sedimentary structures—a conceptual model. In R. A. Davis and R. L. Ethington (eds.), *Beach and Nearshore Sedimentation,* Society for Sedimentary Geology Special Publication 24, Tulsa, Okla., pp. 126–48.

Crowe, B. M., and R. V. Fisher, 1973, Sedimentary structures in base-surge deposits with special reference to cross-bedding, Ubebebe Craters, Death Valley, California. *Geological Society of America Bulletin* 84:663–82.

Dalrymple, R. W., 1984, Morphology and internal structure of sandwaves in the Bay of Fundy. *Sedimentology* 31:365–82.

Dott, R. H., Jr., and J. Bourgeois, 1982, Hummocky stratification: Significance of its variable bedding sequences. *Geological Society of America Bulletin* 93:663–80.

Dott, R. H., Jr., and M. A. Roshardt, 1972, Analysis of cross-stratification orientation in the St. Peter Sandstone in southwestern Wisconsin. *Geological Society of America Bulletin* 83:2589–96.

Duke, W. L., 1985, Hummocky cross-stratification, tropical hurricanes, and intense winter storms. *Sedimentology* 32:167–19. (Also see 1987 Discussions [by G. deV. Klein and K. M. Marsaglia; D. J. P. Swift, and D. Nummedal] and Replies in *Sedimentology* 34:333–51.)

Dzulynski, S., and J. E. Sanders, 1962, Current marks on firm mud bottoms. *Connecticut Academy of Science Transactions* 42:57–96.

Harms, J. C., 1969, Hydraulic significance of some sand ripples. *Geological Society of America Bulletin* 80:363–96.

Harms, J. C., and R. K. Fahnestock, 1965, Stratification, bed forms, and flow phenomena (with an example from the Rio Grande). In G. V. Middleton (ed.), *Primary Sedimentary Structures and Their Hydrodynamic Interpretation,* Society for Sedimentary Geology Special Publication 12, Tulsa, Okla., pp. 84–115.

Harms, J. C., J. B. Southard, D. R. Spearing, and R. G. Walker, 1975, *Depositional Environments as Interpreted from Primary Sedimentary Structures and Stratification Sequences.* Short Course 9, Society for Sedimentary Geology, Tulsa, Okla.

Hunter, R. E., 1977, Basic types of stratification in small eolian dunes. *Sedimentology* 24:361–87.

Ingram, R. L., 1954, Terminology for the thickness of stratification and parting units in sedimentary rocks. *Geological Society of America Bulletin* 65:937–38.

Jones, B. G., and B. R. Rust, 1983, Massive sandstone facies in the Hawkesbury Sandstone, a Triassic fluvial deposit near Sydney, Australia. *Journal of Sedimentary Petrology* 53:1249–59.

Jopling, A. V., 1966, Some applications of theory and experiment to the study of bedding genesis. *Sedimentology* 7:71–102.

Jopling, A. V., and R. G. Walker, 1968, Morphology and origin of ripple-drift cross-lamination, with examples from the Pleistocene of Massachusetts. *Journal of Sedimentary Petrology* 38:971–84.

Kocurek, G., 1988, First-order and super bounding surfaces in eolian sequences—bounding surfaces revisited. *Sedimentary Geology* 56:193–206.

Kocurek, G., and R. H., Dott, Jr., 1981, Distinctions and uses of stratification types in the interpretation of eolian sand. *Journal of Sedimentary Petrology* 51:579–95.

Lindholm, R. L., 1981, Flat stratification: Two ancient examples. *Journal of Sedimentary Petrology* 52:227–31.

McCabe, P. J., 1977, Deep distributory channels and giant bedforms in the Upper Carboniferous of the central Pennines, northern England. *Sedimentology* 24:271–90.

McKee, E. D., 1966, Structure of dunes at White Sands National Monument, New Mexico (and a comparison with structures of dunes from other selected areas). *Sedimentology* 7:1–70.

McKee, E. D., and G. W. Weir, 1953, Terminology for stratification and cross-stratification in sedimentary rocks. *Geological Society of America Bulletin* 64:381–90.

Meckel, L. D., 1967, Tabular and trough cross bedding: Comparison of dip azimuth variability. *Journal of Sedimentary Petrology* 37:80–6.

Miall, A. D., 1974, Paleocurrent analysis of alluvial sediments: A discussion of directional variance and vector magnitude. *Journal of Sedimentary Petrology* 44:1174–85.

Middleton, G. V. (ed.), 1965, *Primary Sedimentary Structures and Their Hydrodynamic Interpretation.* Society for Sedimentary Geology Special Publication 12, Tulsa, Okla., 265p.

Middleton, G. V. (ed.), 1977, *Hydraulic Interpretation of Primary Sedimentary Structures.* Society for Sedimentary Geology Reprint Series 4, Tulsa, Okla.

Middleton, G. V., and J. B. Southard, 1984, *Mechanics of Sediment Movement*, 2d ed. Short Course 3, Society for Sedimentary Geology, Tulsa, Okla.

Nelson, C. H., 1982, Modern shallow-water graded sand layers from storm surges, Bering Shelf: A mimic of Bouma sequences and turbidite systems. *Journal of Sedimentary Petrology* 52:537–45.

Pettijohn, F. J., and P. E. Potter, 1964, *Atlas and Glossary of Primary Sedimentary Structures.* Springer-Verlag, New York, 370p.

Reddering, J. S. V., 1987, Subtidal occurrences of ladder-back ripples: Their significance in palaeo-environmental reconstruction. *Sedimentology* 34:253–7.

Reineck, H.-E., and F. Wunderlich, 1968, Classification and origin of flaser and lenticular bedding. *Sedimentology* 11:99–104.

Sharp, R. P., 1963, Wind ripples. *Journal of Geology* 71:617–36.

Southard, J. B., and L. A. Boguchwal, 1973, Flume experiments on the transition from ripples to lower flat bed with increasing sand size. *Journal of Sedimentary Petrology* 43:1114–21.

Southard, J. B., and L. A. Boguchwal, 1990, Bed configurations in steady unidirectional water flows, Part 2. Synthesis of flume data. *Journal of Sedimentary Petrology* 60:658–79.

Tanner, W. F., 1967, Ripple mark indices and their uses. *Sedimentology* 9:89–104.

Thomas, G. S. P., and R. J. Connell, 1985, Iceberg drop, dump, and grounding structures from Pleistocene glacio-lacustrine sediments, Scotland. *Journal of Sedimentary Petrology* 55:243–9.

Thomas, R. G., D. G. Smith, J. M. Wood, J. Visseer, E. A. Calverley-Range, and E. H. Koster, 1987, Inclined heterolithic stratification—terminology, description, interpretation and significance. *Sedimentary Geology* 53:123–79.

Walker, R. G., 1967, Turbidite sedimentary structures and their relationship to proximal and distal depositional environments. *Journal of Sedimentary Petrology* 37:25–43.

Wells, N. A., 1988, Working with paleocurrents. *Journal of Geological Education* 35:39–43.

Secondary Sedimentary Structures

Arakel, A. V., and D. M. McConchie, 1982, Classification and genesis of calcrete and gypsite lithofacies in paleodrainage systems in inland Australia and their relationship to carnotite mineralization. *Journal of Sedimentary Petrology* 52:1149–70.

Boles, J. R., C. A. Landis, and P. Dale, 1985, The Moeraki Boulders—anatomy of some septarian concretions. *Journal of Sedimentary Petrology* 55:398–406.

Braunstein, J., and G. D. O'Brien (eds.), 1968, *Diapirism and Diapirs.* American Association of Petroleum Geologists Memoir 8, Tulsa, Okla.

Carl, J. D., and G. C. Amstutz, 1958, Three-dimensional Liesegang rings by diffusion in a colloidal matrix, and their significance for the interpretation of geological phenomena. *Geological Society of America Bulletin* 6:1467.

Diller, J. S., 1890, Sandstone dikes. *Geological Society of America Bulletin* 1:411–42.

Elliott, R. E., 1965, A classification of subaqueous sedimentary structures based on rheological and kinematical parameters. *Sedimentology* 5:193–209.

Gregory, M. R., 1968, Sedimentary features and penecontemporaneous slumping in the Waitemata Group, Whangaparaoa Peninsula, North Auckland, New Zealand. *New Zealand Journal of Geology and Geophysics* 12:248–82.

Jones, M. E., and R. M. F. Preston, 1987, *Deformation of Sediments and Sedimentary Rocks.* Geological Society Special Publication 29, Blackwell Scientific Publications, Oxford, 350p.

Kennedy, W. J., R. C. Lindholm, K. P. Helmold, and J. M. Hancock, 1977, Genesis and diagenesis of hiatus- and breccia-concretions from the mid-Cretaceous of Texas and northern Mexico. *Sedimentology* 24:833–44.

Laird, M. G., 1968, Rotational slumps and slump scars in Silurian rocks, western Ireland. *Sedimentology* 10:111–20.

Laird, M. G., 1970, Vertical sheet structure—a new indicator of sedimentary fabric. *Journal of Sedimentary Petrology* 40:428–34.

Lajoie, J., 1972, Slump fold axis orientations: An indication of paleoslope? *Journal of Sedimentary Petrology* 42:584–6.

Lewis, D. W., 1973, Polyphase limestone dikes in the Oamaru region, New Zealand. *Journal of Sedimentary Petrology* 43:1031–45.

Lewis, D. W., D. Smale, and G. J. van der Lingen, 1979, A sandstone diapir cutting the Amuri Limestone, North Canterbury, New Zealand. *New Zealand Journal of Geology and Geophysics* 22:295–305.

Lowe, D. R., 1975, Water escape structures in coarse-grained sediments. *Sedimentology* 22:157–204.

Lowe, D. R., and R. D. LoPiccolo, 1974, The characteristics and origins of dish and pillar structures. *Journal of Sedimentary Petrology* 44:484–501.

McClay, K. R., 1977, Pressure solution and Coble creep in rocks and minerals: A review. *Journal of the Geological Society* 134:57–70.

McConchie, D. M., 1987, The geology and the geochemistry of the Joffre and Whaleback Shale Members of the Brockman Iron Formation, Western Australia. In P. Appel and G. LaBerge (eds.), *Precambrian Iron Formations.* Theophrastus Publications, Athens, pp. 541–601.

McKee, E. D., and M. Goldberg, 1969, Experiments on formation of contorted structures in mud. *Geological Society of America Bulletin* 80:231–44.

Mills, P. C., 1983, Genesis and diagnostic value of soft-sediment deformation structures—a review. *Sedimentary Geology* 35:83–104.

Morris, R. C., 1971, Classification and interpretation of disturbed bedding types in Jackfork flysch rocks (Upper Mississippian), Ouachita Mountains, Arkansas. *Journal of Sedimentary Petrology* 41:410–24.

Moussa, M. T., 1974, Rain-drop impressions? *Journal of Sedimentary Petrology* 44:1118–21.

Park, W. C., and E. H. Schot, 1968, Stylolites: Their nature and origin. *Journal of Sedimentary Petrology* 38:175–91.

Peterson, G. L., 1968, Flow structures in sandstone dikes. *Sedimentary Geology* 2:177–90.

Plummer, P. S., and V. A. Gostin, 1981, Shrinkage cracks: Desiccation or syneresis. *Journal of Sedimentary Petrology* 51:1147–56.

Raiswell, R., 1971, The growth of Cambrian and Liassic concretions. *Sedimentology* 17:147–71.

Ramberg, H., 1955, Natural and experimental boudinage and pinch and swell structures. *Journal of Geology* 63:512–26.

Sanders, J. E., 1960, Origin of convolute lamination. *Geological Magazine* 97:408–21.

Selley, R. C., and D. J. Shearman, 1962, Experimental production of sedimentary structures in quicksands. *Geological Society of America Proceedings* 1599:101–2.

Simpson, J., 1985, Stylolite-controlled layering in a homogeneous limestone: Pseudo-bedding produced by burial diagenesis. *Sedimentology* 32:495–505.

Strakhov, N. M., 1967–70, *Principles of Lithogenesis.* Oliver & Boyd, Edinburgh, 3 vols. (1, 1967; 2, 1969; 3, 1970).

Swarbrick, E. E., 1968, Physical diagenesis: Intrusive sediment and connate water. *Sedimentary Geology* 2:161–75.

Thomson, A., 1973, Soft-sediment faults in the Tesnus Formation and their relationship to paleoslope. *Journal of Sedimentary Petrology* 43:525–8.

Trendall, A. F., and J. G. Blockley, 1970, *The Iron Formations of the Precambrian Hamersley Group, Western Australia, with Special Reference to the Associated Crocidolite.* Geological Survey of Western Australia Bulletin 119, 366p.

Woodcock, N. H., 1979, The use of slump structures as palaeoslope orientation estimators. *Sedimentology* 26:83–99.

Biogenic Structures

Anon., 1991, *Ichnofabric and Ichnofacies. Palaios* (Special Issue) 6(3):197–343.

Basan, P. B. (ed.), 1978, *Trace Fossil Concepts.* Short Course 5, Society for Sedimentary Geology, Tulsa, Okla., 201p.

Bromley, R. G., 1991, *Trace Fossils, Biology and Taphonomy.* Unwin Hyman, London, 280p.

Bromley, R. G., and A. A. Ekdale, 1986, Composite ichnofacies and tiering of burrows. *Geological Magazine* 123:59–69.

Budd, D. A., and R. D. Perkins, 1980, Bathymetric zonation and paleoecological significance of microborings in Puerto Rican shelf and slope sediments. *Journal of Sedimentary Petrology* 50:881–904.

Buick, R., J. S. R. Dunlop, and D. I. Groves, 1981, Stromatolite recognition in ancient rocks: An appraisal of irregularly laminated structures in an early Archaean chert-barite unit from North Pole, Western Australia. *Alcheringa* 5:161–81.

Burne, R. V., and L. S. Moore, 1987, Microbialites: Organosedimentary deposits of benthic microbial communities. *Palaios* 2:241–54.

Burne, R. V., and L. S. Moore, 1988. The ecological and environmental settings of modern microbialites: Significance for the interpretation of ancient examples. *Terra Cognita* 8:225.

Carriker, M. R., L. H. Smith, and E. T. Wilce (eds.), 1969, Penetration of calcium carbonate substrates by lower plants and invertebrates. *American Zoologist* 9:629–1020.

Chamberlain, C. K., 1971, Bathymetry and paleoecology of Ouachita geosyncline of southeastern Oklahoma as determined from trace fossils. *American Association of Petroleum Geologists Bulletin* 55:34–50.

Clifton, H. E., and R. E. Hunter, 1973, Bioturbational rates and effects in carbonate sand, St. John, U. S. Virgin Islands. *Journal of Geology* 81:253–68.

Cornish, F. G., 1986, The trace-fossil *Diplocraterion:* Evidence of animal–sediment interactions in Cambrian tidal deposits. *Palaios* 1:478–91.

Crimes, T. P., 1977, Modular construction of deep water trace fossils from the Cretaceous of Spain. *Journal of Paleontology* 51:591–605.

Crimes, T. P., and J. C. Harper (eds.), 1971, *Trace Fossils.* Geological Journal Special Issue 3, Liverpool, England, 547p.

Crimes, T. P., and J. C. Harper (eds.), 1977, *Trace Fossils 2.* Geological Journal Special Issue 9, Liverpool, England, 351p.

Droser, M. L., and D. J. Bottjer, 1986, A semiquantitative field classification of ichnofabric. *Journal of Sedimentary Petrology* 56: 558–9.

Ekdale, A. A. (ed.), 1978, Trace fossils and their importance in paleoenvironmental analysis. *Palaeogeography, Palaeoclimatology, Palaeoecology* 23:167–373.

Ekdale, A. A., 1988, Pitfalls of paleobathymetric interpretations based on trace fossil assemblages. *Palaios* 3:464–72.

Ekdale, A. A., and D. W. Lewis, 1991*a*, Trace fossils and paleoenvironmental control of ichnofacies in a late Quaternary gravel and loess fan delta complex, New Zealand. *Palaeogeography, Palaeoclimatology, Palaeoecology* 81:253–79.

Ekdale, A. A., and D. W. Lewis, 1991*b*, The New Zealand *Zoophycos* revisited: Morphology, ethology and paleoecology. *Ichnos* 1:183–94.

Frey, R. W., 1973, Concepts in the study of biogenic sedimentary structures. *Journal of Sedimentary Petrology* 43:6–19.

Frey, R. W. (ed.), 1975, *The Study of Trace Fossils.* Springer-Verlag, New York, 562p.

Frey, R. W., J. D. Howard, and W. A. Pryor, 1978, *Ophiomorpha:* Its morphologic, taxonomic, and environmental significance. *Palaeogeography, Palaeoclimatology, Palaeoecology* 19:199–229.

Hantzschel, W., 1975, *Trace Fossils and Problematica,* 2d ed. *Treatise on Invertebrate Paleontology, Part W. Miscellanea, Supplement 1,* C. Teicher (ed.). Geological Society of America and University of Kansas, Boulder, Colo., and Lawrence, Kans., 269p.

Hantzschel, W., F. el-Baz, and G. C. Amstutz, 1969, *Coprolites, an Annotated Bibliography.* Geological Society of America Memoir 108, Washington, D.C., 132p.

Hofmann, H. J., 1973, Stromatolites: Characteristics and utility. *Earth-Science Reviews* 9:339–73.

Lewis, D. W., 1992, Anatomy of an unconformity on mid-Oligocene Amuri Limestone, Canterbury, New Zealand. *New Zealand Journal of Geology and Geophysics* 35:463–75.

Lewis, D. W., and A. A. Ekdale, 1992, Composite ichnofabric of a mid-Tertiary unconformity on pelagic Oligocene Amuri Limestone, Canterbury, New Zealand. *Palaios* 7:222–35.

Logan, B. W., R. Rezak, and R. N. Ginsburg, 1964, Classification and environmental significance of stromatolites. *Journal of Geology* 72:68–83.

Logan, B. W., G. R. Davies, J. F. Read, and D. E. Cebulski, 1970, *Carbonate Sedimentation and Environments, Shark Bay, Western Australia.* American Association of Petroleum Geologists Memoir 13, Tulsa, Okla., 232p.

McCall, P. L., and M. J. S. Tevesz, 1982, *Animal–Sediment Relations: The Biogenic Alteration of Sediments.* Plenum Press, New York, 336p.

Meldahl, K. H., 1987, Sedimentologic and taphonomic implications of biogenic stratification. *Palaios* 2:350–8.

Miller, M. F., A. A. Ekdale, and M. D. Picard (eds.), 1984, Trace Fossils and Paleoenvironments: Marine Carbonate, Marginal Marine Terrigenous and Continental Terrigenous Settings. *Journal of Paleontology* (Special Issue) 58:283–597.

Monty, C. L. (ed.), 1981. *Phanerozoic Stromatolites.* Springer-Verlag, Berlin.

Pryor, W. A., 1975. Biogenic sedimentation and alteration of argillaceous sediments in shallow marine environments. *Geological Society of America Bulletin* 86:1244–54.

Seilacher, A., 1964, Biogenic sedimentary structures. In J. Imbrie and N. Newall (eds.), *Approaches to Paleoecology,* Wiley, New York, pp. 296–316.

Seilacher, A., 1967, Bathymetry of trace fossils. *Marine Geology* 5:413–28.

Walter, M. R., R. Buick, and J. S. R. Dunlop, 1980, Stromatolites 3,400–3,500 Myr old from the North Pole area, Western Australia. *Nature* 284:443–5.

Ward, D. M., and D. W. Lewis, 1975, Paleoenvironmental implications of storm-scoured, ichnofossiliferous mid-Tertiary limestones, Waihao District, South Canterbury. *New Zealand Journal of Geology and Geophysics* 18:881–908.

Wetzel, A., 1991, Ecologic interpretation of deep-sea trace fossil communities. *Palaeogeography, Palaeoclimatology, Palaeoecology* 85:47–69.

5
Texture of Detrital Sediments

Texture refers to the size and shape of grains (sphericity or form and roundness), grain surface features, and grain fabric (packing and orientation, and consequentially porosity and permeability). Pettijohn (1975, chapter 3) provides an excellent expanded discussion on the history of textural studies to 1975. We discuss here the principles of classification and interpretation for sediments comprising detritus derived from other rocks (igneous, metamorphic, and sedimentary); previous chapters in this volume also discuss interpretations involving texture, and **AS** Chapter 7 describes procedures for textural analysis of these sediments and sedimentary rocks. Textural studies of both modern and ancient sediments provide information both on the characteristics of the sediment, and by comparison with what has been observed in nature and in experiments, on the processes involved at the time of accumulation. In addition, of all the information that sedimentologists are asked to provide for project planning and environmental impact assessment, it is probably textural data and interpretations that are most frequently required. Hence, a sound understanding of sediment textures and their relationship to natural processes is essential for any sedimentologist involved in development and environmental management programs.

GRAIN SIZE

The size scale most commonly used for sediments by English-speaking geologists is the Udden 1898 grade scale as slightly modified by Wentworth (1922). This scale is geometric and based on a $\log_2$ scale: each boundary is one-half or twice the millimetre value of the next. To avoid dealing with fractions of millimetres and having to use semi-log graph paper, Krumbein (1934) introduced the more convenient ϕ (phi) scale, where phi = $-\log_2$mm, and each Udden-Wentworth class boundary is a full integer; the most abundant particles—those finer than 1 mm—have positive values. These scales are given in Fig. 5-1; see Tanner (1969) for a comparison with other size scales (and for a discussion of the advantage of this grade scale in sediment studies).

It is common practice to distinguish three major groups on the basis of size: *gravel* (grains larger than 2 mm in diameter), *sand* (grains 2.0–0.0625 mm in diameter), and *mud* (grains smaller than 0.0625 mm in diameter). In the mud fraction, major distinction is usually made between *silt* (0.0625–0.0039 mm), and *clay* (less than 0.0039 mm) because they behave differently in sedimentation (cohesive clays vs. noncohesive silts). Gravels mostly derive from blocky fracturing of the source rocks; sands largely result from the breakdown of rocks into their component crystals (numerous exceptions exist); and muds represent the finest products of disintegration (silt and clay size detritus) and chemical decomposition (clay minerals).

Classification System

Systematic description of rocks is necessary both for the orderly study of rock suites and for the clear and concise communication of observations to others. Many different classifications are in use and different criteria or limits are frequently applied to the same term. Hence, the system of terminology used must be specified in any report or paper, or confusion reigns! The system presented here for detrital sediment texture (Figs. 5-2, 5-3) is that of Folk, Andrews, and Lewis (1970). This classification system is adopted because it provides comprehensive and serviceable textural descriptions and its use is readily mastered.

Use of triangular graphs permits rapid and intelligible plotting of data for sample comparisons. The relative proportions of gravel, sand, and mud are plotted on one triangle; sand, silt, and mud are plotted on the other. Categories are named according to the relative proportions of the grain classes. Only loose sediment names have been given in the illustrations. When the sediment is indurated, substitute con-

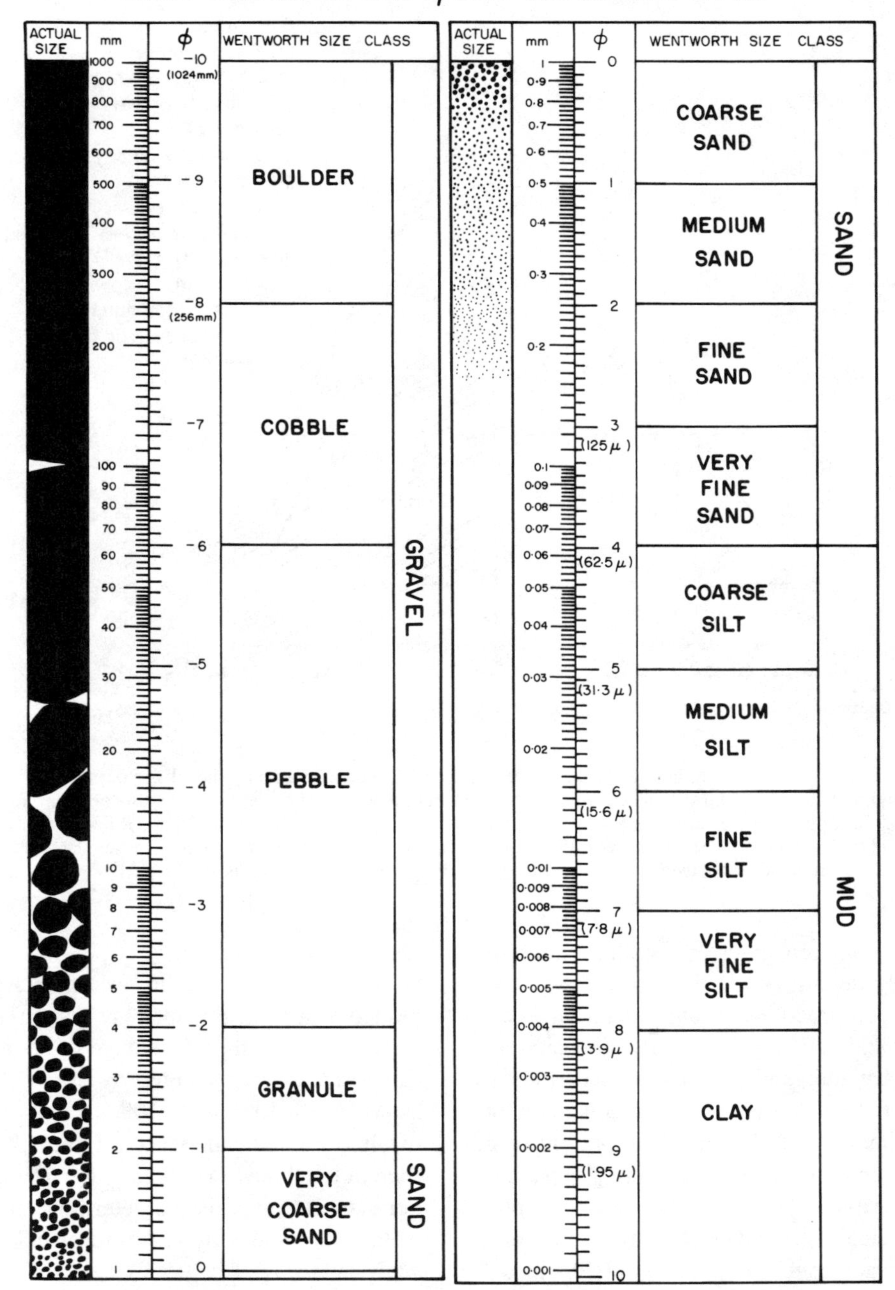

Figure 5-1. The Udden-Wentworth grade scale for grain sizes and the φ/mm conversion chart.

glomerate (or breccia, if particles are angular) for gravel, and substitute sandstone, mudstone, siltstone, and claystone for sand, mud, silt, and clay.

Gravel-Bearing Sediments

Gravel-bearing sediments are subdivided by the ratio of gravel:sand:mud; for sample comparison, percentages are plotted on a triangular diagram (Fig. 5-2), the apices of which represent 100% gravel, 100% sand, and 100% mud, respectively.

The textural class always should be determined in the field because it is difficult to collect representative samples of gravelly sediment for laboratory analysis. First determine the percentage of gravel; five categories are represented by tiers in the triangle: >80%, 30–80%, 5–30%, 0.01 (a trace)–5% (if any gravel at all is present, it is meaningful), and gravel-free. The last category is provided so that the small number of gravel-free samples in a suite of gravelly sediments may be plotted on the same diagram. Next determine the ratio of sand to mud. Boundaries at 9:1 and 1:1 sand to mud subdivide each of the middle tiers of the triangle into three named classes. The bottom (gravel-free) tier is subdivided at 9:1, 1:1, and 1:9 sand to mud to give four textural classes.

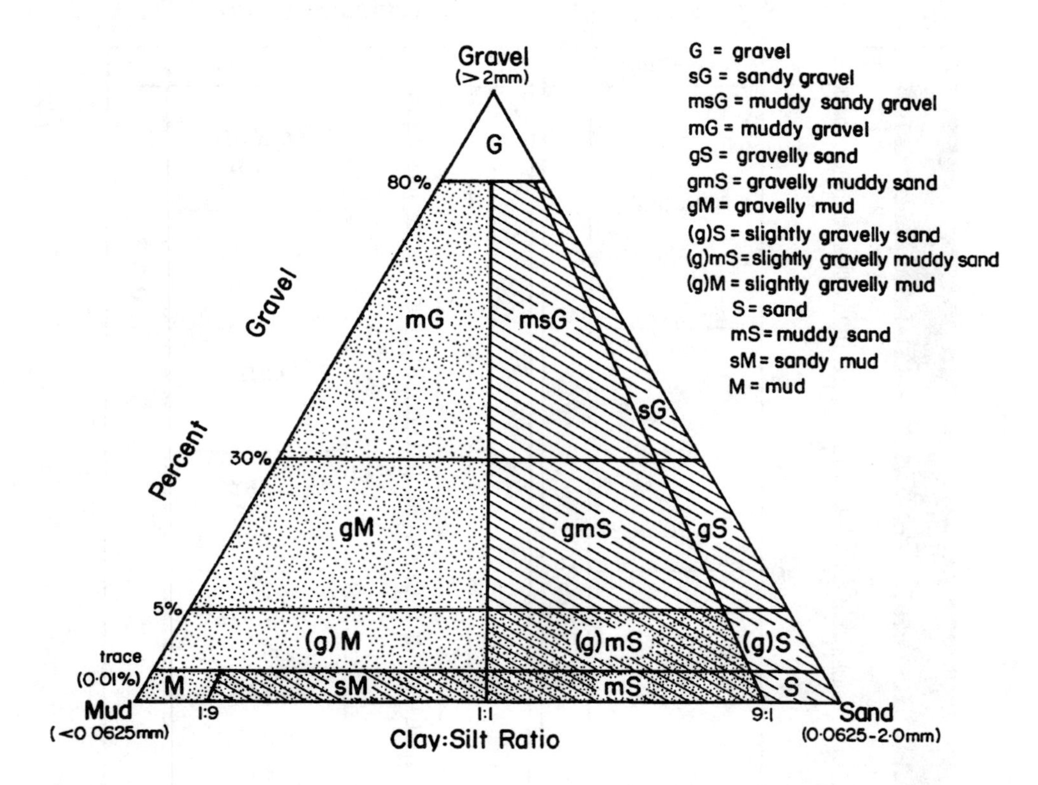

Figure 5-2. Textural terminology for gravel-bearing detrital sediments. Specify modal size of gravel throughout. Specify modal size of sand in cross-hatched area. Where practicable, substitute *silty* or *clayey* for *muddy* in stippled area. Specify *clast-supported* or *matrix-supported* before the textural name if matrix of sand or mud is abundant. If the sediment is lithified, use *conglomerate* (or *breccia* if most clasts are angular) for gravel, *sandstone* for sand, *mudstone* for mud. (After Folk, Andrews, and Lewis 1970.)

The *mode*, or most common grain size, is the grain size parameter most readily determined in the field. The modal size in the gravel and sand fractions always should be specified because it is valuable in determining gross depositional trends in a sedimentary unit and for provenance studies. Specification of the mode also serves to distinguish among the wide range of sediment types that may be represented in any one textural class, and it provides a far more precise impression of the size distribution characteristics. For example, both a granular coarse sandstone and a cobbly fine sandstone belong to the gravelly sandstone class, yet they are distinctly different and have quite different geological significance.

In specifying the modal size, the gravel and sand fractions are considered independently of each other. The modal grain diameter of each is expressed in terms of the equivalent Udden-Wentworth size grade, for example, boulder as in boulder conglomerate. The diameter of the modal size in the gravel fraction is always determined regardless of the proportion of gravel in the sediment. The modal size of the sand fraction is determined for the cross-hatched area of Fig. 5-2.

Where sand or mud is abundant, note whether the gravel is clast-supported or matrix-supported. Table 2-8 lists a commonly used shorthand notation in facies analysis of gravelly deposits. Laznicka (1988) presents a thorough discussion of features and associations of gravelly rocks (mainly breccias), as well as an alternative classification and approach to studying these rocks.

Gravel-Free Sediments

Gravel-free sediments and sedimentary rocks also can be plotted on a sand, silt, and clay triangular diagram (Fig. 5-3). First determine the percentage of sand in the sample; four categories are represented by tiers on the triangle: >90%, 50–90%, 10–50%, and <10% sand. Next determine the ratio of silt to clay; boundaries at 2:1 and 1:2 silt to clay subdivide each of the three lower tiers of the triangle into three classes. For sample comparisons, determine the weight percentages of the size grades by techniques such as sieve and pipette analysis (see **AS** Chapter 7).

The *mode* of the sand fraction is always specified. However, the sand fraction in many sediments contains not one dominant grain size but two abundant grain sizes with very few grains of intermediate size (e.g., Folk 1968; Taira and Scholle 1979). Such *bimodality* provides important clues to the history of a sand. In these cases, insert the word *bimodal* before the term *sand* and specify the two modes in terms of the equivalent Udden-Wentworth size grades. For example, instead of silty sand (zS), the name may be silty bimodal medium and very fine sand, and instead of gravelly sand (gS), the name may be pebbly bimodal medium and very fine sand. This procedure should be adopted only in the textural classes that are cross-hatched in Figs. 5-2 and 5-3. Bimodality in the gravel or the mud fraction does not have the same significance as in the sand fraction and usually does not need to be specified (exceptions exist).

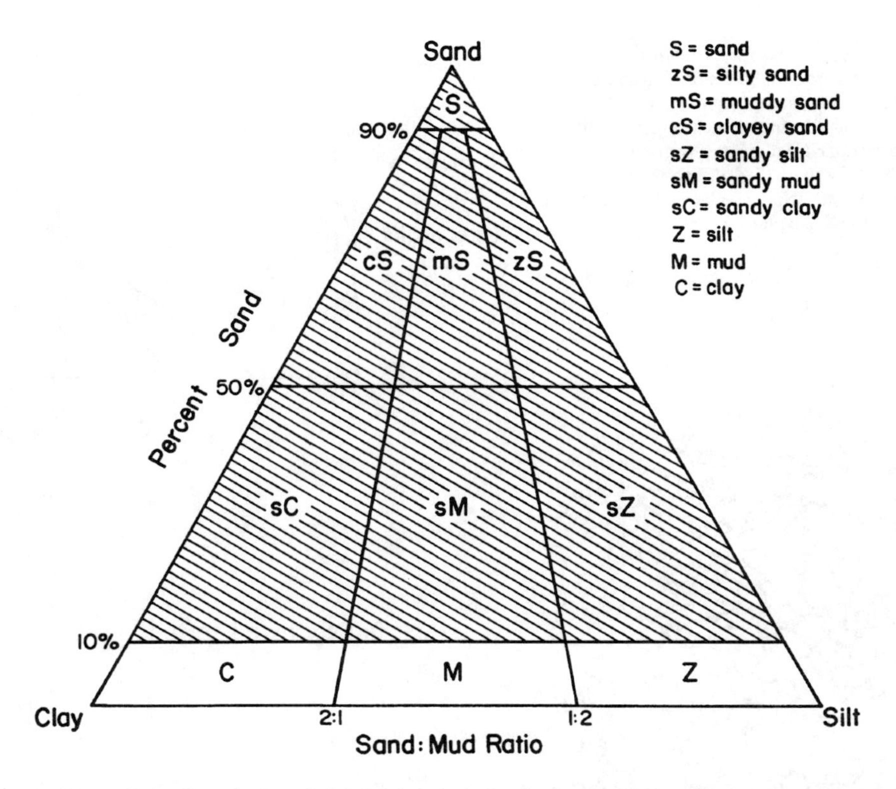

Figure 5-3. Textural terminology for gravel-free detrital sediments. Specify modal size of sand in cross-hatched area. If the sediment is lithified, substitute *sandstone* for sand, *mudstone* for mud, *siltstone* for silt, *claystone* for clay.

Usage

Because only a percentage *range* of a size fraction and not the precise percentage is needed to assign a sample to its textural class, the nomenclature may be applied in the field by using nothing more sophisticated than a hand lens and a grain size comparator (see **AS** Chapter 3). However, only analysis in the laboratory can give precise proportions for each size fraction and enable boundary-line samples to be assigned or samples plotted on the triangles.

Without laboratory analysis, resolution of the silt to clay ratio of the mud fraction is difficult. A good qualitative field test is to grind a tiny sample of mud between your teeth—if silt is present you will sense it (grit vs. smooth paste)! If both silt and clay are present, classify it as *mud*. After laboratory analysis, if the sediment has from one-third to two-thirds silt or clay, use the quantitative term *mud* (*sensu stricto*).

For many purposes a simple field description of sediment texture along the lines described above is sufficient, but when data are available from more refined textural analyses of sediments it is possible to assess the influence of depositional processes, and hence depositional environments, in much more detail. These more detailed approaches and the procedures for obtaining the necessary base data are described and discussed in **AS** Chapter 7.

The terminology of mudrocks has a complex history in the literature, with common usage of *shale* for indurated muds that may or may not show fissility (secondary characteristic imposed by compaction and/or tectonic deformation), and with terms such as *laminated* suggested as a major distinguishing feature between mudrocks (e.g., Flawn 1953, Lundegard and Sanders 1980; Spears 1980). We prefer to avoid the use of shale (=fissile mudstone) and suggest that nontextural characteristics be added as supplementary names to the basic texture term.

Sorting

Sorting, a measure of the range of grain sizes present, generally reflects energy levels in the depositional environment and the stability of those energy conditions over time. It is a particularly useful criterion with which to distinguish between categories of samples falling in the two sand classes S and zS of Fig. 5-3. Three classes of sorting are readily determinable in the field on the basis of the central two-thirds of the grain size range: if that majority of the grains falls within less than the equivalent of one Udden-Wentworth size grade, the sample is *well sorted*; if it ranges over the equivalent of one to two Udden-Wentworth size grades, the sample is *moderately sorted*; if it ranges over more than the equivalent of two Udden-Wentworth size grades, the sample is *poorly sorted*. These three sorting classes are consistent with detailed statistical classifications of standard deviation in grain size for the grain population (see **AS** Chapter 7). As a note of warning for sedimentologists who deal with engineers, when engineers refer to a well-

Figure 5-4. Visual comparison diagrams for estimating sorting in thin section. Figures were prepared from mixtures of Pleistocene dune sands with rounded quartz grains, sieved at 1/4ϕ intervals, blended to produce the distributions illustrated, then cemented into epoxy-resin blocks that were sectioned. Silhouettes from the sections are illustrated at critical intervals for the Folk (1951) textural maturity scale (see subsection of text). (Sigma values are graphical standard deviation measures, see **AS** Chapter 7.)

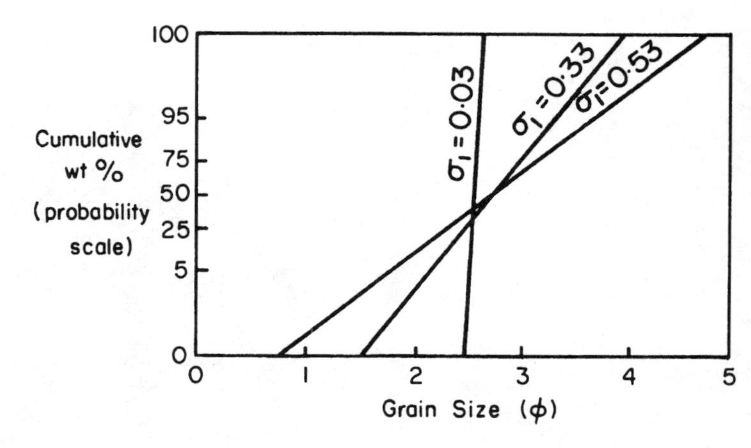

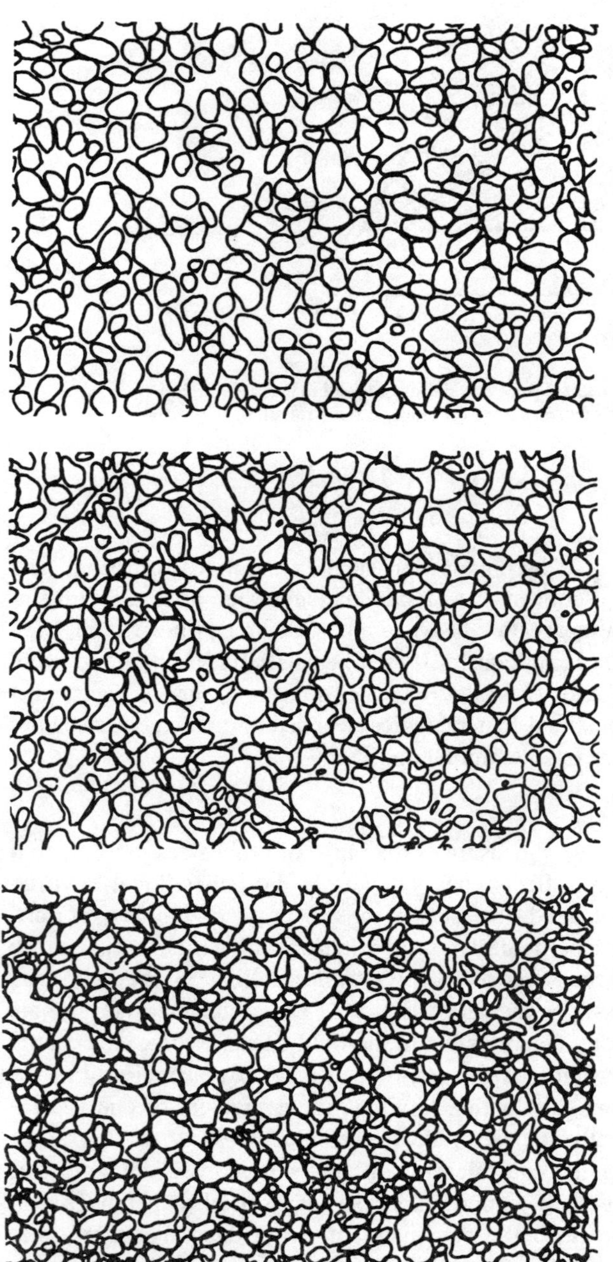

$\sigma_1 = 0.0$ very well sorted

$\sigma_1 = 0.35$

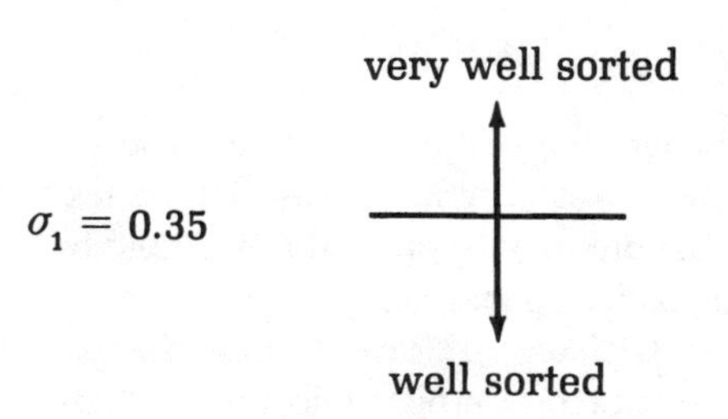

$\sigma_1 = 0.5$

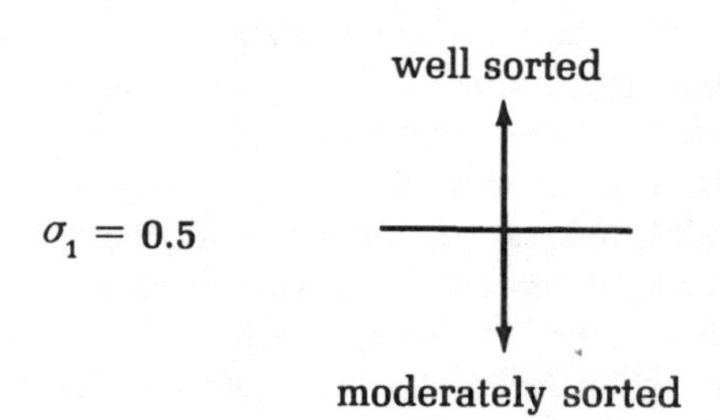

sorted or well-graded sediment, they mean that it has a wide range of grain sizes (more suitable for aggregate)—the opposite of what it means to the sedimentologist!

In determining the size sorting of sand, there is one exception to the above procedure. Some bimodal sands that would be considered poorly sorted as a whole consist of two well-sorted modes. Such sands are well-sorted bimodal sands, as in well-sorted bimodal medium and very fine sand.

In thin sections prepared for rock and soil studies (**AS** Chapter 5) there is a common tendency to underestimate the sorting of sands, because the plane of the section does not cut through the centers of most grains. Accurate grain size analyses are generally impossible, but consistent estimations are possible with experience. Figure 5-4 provides a standard for comparison to help distinguish between the better-sorted sands in thin section.

Except for some granular sediments, gravel-bearing sediments are usually poorly sorted because the high-energy conditions necessary to transport the grains are usually short lived, and as the flow velocity drops rapidly, the fine grains fall out together with the coarse grains. Exceptions exist in which a narrow range of grain sizes was available initially, slightly lower-energy conditions persist after the main depositional event and winnow finer grain sizes, or energy input remains consistently high (e.g., the bed of some rivers and on beaches).

Interpreting Grain Size and Sorting

In conjunction with information derived from other textural attributes (such as shape and roundness), compositional attributes (such as the range of specific gravities of the particles), sedimentary structures, and the vertical (and lateral) stratigraphic relationships, grain size analyses provide powerful guidelines for interpreting depositional conditions. As a generalization that holds true for processes in most depositional environments, the mean grain size in a deposit is largely a function of the energy of the processes controlling transport and deposition: particles are segregated according to their hydrodynamic behavior (e.g., Middleton 1976), which largely depends on their size (although specific gravity and shape are also very influential). In contrast, the degree of sorting of grains in the deposit is a function of the persistence and stability of those energy conditions. The principal exception to this general rule applies when the size of grains that can be deposited is constrained by the size of grains available in the environment (i.e., however high the energy conditions, if no pebbles are available, none will be deposited!). The size distribution in a sediment may also be strongly influenced by source rocks (e.g., a source of sediments already sorted).

Sandy beaches are almost always composed of well-sorted, matrix-free sands because waves continuously winnow fines away, and either concentrate coarse or heavy grains in the highest-energy locales or bury them (exceptions generally indicate rapid sediment supply to a beach, as when shorelines are rapidly retrograding). Most sandy floodplain deposits, in contrast, are less well sorted and contain some muddy matrix because river energy levels fluctuate and deposits are not continuously reworked (exceptions occur with well-sorted sediments in river systems where floodplains are very long). Deposits from turbidity currents tend to be poorly sorted and contain abundant matrix because of rapid sedimentation under low-energy conditions.

Minimum depositional conditions for traction loads of aqueous currents can be inferred from known relationships between flow conditions and sediment size (e.g., the Hjulstrom diagram, Fig. 3-11). However, for most deposits such relationships cannot be used in a rigorous quantitative sense because too many influential variables are unknown (for instance, local availability of particle sizes, volume and depth of the transporting current, degree of turbulence, frictional resistance to flow of the substrate, and total quantity of sediment that was being transported). When the competency (ability to carry coarse, heavy grains) and capacity (ability to carry volume of particles) of currents are high, the coarsest grains that can be transported travel with all the finer particles; if capacity and competency are rapidly lost (that is, if the driving force of the current wanes abruptly), all the sizes will accumulate together. On the other hand, if the capacity and competency wane gradually, grains of only one size will accumulate at one time at any one locality—the finer particles will be carried on. Hence, deposits of floodplains may be well sorted locally, and sorting improves upward in turbidites. These principles can be applied when interpreting trends in sediment characteristics (e.g., Fig. 5-5, and see McLaren and Bowles 1985).

Size distributions are also influenced by the previous history of the sediments—for example, eolian deflation of the fine particles in an earlier stage of the sediment history may leave only coarse grains for transport by a high-energy process. In addition, postdepositional processes may modify the size distribution of a deposit; examples are the biogenic mixing of initially well-sorted layers (e.g., Faas and Nittrouer 1971), the pedogenetic infiltration of colloids and clays (e.g., Brewer and Haldane 1957), and the production of clayey matrix by diagenetic breakdown of unstable minerals (e.g., Whetton and Hawkins 1970).

Much effort has been applied over the past 50 years to the distinction between sedimentary environments on the basis of quantitative grain size analysis (discussed more fully in **AS** Chapter 7). Quantitative analysis serves to characterize grain size distributions precisely and permits statistical distinction of similarities and differences between samples; thus, it can be very effective in relatively local studies where deposits of two or more environments are represented. However, the distinction is actually between *processes* of sedimentation, not environments, and the same processes may operate in different settings. Hence there are no universal rules and relationships that can be applied to all detrital sediments. There are considerable reservations among most sedimentologists today about the utility of grain size analyses from single sedimentary units. The most effective use of

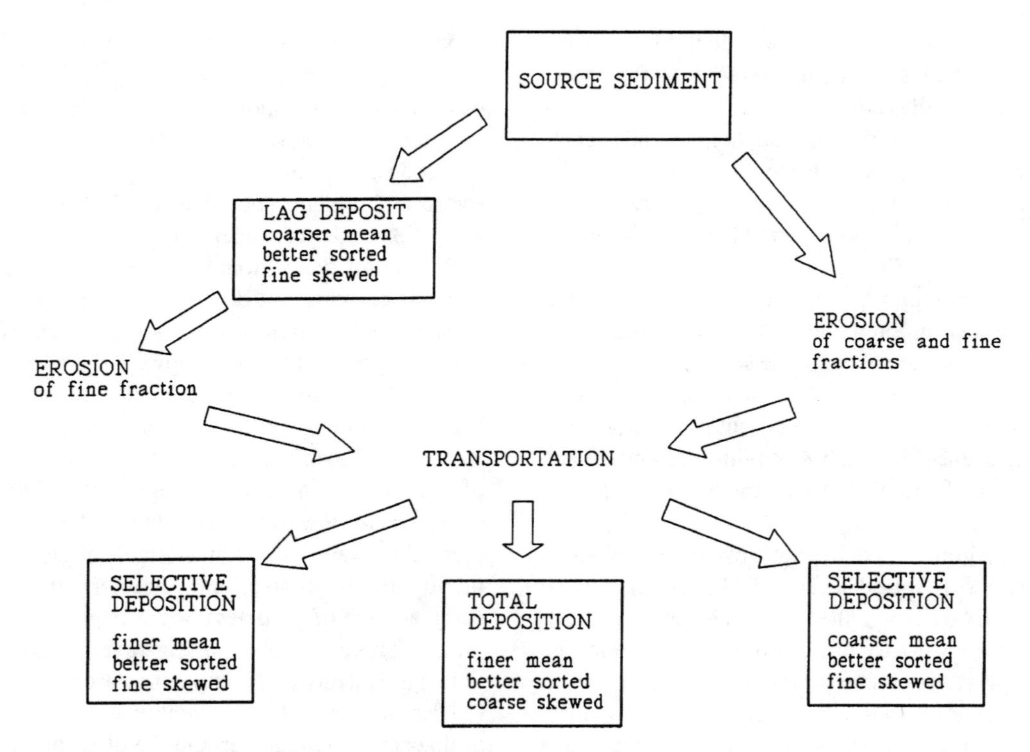

Figure 5-5. Trends in grain texture with transport and deposition. Mean is the arithmetic average of the grain size distribution and skewness is a measure of whether or not the distribution has a non-Gaussian normal tail—either fine or coarse—to the grain size distribution. (After McLaren and Bowles 1985; see that publication or **AS** Chapter 7 for discussion of statistical terms.)

textural studies eventuates when *trends* can be detected—laterally between contemporaneous units or vertically in a sedimentary sequence.

Despite an abundance of studies of the "size" (actually, settling velocity) characteristics of muds, experimental conditions are usually so different from natural conditions that they have little value in interpretation of depositional process. Individually, clay particles settle so slowly that *any* current with an upward component will prevent deposition. However, in natural conditions most clays probably aggregate as fecal pellets (e.g., Haven and Morales-Alamo 1968; Prokopovich 1969) or flocs and aggregates (e.g., Pryor and Vanwie 1971; Feely 1976) and accumulate as relatively coarse particles. Dispersion and deflocculation procedures used in the laboratory for size analysis of muds (see **AS** Chapter 5) thus alter the size distribution of the original sediment and may provide an entirely misleading picture of the depositional setting and processes. Furthermore, muds in compacted sediments, or sedimentary rocks, have developed diagenetic textures and compositions that may be quite unlike those of the original sedimenting population. Nonetheless, size analyses of modern muds, particularly when carried out under natural conditions, are necessary and useful in studies of suspension load transport, particularly because clay minerals adsorb and transport ions and organic complexes that are crucial in dictating the chemical characteristics of environments.

No widely accepted standards exist for quantitative textural studies of conglomerates; because of sampling difficulties, relatively few attempts have been made to study the size distributions (but see Landim and Frakes 1968).

ROUNDNESS

The roundness of gravel and sand clasts should be noted in any rock description, although it is generally not mentioned in the short name for a detrital sediment. Roundness indicates the extent of abrasion the grains have undergone; it reflects overall transport history. Roundness does *not* necessarily reflect the distance the grains have traveled from their source: rounded grains may have been derived locally from a sedimentary rock, or may have been extensively abraded in an environment near the source, such as a beach adjacent to a cliff. Because waves provide consistently high-energy conditions, the active wave zone of beaches is the most effective subaqueous setting for physical rounding of sedimentary grains. Aeolian reworking in beach and inland dunes is even more effective in rounding sand grains because water tends to "muffle" some of the abrasion processes (e.g., impact). In some soils, chemical action rounds grains (e.g., Crook 1968).

Quantitatively, true roundness is generally expressed by the Waddell formula:

$$\text{roundness} = \Sigma \frac{(r/R)}{N}$$

where r is the radius of curvature of grain corners, R is the

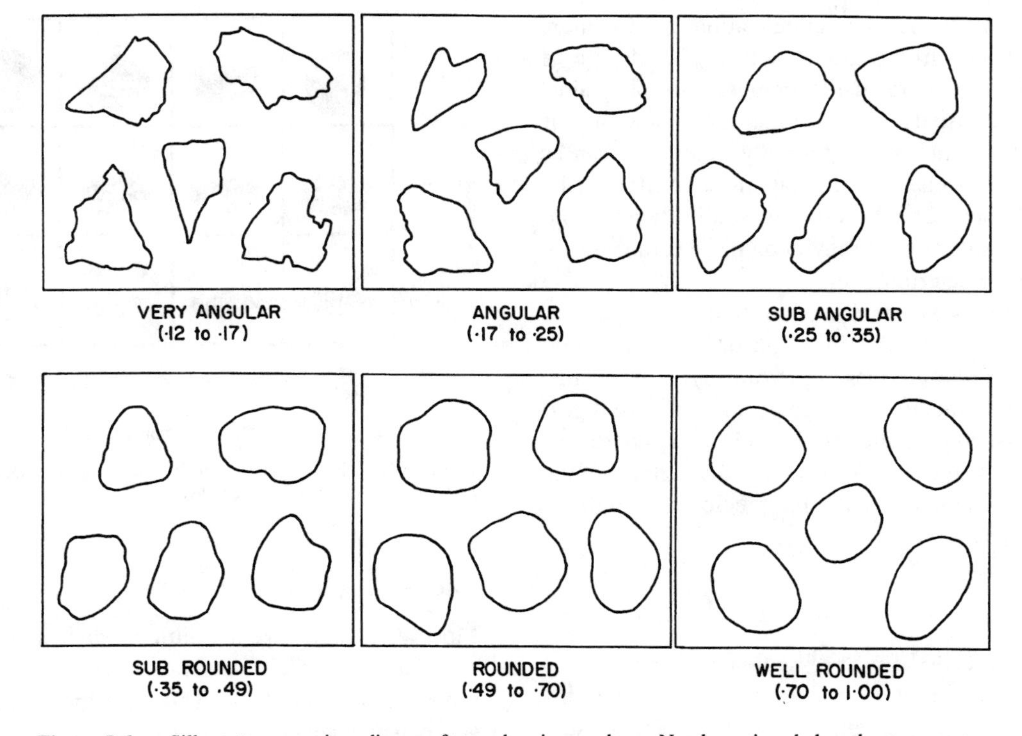

Figure 5-6. Silhouette comparison diagram for sand grain roundness. Numbers given below the names are the quantitative limits to the classes according to the Waddell formula. The boundary between subangular and subrounded corresponds to the borderline between the mature and supermature stages of textural maturity (Folk 1951 and see subsection of text). (After Shepard and Young 1961)

radius of the largest inscribed circle, and *N* is the number of corners. See **AS** Chapter 7 or Barrett (1980) for discussions of practical quantitative measurement procedures.

Unless quantitative analysis is justified by the likely results, practical measures of roundness rely on visual comparison with standard silhouette charts (e.g., Fig. 5-6 for sand grains and see Krumbein 1941 for pebbles). For consistency, silhouettes from which such measurements are made should be taken from grains oriented to show maximum projection sphericity—that is, viewing the plane that includes the long and intermediate axes (easily achieved in studies of loose sediment by gently shaking the tray on which the grains are lying).

Roundness of grains within a sediment will commonly vary even if all grains have been subjected to the same history of abrasion (or solution). Various minerals and rock fragments differ in their physical and chemical properties, e.g., hardness, brittleness, type of internal anisotropism, and solubility. Hence, in comparing samples it is necessary to contrast roundness of the same type of component. Quartz is most commonly used in sandstones because it is abundant, hard, and has relatively isotropic physical and chemical properties. In addition, grains of different size round at different rates as a result of the greater effectiveness of the abrasion processes on coarser grains. Hence, to discover trends among the samples, the same grain size should be examined. Additionally, in any assemblage of grains the degree of rounding is expected to decrease with decreasing grain size. If the coarse quartz grains in a sample are less rounded than many of the fine quartz grains, it is clear that most of the coarse fraction has a different history from the bulk of the fine fraction, irrespective of the grain size distribution of the assemblage. In rare sediments, mixing of grains with different abrasion histories can result in assemblages that are bimodal with respect to roundness. Furthermore, even if the grain assemblage as a whole has the same history, individual grains within the assemblage will have been subjected to somewhat different degrees of abrasion or solution (e.g., in high-energy environments some grains may be fractured by collisions or by being ground between larger grains to produce "broken rounds"). Thus, it is commonly necessary to record the range in roundness as well as the average, and to note any differences correlated with size or composition.

Interpreting Roundness

Roundness is the least indicative texture of depositional environment; it requires a long history of abrasion to round detrital grains substantially. In sedimentary deposits, the two most important characteristics to look for are (1) trends of changing roundness vertically or laterally in a succession, which may indicate respectively the rate of supply of sediments from the source (with consequent tectonic implications) or the direction of transport, and (2) unexpected differences within the local population (such as marked differences in roundness between grains of the same size and composition; finer grains with higher degree of roundness than coarser grains of the same composition; harder, more brittle, or more anisotropic grains with higher degree of roundness than softer, more ductile, or more isotropic grains) that will indicate unusual histories or multiple sources for the sediment. In conjunction with sphericity, roundness can be useful in environmental discrimination (e.g., Krumbein, 1941), although there are relatively few studies in which interpretations rely sufficiently heavily on these parameters to justify quantitative determinations.

An origin that must be considered for some small rounded pebbles found in clusters either in marine or nonmarine environments is as "gizzard stones" (*gastroliths*)—some birds, reptiles, and marine mammals utilize pebbles to assist digestion, and the constant grinding action produces a high roundness. Unfortunately, most workers require the skeleton of the organism to be found in close proximity before they will accept such an explanation!

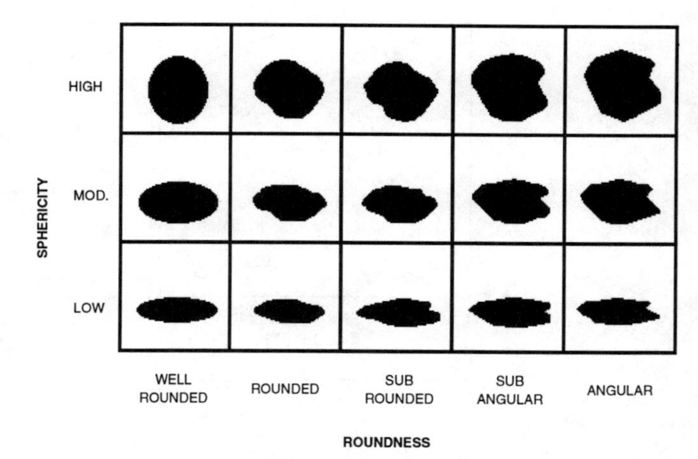

Figure 5-7. Representation of the relationship between sphericity and roundness of sedimentary particles.

SPHERICITY

Shape (sometimes called form) of sedimentary particles is conceptually very different from size, but practically it is difficult to separate their measurement. Sphericity is a measure of how nearly equal the axial dimensions of a particle are—a very different concept to roundness, but commonly confused with that shape attribute (Fig. 5-7). *True sphericity* is the surface area of a grain divided into the surface area of a sphere of the same volume—a rather impractical property to measure! Commonly, sphericity is defined by

$$\text{sphericity} = [(I \cdot S)/L^2]^{1/3}$$

where I is the length of the intermediate axis of a grain, S is the length of the short axis of the grain, and L is the length of the long axis of the grain. The two most widely used verbal classifications and quantitative graphical representations of sphericity are shown in the Zingg diagram and the Sneed and Folk form diagram, both of which are based on ratios of I, S, and L (Fig. 5-8); see also **AS** Chapter 7 and/or Barrett (1980) for practical quantitative measures. Silhouette charts are available for visual comparison in qualitative studies (e.g., Rittenhouse 1943).

Sphericity is strongly influenced by physical anisotropism of the rock and mineral particles (e.g., Moss 1966); hence comparisons must be between clasts with the same structure, texture, and composition. Because gravels are composed of particles that differ substantially in these properties, their sphericity can usefully be included in sediment descriptions. In contrast, sphericity of sand grains is generally not reported unless there is an obvious and abundant population of inequant particles. Sphericity cannot be reliably estimated in two-dimensional views (e.g., in thin section), but hand specimens of indurated rocks should be examined to determine whether a preferred fabric has been imparted by inequant grains; if so, size measurements in thin section may be strongly biased.

Interpreting Sphericity

Most workers believe that little fundamental modification to sphericity occurs after derivation from the source; inherited properties such as internal isotropism are thought to dictate the shape. Nonetheless, particular gravel clast shapes concentrate in particular environments (e.g., discs on beaches, rollers and blades in rivers; e.g., Luttig 1962; Dobkins and Folk 1970; Hobday and Banks 1971; Gale 1990) either because of selective sorting or because of actual modification to grain shape by environmental processes. However, to date insufficient work has been performed to provide firm worldwide generalizations, and sphericity studies always should be integrated with other data (e.g., Goede 1975). An exception is the occurrence of wind-worn shapes of pebbles, ventifacts like *dreikanter* (pebbles with two distinctive sandblasted facets), which are highly diagnostic of eolian environments, and some gastroliths, the stomach or "gizzard" stones of various birds, mammals, and reptiles (e.g., Whitney and Dietrich 1973).

Detailed sphericity analysis of loose sand grains also can prove useful in determination of paleohydrodynamic conditions (see Moss 1972), particularly when combined with studies of size distributions and sedimentary structures (see also Mazzullo and Magenheimer 1987). Certainly both sphericity and roundness affect the behavior of sediment during transport (e.g., Williams 1966).

SURFACE TEXTURE

Surface textures of gravels, determined by visual inspection, can be useful. Glacial processes (and gravitational mass movements) may produce striations, chatter marks, or facets; wind may etch surfaces; beach reworking may produce chink facets; fluvial action may produce percussion marks and spalls; and differential chemical solution (sometimes assisted by pressure at grain contacts) can produce pits.

Detailed study of the surface features of detrital sand grains requires use of high-resolution analytical equipment such as the scanning electron microscope (SEM, see **AS**

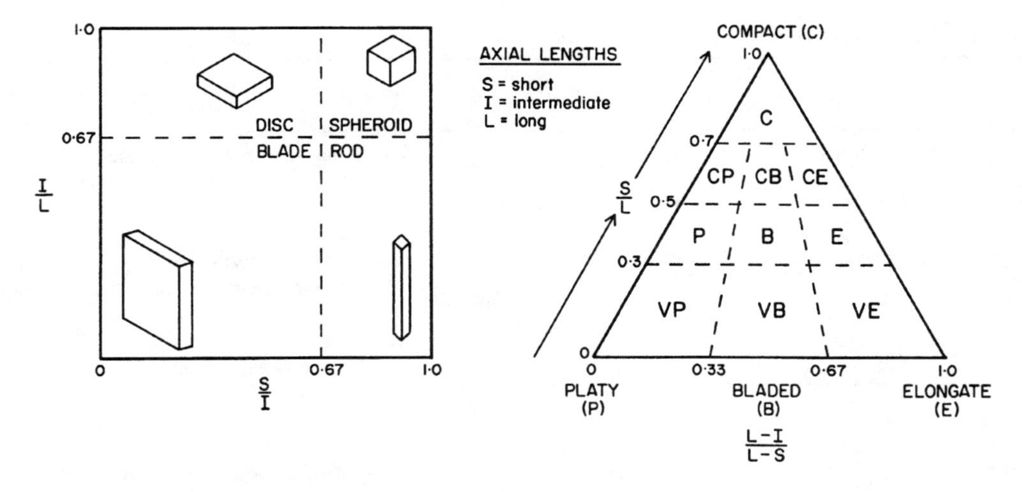

Figure 5-8. Two systems for representing particle shape. *A:* Sphericity classification after Zingg (1935, in Pettijohn 1975). *B:* Sphericity form diagram after Sneed and Folk (1958).

Chapter 7). Various polishes or surface pits and protrusions appear to be caused by different processes, and these processes may be characteristic of certain environments. Hence the surface textures may be used to infer the history of grains or grain assemblages (e.g., Krinsley and Doornkamp 1973; Baker 1976; Whalley and Krinsley 1974). However, there are many complications and limitations to these studies (e.g., Culver et al. 1983), and results are often equivocal; for example, textures may be inherited from the earlier grain history, or early diagenetic processes can substantially modify or obscure them (e.g., Passaretti and Eslinger 1987). Representative sampling is also a major problem.

FABRIC

In gravel deposits, packing and orientation of grains can be studied relatively easily, although problems arise with measurements in well-indurated conglomerates. The most obvious fabric is *imbrication*, where nonspherical (generally discoidal) particles are aligned and overlap each other in an imbricate pattern. It is also important for interpretive purposes to note whether framework clasts are clast or matrix supported, have a preferred orientation (of long axes, or of the planes containing the intermediate and long axes), or show grading (normal, inverse, or symmetrical). Beware of deformation fabrics produced after deposition (even compaction in some cases may produce a preferred orientation).

Fabric of sand grains has been little studied because of the difficulties involved in sampling loose modern sediments and in determining any fabric in three dimensions. However, some useful orientation studies have been made in sedimentary rocks (see Pettijohn 1975), and any obvious fabric should be noted not only because it could prove informative but because it may affect the apparent size distribution of sediments in thin section. Packing of grains (e.g., Fig. 5-9) controls sediment porosity and permeability and the way in which sediments respond to stress. The property of *thixotropy*, which accounts for "quick" sands and muds, results from a relatively open packing array attempting to repack to a closer array more rapidly than interstitial water can escape (grains effectively float in the water, which has no shear strength). *Dilatancy* is the reverse situation, where repacking under a new stress (e.g., foot pressure) moves toward a more open array and water is drawn into the pores (wet sand dries out around the foot).

Porosity is a measure of the total volume between the grains in a sediment or rock; *permeability* is a measure of the interconnectedness of the pores and thus the relative ease of movement of pore fluids. These properties are largely consequential to packing (e.g., Graton and Fraser 1935; Beard and Weyl 1973) and strongly influence diagenetic cementation and dissolution (e.g., Table 5-1). Methods for estimating or quantitatively measuring these parameters are discussed in **AS** Chapter 7.

Fabric of the clayey components in sedimentary rocks also has been little studied by petrologists. Soil scientists have given more attention to this attribute; the descriptive terminology given in Table 5-2 lists part of the much broader system presented by Brewer (1964).

With the development of relatively inexpensive and easily used texture goniometers for X-ray diffraction instruments,

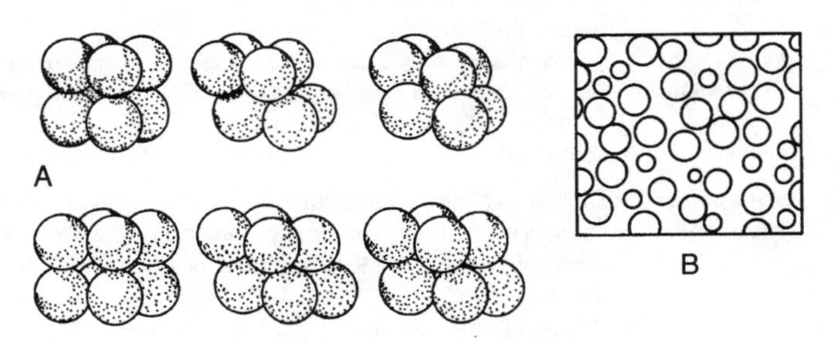

Figure 5-9. *A:* Ideal packing configurations for equal-size spheres. Upper left represents cubic (*open*) packing; lower right rhombohedral represents (*closest*) packing. *B:* Random cut through open-packed arrangement of equal-size spheres. (Both after Graton and Fraser 1935.)

Table 5-1. Some General Types and Origins of Porosity in Sediments

Porosity: Primary	Porosity: Secondary	
Intergranular—between grains; may be submicroscopic pore throats	Fractures	Grain—during compaction, tectonic shocks Rock—during tectonic deformation, pore-fluid over pressuring
Intragranular—within grains such as fossils		
Shelter—below bridges formed between other grains by platy particles	Dissolution	Grains Partial corrosion Complete corrosion (e.g., "moldic"—after skeletal grains) Cement Matrix Stylolites—pressure-solution

the assessment of preferred orientations in sedimentary rock slabs, particularly where very fine-grained mineral components are involved, is becoming more efficient. As a result of this technology, it is likely that assessment of preferred orientation fabrics in sedimentary rocks will be carried out more widely in the future and may produce findings of general applicability.

Interpreting Fabric

Fundamental studies of detrital sediment fabric are rare (but see Martini 1971). Pebble imbrication indicates the direction of the depositional agent (e.g., Rust 1972). In fluvial and marine current deposits, pebbles dip upcurrent; on beaches they dip seaward, and in some gravity flow deposits they dip upslope (see also Walker, 1975; Lindsay 1968; Major and Voigt 1986). Pebble fabrics in glacial sediments have received some attention (e.g., Dowdeswell and Sharp 1986), and apparently preferred orientations of empty shells have intrigued workers (e.g., Jones and Dennison 1970; Clifton 1971). Current lineation fabric in sandstones appears as a streakiness or parting lineation on bed surfaces and indicates the orientation of the depositional current (e.g., Shelton and Mack 1970). Porosity and permeability, together with other

Table 5-2. Terminology for Clayey Fabric in Sediments

Fabric Distribution Patterns Overall	Individual Particles	Relation to Reference Feature
Porphyroskelic (dense groundmass in which grains are set)	Random	Unrelated
	Clustered	Normal (perpendicular to . . .)
	Banded	Parallel
Intertexic (grains touching or linked)	Radial	Inclined (state angle to . . .)
	Concentric	
Granular (no true matrix)		Cutanic (concentrations forming a unit mass adjacent to . . .)

Fabric Orientation Patterns (note magnification used)

Individual Particles

Oriented:	Individuals sufficiently oriented that there is roughly continuous birefringence under crossed nicols (UXN); with rotation of the stage, dark extinction lines or bands move across the aggregation of individuals. Can be subdivided into strongly, moderately, or weakly oriented, although attitude of thin section relative to fabric may influence apparent degree of orientation.
Unresolved:	At magnification used, appears to be some anisotropism within aggregates that are too small to observe in detail; UXN the mass appears to have random fabric.
Unoriented:	At magnification used, fabric is isotropic due to apparent random orientation of individual particles.
Indeterminate:	Fabric appears isotropic because of opacity or crystallographic character.

Domains (aggregations of clayey material, each aggregation with a preferred orientation)

Asepic fabric:	Dominantly anisotropic, with anisotropic domains unoriented with respect of each other.
Sepic fabric:	Various recognizable anisotropic domains with various patterns of preferred orientation (see Brewer 1964 for subdivisions).
Undulic fabric:	Practically isotropic at low magnifications, and weakly anisotropic with faint undulose extinction at high magnifications. Domains not distinct.
Isotic fabric:	Apparently isotropic matrix at all magnifications.
Crystic fabric:	Anisotropic fabric involving recognizable crystals deposited during pedogenesis or diagenesis (see Brewer 1964 for subdivisions).
Strial fabric:	Clayey material as a whole exhibits preferred parallel orientation, giving extinction pattern that is either unidirectional or shows several preferred extinction directions. Common in sedimentary rocks, probably imparted by compaction and other diagenetic processes. May be superimposed on other fabrics, which should be sought when strial fabric is at extinction.

Source: After Brewer 1964.

fabrics imparted by postdepositional processes such as compaction, can provide considerable information on the diagenetic history of sediments (e.g., Maxwell 1964; Rittenhouse 1971). Mud fabrics can provide information on depositional and diagenetic conditions, as demonstrated by a number of publications (e.g., Helig 1970; O'Brien, Nakazama, and Tokuhashi 1980; O'Brien 1987; Kuehl, Nittrouer, and DeMaster 1988).

TEXTURAL MATURITY

The "expected" progression of several textural attributes as more energy is expended on detrital sands with time is provided by a scale of textural maturity devised by Folk (1951; Table 5-3). (See Fig. 5-4 for the critical boundary between "moderately" and "well"-sorted sands in thin section views, and Fig. 5-6 for the critical boundary for "rounded" grains in this scheme.)

The textural maturity stage attained by any sandstone provides a basis for initial interpretation of the sediment history. Inversions or deviations from the listed characteristics are common—perhaps even the rule. However, the principal value of the scheme lies in the recognition of such deviations, whose existence implies that something unusual has happened and requires an explanation for the cause. Two examples are (1) rounded grains in poorly sorted deposits, possibly due to mixing of originally well-sorted layers after initial deposition by storms or bioturbation, or a supply of rounded grains from a sedimentary source rock; (2) well-sorted grains with a clayey matrix, perhaps finally deposited after an energetic history in a very low-energy environment, such as sand dunes or a barrier bar migrating into a lagoon, or due to the pedogenetic infiltration of clays, or due to storm mixing or bioturbation of once segregated layers. Other inversions (such as bimodal grain roundness) and other explanations for them can be imagined. After imagining the alternative possibilities, you will know what additional data to seek from the sediment to explain the cause. As a rule, you should class the sediment according to the lowest stage of textural maturity that it shows (thought to be indicative of the last set of processes acting on it), qualified by the kind of textural inversion present.

Figure 5-10 illustrates the range of textural maturity that is most commonly represented in various environments. Low-energy environments, where winnowing and selective sorting during deposition are not active processes, tend to have a low textural maturity (and the converse), but there are exceptions where, for example, the available grain sizes are limited or wind has deflated the clays.

Table 5-3. Stages of Textural Maturity

Immature	> 5% clays	poorly sorted	grains angular
Submature	< 5% clays	poorly sorted	grains angular
Mature	< 5% clays	well sorted	grains angular
Supermature	< 5% clays	well sorted	grains rounded

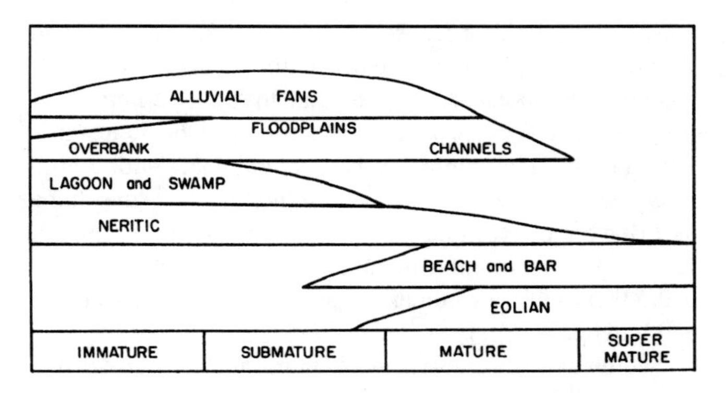

Figure 5-10. Expected relationships between environments and textural maturity. Exceptions exist, particularly if the sands are recycled. The exceptions (textural inversions) are important aids to thinking about the origin of the sediment. (Modified from a sketch by R. L. Folk.)

SEDIMENT TEXTURE IN ENVIRONMENTAL IMPACT ASSESSMENT

Because the texture of a sediment is a function of the particular type, magnitude, and persistence of the processes that acted to produce it, any change in the relationship between a sediment and the processes it experiences is likely to result in reworking of the sediment until a new equilibrium relationship is achieved. For example, if muddy sediment is dredged from calm harbor waters and dumped into a higher-energy, sand-dominated area outside the harbor entrance, the fine sediment will be reworked, and unless the prevailing current conditions have been calculated when selecting the dump site, it is possible that the dumped sediment will wash straight back into the harbor. Similarly, if fine sand is added to a beach consisting of well-sorted medium sand as part of a beach nourishment program, it should be expected that the fine sand will rapidly be winnowed away, since the wave and current regime remain unchanged: the beach-nourishment operation will have achieved nothing. In situations where this type of sediment remobilization is likely, it is necessary to assess hydrodynamic conditions to determine where the remobilized sediment is likely to be redeposited, as well as how the area from which the sediment was removed will change in its own profile of equilibrium.

In addition, the physical disturbance of a sediment caused by subaqueous dredging and dumping operations almost always produces a plume of fine resuspended particles that will move with the prevailing currents until the fine particles settle out. The rate at which these particles settle out, hence the persistence of the plumes, will be determined largely by the grain size distribution in the dredged sediment and the existent hydrodynamic conditions. Because a persistent high-suspension load can have an adverse impact on aquatic ecosystems, as well as an aesthetic impact, planning for dredging operations must include an assessment of the sediment texture in relation to the water depth and hydrodynamic conditions at potential dump sites. Furthermore, because most pollutants are bound to the finest sedimentary particles (clays and colloids because of their high surface area and high charge to mass ratio), redistribution of any pollutant elements or compounds can accentuate the adverse impacts of the plume itself (or the plume may remobilize the pollutants). For example, if fine-grained harbor sediments with a high heavy metal load are dumped where currents cause the plume to drift over an oyster culture site, even if the oysters are not killed, bioaccumulation of metals from the sediment may make them unfit for human consumption (e.g., McConchie and Lawrence 1991).

Sediment textures, particularly fabric and grain size distribution, also have a major influence on the drainage characteristics of a sediment or soil. This influence needs to be evaluated in a wide range of civil engineering projects where water retention could lead to site instability, or where water retention is required (e.g., in reservoirs).

SELECTED BIBLIOGRAPHY

General

Bagnold, R. A., 1968, Deposition in the process of hydraulic transport. *Sedimentology* 10:45–56.

Bowler, J. M., 1973, Clay dunes: Their occurrence, formation and environmental significance. *Earth-Science Reviews* 9:315–38.

Brewer, R., and A. D. Haldane, 1957, Preliminary experiments on the development of clay orientations in soils. *Soil Science* 84:301–9.

Faas, R. W., and C. A. Nittrouer, 1976, Post depositional facies development in the fine grained sediments of the Wilkinson Basin, Gulf of Maine. *Journal of Sedimentary Petrology* 4B:337–44.

Feely, R. A., 1976, Evidence for aggregate formation in a nepheloid layer and its possible role in the sedimentation of particulate matter. *Marine Geology* 20:M7–14.

Flawn, P. T., 1953, Petrographic classification of argillaceous sedimentary and low grade metamorphic rocks in subsurface. *American Association of Petroleum Geologists Bulletin* 37:560–5.

Folk, R. L., 1951, Stages of textural maturity, *Journal of Sedimentary Petrology* 21:127–30.

Folk, R. L., 1968, Bimodal supermature sandstones: Product of the desert floor. *Proceedings XXIII International Geological Congress, Prague* 8:9–32.

Folk, R. L., P. B. Andrews, and D. W. Lewis, 1970, Detrital sedimentary rock classification and nomenclature for use in New Zealand. *New Zealand Journal of Geology and Geophyics* 13:937–68.

Haven, D. S., and R. Morales-Alamo, 1968, Occurrence and transport of faecal pellets in suspension in a tidal estuary. *Sedimentary Geology* 2:141–51.

Krumbein, W. C., 1934, Size frequency distributions of sediments. *Journal of Sedimentary Petrology* 4:65–77.

Krumbein, W. C., 1941, Measurement and geological significance of shape and roundness of sedimentary particles. *Journal of Sedimentary Petrology* 11:64–72.

Landim, P. M. B., and L. A. Frakes, 1968, Distinction between tills and other diamictons based on textural characteristics. *Journal of Sedimentary Petrology* 38:1213–23.

Laznicka, P., 1988, *Breccias and Coarse Fragmentites: Petrology, Environments, Associations, Ores.* Developments in Economic Geology 25, Elsevier, New York, 832p.

Lundegard, P. D., and N. D. Sanders, 1980, Field classification of fine-grained sedimentary rocks. *Journal of Sedimentary Petrology* 50:781–6.

McConchie, D. M., and L. M. Lawrence, 1991, The origin of high cadmium loads in some bivalve molluscs from Shark Bay, Western Australia: A new mechanism for cadmium uptake by filter feeding organisms. *Archives of Environmental Contamination and Toxicology* 21:303–10.

McLaren, P., and D. Bowles, 1985, The effects of sediment transport on grain-size distributions. *Journal of Sedimentary Petrology* 55:457–70.

Maxwell, J. C., 1964, Influence of depth, temperature, and geologic age on porosity of quartzose sandstones. *American Association of Petroleum Geologists Bulletin* 48:697–709.

Middleton, G. V., 1976, Hydraulic interpretaion of sand size distributions. *Journal of Geology* 84:405–26.

Moss, A. J., 1972, Bed-load sediments. *Sedimentology* 18:159–219.

Pettijohn, F. J., 1975, *Sedimentary Rocks.* Harper & Row, New York, 628p.

Prokopovich, N. R., 1969, Deposition of clastic sediment by clams. *Journal of Sedimentary Petrology* 39:891–901.

Pryor, W. A., and W. A. Vanwie, 1971, The "Sawdust Sand"—an Eocene sediment of floccule origin. *Journal of Sedimentary Petrology* 41:763–69.

Risk, M. J., and J. S. Moffat, 1977, Sedimentological significance of fecal pellets of Macoma bathica in the Minas Basin, Bay of Fundy. *Journal of Sedimentary Petrology* 47:1425–36.

Rittenhouse, G., 1971, Pore-space reduction by solution and cementation. *American Association of Petroleum Geologists Bulletin* 55:80–91.

Spears, D. A., 1980, Towards a classification of shales. *Geological Society of London Quarterly Journal* 137:125–9.

Taira, A., and P. A. Scholle, 1979, Origin of bimodal sands in some modern environments. *Journal of Sedimentary Petrology* 49:777–86.

Tanner, W. F., 1969, The particle size scale. *Journal of Sedimentary Petrology* 39:809–12.

Wentworth, C. K., 1922, A scale of grade and class terms for clastic sediments. *Journal of Geology* 30:377–92.

Whetten, J. T., and J. W. Hawkins, Jr., 1970, Diagenetic origin of greywacke matrix minerals. *Sedimentology* 15:347–61.

Shape

Barrett, P. J., 1980, The shape of rock particles, a critical review. *Sedimentology* 27:291–303.

Crook, K. A. W., 1968, Weathering and roundness of quartz sand grains. *Sedimentology* 11:171–82.

Dobkins, J. E., Jr., and R. L. Folk, 1970, Shape development on Tahiti–Nui. *Journal of Sedimentary Petrology* 40:1167–203.

Gale, S. J., 1990, The shape of beach gravels. *Journal of Sedimentary Petrology* 60:787–9.

Goede, A., 1975, Downstream changes in the pebble morphometry of the Tambo River, Eastern Victoria. *Journal of Sedimentary Petrology* 45:704–18.

Hobday, D. K., and N. L. Banks, 1971, A coarse-grained pocket beach complex, Tanafjord (Norway). *Sedimentology* 16:129–34.

Krumbein, W. C., 1941, Measurement and geological significance of shape and roundness of sedimentary particles. *Journal of Sedimentary Petrology* 11:64–72.

Luttig, G., 1962, The shape of pebbles in the continental, fluviatile, and marine facies. In *International Association of Scientific Hydrology Publication 59*, pp. 253–8.

Mazzullo, J., and S. Magenheimer, 1987, The original shapes of quartz sand grains. *Journal of Sedimentary Petrology* 57:479–87.

Mazzullo, J., D. Sims, and D. Cunningham, 1986, The effects of eolian sorting and abrasion upon the shapes of fine quartz sand grains. *Journal of Sedimentary Petrology* 56:45–56.

Moss, W. J., 1966, Origin, shaping and significance of quartz sand grains. *Journal of the Geological Society of Australia* 13:97–136.

Passaretti, M. L., and E. V. Eslinger, 1987, Dissolution and relic textures in framework grains of Holocene sediments from the Brazos River and Gulf Coast of Texas. *Journal of Sedimentary Petrology* 57:94–7.

Rittenhouse, G., 1943, A visual method for estimating two-dimensional sphericity. *Journal of Sedimentary Petrology* 13:79–81.

Rittenhouse, G., 1946, Grain roundness—a valuable geologic tool. *American Association of Petroleum Geologists Bulletin* 30:1192–7.

Shepard, F. P., and R. Young, 1961, Distinguishing between beach and dune sands. *Journal of Sedimentary Petrology* 31:196–214.

Sneed, E. D., and R. L. Folk, 1958, Pebbles in the lower Colorado River, Texas—a study in particle morphogenesis. *Journal of Geology* 66:114–50.

Waddell, H., 1932, Volume, shape, and roundness of rock particles. *Journal of Geology* 40:443–51.

Whitney, M. I., and R. V. Dietrich, 1973, Ventifact sculpture by windblown dust. *Geological Society of America Bulletin* 84:2561–82.

Williams, G. P., 1966, Particle roundness and surface texture effects on fall velocity. *Journal of Sedimentary Petrology* 36:255–9.

Fabric

Beard, D. C., and P. K. Weyl, 1973, Influence of texture on porosity and permeability of unconsolidated sand. *American Association of Petroleum Geologists Bulletin* 57:349–69.

Brewer, R., 1964, *Fabric and Mineral Analysis of Soils.* Wiley, New York. (Reprinted in 1976 by Robert E. Kreiger Publishing, Huntington, New York, 482p.)

Clifton, H. E., 1971, Orientation of empty pelecypod shells and shell fragments in quiet water. *Journal of Sedimentary Petrology* 41:671–82.

Dowdeswell, J. A., and M. J. Sharp, 1986, Characterization of pebble fabrics in modern terrestrial glacigene sediments. *Sedimentology* 33:699–710.

Graton, L. C., and H. J. Fraser, 1935, Systematic packing of spheres—with particular relation to porosity and permeability. *Journal of Geology* 43:785–909.

Heling, D., 1970, Micro-fabrics of shales and their rearrangements by compaction. *Sedimentology* 15:247–60.

Jones, M. L., and J. M. Dennison, 1970, Oriented fossils as paleocurrent indicators in Paleozoic lutites of southern Appalachians. *Journal of Sedimentary Petrology* 40:642–9.

Kuehl, S. A., C. A. Nittrouer, and D. J. DeMaster, 1988, Microfabric study of fine-grained sediments: Observations from the Amazon subaqueous delta. *Journal of Sedimentary Petrology* 58:12–23.

Lindsay, J. F., 1968, The development of clast fabric in mudflows. *Journal of Sedimentary Petrology* 38:1242–53.

Major, J. J., and B. Voight, 1986, Sedimentology and clast orientations of the 18 May 1980 southwest-flank lahars, Mount St. Helens, Washington. *Journal of Sedimentary Petrology* 56:691–705.

Martini, I. P., 1971, An analysis of the interrelationsips of grain orientation, grain size and grain elongation. *Sedimentology* 17:265–75.

O'Brien, N. R., 1987, The effects of bioturbation on the fabric of shale. *Journal of Sedimentary Petrology* 57:449–55.

O'Brien, N. R., K. Nakazama, and S. Tokuhashi, 1980, Use of clay fabric to distinguish turbiditic and hemipelagic siltstones and silts. *Sedimentology* 27:47–61.

Rust, B. R., 1972, Pebble orientation in fluvial sediments. *Journal of Sedimentary Petrology* 42:384–8.

Shelton, J. W., and D. E. Mack, 1970, Grain orientation in determination of paleocurrents and sandstone trends. *American Association of Petroleum Geologists Bulletin* 545:1108–19.

Walker, R. G., 1975, Generalized facies models for resedimented conglomerates of turbidite association. *Geological Society of America Bulletin* 86:737–48.

Grain Surface Textures

Baker, H. W., Jr., 1976, Environmental sensitivity of submicroscopic surface textures on quartz sand grains—a statistical evaluation. *Journal of Sedimentary Petrology* 46:871–80.

Culver, S. J., P. A. Bull, S. Campbell, R. A. Shanesby, and W. B. Whalley, 1983, Environmental discrimination based on quartz grain surface textures: A statistical investigation. *Sedimentology* 30:129–36.

Krinsley, D. H., and J. C. Doornkamp, 1973, *Atlas of Quartz Grain Surface Textures.* Cambridge University Press, Cambridge, U.K., 91p.

Whalley, W. B., and D. H. Krinsley, 1974, A scanning electron microscope study of surface textures of quartz grains from glacial environments. *Sedimentology* 21:87–105.

6
Composition of Detrital Sediments

Composition of sediments refers to the chemical and mineralogical makeup of the constituent particles, which may be individual minerals, combinations of minerals as in rock fragments, organic particles or constituents chemically precipitated in the depositional or diagenetic environment. In this chapter we treat sediments comprising more than 50% detritus from preexisting igneous, metamorphic, or sedimentary rocks. Some authors call these *clastic* sediments, but since most limestones are composed largely of fossil clasts (broken fragments), they too are properly clastic; hence the word *detrital* (or *siliciclastic)* is more definitive. The composition of detrital gravel and sand is dictated by the composition of the source rocks and is influenced by climatic and diagenetic processes (and rarely, by environmental sorting processes); the tectonics of the source area and basin of deposition are also influential insofar as they determine the kinds of source rocks exposed and the rate at which the sediments are buried (see Chapter 2); time is implicit as a factor as well. In contrast, the composition of clay minerals is generally related to climatic and diagenetic processes, and although many clay minerals are derived from the alteration of detrital minerals, it is rare that the original mineral can be definitively identified; many clay minerals also form in the depositional or diagenetic environment by chemical precipitation. Hence only clay minerals derived from preexisting sedimentary rocks are truly detrital, and it is appropriate to treat clays with colloids separately in Chapter 7.

Gravel or conglomerate composition is best studied in the field; representative samples are difficult to collect for laboratory analysis. Grids are commonly laid out on exposure surfaces and pebbles at the intersection points are counted (see **AS** Chapter 4). Sand composition is most commonly identified in thin section or loose grain mounts (**AS** Chapter 5), using petrographic and binocular reflected-light microscopes and point-counting techniques with a mechanical stage (**AS** Chapter 8). Clay-size particles generally require identification by X-ray diffraction, differential thermal analysis, infrared analysis, or other specialized techniques (see Chapter 7 and **AS** Chapter 8).

No single classification is in universal use for any of the main classes of detrital sediment, which are established on the basis of their textural characteristics: gravels (conglomerates), sands (sandstones), and muds (mudstones). Each scheme has its own advantages and disadvantages; the most suitable one may depend on the situation encountered. However, in any formal report it is important that the system of classification used be clearly cited and used consistently in the report to avoid potential confusion among readers.

Following Lyell (1837, pp. 456–457), a sandstone is "any stone which is composed of an agglutination of grains of sand, whether calcareous, siliceous, or of any other mineral nature." Many limestones and some other sediments with components formed in the depositional environment have a texture of sand-size particles. As cogently argued by Folk (1954), mineralogical and textural classifications should be distinct because they can vary independently and because the possible interpretations from each kind of property are different. Controls for sediment texture are very different from those for composition; hence it is desirable to use different terms when discussing each property. The term *arenite* (from Latin *arena*, meaning sand) is used here when referring to the composition of a detrital sand or sandstone. In parallel with this usage, detrital muds or mudrocks are *lutites* (from *lutum* for mud), and detrital gravels, conglomerates, or breccias are *rudites* (from *rudus* for rubble).

DETRITAL SANDSTONES

Many classification systems have been proposed for detrital sandstones (see Klein 1963 for a good critical review of many suggested to that date; few new schemes have been proposed since the early 1970s). Many redefine the same terms, such as *arkose* (see Pettijohn 1943; Oriel 1949) and *greywacke* (see Boswell 1960; Dott 1964). The classification for detrital sandstone composition presented here is that of Folk, Andrews, and Lewis (1970). It employs terminology that is self-explanatory, avoids terms that have been extensively redefined in previous schemes, and places a high value on flexibility. Because the classification scheme is basically objective and does not carry pronounced genetic implications, sediment descriptions based on it will retain their value even though genetic models may change in the future.

Matrix (arbitrarily set for arenites as detritus and clay minerals finer than 0.03 mm) and the gravel fraction coarser than 4 mm are taken into account in the textural classification but are not named in the arenite compositional classification. Although arenites technically incorporate only sand-size grains (2.0–0.06 mm in diameter), extension of the limits through coarse silt at one end (down to 0.03 mm) and granules (up to 4 mm) at the other is both practical and convenient for purposes of compositional classification.

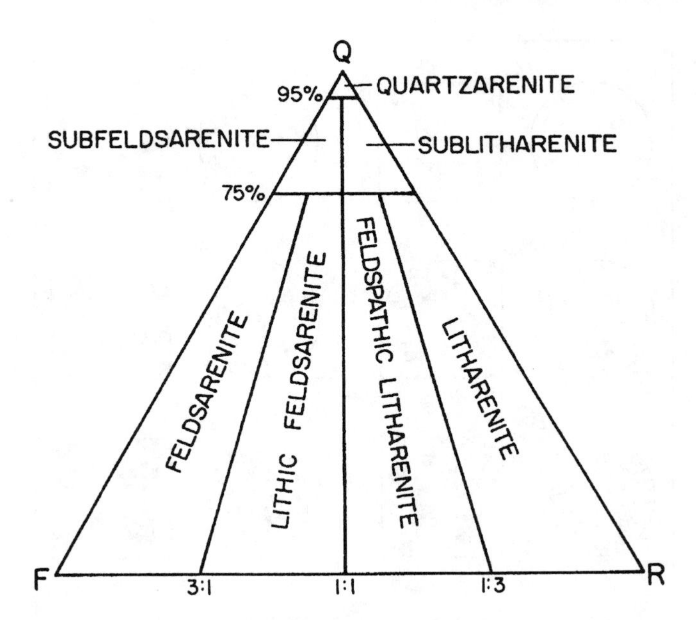

Figure 6-1. Parent triangle for arenite classification of Folk, Andrews, and Lewis 1970. Q = monocrystalline and polycrystalline quartz (excluding chert). F = monocrystalline feldspar. R = rock fragments (igneous, metamorphic, and sedimentary, including chert).

The Classification

Most detrital sand grains are either quartz (Q), feldspar (F), or rock fragments (R; commonly denoted as L for lithics). Thus, a ternary diagram is constructed in which the QFR components of any arenite may be quantitatively plotted (Fig. 6-1). Other compositional attributes of the arenite are stated separately. The compositional classification of an arenite would thus follow the system:

(cement) (prominent nondetrital grain type) (prominent detrital non-QFR grain type) (QFR compositional term)

At the Q pole are grouped all monocrystalline and polycrystalline quartz grains excluding detrital chert (a rock fragment). Whereas all polycrystalline grains might be considered rock fragments, there are no criteria for consistent distinction between such grains derived from plutonic igneous or metamorphic rocks, silica-cemented quartz-rich arenites, and metaquartzites (e.g., Blatt 1967*a*, 1967*b*), and most monocrystalline quartz grains are derived from the same kinds of source rocks; there is also only a small difference in the mechanical and chemical stability of these components. Thus, it is reasonable to group them all at one pole.

At the F pole are grouped all monocrystalline feldspars; polycrystalline feldspars are placed in the R category because the size of the component crystals usually permits identification of the derivation of these grains as either volcanic (small lathlike or euhedral crystals set in aphanitic groundmass) or plutonic (large anhedral crystals). At the R pole are grouped all recognizable rock fragments (igneous, metamorphic, and sedimentary, including chert).

Although subdivision of the QFR triangle is largely arbitrary, the various divisions reflect an attempt to group arenites with similar source rock lithologies and histories. For example, many arenites, especially in continental regions, are composed almost exclusively of quartz, either because of prolonged weathering that results in elimination of unstable minerals prior to deposition, or because most of the grains are recycled from other quartz-rich sedimentary rocks (see Blatt 1967*a*). It is therefore useful to distinguish *quartzarenites* from other quartz-rich arenites by a boundary at 95% Q. The remainder of the triangle is divided into two equal parts at a ratio of 1:1 F to R grains.

Previous petrological studies of arenites (e.g., see Klein 1963) have shown that a boundary of 25% F grains is a useful limit for rocks of intermediate composition between quartz-rich (*quartzarenite*) and feldspar-rich (*feldsarenite*) sediments—the *subfeldsarenites*. It is convenient to take the same limit for these arenites as for the intermediate-R arenites (the *sublitharenites)*. Purely as a choice of convenience and to permit further refinement, the two wide fields for arenites rich in F and R grains are split into equal halves at ratios of F:R = 3:1 and 1:3.

The QFR triangle is not utilized in naming the geologically less common arenites that have a predominance of detrital mineral grains of neither Q, F, nor R; they are best treated as *(mineral)-arenites*, for example, magnetite-arenite. If these non-QFR minerals constitute less than the QFR components by volume, they appear only as a preceding term, for example, magnetite quartzarenite.

For plotting sample compositions onto the QFR triangle, the percentage of the Q, F, and R components alone are re-

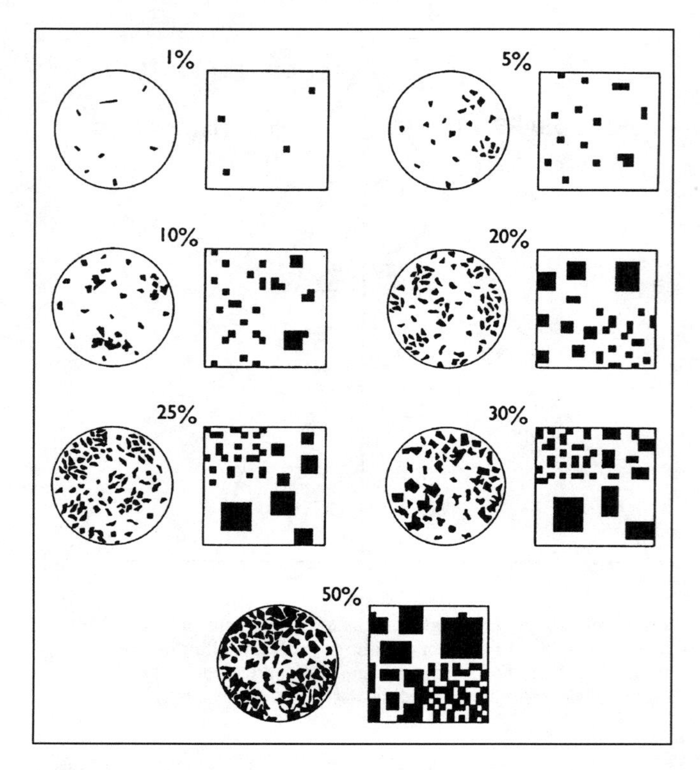

Figure 6-2. Percentage estimation comparison charts. (Reproduced from Folk, Andrews, and Lewis 1970, *New Zealand Journal of Geology and Geophysics*.)

calculated to 100%. Percentages of components are precisely determined by point-counting techniques (see **AS** Chapter 8), but they can be estimated with reasonable precision after some experience with a comparison chart such as Fig. 6-2.

Refinement of Nomenclature

The relative proportion of igneous to metamorphic and/or sedimentary rock fragments, or of specific varieties of these rock fragments, is important for interpreting *provenance* (source rock character) and sediment history. In some sample suites, it is also useful to refine the quartzarenite and feldsarenite classes. The level of refinement desired depends on the particular suite of arenites. Refinement in nomenclature for the litharenites is achieved by appending the name of the most common rock fragment type to litharenite or sublitharenite (in the case of lithic feldsarenite, replace lithic with the predominant rock fragment type), e.g., sedimentary litharenite (sedarenite), igneous litharenite, volcanic litharenite (volcarenite), plutonic litharenite, volcanic feldsarenite. Phyllarenite is probably the most unequivocal name where phyllitic or micaceous metamorphic rock fragments predominate; use of metamorphic (or schistose) litharenite could mislead in implying a litharenite that has been metamorphosed. Where detrital carbonate rock fragments predominate, the term *limestone litharenite* or *calclitharenite* must be used; calcarenite has the preemptive connotation in limestone classifications of a rock composed mainly of carbonate sand particles that have been derived from within the basin of deposition.

To show similarities and differences in suites of arenite subvarieties, second- and third-order triangles may be devised as in Fig. 6-3. Any end members that are useful for the particular suite of sediments may be selected; only one set of examples is shown in the figure. For quartzarenite suites, it may be useful to devise a second-order triangle at the Q pole, with poles for polycrystalline quartz, monocrystalline quartz of undulatory extinction, and monocrystalline quartz of straight extinction, because the first two grain types are preferentially eliminated, in the order given, by prolonged action of sedimentary processes. Whenever plotting on a second- or third-order triangle, the constituents represented by the poles of the triangle are recalculated to 100% before plotting.

It may be appropriate to go directly to different triangles for determination of source rocks, climatic modifications to composition, tectonic settings, or other specific goals. In such cases, the end members selected will depend on the purpose of the study and may comprise specific varieties of Q, F, or R components, such as Qp (polycrystalline quartz), Lv fine-grained volcanic rock fragments), Ls (fine sedimentary–metasedimentary rock fragments). Examples of such triangles are provided later in this chapter. Such applications of graphic plots require precise quantitative data, achieved only by point counting of thin sections (see **AS** Chapter 8); such procedures may involve the Gazzi-Dickinson method (see Ingersoll et al. 1984), in which all crystals or grains of sand size that are included in larger rock fragments are counted as those specific crystal or grain types rather than as the larger rock fragment (this procedure helps to minimize differences in composition that relate to differences in grain size, but obscures important information about source rocks).

Nondetrital Components

Separate classifications exist for sandstones that contain more than 50% nondetrital grains by volume (it may be appropriate to give both a detrital and nondetrital name to a sediment with substantial components of each). However, in detrital sediment nomenclature, the presence of small quantities of nondetrital grains in a sediment should be indicated by adding the name of the grain type as an adjectival modifier e.g., *fossiliferous* or *glauconitic named-arenite*. Workers may decide for themselves at which percentage limits these grain types should be included in the name. Postdepositional precipitates that act as cements are another modifier of the arenite name e.g., carbonate-cemented, calcite-cemented, iron oxide-cemented, pyrite-cemented.

Example of Classifying an Arenite

Analysis shows a composition of 45% quartz, 18% feldspar, 5% rock fragments, 3% mica, 15% glauconite, and 14% carbonate cement. The Q component is 100 × [quartz/(quartz + feldspar + rock fragments)] = 100 × [45/(45 + 18 + 5)] = 66.2%. The point representing this arenite would thus lie on the line of 66.2% Q (parallel to the base of the triangle and

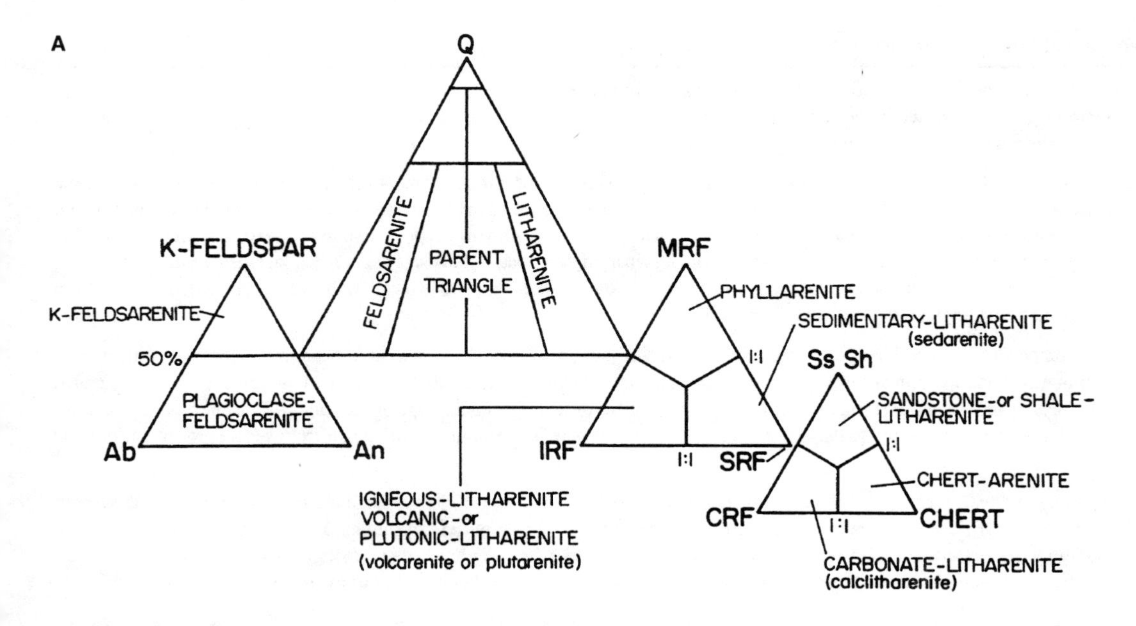

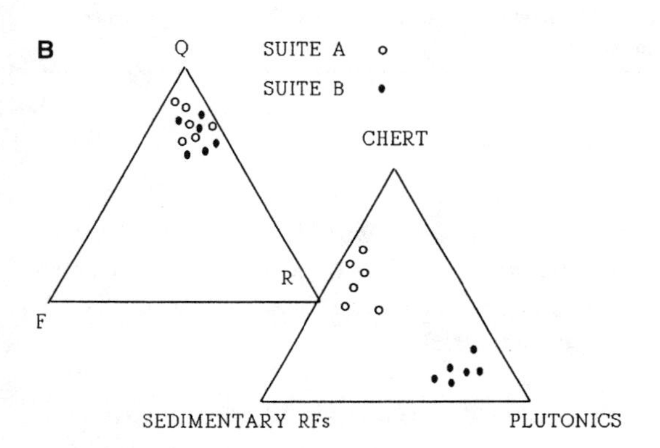

Figure 6-3. Example of daughter triangles for arenite classification. *A:* A set of possible daughter triangles; workers may devise different combinations depending on the composition of the arenites. MRF = metamorphic rock fragments; IRF = igneous rock fragments; SRF = sedimentary rock fragments; CRF = carbonate rock fragments. *B:* Example of distinction that can be made between sample suites using daughter triangles. (Part *A* after Folk, Andrews, and Lewis 1970.)

two-thirds of the way toward the *a* apex). The F:R ratio is 3.6:1; thus, the point lies in the feldsarenite field. For precise plotting of the point, the F component is (100 × 18/68), or 26.5%, and the R component is (100 × 5/68), or 7.4%. The sample point lies at the intersection of the 7.4% R line (parallel to the QF side of the triangle) and/or of the 26.5% F line (parallel to the QR side) with the 66.2% Q line. (If the three percentages do not intersect at the same point, there is an error in calculation!) Assuming other characteristics of the sediment, the final name might be *pale greenish-gray friable, cross-laminated, poorly sorted bimodal medium and fine sandstone: calcite-cemented glauconitic mica subfeldsarenite*. This name combines all characteristics of the sediment (cf. Chapter 1), of which only the latter part (after the colon) refers to composition. Such a classification may seem so lengthy as to be cumbersome, but it does provide distinctive and important characteristics of a deposit for both description and interpretation. Not even this sequence of terms comprehensively describes a specific rock; further details would be necessary on the cross-lamination, cementation, types of feldspar, roundness of grain types, and other properties.

Common Arenite Components

Matrix and Cement

Mineral matter occupying interstitial spaces between framework grains in sediments is either *matrix*, deposited mechanically at the time of sedimentation, or *cement*, chemically precipitated from solution after deposition. Characteristics of these components must be studied under the petrographic microscope (a variable-power microscope capable of polarizing light in two directions and used to study thin sections or slices or rock; see **AS** Chapter 8). The definitions of these components and their derivatives (Table 6-1) are genetic. Objective distinction would seem to be feasible: cement comprises a visibly crystalline mineral, whereas matrix comprises a mixture of discrete fine particles (in arenites, detritus of mixed clay and silt components). Distinction may be difficult, however, because finely crystalline cement with impurities (especially iron oxide stains) may resemble densely packed fine silts or clays with or without minor cement. Impurities also may mask completely the character of the interstitial material or, because the average thickness of a thin section is

Table 6-1. Definitions of Cement and Matrix Components of Arenite

Primary cement: mineral matter that has filled original interstitial voids (i.e., the cement has not replaced other material). Common attributes include development as overgrowths on detrital parent grains (indicated by concentrations of dust on the rims); transparent crystals without matrix inclusions; evidence of radial inward growth of crystals from the edges of surrounding framework grains.

Secondary cement: crystalline mineral matter filling interstitial spaces previously occupied by matrix or primary cement. The cement may be replacive (expect remnant matrix inclusions) or displacive (expect distorted or compressed surrounding matrix fabrics). Somewhere in the rock unit there should be textural evidence of intermediate stages in the development of secondary cement, but secondary cement is sometimes pervasive (as in some dolomitization). The secondary nature of the cement may be indicated where framework grains are "floating" in a single crystal—the result of grain displacement as the crystal grew (appears as grains isolated in cement—not possible initially because grains are always in contact when they are deposited); however, beware: precompaction primary cements can *appear* to surround grains in the two-dimensional slices of thin sections.

Primary matrix: fine clastic (any compositon) mineral matter, interstitial to the framework grains, deposited in the environment of sedimentation penecontemporaneously with the framework grains. Primary matrix may have been deposited concurrently with the framework grains or have been blended with them soon afterward (e.g., because of bioturbation). Primary matrix may recrystallize or undergo mineralogical changes (for instance, smectite clays commonly change to illite and/or chlorite and/or chert in late diagenesis), in which cases it may be difficult to distinguish from secondary matrix or even cement.

Secondary matrix: fine interstitial mineral matter (essentially clay minerals) deposited in void spaces during pedogenesis or diagenesis from colloidal suspensions (such as cutans) or from solution (e.g., where unstable framework grains have been extensively altered and the alteration products have reprecipitated). It is difficult to distinguish secondary matrix objectively from primary matrix, particularly where the latter has been recrystallized or mineralogically transformed. In the case of precipitation from solution, the distinction from cement is arbitrarily based on the clayey nature of the material.

Pseudomatrix: discontinuous interstitial paste formed by the deformation of weak framework grains. Examples are squashed and deformed detrital micas that have altered to clay minerals, glauconite pellets, and argillaceous rock fragments. Pseudomatrix is perhaps most difficult to distinguish in volcaniclastics that have been deformed. Wispy extensions from deformed grains into narrowing interstices between framework grains are commonly used as evidence for pseudomatrix. Pseudomatrix should be recorded as part of the original framework grain population.

0.03 mm, finer particles can be superimposed and produce indistinct aggregate optical properties.) In addition:

1. Material resembling primary matrix can be produced by the breakdown of unstable detrital minerals (e.g., feldspars, particularly the calcic plagioclases). Intermediate stages in this breakdown often can be observed in thin section or a suite of thin sections from the rock unit; these intermediate stages may provide the only clue to the diagenetic transformations that have taken place. Authigenic clays may precipitate and/or grow from solution, as well as form by recrystallization of finer detrital phyllosilicates. Clays and the humic materials in soils also may be translocated by groundwater during pedogenetic eluviation or illuviation. Locally such clay particles may pack tightly parallel to framework grain boundaries to form a birefringent *cutan*, often with parallel banding and showing a sinuous band of extinction that moves across the cutan when the stage is rotated under Crossed Polars (CPL); these features may be reorganized during later diagenesis to resemble a primary matrix. Microcrystalline calcite resembling lime mud that forms in the depositional environment (see Chapter 8) can also precipitate during diagenesis or form by recrystallization of carbonate grains or of carbonate cement.

2. Material resembling precipitated void-filling cement may be produced by diagenetic recrystallization of a primary matrix (e.g., lime mud may transform into calcite cement) or pervasive replacement (e.g., silicification during silcrete formation) may destroy all evidence of original matrix. Chert, a common void-filling cement, may also form during diagenesis from primary detrital matrix with the concurrent formation of fine chloritic material, illite, or sericite. Overgrowths on quartz, feldspar, and carbonate grains may also absorb and replace a primary matrix.

Thus, the problem of discriminating between matrix and cement can be major, and accurate distinction depends heavily on the operator's skill in evaluating the diagenetic history of the sediment.

Quartz

Because quartz is the most common detrital mineral in arenites and because it is virtually the only one in mineralogically mature arenites, subvarieties have been distinguished for both descriptive and interpretive purposes (e.g., Blatt 1967*b*; Folk 1974; Basu et al. 1975; Young 1976; Sanderson 1984; Basu in Zuffa 1985). Descriptive varieties of quartz are:

1. *Unstrained*: grains are single crystals that extinguish as a unit when using a polarizing microscope under CPL.

2. *Strained*: grains are single crystals that never wholly extinguish CPL. An irregular band of extinction migrates across the crystal as the stage is rotated. Most grains that require rotation of the microscope stage for more than 5° to obtain extinction in different parts of the grain are probably derived from low-rank metamorphic rocks; those requiring less than 5° are probably derived from plutonic rocks (Basu et al. 1975). However, care must be taken to ensure strain did not develop with fold or fault deformation after deposition (many grains will then show a similar extinction pattern).

3. *Polycrystalline*: single grains composed of more than one crystal, each of which is strained or unstrained. Crystal boundaries may be obscure (due to recrystallization), but extinction characteristics will generally indicate the polycrystalline character. Folk (1980) classed grains with obscure crystal boundaries and similar optic orientations in the crystals as *semi-composite*. Individual crystal sizes increase with an increase in the source rock's metamorphic grade. More polycrystalline quartz grains in the medium sand range are initially supplied from low-rank metamorphic rocks (around 50%) instead of from middle- and upper-rank metamorphic (about 30%) or from plutonic rocks (about 15%) (Basu et al. 1975). The number of polycrystalline grains is also a function of grain size—the finer the size, the fewer the polycrystalline grains.

Genetic varieties of quartz (after Folk 1980) include:

1. *Volcanic*: commonly water clear, generally unstrained, mostly monocrystalline (cagey generalizations!). Hexagonal-bipyramidal (A-quartz) shape is definitive. Embayments are common due to corrosion prior to extrusion; conchoidal fractures are common; shards occur in fresh volcaniclastics but are generally rare.

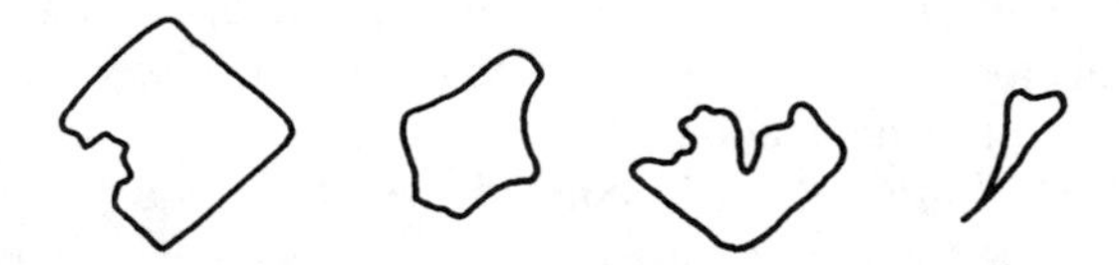

2. *Vein*: generally abundant vacuoles (which account for the milky appearance of large grains), unstrained or slightly strained. Microlites are very rare. Vein varieties may be monocrystalline or polycrystalline (commonly with obscure crystal boundaries). They may show comb structure (appressed crystals that grew perpendicular to the vein walls) and/or contain vermicular chlorite microlites.

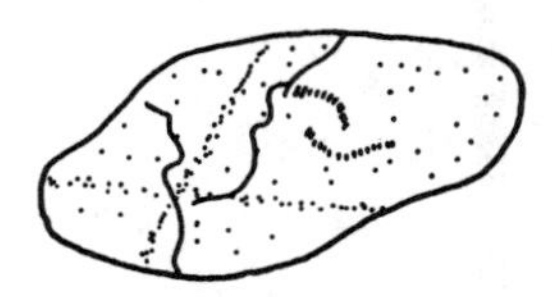

3. *Recycled sedimentary*: may dominate in almost any sediment, but most grains will reflect the characteristics of the ultimate source rather than of the immediate sedimentary source. The presence of rounded or worn overgrowths is the main direct indicator. If only a few grains have overgrowths, and these do not interlock with other overgrowths, they are probably recycled. (Because overgrowths are optically continuous with the parent, concentrations of minute dust along the margin of the parent crystal are commonly the only distinguishing characteristic. Microlites will be restricted to the parent, but vacuoles may be present in both parent and overgrowth; strain may be duplicated in the overgrowth.)

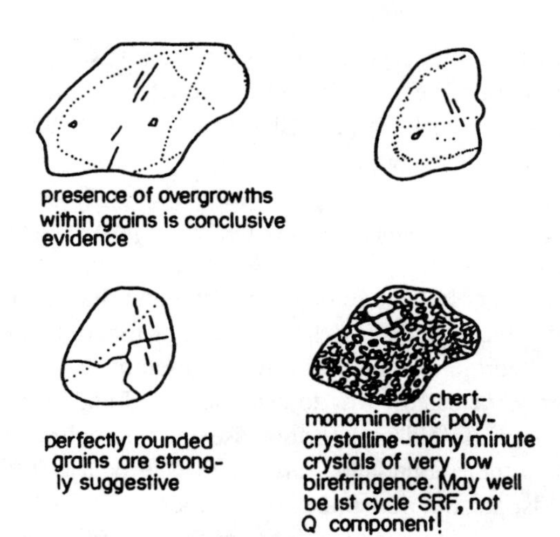

4. *Metamorphic*: often indistinguishable from primary plutonic igneous quartz (see item 5, common quartz). The most obvious metamorphic quartz grains are polycrystalline varieties with elongate internal crystals showing smooth, crenulated, or granulated boundaries and a subparallel optic orientation. They can be classed as MRFs if their metamorphic origin is convincing. Polycrystalline medium sand grains with 10 or more individual crystals appear to be largely of metamorphic origin, but multiple crystals are also characteristic of other siliceous rock fragments. Some polycrystalline grains have several parallel micas (remnants of schistosity) or microlites of metamorphic minerals; these are polymineralic and can be referred to as MRFs.

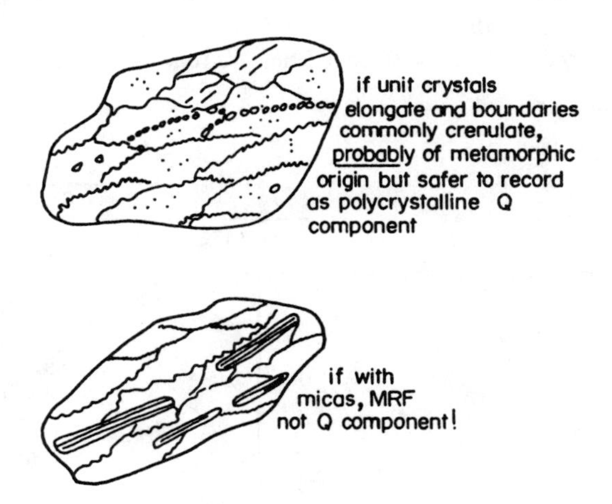

5. *Common*: Most quartz grains do not have any clear characteristic indicative of their origin and probably have a plutonic igneous/metamorphic ultimate source, but they may have been recycled. They are either strained or unstrained, mono- or polycrystalline, generally with few scattered vacuoles and/or microlites. Polycrystalline varieties generally have very inequant individual crystals.

Inclusions may be present in any quartz grain variety. These consist of vacuoles (minute bubbles, often partly filled with liquid; difficult to resolve even under high power) or microlites (minute crystals of other minerals, such as rutile needles). Microlite varieties may characterize quartz from specific source rocks (e.g., Keller and Littlefield 1950).

Feldspar

Feldspar varieties can prove very useful for provenance studies (see Fig. 6-4; e.g., Pittman 1970), and their relative degree of alteration is indicative of weathering and diagenetic history. **AS** Chapter 8 discusses methodology for studying feldspars in sediments and lists relevant literature. Staining is commonly the easiest way to estimate the abundance of the potassium or K-feldspars (orthoclase and microcline) and to determine their abundance relative to plagioclase (Ca- and Na-) feldspars, but note that clays may also take up the stain. Commonly distinguished varieties of feldspar in thin-section studies of arenites are:

1. *Microcline*: shows characteristic cross-hatch twinning. (Anorthoclase, with a finer pattern, may also be distinguished.) Microcline feldspars are the most stable of the feldspars in the sedimentary environment; thus, they are commonly fresh and may show overgrowths (often by untwinned K-feldspar).

2. *Orthoclase*: generally untwinned; refractive index (RI) is less than quartz and balsam; biaxial optical figure; grains are often slightly altered (unlike quartz) and often show overgrowths in originally permeable arenites. Where quartz and feldspar grains are touching, their differential relief will permit distinction under low to medium magnifications when the substage diaphragm is closed and the field of view is slightly out of focus. (Becke lines are rendered obvious).

3. *Plagioclase*: generally identified by polysynthetic twinning; mineral species in this group generally require identification with the universal stage (in thin section) or by index-oil identification (with loose grains). In thin section, the RI of calcium varieties is greater than that of quartz and balsam. The main problem is in distinguishing untwinned varieties of sodium-rich plagioclase (albite and oligoclase, which are common in metamorphic rocks) from quartz and orthoclase. The RI of albite is only slightly lower than that of quartz; the RI of oligoclase overlaps that of quartz.

4. *Altered*: a general category for all feldspars that have been so extensively altered that their original composition is indeterminate. If all feldspars show the same degree of alteration, then diagenetic processes are probably to blame; if varying degrees of alteration are evident in grains of the same composition, predepositional weathering processes are probably to blame.

Three types of feldspar alteration are commonly distinguished:

1. *Vacuolization*: a multitude of minute bubbles, often partly filled with water, probably a result of hydration/hydrolysis.
2. *Kaolinization*: very low birefringent microflakes, generally too fine to distinguish; at low magnifications may resemble vacuolization.
3. *Sericitization* (strictly should be *illitization* because sericite tends to be a late diagenetic or metamorphic product): an alteration to clayey microflakes that are generally discernible by first-order white to yellow birefringence.

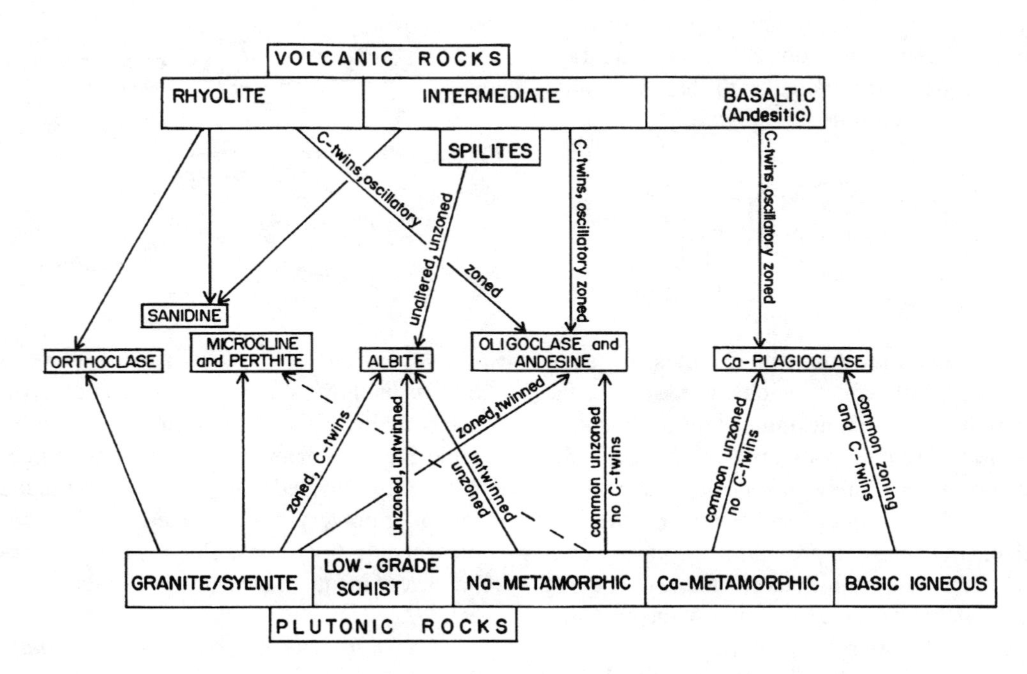

Figure 6-4. Relationships of feldspar varieties to source rocks. (After Pittman 1970 and Shelley, University of Canterbury, pers. comm.)

These alteration products may be disseminated throughout the feldspars or concentrated along cleavages; authigenic feldspar overgrowths will not show them unless alteration is late-stage diagenetic or due to weathering after emergence of the rock.

Rock Fragments

Rock fragments are the best indicators of provenance and should be carefully sought in thin section. Identification of rock fragments is dependent on the size of the sand grains as well as the crystallographic character of the source rock (see Fig. 6-5; Boggs 1968). Although volcanic glass shards are in a sense monomineralic, when present in a sediment they should be classed as rock fragments (e.g., McConchie 1987). Because of their importance for interpretation, all varieties present in the rock should be carefully identified in qualitative study before quantitative analysis (see **AS** Chapter 8). Identification is not always easy, and fine-grained small rock fragments may be exceedingly difficult to distinguish from altered feldspars (Fig. 6-6); hence precise descriptions of texture and composition are desirable in case an initial identification requires review after further study of associated rocks.

In plane-polarized light (PPL), most rock fragments may appear colorless or a murky pale brown, but low-magnification PPL views are essential to distinguish grain outlines where muddy or cherty matrix is (or appears to be) present. The main problem in identifying rock fragments lies in distinguishing among the very fine-grained/finely crystalline varieties, which can resemble each other very closely, particularly after some diagenetic recrystallization (see Fig. 6-6). No hard and fast rule can be applied; indications as to the character of the fragments are the other rock fragments and minerals in the sediment, a knowledge of the general geological setting, and/or likely provenance. In general, under CPL, *chert* appears as a mosaic of roughly equal-sized anhedral crystals (smaller than 20 microns) with very low birefringence (gray colors). *Siltstone* clasts usually have more variably sized quartz crystals that are most commonly angular and do not form a good mosaic texture (unless overgrowth interlocking or metamorphic recrystallization has occurred). *Volcanic* clasts commonly contain some lathlike feldspars and exhibit a porphyritic texture; shard outlines may be present in tuffs. A foliate fabric and presence of abundant phyllosilicates are diagnostic of the *metamorphosed*

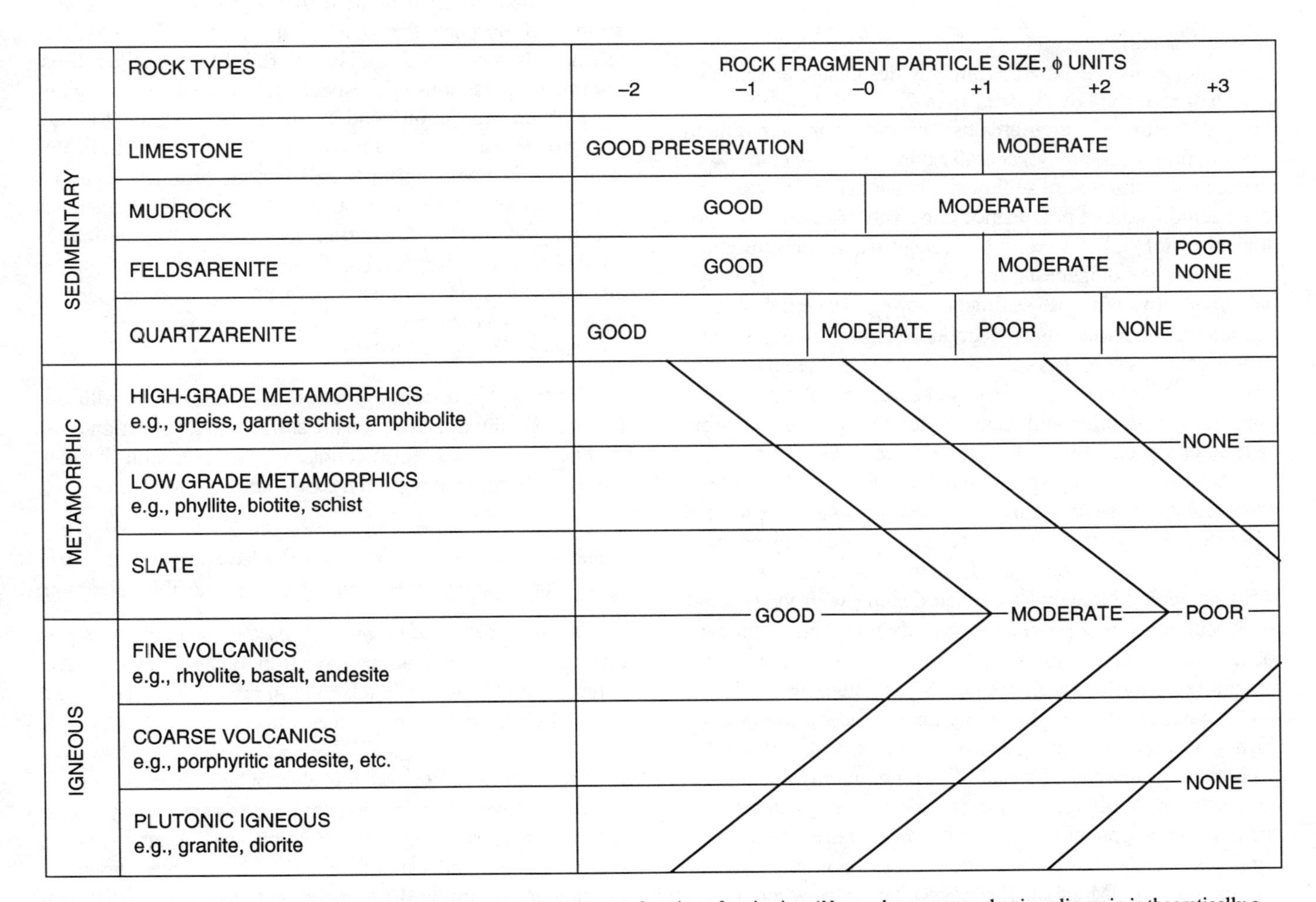

Figure 6-5. Relative degree of preservation of parent-rock textures as a function of grain size. (*Note:* whereas any polymineralic grain is theoretically a rock fragment, grains with small microlites present as inclusions that originated during magmatic crystallization or metamorphic recrystallization do not normally count as rock fragments; however, they may if the worker is convinced of their origin and is consistent.) (After Boggs 1968.)

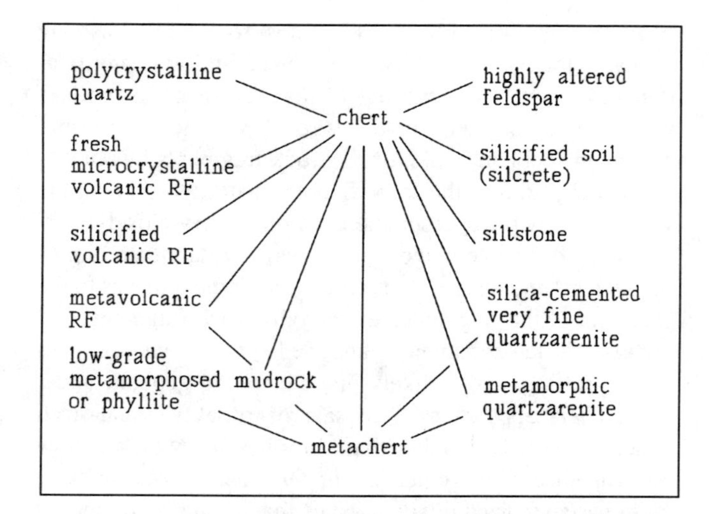

Figure 6-6. Arenite components that can resemble chert in internal texture. RF = rock fragment. (Modified after Wolf 1971.)

mudstones. Finely crystalline low-grade metamorphic rock fragments may be impossible to distinguish from mudstones.

Heavy Minerals

Heavy minerals (HMs) are arbitrarily designated as minerals with a specific gravity greater than 2.85. Whereas there are many such minerals, most are unstable under normal sedimentary conditions and they generally constitute less than 1% of the sand-size fraction of sediments. Nonetheless, they are important indicators of provenance (e.g., Table 6-2) and the geological history of the sediments (weathering, transportation, deposition, and diagenesis; see later discussion). Some heavy minerals grow in the sediment during diagenesis; these *authigenic* minerals may occur as overgrowths on detrital grains or may occur as totally new crystals that are usually of very fine sand size and show euhedral (excellent) crystal shapes. As with other authigenic minerals, they provide information on the chemistry of the diagenetic environment.

Because they are so sparse in most sediments, they are not representatively sampled in thin sections or small assemblages of loose grains; they usually must be concentrated by panning or more specialized techniques from much larger volumes (see **AS** Chapter 8). Nonetheless, when dealing with the original bulk sediments, it is possible to identify many under the binocular microscope or in thin section, and thin-section studies are vital to determine their relationships to other materials in a rock. Standard optical mineralogy texts or books specializing in heavy minerals are necessary for identification until the worker has experience. Crystal habit is commonly useful, particularly with authigenic minerals and where grains are too small for good optical figures. It may prove worthwhile to recognize several *varieties* of specific minerals (e.g., zircon or tourmaline; e.g., Marshall 1967; Krynine 1946) when undertaking detailed provenance or diagenetic studies in a particular petrographic province—varieties based on color, habit, inclusions, or other distinctive characteristics. In some studies, radiometric age determinations from HMs (e.g., lead/lead dating of zircons) can provide a powerful tool in provenance determinations.

Opaque heavy minerals generally require reflected-light microscopy on polished sections for identification (see **AS** Chapter 8). Their potential for provenance determination has only recently received attention (e.g., Basu and Molinaroli 1989; Grigsby 1990). A few can be distinguished under the microscope by illuminating the top of a normal thin section with a microscope lamp:

1. *Leucoxene* (a mixture of Ti-oxides and hydroxides): indicated by a bright, white reflection (do not confuse with the adamantine luster of some translucent heavies). A common diagenetic alteration product of titanium-bearing minerals (such as ilmenite and rutile), it may recrystallize to form tiny euhedral crystals of anatase, brookite, or sphene.

2. *Hematite and limonite* (a mixture of Fe-oxides and hydroxides): indicated by red (hematite) and reddish to yellowish brown (limonite) reflected colors. These are common weathering and diagenetic alteration products of iron-bearing minerals. A small amount may produce marked coloration and obscure the optical properties of grains (for example, hematite in altered biotite flakes). Hematite is most commonly present as microspecks (distinguishable only under high magnification), and its distribution relative to grain overgrowths and alteration products commonly indicates whether it is a weathering or a diagenetic product. Limonite appears to be mobile in interstitial fluids during diagenesis and may appear as discontinuous stringers or continuous bands (such as Leisegang bands) in the rock; it commonly is an obscuring stain on other minerals in thin section (such as carbonates, where it may have been exsolved during the conversion of siderite to calcite) .

3. *Pyrite and marcasite*: give a brassy yellow, metallic reflection. Distinction between the sulfides in thin section must be based on crystal form (cubic vs. orthorhombic habits). Both commonly show partial oxidation to limonite.

4. *Magnetite, ilmenite, and many other metallic minerals*: cannot easily be distinguished with reflected light in thin section. Many opaque minerals give a similar metallic reflection.

5. *Carbonaceous and bituminous matter*: can commonly be distinguished from opaque heavy minerals by nonreflective, generally dull brown to black color in reflected light (the coal maceral vitrinite is a shiny exception).

In HM studies, the first step should be to identify the entire suite present. For detailed studies, it is commonly necessary to concentrate on several size grades (the same for each sample). One grade should be of uniform size for every sample (to minimize differences that may reflect different grains in different sizes). The other should be at the next finest size to the mode of the overall grain size distribution (to

Table 6-2. Provenance Indicators

Sedimentary Provenance		
mudstone/sandstone/limestone fragments	zircon	
chert	rutile	particularly if
quartz (particularly if abraded overgrowths)	tourmaline	rounded
altered feldspars (rounded)	sphene	
abraded glauconite	other rounded hard/tough heavy minerals	
	leucoxene(?)	
Low-Rank Metamorphic Provenance		
slate/phyllite/quartzite fragments	leucoxene	
muscovite	tourmaline (particularly small euhedral brown crystals)	
chlorite	other stable HMs (as for sedimentary source)	
quartz (especially metaquartzite types)		
altered feldspars		
Higher-Rank Metamorphic Provenance		
schist/gneiss fragments	garnet	
muscovite biotite	epidote/zoisite	
chlorite	staurolite	
feldspars	kyanite/sillimanite/andalusite	
quartz (especially metaquartzite types)	magnetite/ilmenite	
	sphene	
	zircon	
Acid Igneous		
acid volcanics or granitic syenitic fragments	monazite/sphene rutile	
quartz (common or volcanic types)	tourmaline	
K-feldspar	zircon	
Na-plagioclase	magnetite	
biotite muscovite	apatite	
hornblende		
Basic/Intermediate Igneous		
basic volcanic (less commonly plutonic) fragments	ilmenite/magnetite	
olivine (often serpentinized)	anatase/brookite	
calcic plagioclase	rutile	
pyroxenes	chromite	
Pegmatite		
quartz (common type)	cassiterite	
orthoclase	tourmaline	
microcline	beryl	
Na-plagioclase	topaz	
muscovite	monazite	
	fluorite	
	(other uncommon ones)	

Notes: This table has been compiled from various sources (for example, Pettijohn 1975).

Only rock fragments are conclusive by themselves; in general it will be the *suite* of minerals present that will be indicative.

minimize differences due to transport fractionation and because this is the size class in which the mode of the denser heavy minerals is likely to occur). **AS** Chapter 8 discusses analytical procedures more fully.

Interpretation of Arenite Composition

The composition of detrital sediments generally reflects the major controls of source rock character, or provenance (e.g., Blatt 1967*a*; a number of papers in Zuffa 1985); tectonism, which dictates what source rocks are uplifted into the zone of erosion (e.g., Krynine 1942; Dickinson in Zuffa 1985; Ingersoll 1990); and the combination of climate (e.g., Crook 1967; Ataman and Gokcen 1975; Basu in Zuffa 1985; Girty 1991), physiography (e.g., Johnsson and Stallard 1989), and time, which controls the extent of chemical weathering and the selective destruction of minerals that occurs on the way from the source to their site of deposition. Temperature, pressure, and chemistry of the diagenetic environment may also play an important role (but not as great as with nondetrital sediments): alterations to detrital minerals or authigenesis of new minerals indicates geochemical aspects of the early or late diagenetic environment (e.g., McBride in Zuffa 1985; Dutta and Suttner 1986). Composition at particular localities may also reflect the effects of selective sorting during transport and of those processes acting in the final depositional environment (e.g., Davies and Ethridge 1975). Lateral or vertical stratigraphic trends of changing composition are particularly powerful interpretive tools.

Composition of arenite components should not be studied

in isolation from texture: grain size and roundness are important properties that may provide essential information for interpretation of composition. For example, the relative abundance of different components depends on grain size in the first instance (Figs. 6-5, 6-7); hence two sediments might differ in composition purely as a result of one component not being available in the size range represented in the other sediment. Unexpected differences in roundness (finer grains should be less rounded than coarser grains of the same composition; softer and isotropic minerals should be better rounded than harder and anisotropic minerals) can be informative of different source areas contributing sediment or of different grain histories encountered by the different grain suites.

Provenance

Composition of the rocks in the source area is of paramount importance in dictating the composition of arenites (e.g., Fig. 6-8). Provenance may be simple and comprise only one dominant rock type, or it may be complex where there is more than one main rock type in the source area, there is more than one source area, and/or sediments are recycled from other sediments (e.g., Fig. 6-9). Textural attributes such as the roundness may indicate whether grains came directly from the *ultimate source* (the original igneous or metamorphic rock) or from an *immediate source* comprising a sedimentary rock that derived its minerals from the ultimate source. Lateral changes in composition within a sedimentary unit usually reflect different source rocks or source areas (although there are cases in which they reflect different subenvironments; e.g., Mack 1978), whereas vertical changes in a sequence of rock units may reflect either a change in dominant source area supplying the region, *or* a change in the source rock character in the same source area due to tectonic uplift and progressive unroofing of basement rocks, *or* a change in climate or the depositional environment.

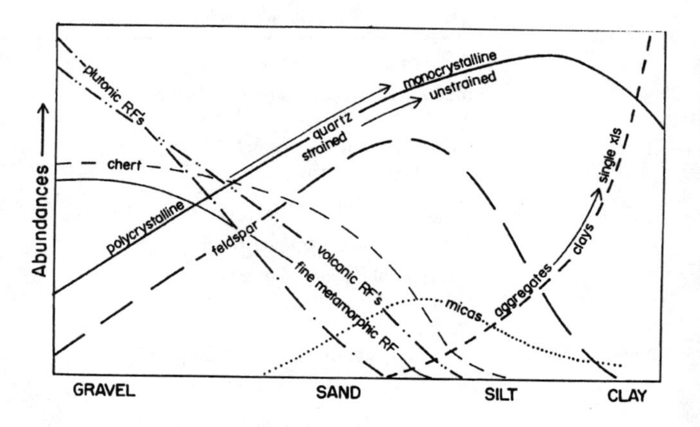

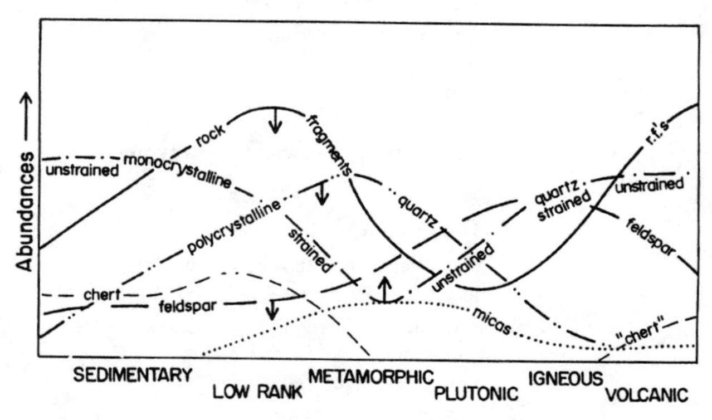

Figure 6-7. Two controls on the abundance of detrital sedimentary components. *A:* Generalization on the abundance of common detrital components in relation to grain size. Provenance and selective weathering are not the only controls on the abundance of components in a sediment. *B:* Expected abundance of sand size components as a function of provenance; arrows show whether constituent increases or decreases with weathering. Each curve should be considered separately—it is not intended that relative abundances of different components be represented. There are many exceptions to the generalizations.

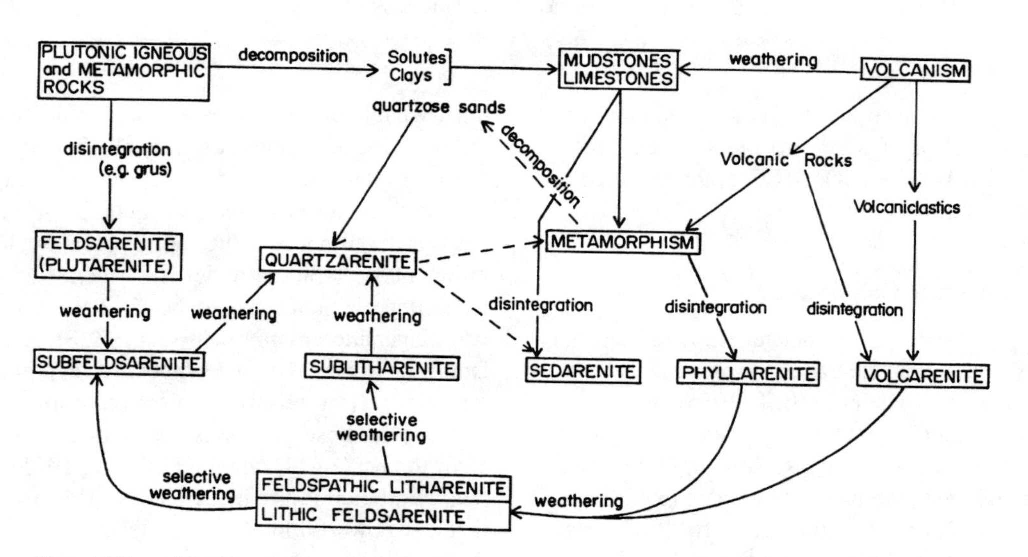

Figure 6-8. General evolutionary routes for arenites (cf. Figs. 6-1 and 6-3 for definition of arenite names).

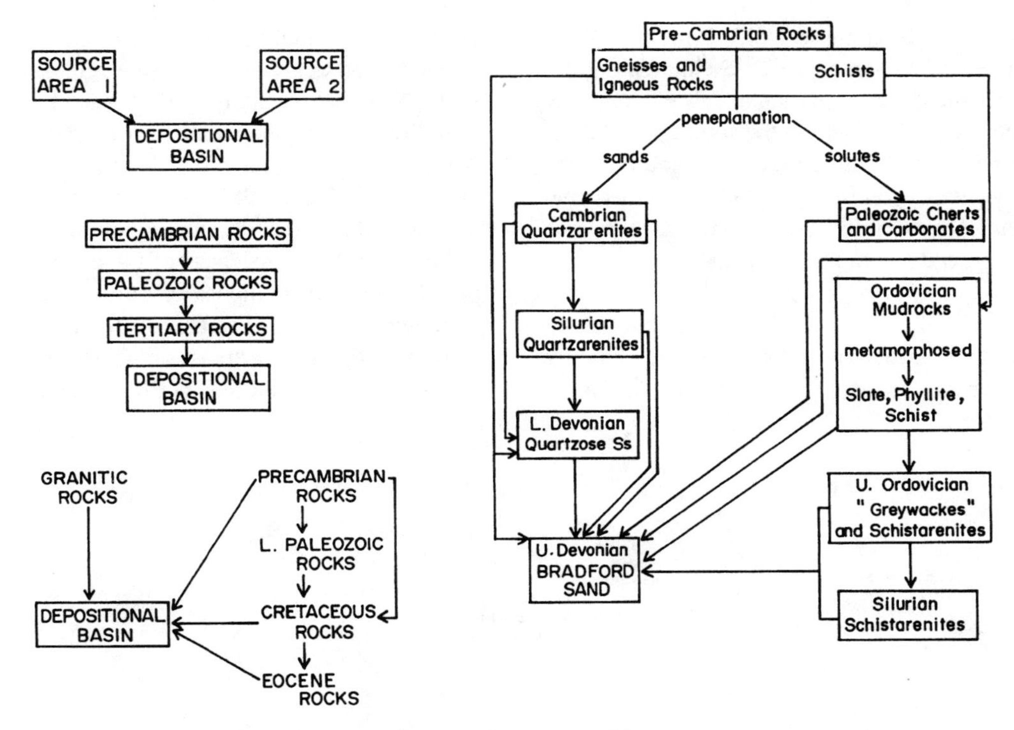

Figure 6-9. Four examples of the possible derivation and evolution of arenites. The three on the left show increasing levels of complexity; the example on the right is an actual interpretation of the evolution of a sedimentary unit in Pennsylvania, deduced by P. D. Krynine (1940).

The most informative components of arenites for provenance interpretations are rock fragments and heavy minerals (Table 6-2). Feldspars may also be very useful (Fig. 6-4). The most common component of arenites—quartz (see discussion above and Fig. 6-10)—generally requires intensive study and comparison in related sample suites to have any hope of obtaining useful information relevant to provenance or the previous history of the assemblage.

Heavy minerals are also indicative of provenance because they include a much wider range of minerals than the abundant light grain fraction, and they are particularly useful for regional paleogeographic reconstruction when multiple sources contribute to single sedimentary units (Rittenhouse 1943; Rice, Gorsline, and Osborne 1976; Morton in Zuffa 1985). Provenance indicators may be masked because of the differential chemical and physical stability of the various minerals, but these effects may prove useful in determining paleoclimate (e.g., Hubert 1962) and diagenetic history (e.g., Pettijohn 1941; van Andel 1959) and by integrating the results with interpretations from light mineral studies. In HM studies it is important to remember that because of their high specific gravity, any HMs transported with the more common sedimentary minerals will tend to be finer than those less dense grains. The hydraulic equivalence factor also varies between HMs—for instance, zircons, (s.g. 4.88) of one size grade will be associated with larger tourmalines (s.g. 3.0–3.25); applications using the principle can provide important additional information on hydrodynamic processes

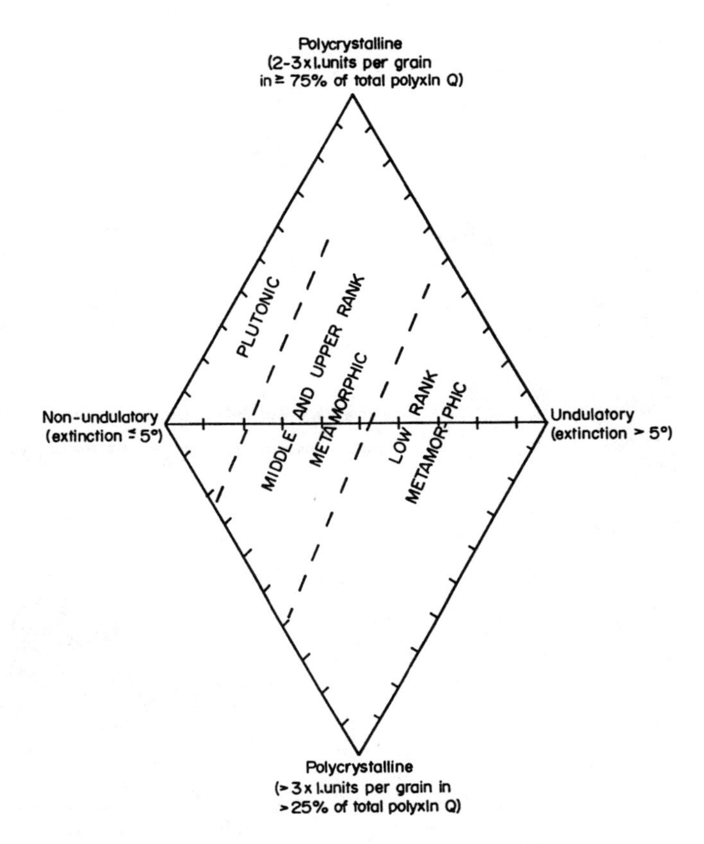

Figure 6-10. Graph for plotting quartz varieties in arenites, showing fields characteristic of particular source rocks. (After Basu et al. 1975; consult that source for limitations in its application.)

of sedimentation (e.g., Rittenhouse 1943; Lowright, Williams, and Dachille 1972; Stapor 1973; Steidtmann 1982; Slingerland 1984). In addition, particular HMs commonly occur in a restricted size range within the source rock. Thus, relative HM abundances or ratios may vary greatly in the same sediment, depending on the size grade(s) studied, and may give false indications of differences in provenance (see **AS** Chapter 8 for procedures in HM analysis).

Tectonism

The history of earth deformation in the source area broadly dictates the kind of rocks *exposed* to erosion (Chapter 2), but because different deformation histories can be superimposed in one area, interpretation of paleotectonism can be very difficult. Nonetheless, many workers attempt to relate composition directly to plate tectonic settings (Figs. 2-3, 2-4; Table 2-1; Fig. 6-11; and see Dickinson and Suczec 1979; Dickinson in Zuffa 1985). This approach is unlikely to succeed without extensive sampling of both vertical and lateral sedimentary facies, because of the extensive overlap that is possible between compositions of source rocks in different plate-tectonic settings (e.g., Fig. 6-12), and because of the possibility that source areas contain rocks of both the present and older tectonic settings. Furthermore, a spurious interpre-

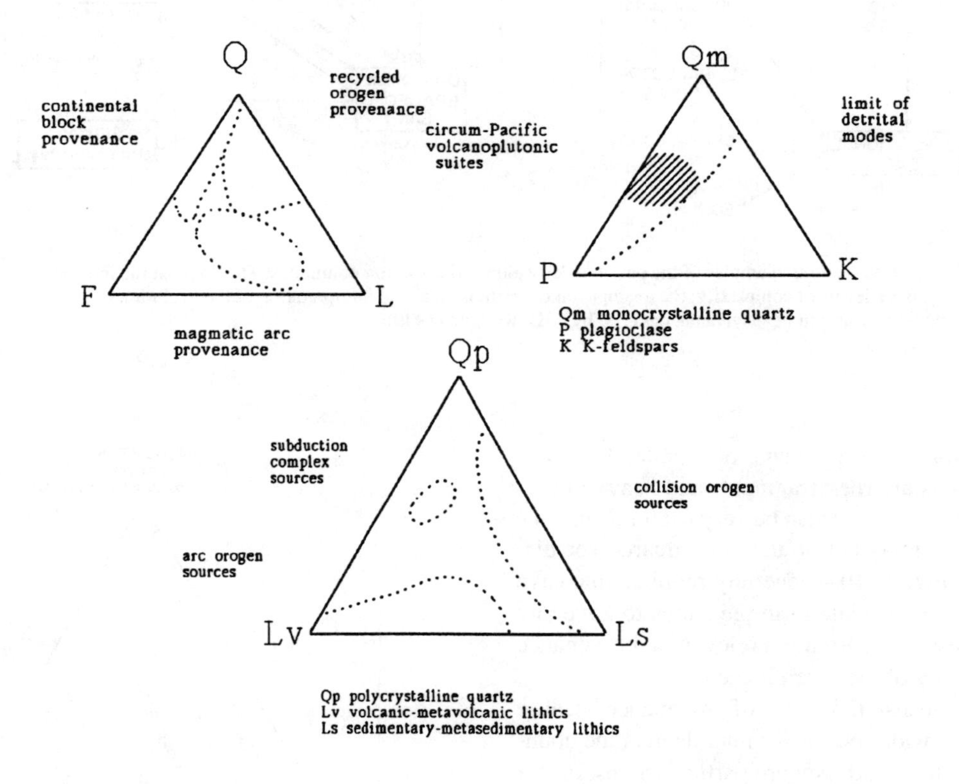

above - reported distributions in the literature

below - fields for different tectonic sources, generalized from the above

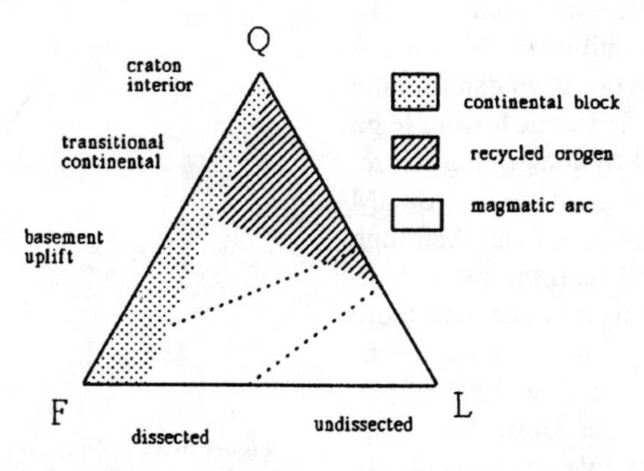

Figure 6-11. Graphical representation of generalized relationships between arenite composition and tectonic setting. (After Dickinson 1985, in Zuffa 1985.)

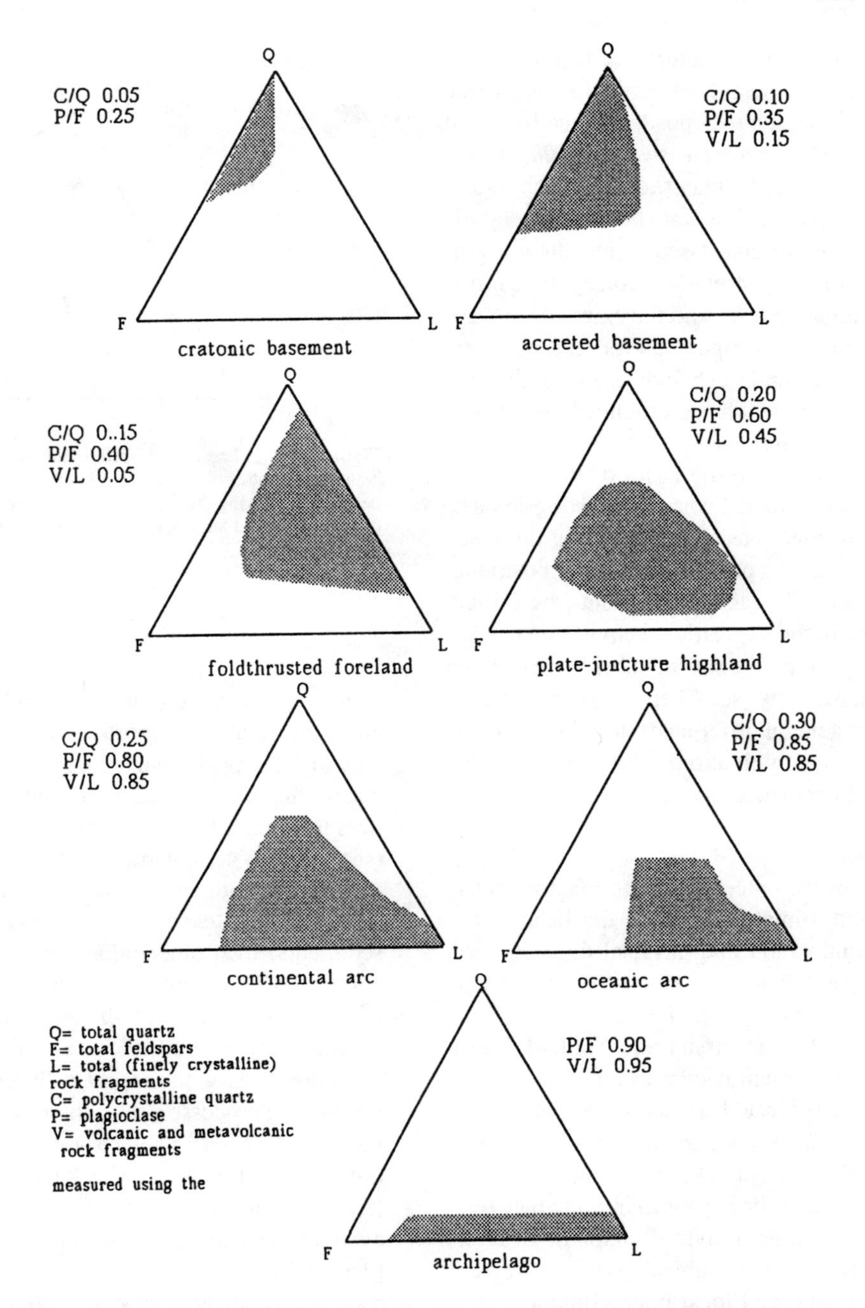

Figure 6-12. Approximate fields for the provenance of modern marine sands, showing the substantial overlap that exists; slightly better resolution is obtained when grain ratios of C/Q, P/F, and V/L are calculated. Components used in the plots are: Q = mono- and polycrystalline quartz plus chert; P = plagioclase; V = volcanic (and metavolcanic) rock fragments; L = total rock fragments of all kinds; C = polycrystalline quartz; F = total feldspars. (After Valloni 1985, in Zuffa 1985.)

tation may result whatever the sample suite because different combinations of climate-, provenance-, or environment-controlled factors can override the tectonic influence (e.g., see Mack 1984; Velbel 1985; Girty and Armitage 1989). In addition, plate tectonic regimes are on such a large scale that their interpretation may not greatly assist in interpretations of smaller-scale tectonics (e.g., of individual mountain ranges). As demonstrated by Molinaroli, Blom, and Basu (1991), none of the current schema for classifying arenites according to tectonic provenance achieve more than 85% precision in correctly identifying their own data; the geologist must always integrate additional information.

Direct relationships exist only between arenite composition and provenance; the first step in reconstructing a *paleotectonic history* is to identify the *immediate* source rocks (e.g., deep-seated metamorphic rocks), then to interpret *their* relation to the intermediate-level tectonic regime (e.g., derived from the Appalachian mountain range), and only then extend the inferences to the plate tectonic regime (e.g., convergent margin suture belt; e.g., see Ingersoll

1990). In the rock record, the paleotectonic history of the source area can be deduced best from vertical stratigraphic trends of changing composition in deposits derived from that area, which commonly reflect *progressive unroofing* as uplift occurs (e.g., oldest sediments may show a recycled sedimentary source, younger sediments show a low-grade metamorphic source, and youngest sediments show a plutonic igneous/metamorphic source; e.g., Dorsey 1988). It is also important to examine textures and the extent of mineral alteration in association with composition for tectonic inferences; for example, angular and fresh feldspars are indicative of a primary, proximal source whereas rounded and altered feldspars may reflect recycling.

The influence of tectonics in the basin of deposition is commonly inadequately considered when interpreting arenites. Tectonics of the depositional basin exert a strong influence on the rate at which sediments pass through the depositional interface and are buried. The slower sediments are buried, the less time processes in the depositional environment have to act on them. Consequently, the rate of subsidence exerts a control on the textural maturity (see Chapter 5) of sediments. Table 6-3 summarizes general differences in sediment suites resulting from different combinations of tectonics in the source area and site of deposition.

Climate

The relative intensity of tectonic and climatic factors determines relief; relief and climate determine the balance between chemical decomposition and physical disintegration of rocks and minerals, and thus the products that *can be supplied* to the depositional basin (e.g., Fig. 2-5; Basu 1976; Grantham and Velbel 1988; Johnsson and Stallard 1989). The time factor is also important insofar as it determines the duration that detritus is subjected to any set of weathering processes. Detrital clasts and secondary minerals (e.g., clays) may move directly to the final site of deposition, but in most settings they pause for varied time spans in intermediate resting places en route (e.g., in an alluvial floodplain or desert basin), in which case they may be subjected to lengthy periods of exposure to climatic and local microclimatic conditions that differ from those in the source area (e.g., Johnsson and Meade 1990). Although in some areas the climatic impress on compositional characteristics is considered by many to be less important than the tectonic impress (e.g., Suttner and Dutta 1986; Johnsson 1990), there is no doubt that it can play an influential role or that intensive and/or prolonged chemical weathering can effectively eliminate all minerals other than quartz and/or secondary minerals such as clays and iron oxides (e.g., the laterite soils; e.g., Suttner, Basu, and Mack 1981). Figure 6-13 shows an example of the kinds of differences in composition that can be expected to result from climatic effects on the same source rocks (see also Basu in Zuffa 1985).

The major effect of climate-related processes is the *selective elimination* of chemically unstable or mechanically nonresistant minerals from the original suite available in the source rock. Once the source rock character has been discovered, the relative proportion of minerals destroyed on the way to the depositional site can be used to infer the extent of weathering. A knowledge of relative mineral stabilities permits this kind of paleoclimatic (or diagenetic) interpretation (see Table 6-4). Quartz is the most stable of the common rock-forming minerals. It is usually less common in potential source rocks (except other sediments and some metasediments) than other minerals such as feldspar. Thus, the approach of an arenite to a pure quartz composition can be taken as a measure of the *mineralogical maturity* of the arenite. Zircon, tourmaline, and rutile are the most stable common HMs with respect to both physical and chemical weathering processes; thus, their ratio relative to other heavy minerals in a sediment provides a more refined index for mineralogical maturity (the *ZTR index*, e.g., Hubert 1962). (Another ultrastable HM is monazite, but it is relatively rare, even though economically important if sufficiently concentrated.) Some minerals, such as apatite and garnet, are quite resistant to physical processes, but they can be rapidly destroyed under some conditions of chemical weathering or diagenesis. Clay minerals may provide an excellent guide to paleoclimate if diagenetic changes have not been extensive (see Chapter 7).

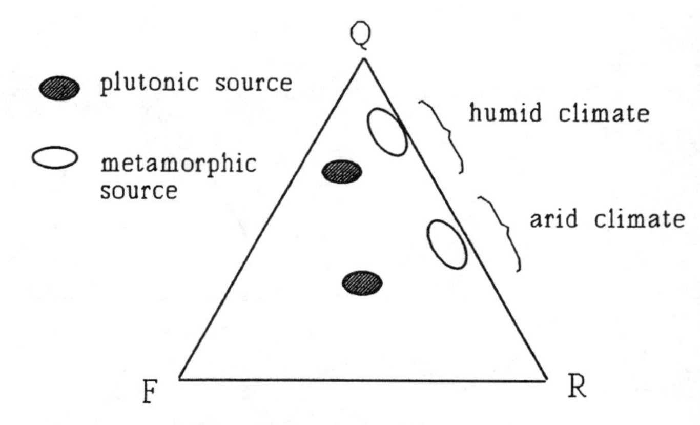

Figure 6-13. Example of differentiation in first-cycle medium sands between climatic regions where provenance is known (had the samples been counted using the Gazzi-Dickinson method, much of the clear distinction would have been lost). (After Basu 1985.)

Diagenesis

Diagenetic chemical processes may selectively alter, destroy, and/or create minerals (e.g., McBride in Zuffa 1985). In general, the changes affect detrital minerals that are present only in small quantities, but heavy minerals or feldspars that are particularly useful for interpretive purposes may be involved. To evaluate the role of diagenetic changes, compare compositions in originally permeable sediments with those of adjacent impermeable ones (e.g., early concretions or primary clay-rich layers) wherein changes are likely to be much less extensive. Relative stabilities of minerals to diagenetic conditions are

Table 6-3. Empirically Expected Relationships Between Tectonics and Sedimentary Basins

Source Area Tectonics	Tectonics of Depositional Basin		Results
Orogeny (Fold or Fault) (O) Rapid supply of mineralogically immature sediments	Rapid subsidence (R) Rapid burial of sediments, little time for environmental impress, blurred boundaries between paleoenvironments	O+R	Initial recycling of sediments, then thick deposits of mineralogically and texturally immature detrital sediments. Deep and/or shallow and/or emergent paleoenvironments, depending on balance and/or intensity of O to R (and climate). Alluvial and submarine fans and deltas with mass flow deposits. Breccias, conglomerates, and arenites proximally, with muds intermixed to dominant distally. Rapid lateral thickness variations.
	Slow subsidence (S) Slow burial of sediments, environmental processes with strong effect on sediment characteristics; sharp boundaries between paleoenvironments, which cover broad areas	O+S	Mineralogically immature but may be texturally mature due to reworking. Relatively thin, varied detrital deposits. Most sediment bypassed to more distal depocenters (e.g., via channelized routes as mass flows); typically coarse texturally immature deposits (conglomerates and arenites), with massive homogeneous muds distally.
Epeirogenic uplift (Eustatic fall) (E) Supplies of detritus lacking most unstable components; paleoclimate and time spent in intermediate environments dictate compositional characteristics		E+R	(rare) Texturally immature (due to rapid final deposition) to mature (due to previous history); mineralogical maturity depends on previous history. High proportion of nondetrital sediment, e.g., bioclastic limestones; reefs along hingeline. Arenites and mudstones commonly in deltas and/or submarine fans.
		E+S	Texturally mature detrital sediments of distal alluvial to shallow marine paleoenvironments. Cyclothems. Arenites proximally to mudstones distally.
Quiescence (Q) Minor supplies of mineralogically mature detritus		Q+R	Proximal texturally and mineralogically mature to supermature arenites. Evaporites where arid climate and restricted basinal circulation; black shales where temperate and restricted circulation. Hinge-line reefs, some mass flow deposits seaward of hinge line. Starved basins with deep-water carbonate and siliceous oozes.
		Q+S	Texturally and mineralogically supermature arenites. Common siltstones and claystones. Abundant carbonates, either bioclastic or lime mudstones. Shallow to moderate-depth marine deposits (shelf-type).

Note: All else being equal, tectonics of the source area tend to dictate mineralogical maturity, whereas tectonics of the depositional basin tend to dictate textural maturity.

similar to those for weathering conditions (Table 6-4) at temperatures below about 150°C, but as burial temperatures rise, diagenesis can involve reactions that do not occur under normal weathering (e.g., between iron carbonate siderite and hematite to produce magnetite in banded iron formations; McConchie 1987). Complex (*paragenetic*) histories of intrastratal solution, alteration, and authigenesis can be deduced from occurrences or local absences of particular minerals (e.g., alteration of original ilmenite to leucoxene, followed by authigenic growth of sphene in the same position).

Table 6-4. Mineral Stability Relationships under Sedimentary Conditions

Goldich Stability Series (generalization for common rock-forming minerals, effectively the inverted Bowen's Reaction Series without reactions)

	Quartz		
	Muscovite		
	K-feldspar		↑ Stability
Biotite		Na-plagioclase	
Hornblende		Na-Ca plagioclase	
Augite		Ca-Na plagioclase	
Olivine		Ca plagioclase	

Relative Chemical Stability of Some Minerals[a] (see Brewer 1976 for sources and discussion)

			Graham	Marel
Pettijohn (1975)		Weyl (1952)	(in Jackson and Sherman 1953)	
Anatase	More common in ancient rocks	Rutile, zircon	Quartz, muscovite,	Tourmaline
Muscovite	than modern sediments, some	Tourmaline, sphene	Rutile, zircon,	Rutile
Rutile	are authigenic	Magnetite	Tourmaline,	Staurolite
			ilmenite, andalusite,	Zircon
Zircon		Kyanite, andalusite	kyanite, sphene, magnetite	Garnet
Tourmaline		Sillimanite		Muscovite
Monazite		Epidote, garnet	Staurolite	Epidote
Garnet				Amphibole
Biotite		Augite, hornblende	Biotite, epidote, garnet, augite,	Augite
Apatite		Olivine	hornblende	Biotite
Ilmenite				Hypersthene
Magnetite			Apatite, olivine	Olivine
Staurolite				
Kyanite				
Epidote				
Hornblende				
Andalusite				
Topaz				
Sphene				
Ziosite				
Augite				
Sillimanite				
Hypersthene				
Diopside				
Actinolite				
Olivine				

[a]Significant differences shown by these workers between the relative stabilities of some minerals probably reflects different chemical conditions in the settings studied. Hence, depending on the pedogenetic/diagenetic conditions, relative positions of some minerals change.

Compositional Classes

Quartzarenites

Quartz-rich arenites result from preferential destruction of other minerals during the sedimentary cycle. Quartzarenites therefore reflect intensive chemical weathering of source rocks or of the grain assemblage after derivation from the source, as might be expected in a warm humid climate (e.g., Akhtar and Ahmad 1991); alternatively a prolonged time of weathering in a rigorous climate or recycling from sedimentary source rocks might produce similar results (e.g., Suttner, Basu, and Mack 1981). Grain roundness sometimes can help to distinguish between the intense or prolonged weathering/recycling alternatives; better rounding commonly results from more mechanical abrasion, which is likely in the latter cases. Most quartzarenites, particularly those with *well*-rounded grains, reflect recycling from source rocks of older quartz-rich sediment. Quartz requires a truly extensive history of abrasion to become well rounded, and this is rarely achieved in one sedimentary cycle, however long (localized exceptions may exist in the case of quartz-rich sands subjected to prolonged eolian or beach processes). Unfortu-

Relative Resistance to Abrasion of Some Minerals (see Pettijohn 1975 for sources and discussion)

after Friese (1931)		after Thiel (1940)	
Tourmaline	Epidote	Quartz	Rutile
Pyrite	Olivine	Tourmaline	Hypersthene
Staurolite	Apatite	Microcline	Apatite
Augite	Kyanite	Staurolite	Augite
Topaz	Andalusite	Sphene	Hematite
Magnetite	Orthoclase	Garnet	Kyanite
Garnet	Monazite	Epidote	Fluorite
Ilmenite	Hematite	Zircon	Siderite
		Hornblende	Barite

Fieldes and Swindale (1954)	Smithson (1941)	Dryden and Dryden (1946)	Clay-Size Minerals (after Jackson and Sherman 1953)
Quartz	Zircon, rutile, tourmaline, apatite	Zircon	Anatase, zircon, rutile, ilmenite, leucoxene, corundum, etc.
		Tourmaline	
Feldspars, acid	Monazite	Sillimanite	Hematite, geothite, limonite, etc.
Volcanic glass		Monazite	
	Garnet, staurolite, kyanite	Chloritoid	Gibbsite, boehemite, allophane, etc.
Muscovite, biotite		Kyanite	Kaolinite, halloysite, etc.
	Mafic minerals	Hornblende	
Zeolites, basic volcanic glass		Staurolite	Montmorillonite, beidellite, saponite, etc.
		Garnet	Mixed 2:1 layer clays and vermiculite
Augite, hornblende, hypersthene, olivine		Hypersthene	
			Muscovite, sericite, ilite
			Quartz, crisotbalite, etc.
			Albite, anorthite, stilbite, micro-cline, orthoclase, etc.
			Biotite, glauconite, Mg-chlorite, antigorite, nontronite, etc.
			Olivine, hornblende, pyroxenes, diopside, etc.
			Calcite, dolomite, aragonite, apatite, etc.
			Gypsum, halite, Na-nitrate, ammonium chloride, etc.

nately, chemical weathering can also round grains: corners and edges of grains have a much higher surface area to volume ratio than flat faces and solution concentrates on these more reactive areas (finer grains tend to round faster than coarser grains, in contrast to the effects of abrasion). Chemical rounding usually produces embayments that are lacking on mechanically rounded grains, but it may require roundness comparison between different sizes and minerals (i.e., of different chemical and physical resistance) in the same sediment to distinguish between chemical and mechanical rounding effects. The presence of abraded overgrowths would indicate the recycled character of the grains. A heavy mineral assemblage containing only the ultrastable minerals (e.g., a high ZTR index) would also be more likely in recycled than first-cycle sediments; survival of other minerals that are resistant to physical abrasion but less resistant to chemical weathering suggests prolonged exposure under conditions of relatively low-intensity chemical weathering (Table 6-4). For example, the presence of rounded feldspars suggests an arid paleoclimate.

The relative abundance of polycrystalline strained quartz to monocrystalline unstrained quartz may be indicative of abrasion history, because the former is preferentially destroyed; however, its original abundance and character are dictated by provenance. Trends within, or differences between, units in stable versus unstable polycrystalline quartz grains (see Young 1976; Mack 1978) or of undulose (extinction >5° stage rotation) versus nonundulose (extinction <5°) versus polycrystalline (2–3 vs. >3 crystal units per grain) may also prove useful in provenance determinations (see Basu et al. 1975 and Fig. 6-10). (*Note:* When studying the relative proportions of these quartz varieties, comparisons must be made within the same size fractions. In addition, when provenance is uncertain, interpretations usually must be based on vertical stratigraphic trends.)

Quartzarenites are expected to form under stable, quiescent tectonic settings in, or on the edges of, cratons. A sheet geometry, reflecting braidplain sedimentation (e.g., Fedo and Cooper 1990) or nearshore deposits of marine transgression during epeirogenic subsidence (or occasionally, regression during tectonic rejuvenation), is typical, but some occur as shoestring beach or bar deposits in more active tectonic regimes. Most of these deposits are texturally mature or supermature, but textural inversions such as bimodal grain roundness, bimodal size distributions, or more than 5% clayey matrix are also common. Diagenetic destruction of accessory minerals and development of grain overgrowths are common. In some strata, original high-permeability subfeldsarenites are converted to quartzarenites by the diagenetic (or epigenetic) destruction of feldspars (e.g., McBride 1987); indication of this change may be an irregular distribution and size of voids or the presence of disseminated clay patches (often lost during thin-section preparation). Cements in quartzarenites are most commonly silica (usually as grain overgrowths), carbonates, or iron oxides. Disseminated specks of hematite may obscure the nature of clayey matrix or carbonate cement and impart a distinctive red color to the rock; this hematite may form in the depositional environment (specks concentrated on the rims of the detrital grains) or as a result of diagenetic or epigenetic processes (specks concentrate outside the rim of overgrowths).

Feldsarenites

Feldspar is the commonest mineral in most ultimate source rocks. Whenever chemical weathering is restricted, either due to arid climatic conditions or because too little time is available, feldspars should be abundant in the sand size fraction. All gradations from feldsarenites to subfeldsarenites are common.

Tectonic feldsarenites are produced when the rate of sediment supply is too rapid for chemical destruction of the feldspars (i.e., mechanical fragmentation dominates over chemical decomposition). Characteristically, tectonism will have created a high relief, such that physical processes of erosion dominate and grains are rapidly moved from the source rocks into the site of deposition and rapidly buried there. Block faulting, resulting in the exposure of originally deep-seated igneous and metamorphic rocks, is a typical cause. The overall geometric pattern for feldsarenites and associated facies is characteristically wedge-shaped (as in alluvial fans). Nonmarine deposits are generally abundant and are commonly red because of deposition and diagenesis under oxidizing conditions; more distal deposits commonly show drab colors because these environments are marine or have a consistently high groundwater table that inhibits oxidation. Conglomerates are characteristically associated with tectonic feldsarenites, and the gravel clasts therein commonly show the range of rock types in the source area. Such coarse deposits may be sparse in the more distal or stratigraphically highest parts of the succession, but in there the coarser sand grades commonly contain diagnostic rock fragments (especially polycrystalline feldspars or feldspars plus quartz or mica). Chemically unstable minerals (such as biotite or amphibole) are commonly present. Feldspar grains are typically fresh, but some may be partly weathered (particularly the calcium plagioclases, but see Todd 1968); frequently, both fresh and partly weathered feldspars of the same species are present, because different grains have been exposed to different degrees of chemical weathering. Grains are typically angular and most deposits are texturally immature. Muds are common as a matrix to the sandstones, and as silty interbeds.

Climatic feldsarenites are produced when climatic conditions are sufficiently arid or cold that chemical weathering is inhibited. Minerals chemically more unstable than the alkaline feldspars are generally lacking (although chemical weathering processes are relatively restricted, the sediments have been exposed to them long enough for widespread destruction of the less stable components). The ultrastable minerals commonly dominate the heavy mineral suites, but less stable minerals (such as garnet and apatite) should also be present. Microcline is typically fresh, orthoclase is fresh to weathered, and most plagioclase is destroyed or altered beyond the level of species identification. Authigenic mineral growths (such as quartz and feldspar overgrowths) are generally abundant, and diagenetic destruction of chemically unstable minerals may be extensive (as with quartzarenites, most deposits are highly porous and permeable). Tectonic quiescence or slow broad-scale movements prevail and the grains commonly are subject to extensive physical reworking before they are finally buried; hence grains are commonly rounded and the overall geometry is characteristically sheetlike. Most arenites are texturally mature or supermature, but bimodal size distributions are common. Deposits are commonly gray (or red) and can be mistaken for quartzarenites in the field. Conglomerates are sparse, are relatively finegrained, and consist of stable minerals and rock fragments such as chert, polycrystalline quartz, and silicified acidic volcanics. Muds are generally rare and well segregated (probably due to deflation), whereas gravels commonly interfinger with the sands.

Whereas final accumulation may occur under transgressive (or regressive) marine conditions, the environments in which climatic feldsarenites acquire many of their characteristics were probably analogous to deserts like those of central

Australia today. However, prior to the development of extensive terrestrial vegetation (which acts to bind sediment and retain water as well as to contribute powerful chemical weathering agents via root action and decay products), it is probable that suitable conditions for climatic feldsarenite accumulation existed in more temperate and humid climates than today. Certainly such deposits are much more common in Precambrian and Lower Paleozoic sequences than in younger strata (possibly also because of the limited possibilities of recycling in that early part of Earth history).

Recycled feldsarenites are produced when tectonism uplifts poorly indurated feldspathic sediments and/or volcanics, either in an arid climatic setting where chemical weathering is inhibited or where transport and deposition are very rapid. Feldspars are typically altered. Characteristically, there will be sedimentary (and volcanic) rock fragments present—all gradations to, and associations with, sublitharenites, lithic feldsarenites, and sedlitharenites are likely. Distinction between altered feldspars, volcaniclasts, and sedimentary rock fragments may be difficult. The tectonic setting is commonly dominated by fold deformation, which creates highs that are continuously eroded during uplift and depositional troughs that are continually sinking to receive the detritus; however, early deposits from fault-block movements or from rapid epeirogenic uplift may fall into this category. Textural characteristics are varied, depending on the previous sedimentary history of the grains and the depositional locale, but grains are typically subround and deposits are immature. Because of the intermediate character of the tectonic setting (neither broad, slow uplift nor local extensive vertical uplift), the geometric characteristics of the deposits are intermediate between the tectonic and climatic feldsarenites (thinner and broader than the former, thicker and more localized than the latter). Deltaic and submarine fan deposition of thick alternating sandstones and mudstones is most likely. Conglomerates, with dominant sedimentary and/or volcanic rock fragments, may be either common or sparse.

Igneous feldsarenites are produced essentially in association with intermediate to basic igneous activity, hence form mainly near island arcs or plate boundaries. These sediments may be derived by winnowing from extrusive volcanic deposits, in which case they show gradation to, and association with, volcarenites; or they may be derived from granodiorites and more basic intrusives, in which case plutonic rock fragments (such as polycrystalline feldspars) are generally present in the coarser grain sizes. Distribution and characteristics of these deposits are not well known: they are usually found in sequences that have been extensively deformed (most have probably been metamorphosed out of the sedimentary class). These sediments are mainly associated with volcarenites and recycled feldsarenites; hence their history appears intimately interrelated. Compositionally they are characterized by a dominance of calcic plagioclases, and because of the ease with which these minerals are destroyed by chemical weathering, rapid transport and burial are necessary for preservation. Grains are thus angular, and texturally immature deposits are most common; however, well-sorted igneous feldsarenites are known (e.g., in the Mesozoic rocks of New Zealand; DWL personal observations). Alteration of feldspars to various clay minerals commonly obscures their differentiation from clayey matrix or fine-grained rock fragments.

Cements of the feldsarenites vary, depending on texture as much as composition. In well-sorted deposits, both quartz and alkali feldspar overgrowths tend to be abundant; feldspar overgrowths generally appear to develop first, because euhedral crystal outlines are much more common than with quartz overgrowths, suggesting that there was less interference to their growth on the parent grains. Where matrix content is high, or there are abundant micas (or relatively soft rock fragments) that deform easily and are squeezed between framework grains during compaction, cements are poorly developed (many tectonic feldsarenites). Where volcanic debris is involved, cements may comprise zeolites, which require refined analytical techniques to identify precisely.

Litharenites

Most litharenites are deposited during times of orogenic activity. Physical weathering processes are as effective as chemical weathering in destroying rock fragments (and are more effective in the case of most sedimentary and metamorphic clasts; e.g., Cameron and Blatt 1981). Abrasion stresses the internal planes of weakness between crystals in the fragments, which progressively disintegrate into smaller grains until monocrystalline mineral fragments remain (coarsely crystalline parent rocks are lost by the time grains are in the +2 to +3 ϕ range; see Fig. 6-5). Hence, coarse-grained litharenites commonly are associated with fine-grained sublitharenites; both result from the same geological history. In some litharenites, differentiation of fine-grained rock fragments from clayey matrix is difficult, which is a problem that is commonly increased by the creation of clayey matrix during diagenetic breakdown of unstable accessory minerals and feldspars.

Volcarenites are common in active continental margin plate tectonic settings. Volcanic clasts may be sparse or dominate in an arenite, depending both on proximity to extrusive centers and on geologic history; wherever they are present, much larger volumes of volcanic ash (probably indistinguishable from sedimentary muds) are likely to occur nearby (as matrix or in lateral facies). When volcanic clasts dominate, there is a possibility that the deposit represents a direct airfall volcanic tuff; sedimentary structures and textures (such as grain roundness and sorting) can be used to distinguish between direct air fall and exclusively sedimentary deposits, although all gradations exist (e.g., see Smith 1988). Resedimented volcarenites are commonly difficult to differentiate from direct air fall or pyroclastic flow and surge deposits (cf. Table 3-9). Because of the fine interlocking fabric of the parent rock, many volcanogenic clasts can survive considerable abrasion and retain recognizable textures into the fine sand grades. However, chemical weathering and diagenetic alteration are effective in destroying the intermediate and basic volcaniclasts, and they commonly obscure the original character of clasts such that they become difficult to distinguish from other finely textured clasts (such as mudstones, chert, and phyllitic metamorphics; see Fig. 6-6). Mineralized vesicles, devitrified glass texture, and remnant lathlike feldspars are the main recognition criteria.

Sedlitharenites are characteristic of the early phases of both block faulting and fold deformation; in the former they are commonly overlain by plutonic litharenites and associated with conglomerates, whereas in the latter, conglomerates may be few and overlying arenites may have progressively greater proportions of low-grade metamorphic detritus (which may be difficult to distinguish in thin section from sedimentary rock fragments). Slow, broad-scale tectonic uplift may also produce sedlitharenites, but in such settings erosion is usually slow enough to disarticulate the clasts into their original monocrystalline particles.

Sandstone litharenites are rare because clasts commonly break down into monocrystalline sand grains. However, when the source rock is a silica-cemented quartzarenite, clasts may be common; problems may arise in distinguishing such clasts from plutonic polycrystalline quartz and metaquartzite fragments (see discussion above and Fig. 6-6) unless overgrowths developed during cementation are evident.

Mudrock litharenites are common in some sequences. Mudrock clasts are little affected by chemical weathering (there may be oxidation rims) and may be very tough (as when they are silicified). The main problem in thin section becomes one of distinguishing such clasts from altered volcanics and low-grade metamorphics (see Fig. 6-6), or from any muddy matrix that may be present in the litharenite (hold the thin section up to a light source to distinguish boundaries and look at hand specimens: subtle changes in the textural pattern formed by fine clay minerals or the distribution of very fine opaque components—e.g., hematite or carbonaceous matter—commonly provide good clues to aid this distinction). Evidence must also be sought to distinguish detrital from perigenic (locally derived) mudstone clasts.

Calclitharenites are produced only where erosion, transportation, and deposition are very rapid: hence they are characteristic of block-fault or similar tectonic settings associated with rapid uplift. Carbonates are very unstable in the terrestrial weathering environment (rainwater and most groundwaters are acidic), and carbonate minerals are soft (hardness of 3–4). Calclithrudites are generally intimately associated with these rocks and often dominate volumetrically; they may also be produced—albeit rarely—in the submarine environment, whereas submarine production of calclitharenites is most unlikely. Care must be taken in order to distinguish detrital from perigenic carbonate clasts (intraclasts; see Chapter 8).

Chert arenites are uncommon, but chert sublitharenites are not. Chert is not common as a source rock type, but it is an important component of some local successions (mostly associated with limestones), and it can form thick bedded deposits. Chert is a product of diagenetic (or pedogenetic) silicification of mud, carbonate, or other material and it also can form as a primary precipitate (see Chapter 9). The necessary silica is commonly remobilized from organic hard parts (diatoms, radiolaria, sponge spicules, and the like) or derived during diagenesis from the alteration of unstable silicate minerals or the pressure solution of grains. Because chert is chemically and physically very stable and resistant, it survives into even the finest sand sizes. Clasts from original chert source rocks may be difficult to distinguish from other fine silicious or silicified clasts (Fig. 6-6).

Phyllarenites are produced when high rates of uplift and erosion of low-grade metamorphic rocks enable rapid transport of sediment to a site of deposition. Because of the micaceous foliation, clasts are relatively easily broken down by physical abrasion; conglomerates of these clasts are uncommon, although fine to very fine sand particles may be abundant. The common original chloritic composition of many such clasts results in relatively efficient chemical alteration of these components to sedimentary clay minerals; either weathering or diagenetic alteration can render the clasts very difficult to distinguish in thin section from mudstone or volcaniclasts (Fig. 6-6), as well as from primary clayey matrix. Phyllarenites may be produced in the penultimate stages of complex fold-dominated orogeny, in which case they will stratigraphically overlie recycled sediments (such as recycled feldsarenites and sedlitharenites) and underlie plutonic litharenites; or they may be produced by fault-block rejuvenation of regionally metamorphosed sequences. Subphyllarenites are more common than phyllarenites.

Plutarenites (plutonic litharenites) are produced when crystalline plutonic rocks are rapidly uplifted and eroded. Hence plutarenites are generally coarse grained and are characteristic of rapid supply from block-faulted areas or areas in the final phases of complex orogeny. In the former case, they commonly occur in wedge-shaped units (such as alluvial fans), and stratigraphically above sedlitharenites or quartzarenites, reflecting progressive unroofing of the basement, but in cratonic regions there may be no sedimentary cover. In the latter case, they commonly overlie a thick succession of recycled feldsarenites and sedlitharenites and/or phyllarenites. All gradations to feldsarenite and sublitharenite are common, and they are generally texturally immature, although some well-sorted deposits do occur. Because the mean grain size of minerals in plutonic rocks is coarser than most sand sizes, plutarenites are rare and difficult to recognize (plutarudites—see next section—are much more common).

RUDITES

For a qualitative description compatible with the arenite classification discussed previously, the most common mineral or rock fragment name is appended to the suffix *rudite*. A number of names may be presented, with the convention that the one nearest the suffix is the most common, for example, chert-granite-rudite. Contractions such as ignirudite, plutarudite, volcorudite, and phyllarudite are necessary to avoid connotations of origin that arise with the terms igneous rudite

A

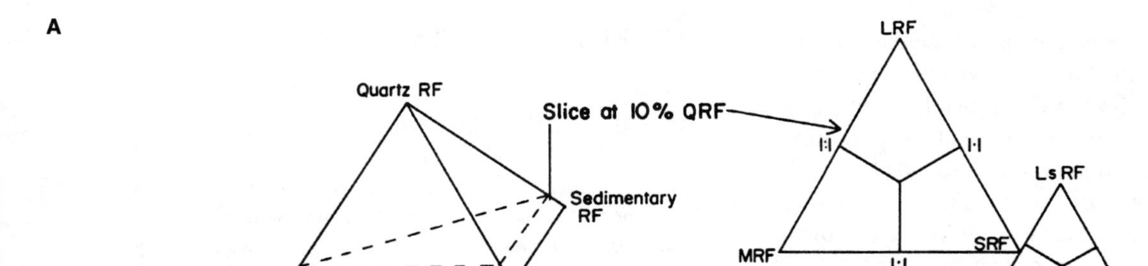

B

EPICLASTIC	EXTRAFORMATIONAL (detrital clasts from outside depositional basin)	ORTHOCONGLOMERATES (matrix <15%)	Metastable clasts <10%	ORTHOQUARTZITIC (OLIGOMICT) CONGLOMERATE (pebbles of one type only)
			Metastable clasts >10%	PETROMICT CONGLOMERATE (specify dominant clast type; e.g., petromict limestone cgl.)
		PARACONGLOMERATES (matrix >15%) also termed diamictites	laminated matrix	LAMINATED CONGLOMERATIC MUDSTONE OR ARGILLITE
			nonlaminated matrix	TILLITE (glacial)
				TILLOID (not glacial); pebbly mudstones, olistostromes, etc.
	INTRAFORMATIONAL (clasts from sediments *within* the depositional basin)	INTRAFORMATIONAL CONGLOMERATES AND BRECCIAS (specify dominant clast type)		
PYROCLASTIC	VOLCANIC BRECCIAS (derived from previously deposited volcanics) VOLCANIC AGGLOMERATES (formed of lava solidified in flight)			
CATACLASTIC	LANDSLIDE AND SLUMP BRECCIAS (not involving fluidlike flow, which generates tilloids)			
	FAULT AND FOLD BRECCIAS (i.e., direct in situ results of tectonism)			
	COLLAPSE AND SOLUTION BRECCIAS (resulting from solution of underlying material: salt, limestone, ?)			
METEORIC	IMPACT OR FALLBACK BRECCIAS			

Figure 6-14. Alternative schemes for conglomerate: rudite classification. *A:* Example of a compositional classification, based on the Folk, Andrews, and Lewis 1970 system, for a rudite suite with 10% or less quartz clasts. Any triangular slice from the tetrahedron could be used for quantitative plots and daughter triangles constructed as in the example. Nomenclature (see text) would follow the textural name of the conglomerate or breccia. *B:* The classification of conglomerates and breccias of Pettijohn (1975), which combines textural and compositional characteristics for "common" groupings.

or metamorphic rudite. Cements and nondetrital grains should be treated the same as they are in the arenite classification system.

For quantitative plots, triangular diagrams are desirable because they are easy to construct and permit graphical distinction within multicomponent systems. However, rudites are predominantly composed of *four* common components: quartz, sedimentary, igneous, and/or metamorphic rock fragments. Thus, to obtain a triangle, one must take a slice from a tetrahedron at a stated value (or restricted range of values) of one component (e.g., Fig. 6-14*a*). This classification procedure parallels that for arenites, but has not received much practical testing.

A widely recognized rudite classification is that of Pettijohn (1975, see Fig. 6-14*b*). It has a major drawback of combining textural, compositional, and genetic terms. However, some of the terms are widely used independently of the remainder of the classification. For example, the auxiliary term *polymictic* (see Pettijohn 1975) is useful where clasts of more than one lithologic type are present. A more modern approach to describing and interpreting rudites (particularly breccias) is provided by Laznicka (1988).

Gravel-size fragments of muddy sediments are commonly generated within a basin of deposition by erosion of semi-consolidated deposits (e.g., Williams 1966; Smith 1972). These fragments may be termed *intraformational* (as in the Pettijohn 1975 scheme), but to many the name implies mass movement rather than fluid flow during emplacement of the clasts. Alternative terms are *penecontemporaneous* or *perigenic* (in contrast to *allogenic* or *authigenic*; see Lewis 1964). The term *intraclast* is also sometimes used in this way, but has the preemptive connotation of a locally derived carbonate clast.

Interpretations from rudites are like those from arenites (e.g., Twenhofel 1947): provenance, tectonics (e.g., Wilson 1970), climate (by its influence on chemical destruction of unstable components), and the history of abrasion (e.g., Abbott and Peterson 1978).

LUTITES

For qualitative description, it would seem appropriate to append the component mineral names to the suffix *lutite*, with the convention that the name nearest the suffix is the most common, e.g., clay-quartz-lutite, quartz-illite-lutite, and so on. Cements and nondetrital components can be treated as they are with arenites.

Most lutites are predominantly composed of quartz, feldspar, and clay minerals (see Chapter 7; Yaalon 1962; Shaw and Weaver 1965). A primary triangle with these three poles can be subdivided into three fields with boundaries at ratios of 1:1 between each component. Secondary, and even tertiary, triangles may be derived from each apex to suit the individual worker's requirements; for example, at the clay apex, a secondary triangle could be formed with end members of illite, montmorillonite, kaolinite. Quantitative plotting will generally require determination of composition by specialized laboratory analysis (e.g., X-ray diffraction; see **AS** Chapter 8).

Note that the term *shale* is widely used for mudrocks (generally excluding siltstones; see Tourtelot 1960); implications of fissility (a secondary feature) confuse its application and render it unsatisfactory when the primary sedimentary characteristics of the rock are being described. The term *argillite* is also applied by some to a mudstone or claystone that is intermediate between a shale and a slate.

Few studies of the mineralogical composition of mudrocks in detail have been attempted, particularly of the nonclay mineral components, which average about 40% (Shaw and Weaver 1965). See Potter, Maynard, and Pryor (1980), Hesse (1984), Weaver (1989), and Chamley (1989) for general reviews of knowledge on mudrock sedimentology, and Blatt (1985) for a discussion of the potential for using the nonclay mineral components of mudrocks for interpretations.

SELECTED BIBLIOGRAPHY

Arenites—General

Boswell, P. G. H., 1960, The term greywacke. *Journal of Sedimentary Petrology* 30:154–7.

Brewer, R., 1976, *Fabric and Mineral Analysis of Soils.* Robert E. Krieger Publishing, Huntington, New York, 482 p.

Cameron, K. L., and H. Blatt, 1971, Durabilities of sand-size schist and "volcanic" rock fragments during fluvial transport, Elk Creek, Black Hills, South Dakota. *Journal of Sedimentary Petrology* 41:565–76.

Carozzi, A. V., 1960, *Microscopic Sedimentary Petrography.* Wiley, New York, 405p.

Davies, D. K., and F. G. Ethridge, 1975, Sandstone composition and depositional environment. *American Association of Petroleum Geologists Bulletin* 59:239–64.

Dickinson, W. R., 1970, Interpreting detrital modes of graywacke and arkose. *Journal of Sedimentary Petrology* 40:695–707.

Dott, R. H., Jr., 1964, Wacke, greywacke and matrix—what approach to immature sandstone classification? *Journal of Sedimentary Petrology* 34:625–32.

Dryden, L., and C. Dryden, 1946. Comparative rates of weathering of some common heavy minerals. *Journal of Sedimentary Petrology* 16:91–96.

Fedo, C. M., and J. D. Cooper, 1990, Braided fluvial to marine transition: The basal Lower Cambrian Wood Canyon Formation, southern Marble Mountains, Mojave Desert, California. *Journal of Sedimentary Petrology* 60:220–34.

Fieldes, M., and L. D. Swindale, 1954, Chemical weathering of silicates in soil formation. *New Zealand Journal of Science and Technology* 36B:140–154.

Fisher, R. V., and H.-U. Schmincke, 1984, *Pyroclastic Rocks.* Springer-Verlag, New York, 472p.

Folk, R. L., 1954, The distinction between grain size and mineral composition in sedimentary rocks. *Journal of Geology* 62: 344–59.

Folk, R. L., 1980, *Petrology of Sedimentary Rocks.* Hemphill, Austin, Tex., 182p.

Folk, R. L., P. B. Andrews, and D. W. Lewis, 1970, Detrital sedimentary rock classification and nomenclature for use in New Zealand. *New Zealand Journal of Geology and Geophysics* 13: 937–68.

Friese, F. W., 1931, Untersuchungen von mineralal auf abnutzbarkeit bei verfrachtung im wasser. *Mineralogische und Petrographische Mitteilungen* 41:1–7.

Ingersoll, R. V., T. F. Bullard, R. L. Ford, J. P. Grimm, J. D. Pickle, and S. W. Sares, 1984, The effect of grain size on detrital modes: A test of the Gazzi–Dickinson point-counting method. *Journal of Sedimentary Petrology* 54:103–16.

Jackson, M. L., and G. D. Sherman, 1953, Chemical weathering of minerals in soils. *Advances in Agronomy* 5:219–318.

Klein, G. deV., 1963, Analysis and review of sandstone classifications in the North American geological literature, 1940–1960. *Geological Society of America Bulletin* 74:555–75.

Lewis, D. W., 1964, "Perigenic": A new term. *Journal of Sedimentary Petrology* 34:875–6.

Lewis, D. W., 1971, Qualitative petrographic interpretation of Potsdam Sandstone (Cambrian), Southwest Quebec. *Canadian Journal of Earth Sciences* 8:853–82.

Lyell, C., 1837, *Principles of Geology,* vol. 1, 5th ed. John Murray, London, 462p.

McBride, E. F., 1987, Diagenesis of the Maxon Sandstone (early

Cretaceous), Marathon region, Texas: A diagenetic quartzarenite. *Journal of Sedimentary Petrology* 57:98–107.

McConchie, D. M., 1987, The geology and geochemistry of the Joffre and Whaleback Shale Members of the Brockman Iron Formation, Western Australia. In P. Appel and G. LaBerge (eds.), *Precambrian Iron Formations.* Theophrastus Publications, pp. 541-601.

Mack, G. H., 1978, The survivability of labile light mineral grains in fluvial, aeolian, and littoral marine environments: The Permian Cutler and Cedar Mesa Formations, Moab, Utah. *Sedimentology* 25:587–604.

Oriel, S. S., 1949, Definitions of arkose. American Journal of Science 247:824–29.

Pettijohn, F. J., 1943, Archean sedimentation. *Geological Society of America Bulletin* 54:925–72.

Pettijohn, F. J., 1963, Chemical composition of sandstones—excluding carbonate and volcanic sands. *U.S. Geological Survey Professional Paper 440-S,* 21p.

Pettijohn, F. J., 1975, *Sedimentary Rocks,* 3d ed. Harper & Row, New York, 628p.

Runkel, A. C., 1990, Lateral and temporal changes in volcanogenic sedimentation: Analysis of two Eocene sedimentary aprons, Big Bend region, Texas. *Journal of Sedimentary Petrology* 60: 747–60.

Scholle, P. A., 1979, A color illustrated guide to constituents, textures, cements and porosities of sandstones and associated rocks. *American Association of Petroleum Geologists Memoir 28,* Tulsa, Okla., 201p.

Smithson, F., 1941, The alteration of detrital minerals in the Mesozoic rocks of Yorkshire. *Geological Magazine* 78:97–112.

Thiel, G. A., 1940, The relative resistance to abrasion of mineral grains of sand size. *Journal of Sedimentary Petrology* 10:103–124.

Weyl, R., 1952, Studies of heavy minerals in soil profiles. *Zeitschrift Pflernahr Dung., Bodenkunde* 57:135–141.

Wolf, K. H., 1971, Textural and compositional transitional stages between various lithic grain types (with a comment on "interpreting detrital modes of greywacke and arkose"). *Journal of Sedimentary Petrology* 41:889.

Arenites—Provenance

Basu, A., S. W. Young, L. J. Suttner, W. C. James, and G. H. Mack, 1975, Re-evaluation of the use of undulatory extinction and polycrystallinity in detrital quartz for provenance interpretation. *Journal of Sedimentary Petrology* 45:873–82.

Blatt, H., 1967*a*, Provenance determinations and recycling of sediments. *Journal of Sedimentary Petrology* 37:1031–44.

Blatt, H., 1967*b*, Original characteristics of clastic quartz grains. *Journal of Sedimentary Petrology* 37:401–24.

Boggs, S. J., 1968, Experimental study of rock fragments. *Journal of Sedimentary Petrology* 38:1326–39.

Gilligan, A., 1919, The petrology of the Millstone Grit of Yorkshire. *Quarterly Journal of the Geological Society of London* 75:251–92.

Keller, W. R., and R. F. Littlefield, 1950, Inclusions in the quartz of igneous and metamorphic rocks. *Journal of Sedimentary Petrology* 20:74–84.

Krynine, P. D., 1940, *Petrology and Genesis of the Third Bradford Sand.* Pennsylvania State College Min. Ind. Experimental Station Publ. No. 29, 134p.

Pittman, E. D., 1970, Plagioclase feldspar as an indicator of provenance in sedimentary rocks. *Journal of Sedimentary Petrology* 40:591–8.

Sanderson, I. D., 1984, Recognition and significance of inherited quartz overgrowths in quartz arenites. *Journal of Sedimentary Petrology* 54:473–86.

Van Der Plas, L., 1966, *The Identification of Detrital Feldspars.* Developments in Sedimentology 6, Elsevier, New York, 305p.

Young, S. T., 1976, Petrographic textures of detrital polycrystalline quartz as an aid to interpreting crystalline source rocks. *Journal of Sedimentary Petrology* 46:595–603.

Zuffa, G. G. (ed.), 1985, *Provenance of Arenites.* D. Reidel Publishing Co., Dordrecht, Holland, 408p.

Arenites—Tectonics

Dickinson, W. R., and C. A. Suczek, 1979, Plate tectonics and sandstone compositions. *American Association of Petroleum Geologists Bulletin* 63:2164–82.

Dorsey, R. J., 1988, Provenance evolution and unroofing history of a modern arc–continent collision: Evidence from petrography of Plio-Pleistocene sandstones, eastern Taiwan. *Journal of Sedimentary Petrology* 58:208–18.

Girty, G. H., and A. Armitage, 1989, Composition of Holocene Colorado River sand: An example of mixed-provenance sand derived from multiple tectonic elements of the Cordilleran continental margin. *Journal of Sedimentary Petrology* 59:597–604.

Krynine, P. D., 1942, Differential sedimentation and its products during one complete geosynclinal cycle. *1st Pan American Congress of Mining and Engineering Geology Annals (Mexico)* 2:537–61.

Ingersoll, R. V., 1990, Actualistic sandstone petrofacies: Discriminating modern and ancient source rocks. *Geology* 18:733–6.

Mack, G. H., 1984, Exceptions to the relationship between plate tectonics and sandstone composition. *Journal of Sedimentary Petrology* 54:212–20.

Molinaroli, E., M. Blom, and A. Basu, 1991, Methods of provenance determination tested with discriminant function analysis. *Journal of Sedimentary Petrology* 61:900–8.

Ruxton, B. P., 1970, Labile quartz-poor sediments from young mountain ranges in northeast Papua. *Journal of Sedimentary Petrology* 40:1262–70.

Smith, G. A., 1988, Sedimentology of proximal to distal volcaniclastics dispersed across an active fold belt: Ellensburg Formation (late Miocene), central Washington. *Sedimentology* 35:953–77.

Valloni, R., and G. Mezzadri, 1984, Compositional suites of terrigenous deep-sea sands of the present continental margins. *Sedimentology* 31:353–64.

Velbel, M. A., 1985, Mineralogically mature sandstones in accretionary prisms. *Journal of Sedimentary Petrology* 55:685–90.

Arenites—Climate

Akhtar, K., and A. H. M. Ahmad, 1991, Single-cycle cratonic quartzarenites produced by tropical weathering: The Nimur Sandstone (Lower Cretaceous), Narmada Basin, India. *Sedimentary Geology* 71:23–32.

Ataman, G., and S. L. Gokcen, 1975, Determination of source and palaeoclimate from the comparison of grain and clay fractions in sandstones: A case study. *Sedimentary Geology* 13:81–107.

Basu, A., 1976, Petrology of Holocene fluvial sand derived from plutonic source rocks: Implications to paleoclimatic interpretation. *Journal of Sedimentary Petrology* 46:694–709.

Basu, A., 1985, Influence of climate and relief on compositions of sands released at source areas. In G. G. Zuffa (ed.), *Provenance of Arenites,* D. Reidel Publishing Co., Dordrecht, Holland, pp. 1–18.

Crook, K. A. W., 1967, Tectonics, climate, and sedimentation, 7th Sedimentological Congress.

Dutta, P. K., and L. J. Suttner, 1986, Alluvial sandstone composition and paleoclimate, II. Authigenic mineralogy. *Journal of Sedimentary Petrology* 56:346–58.

Girty, G. H., 1991, A note on the composition of plutoniclastic sand produced in different climatic belts. *Journal of Sedimentary Petrology* 61:428–32.

Grantham, J. H., and M. A. Velbel, 1988, The influence of climate and topography on rock-fragment abundance in modern fluvial sands of the southern Blue Ridge Mountains, North Carolina. *Journal of Sedimentary Petrology* 58:219–27.

Johnsson, M. J., 1990, Tectonic versus chemical-weathering controls on the composition of fluvial sands in tropical environments. *Sedimentology* 37:713–26.

Johnsson, M. J., and R. H. Meade, 1990, Chemical weathering of fluvial sediments during alluvial storage: The Macuapanim Island point bar, Solimoes River, Brazil. *Journal of Sedimentary Petrology* 60:827–42.

Johnsson, M. J., and R. F. Stallard, 1989, Physiographic controls on the composition of sediments derived from volcanic and sedimentary terrains on Barro Colorado Island, Panama. *Journal of Sedimentary Petrology* 59:768–81.

Mack, G. H. L., and L. J. Suttner, 1977, Paleoclimate interpretation from a petrographic comparison of Holocene sands and the Fountain Formation (Pennsylvanian) in the Colorado Front Range. *Journal of Sedimentary Petrology* 47:89–100.

Suttner, L. J., A. Basu, and G. H. Mack, 1981, Climate and the origin of quartz arenites. *Journal of Sedimentary Petrology* 51:1235–46.

Suttner, L. J., and P. K. Dutta, 1986, Alluvial sandstone composition and paleoclimate, I. Framework mineralogy. *Journal of Sedimentary Petrology* 56:329–45.

Todd, T. W., 1968, Paleoclimatology and the relative stability of feldspar minerals under atmospheric conditions. *Journal of Sedimentary Petrology* 38:832–44.

Arenites—Heavy Minerals

Basu, A., and E. Molinaroli, 1989, Provenance characteristics of detrital opaque Fe-Ti oxide minerals. *Journal of Sedimentary Petrology* 59:922–34.

Force, E. R., 1980, The provenance of rutile. *Journal of Sedimentary Petrology* 50:485–8.

Grigsby, J. D., 1990, Detrital magnetite as a provenance indicator. *Journal of Sedimentary Petrology* 60:940–51.

Hubert, J. F., 1962, A zircon–tourmaline–rutile maturity index and the interdependence of the composition of heavy mineral assemblages with the gross composition and texture of sandstones. *Journal of Sedimentary Petrology* 32:440–50.

Krynine, P. D., 1946, The tourmaline group in sediments. *Journal of Geology* 54:65–88.

Lowright, R., E. G. Williams, and F. Dachille, 1972, An analysis of factors controlling deviations in hydraulic equivalence in some modern sands. *Journal of Sedimentary Petrology* 42:635–45.

Marshall, B., 1967, The present status of zircon. *Sedimentology* 9:119–36.

Pettijohn, F. J., 1941, Persistence of heavy minerals in geologic time. *Journal of Geology* 49:610–25.

Raeside, J. D., 1959, Stability of index minerals in soils with particular reference to quartz, zircon, and garnet. *Journal of Sedimentary Petrology* 29:493–502.

Rice, R. M., D. S. Gorsline, and R. H. Osborne, 1976, Relationships between sand input from rivers and the composition of sands from the beaches of Southern California. *Sedimentology* 23:689–703.

Rittenhouse, G., 1943, Transportation and deposition of heavy minerals. *Geological Society of America Bulletin* 54:1725–80.

Saxena, S. K., 1966, Evolution of zircons in sedimentary and metamorphic rocks. *Sedimentology* 6:493–502.

Slingerland, R., 1984, Role of hydraulic sorting in the origin of fluvial placers. *Journal of Sedimentary Petrology* 54:137–50.

Stapor, F. W., Jr., 1973, Heavy mineral concentrating processes and density/shape/size equilibria in the marine and coastal dune sands of the Apalachicola, Florida, region. *Journal of Sedimentary Petrology* 43:396–407.

Steidtmann, J. R., 1982, Size-density sorting of sand-size spheres during deposition from bedload transport and implications concerning hydraulic equivalence. *Sedimentology* 29:877–83.

van Andel, T. J. H., 1959, Reflections on the interpretation of heavy mineral analyses. *Journal of Sedimentary Petrology* 29:153–63.

Rudites

Abbott, P. L., and G. L. Peterson, 1978, Effects of abrasion durability on conglomerate clast populations: Examples from Cretaceous and Eocene conglomerates of the San Diego area. *Journal of Sedimentary Petrology* 48:31–42.

Laznicka, P., 1988, *Breccias and Coarse Fragmentites: Petrology, Environments, Associations, Ores.* Developments in Economic Geology 25, Elsevier, New York, 832p.

Smith, N. D., 1972, Flume experiments on the durability of mud clasts. *Journal of Sedimentary Petrology* 42:378–83.

Twenhofel, W. H., 1947, The environmental significance of conglomerates. *Journal of Sedimentary Petrology* 17:99–128.

Williams, G. D., 1966, Origin of shale–pebble conglomerate. *American Association of Petroleum Geologists Bulletin* 50:573–7.

Wilson, M. D., 1970, Upper Cretaceous–Paleocene synorogenic conglomerates of southwestern Montana. *American Association of Petroleum Geologists Bulletin* 54:1843–67.

Lutites

Anderson, D. W., and M. D. Picard, 1971, Quartz extinction in siltstone. *Geological Society of America Bulletin* 82:181–6.

Blatt, H., 1985, Provenance studies and mudrocks. *Journal of Sedimentary Petrology* 55:69–75.

Chamley, H., 1989, *Clay Sedimentology.* Springer-Verlag, Berlin, 623p.

Hesse, R. (ed.), 1984, *Sedimentology of Siltstone and Mudstone. Sedimentary Geology* (special issue) 41:113–300.

Potter, P. E., J. B. Maynard, and W. A. Pryor, 1980, *Sedimentology of Shale.* Springer-Verlag, New York, 306p.

Shaw, D. B., and C. E. Weaver, 1965, The mineralogical composition of shales. *Journal of Sedimentary Petrology* 35:213–22.

Tourtelot, H. A., 1960, Origin and use of the word shale. *American Journal of Science* 258A:335–43.

Weaver, C. E., 1989, *Clays, Muds, and Shales.* Developments in Sedimentology 44, Elsevier, Amsterdam, 819p.

Yaalon, D. H., 1962, Mineral composition of the average shale. *Clay Minerals Bulletin* 5:31–6.

7
Clays and Colloids

Clay minerals and colloids are the most abundant sediment; clay minerals comprise the majority of virtually all detrital mudrocks. Because of their minute size (most clays are <0.002 mm in diameter and colloids range down in size to 10^{-6}mm) and because most clay minerals have similar optical properties (biaxial negative with a small 2V and length slow), they are difficult to study and differentiate in thin section, particularly because their properties are obscured by aggregation effects (such as preferred fabrics or stacking of flakes in the rock slice) or mixing (of clay mineral varieties, or with cements and iron oxides). Hence, X-ray diffraction (XRD), infrared spectroscopy (IR), or differential thermal analysis (DTA) must generally be applied to determine the mineralogy of the clays (see **AS** Chapter 8). It is important to distinguish between *clay size* and *clay mineral* (see below) because grains between 0.002 (9ϕ) and 0.004 mm (8ϕ—the upper limit of clay-size particles according to the Wentworth scale) are mostly not clay minerals. Clay minerals form as alteration products of silicate minerals during initial weathering or diagenesis, and they may be precipitated directly from aqueous solutions or suspensions during diagenesis (e.g., Wilson and Pittman 1977). During weathering and diagenesis, one clay mineral may transform to another and trends of increasing maturity may be discernible (e.g., Johnsson and Meade 1990), and during diagenesis, clay minerals commonly recrystallize and grow to form larger authigenic micas or chlorites, which can be difficult to distinguish from detrital flakes. Hence, not only are these sediments difficult to study but their origin and history are also difficult to interpret. Excellent general discussions on most aspects of these minerals are found in Grim (1968), Millot (1970), Weaver (1989), and Chamley (1989).

Deposition of clay minerals is generally thought to occur only in very low-energy zones; almost any current is sufficient to maintain such fine particles in suspension, although once they have accumulated, interparticle cohesion resists erosion. In fact, most clay minerals are probably deposited as aggregates of particles: as flocs (Pryor and Vanwie 1971; Krank 1973, 1975; Gibbs 1983; Gibbs et al. 1989), as pedogenetic aggregates (Rust and Nanson 1989), or as fecal and feeding pellets of organisms (e.g., Pryor 1975 and see discussion in Chapter 5). Also clays may infiltrate coarser previous deposits (e.g., Matlack, Houseknecht, and Applin 1989) as well as originate at a later stage in diagenesis (e.g., Whetton and Hawkins 1970).

Clay minerals are the basis of the ceramics industry (different combinations have different plasticities and respond differently to glazing), and are important in many other industries, including quality paper making (as fillers and coatings); plastic, paint, and rubber manufacture; pharmaceutical preparation; and molding sand mixes. The properties of most soils are dictated by the type and abundance of clay minerals and colloids, and because the properties of a single soil can vary substantially depending on the chemical (e.g., cation concentrations) and physical (e.g., quantity of water present) conditions, clay mineral studies are important to agriculture (e.g., for controlling permeability and nutrient availability) and to engineering geology (e.g., foundation studies). The primary swelling clays also are important in the petroleum industry, where they form the component of drilling muds, because their properties (e.g., density and viscosity) can be artificially controlled (also, these clays can produce major problems when encountered in drilling). In addition, the affinity of swelling clays for organic compounds probably is important in the development of petroleum source rocks, from which relatively late-stage flushing of the hydrocarbons is achieved by the dehydration of these clays. Clay minerals (and organo-clays, which are formed by tight bonds between specific clays and organic complexes) have been widely used in various chemical industries as filters or catalysts. As a result of their unusual chemical and physical properties, clays may also have played an important role in the origin of life on Earth (e.g., Cairns-Smith 1985; Cairns-Smith and Hartman 1986). In the modern environment, clay minerals and colloids are important natural trapping agents for heavy metals, some organic compounds, and radioactive isotopes liberated by human activities; the use of pelletized clays as an aid in cleaning up toxic wastes is a new and expanding application of their physical and chemical properties.

MINERALOGY

Clay minerals are hydrated aluminophyllosilicates. The different groups of clay minerals consist of different combinations of three main structural units (Fig. 7-1): silica tetrahedra, alumina octahedra, and brucite sheets (also with octahedral coordination). Figs. 7-2 and 7-3 illustrate the crystallography, composition, and classification of clay minerals (see also Grim 1968; Millot 1970; Brindley and Brown 1980; Chambley 1989; and Weaver 1989 for detailed descriptions of varieties and subvarieties). Note that *smectite* (vs. a common usage of *montmorillonite* in the older literature) is the currently accepted name for the broad group of expandable clay minerals that includes the subgroups *montmorillonite* (Mg substituting for Al in the octahedral sheet) and *beidellite* (Al substituting for Si in the tetrahedral sheet). Allophane is an amorphous clay (e.g., Fieldes 1966; Fieldes and Furkert 1966) that is widespread in some soils and sediments but rarely recognized because of the difficulty in identifying it (it is isotropic and generally requires IR analysis or a chemical test for identification; see **AS** Chapter 8). The double-chain clay minerals (Palygorskite Group) also are very difficult to distinguish in thin section (e.g., from zeolite minerals) but are generally rare (meerschaum is an example of this group, which has very different properties than most clays). Many types of *mixed-layer clays* are common and cannot be distinguished without XRD, IR, or DTA analysis for their crystallographic ordering (the common classes of ordering are shown in Fig. 7-2).

All clay minerals are crystallographically disordered compared with most other minerals, and crystal lattices are distorted because of common cation substitutions in the tetrahedral (Al^{3+} for Si^{4+}) and octahedral (Mg^{2+} and some Fe^{3+} for Al^{3+}, Fe^{2+} and many other ions for Mg^{2+}) sheets; in addition, some OH^- generally replaces some O^{2-} (relatively minor substitutions compared with cation substitutions). Hence the clay crystals have various levels of charge deficiency, which results in their attracting other ions (metallic, organic, even polar water molecules) onto the *outside* of the clay particles. The small particle size of many clay crystallites means that there will be a high proportion of broken bonds per unit mass of clay; these broken bonds provide additional sites where ions can be adsorbed on the outside of the crystal lattice. These external sites are the positions for *cation exchange* (see further discussion in the next section). Because of the various substitutions *within* the clay lattices, there are countless chemical species of mineral within each major group, and names are pointless beyond a relatively coarse level of discrimination; the ordering of the minerals plays an important role in their species classification. Crystallographic ordering has a close relationship to clay crystallite size and is reflected in shape of X-ray diffraction peaks used to calculate the disorder coefficient (DC; McConchie and Lewis 1980) and the Weaver and Kubler indices (WI and KI; see Weaver 1989), which are finding increasing application in studies of diagenesis and low-grade metamorphism. Ordering and crystallite size generally increase with increasing temperature and time in diagenesis.

CATION EXCHANGE

Many of the important properties of clay minerals are related to the positive charge deficiency in the minerals resulting from broken bonds at the edges of the tiny crystals and isomorphous substitutions *within* the crystal lattices (e.g., as when Si^{4+} is replaced by Al^{3+}). This deficiency is balanced by cations (e.g., Ca^{2+}, Mg^{2+}, K^+, Na^+, H^+) held in the interlayer zone in the clay crystals (see Fig. 7-2). The net charge deficiencies in the major clay mineral groups are:

Micas > 2/unit cell
Sericites 1.5–2/unit cell
Illite 1–1.5/unit cell
Smectites 0.2–0.8/unit cell
Kaolinites about 0/unit cell

In minerals with a high charge deficiency (> ca. 0.8/unit cell) the interlayer ions are tightly held and tend not to exchange with other ions, but when the charge deficiency is between about 0.2 and 0.7/unit cell the bonding is weak and *cation exchange* commonly results when there is a change in the relative ion concentrations in the water surrounding the minerals (see Tables 3-1 and 3-2). When the charge deficiency is low and the crystals are relatively coarse in size there is a low exchange capacity because there are few charged sites on which the exchange can take place.

The exchangeability of ions depends on several factors including the type of clay, the size of the clay crystallites, the type and availability of ions, the Eh and pH condition of the solution, temperature, and time. Common exchange cations include Ca^{2+}, Mg^{2+}, K^+, Na^+, NH_4^+, and H^+, but a variety of trace elements and amino acids can also occupy exchange sites. In general, the smaller the hydrated ion size, the more tightly the cation will be held by the clay, but this rule does not always strictly apply because some cations that fit very neatly into an interlayer position may be very difficult to remove. Cations with a higher charge, e.g., Ca^{2+} (cf. Na^+), tend to be more tightly held, but where there is a conflict the size factor seems to dominate. The type of exchangeable ions present on modern clays can be a useful paleoenvironmental indicator, but with diagenesis equilibrium tends to be established with interstitial waters that commonly circulate, and

Silica tetrahedron
Si^{4+} central and equidistant from 4 O^{2-} ions; units link laterally to form a sheet 4.93Å thick (Al^{3+} substitutes for some Si^{4+} in most clays)

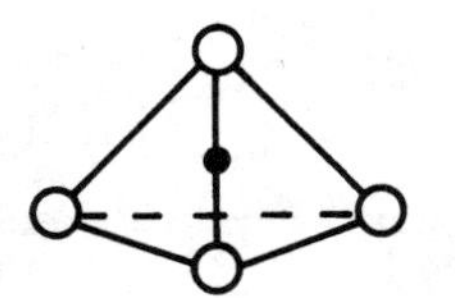

linked verticaly with sheet of

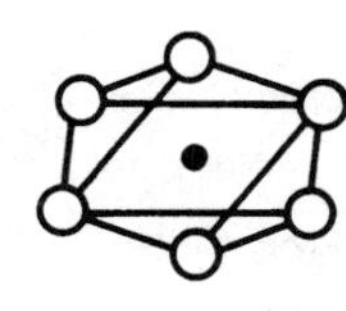

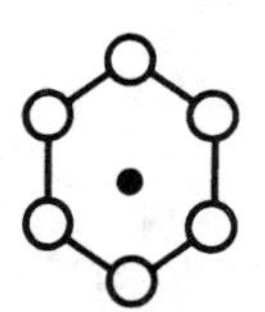

Alumina octahedra
Al^{3+} central and equidistant from 6 O^{2-} ions, units link laterally to form a sheet 5.05Å thick (Mg^{2+} substitutes for some Al^{3+}, and other bivalent ions for Mg^{2+}; $(OH)^-$ substitutes for some O^2)

Brucite sheets: octahedral units of $Mg_2(OH)_6$, laterally linked to form separate sheets in chlorites.

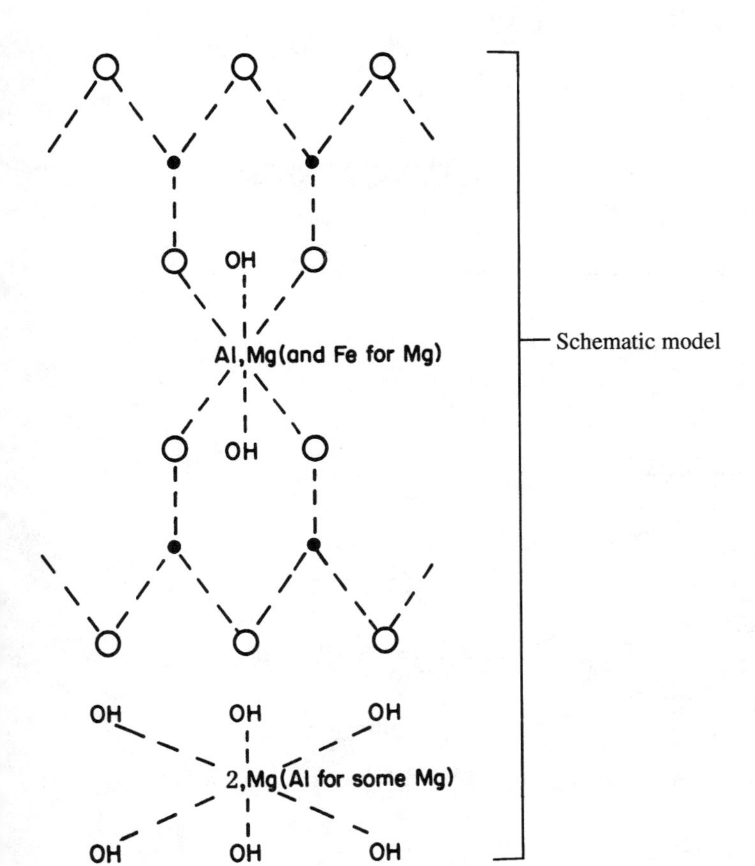

The three fundamental criteria for classification of the major clay minerals:

1. *Combination of structural units—unit layers of clays formed of:*

   1(silica tetrahedral sheet):1(alumina octahedral sheet)—Kaolinite Group
   1:2-Montmorillonite and Illite Groups
   1:2 + 1(brucite sheet)—Normal (sedimentary) Chlorite Group
   (1:1 + 1—Septechlorite Group, of metamorphic origin)

2. *Cation content of octahedral sheet:*

   Dioctahedral—two-thirds of all possible cation lattice positions filled, as for gibbsite $Al_2(OH)_6$ (glauconite, most illite and montmorillonite, muscovite)
   Trioctahedral—all possible cation lattice positions filled, as for brucite $Mg_3(OH)_6$ (vermiculite, normal chlorites, phlogopite, biotite)

3. *Manner and perfection of stacking layers:*

   The vertical stacking array of clay layers, such as kaolinite-illite-kaolinite-illite-etc. vs. illite-montmorillonite-chlorite-montmorillonite-illite-etc.
   The crystallographic symmetry of the vertical array of clay layers, such as 2M (two-layer monoclinic configuration), 1M (one-layer monoclinic), 1 md (one-layer monoclinic disordered), and 3T (three-layer trigonal)

Figure 7-1. Fundamental crystallographic units that constitute the structure of most clay minerals; see text for expanded discussion. Exceptions exist in the case of the amorphous clays (*allophanes*), where there are random associations of the tetrahedral and octahedral units, and double-chain minerals (the *Palygorskite Group*), where chains of silica tetrahedra are linked with octahedral groups (similar to the structure of the amphibole minerals).

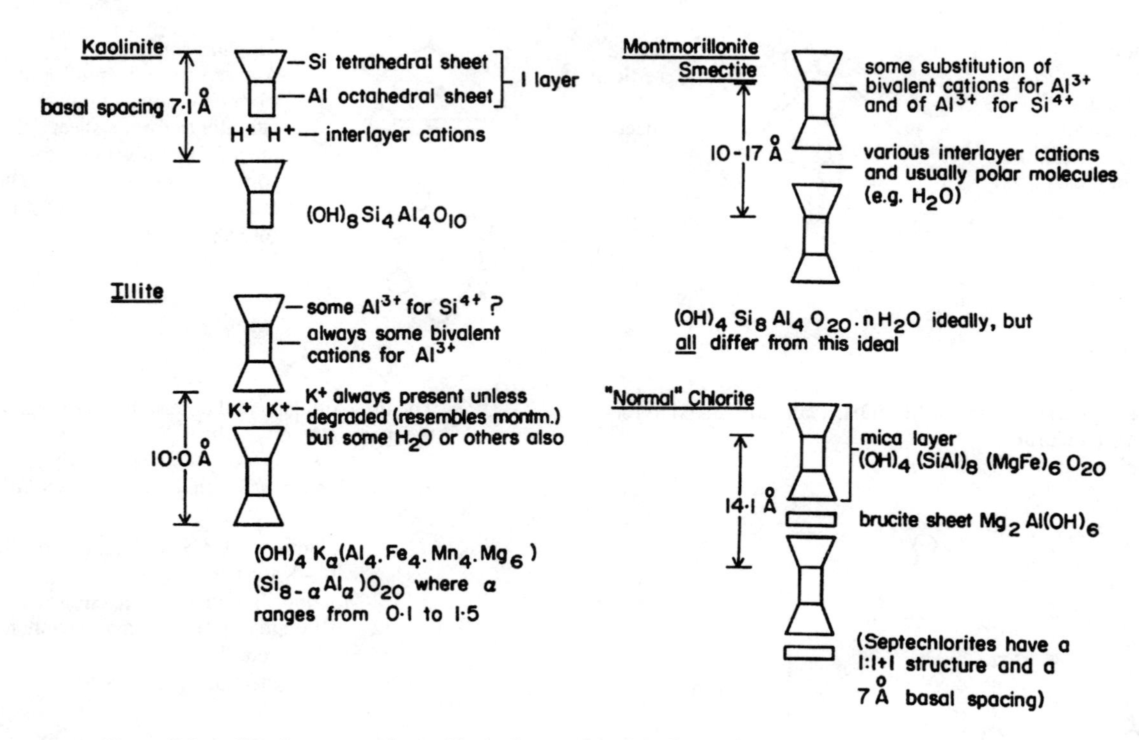

Figure 7-2. Sketch representations of the basic structure of the four major clay groups.

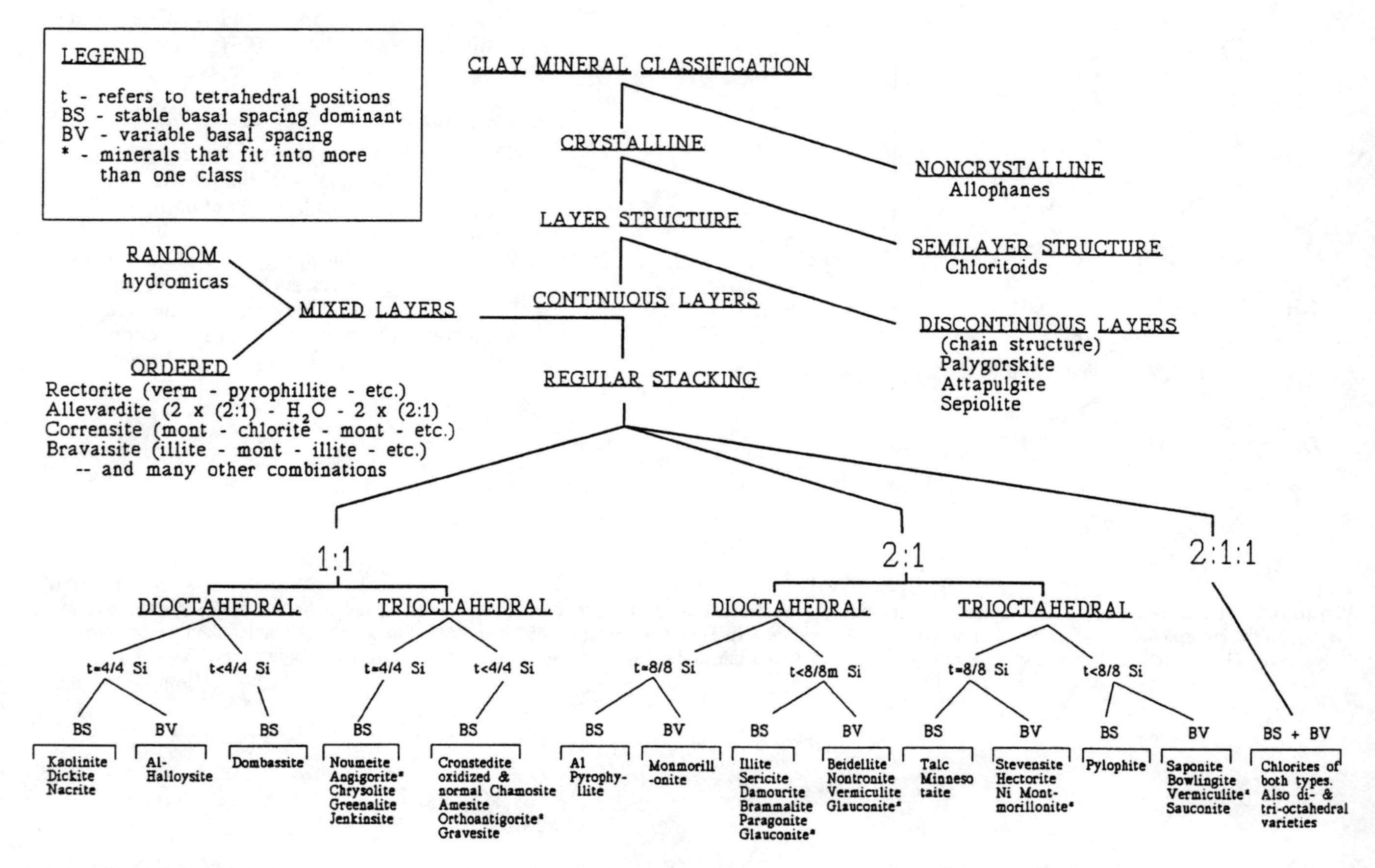

Figure 7-3. An example of a clay mineral classification scheme showing the relationship of many clay mineral species and the tree structure required for refinement of clay mineral classification down to species level.

cations in the exchange positions have only limited use in differentiating diagenetic histories.

Clays with charge deficiencies in the range of 0.25 and 1.0/unit cell also commonly show a tendency to swell (e.g., smectites) in the presence of water or other polar molecules attracted to the charged surface and drawn into the interlayer position. The amount of expansion resulting from polar liquids in the interlayer positions will vary with the type of clay, the nature of interlayer cations, the charge deficiency, and the type (e.g., size and polarity) of the polar liquid molecule. Smectites swell most, with Na- or Li-montmorillonites swelling more than K- or Ca- or Mg-montmorillonites. The surface charges on clay particles are also the major reason why clays transported in a freshwater river tend to flocculate (to form aggregates or flocs) when the ionic strength of the water rises as seawater is encountered (e.g., in an estuary).

Clays that normally do not swell very much, such as illites and chlorites, may have their normally firmly held interlayer ions stripped under some weathering/pedogenetic conditions; the *degraded* clay products then behave as swelling clays (and in fact can be very difficult to distinguish from smectites in all ways) until they reenter an environment from which they can recover their initial ions and reconstitute (or *aggrade*). Overall, as a result of the charges they induce, the isomorphous substitutions in clay mineral crystallites have a major influence on the behavior of clays in transport and deposition (e.g., flocculation effects), agriculture (e.g., the nutrient and trace element retention capacity of soils), and engineering (e.g., the effect of swelling on slope stability and foundation strength).

Table 7-1. General Summary of Environmental Factors Favoring the Formation of the Four Major Types of Clay Minerals

Kaolinite
Low pH
Removal of Ca^{2+}, Mg^{2+}, $Fe^{2+\text{ and }3+}$, Na^{+}, K^{+}
Precipitation > evaporation
Prolonged leaching (no stagnant water, permeable strata), nonmarine conditions
Oxidation of Fe^{2+} to Fe^{3+}
High Al:Si ratio.
Acidic parent material (e.g., granites, rhyolites, etc.)
Illite
pH near neutral or mildly alkaline
Retention of alkali cations, particularly K^{+}
Moderate rainfall with cyclical wetting and drying
Both marine and nonmarine conditions can be favorable ($CaCO_3$ often present)
High Si:Al ratio
Feldspars, amphiboles, and micas are ideal parent materials; N + 1 cycle sedimentary illite is common
Smectite
High pH
Retention of Mg^{2+}, Ca^{2+}, and Na^{+}
Evaporation > precipitation (semi-arid conditions are ideal) for formation in soils
Stagnant water, poor leaching (e.g., standing water in lakes and swamps)
Both marine and nonmarine conditions can be favorable
High Si:Al ratio and silica retention
Alkaline parent rocks (e.g., basalts and gabbros; ferromagnesian silicates and volcanic ash are highly susceptible to alteration)
Chlorite
pH near neutral or alkaline
Retention of Mg^{2+} and $Fe^{2+\text{ and }3+}$
Poor leaching (e.g., standing water in lakes or seas)
Marine conditions are most favorable
Reduction of Fe^{3+} to Fe^{2+} in the genetic environment is desirable
High Si:Al and Fe:Al ratios
Alkaline parent rocks (e.g., basalts and gabbros with a high ferromagnesian mineral content are ideal)

Note: The listed factors are those that generally favor the formation of each type of clay but not all need apply to the formation of a particular clay and the absence of any factor may not preclude the formation of a particular clay.

ORIGIN

Which clay mineral forms where and when depends on the chemistry of the weathering or diagenetic environment, influenced by source rock composition and both macro- and microclimatic factors (e.g., Fig. 7-4, Table 7-1, and see discussion in Singer 1980). Consequently, clays can be useful environmental indicators. However, they need to be interpreted with caution because clays found in a given environment may not have formed there, different clays may form in one environment depending on controls such as source rock type and drainage, and one type of clay mineral can form in a variety of environments when parent rock types and drainage conditions are similar.

Hydrolysis is the major process involved in the breakdown of primary silicates to clay minerals. Hence, the amount of precipitation, circulation of water among the primary minerals, and temperature (as it relates to rates of chemical reactions) are of prime importance. The influence of biota (e.g., Fig. 7-5) and organic decomposition products are also commonly great: bacterial activities, degradation of minerals as they pass through the alimentary canals of sediment feeders, or effects of organic chelating compounds, acids, and bases. Also critical are the type and concentration of inorganic ions in the environment (derived from the decomposition of minerals and supplied by migrating interstitial waters). In addition, the time factor is important: products may be similar (irrespective of the parent rock composition) when chemical activity is intense over a short time span or when it is sluggish but very prolonged. Soil characteristics are particularly important, and since these may differ between adjacent areas (e.g., sites at which clays "pause" during transportation) or can change physically and chemically in one place with time, clay mineralogy may change between the time the clays are initially produced and the time they are buried at their final depositional site.

In general, clay mineralogy at the time of sedimentation tends to reflect the intensity of weathering in the adjacent landmasses; hence studies may indicate the climatic setting in the source area. Temperate humid climates tend to produce mixed-layer smectites and illites and medium crystalline

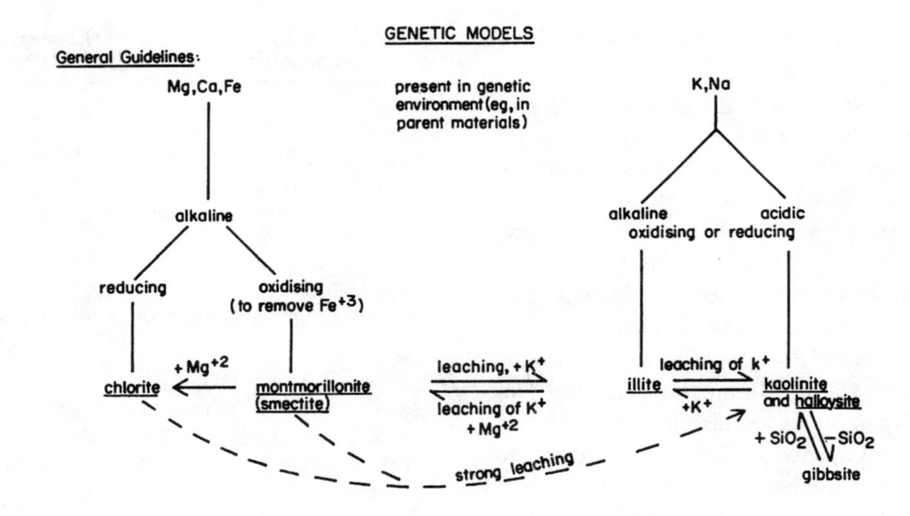

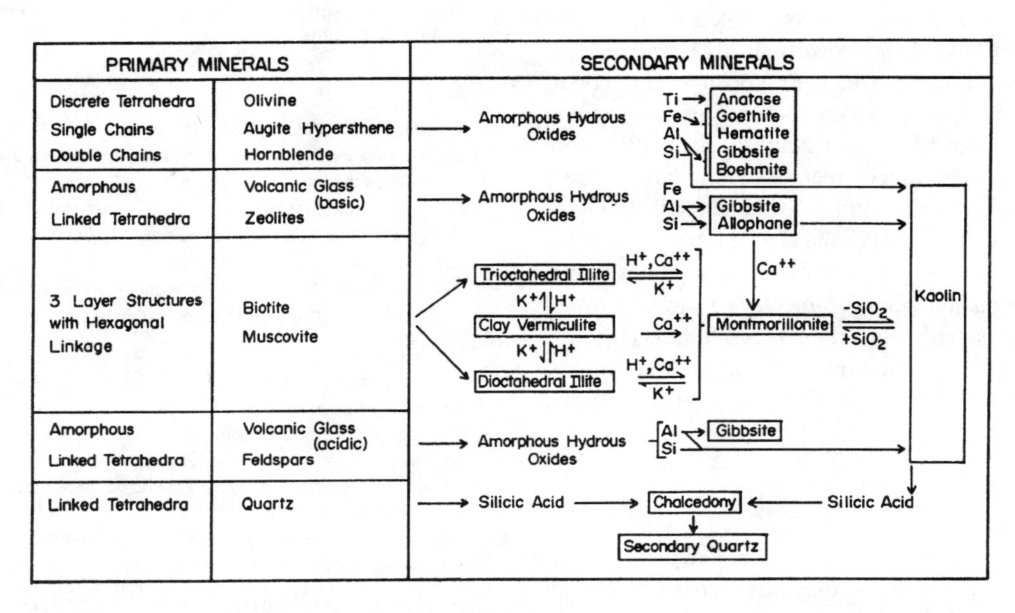

Figure 7-4. Examples of routes for the origin of clay minerals. *Primary clays of neoformation* are created from nonclay minerals (or volcanic glass, ions in solution, or nonclay colloids in suspension). *Transformation clays* are derived from fundamental crystallographic reorganizations of other clay minerals. (Tabular portion after Fieldes and Swindale 1954.)

smectites; in cool temperate climates, mixed-layer vermiculites; in warm temperate, humid climates, degraded smectites (Chamley 1989). The wet tropics, where leaching is at its maximum, produce kaolinites or gibbsites; leaching beneath peats or where abundant organic complexes have flushed through the system also favors kaolinite formation (e.g., Staub and Cohen 1978). In cold or arid climates, where weathering is slow, source rock characteristics may be represented by the clay mineralogy. In strongly alternating wet or dry climates, ions derived from the source rocks are not fully removed from the weathering system and well-crystallized smectites result (calcretes are usually associated with Mg-smectites and occasionally palygorskite or sepiolite; Chamley 1989). Changes in clay mineralogy in the stratigraphic column can be indicative of even slight changes in the tectonics of the source area through the tectonic effect on relief and consequential drainage (e.g., Chamley 1989).

Clay minerals may be produced or modified in the marine environment as well as in nonmarine settings; the term *halmyrolysis* is used for the equivalent of the weathering process on the seafloor. Only clays of the Kaolin Group cannot be formed in the marine setting, because they require acidic leaching conditions (although there is a diversity of origins within this restriction, e.g., Keller 1982). Beidellite, an aluminum-rich smectite, is also unlikely to form in magnesium-rich seawater. Chlorite and smectite can form because they require alkaline conditions where leaching is restricted and where there are supplies of Mg (and, to a lesser degree, Fe and Ca) ions (e.g., from ferromagnesian minerals). However, most sedimentary chlorites appear to be products of later di-

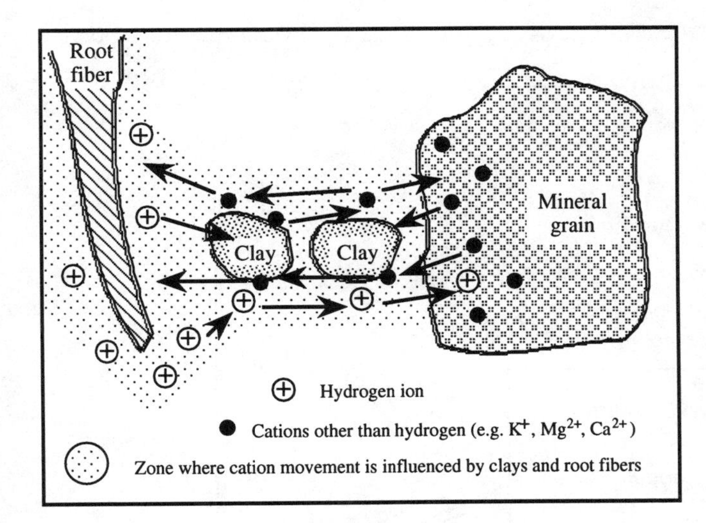

Figure 7-5. Diagrammatic representation of the way in which clays provide a pathway connecting plant roots to mineral grains. Cations from decomposing mineral grains can be transferred along this "bridge" to the plant root, and H^+ ions released by the plant root can be transferred back along the pathway, with consequential acceleration of mineral decomposition. The clay particles may be products of earlier decomposition of the primary mineral, or they may have a different origin.

agenesis, often from transformation of smectites. Illite forms where there are supplies of potassium ions, and since such ions may be supplied from the breakdown of minerals such as K-feldspars and micas, it is usually formed from weathering/diagenetic products derived from acidic igneous/metamorphic rocks, whereas the chlorites/montmorillonites are usually products of more basic rocks (but can be derived from acidic rocks from which the alkalis have been leached).

Clay minerals also are common products of hydrothermal alteration; hence in volcanic settings hydrothermal clays are often reworked into sedimentary environments. Unfortunately, there is no consistent way in which a hydrothermal origin can be deduced from the character of clays found in sediments—the same variety is produced as from rocks in sedimentary weathering environments. When found at the site of origin, the distribution of clay mineral varieties in the alteration haloes (which vary widely in size) normally can be found to change both laterally and vertically, reflecting cooling trends and pH changes associated with increasing distance traveled by the hydrothermal solutions. Commonly, an inner sericite zone is surrounded by a kaolinite zone, then a mixed-layer illite/montmorillonite zone, and finally an outer chlorite zone, but zone boundaries are gradational. Studies of the clay minerals in alteration zones can provide valuable information for hydrothermal mineral exploration programs.

Clay minerals may be *transformed* from one variety to another during diagenesis (e.g., de Segonzac 1970; Velde and Nicot 1985). Very early diagenetic crystallographic transformations are rare. Physical reorganization occurs with compaction and cation exchange reactions commonly take place, but there tends to be no change in crystal lattice structure; progressive lateral changes in clay mineralogy, such as in an offshore direction as noted within some modern sediments (e.g., see Carson and Arlaro 1983), often reflect merely selective sorting of the different clays that have different crystallite sizes and/or flocculate in different places

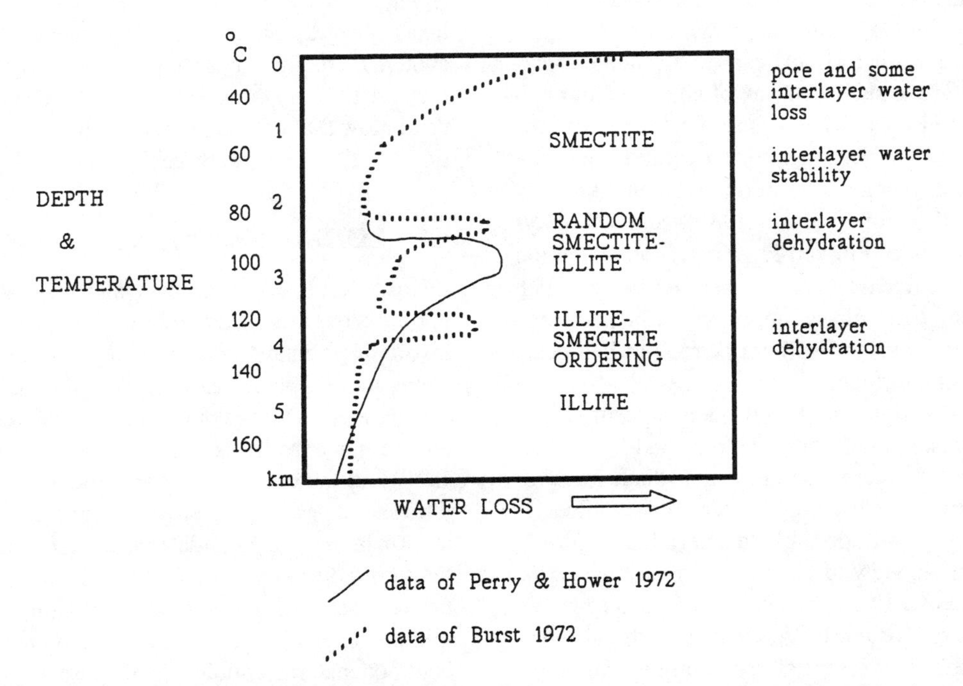

Figure 7-6. Graphical representation of the relationship between depth, temperature, character of the clay mineral assemblage (changing through transformations), and water escape (maximum water escapes during shallow compaction, but additional quantities are provided by interlayer dehydration during transformations to well-ordered illite). (After Chamley 1989, who cites the sources for the curves.)

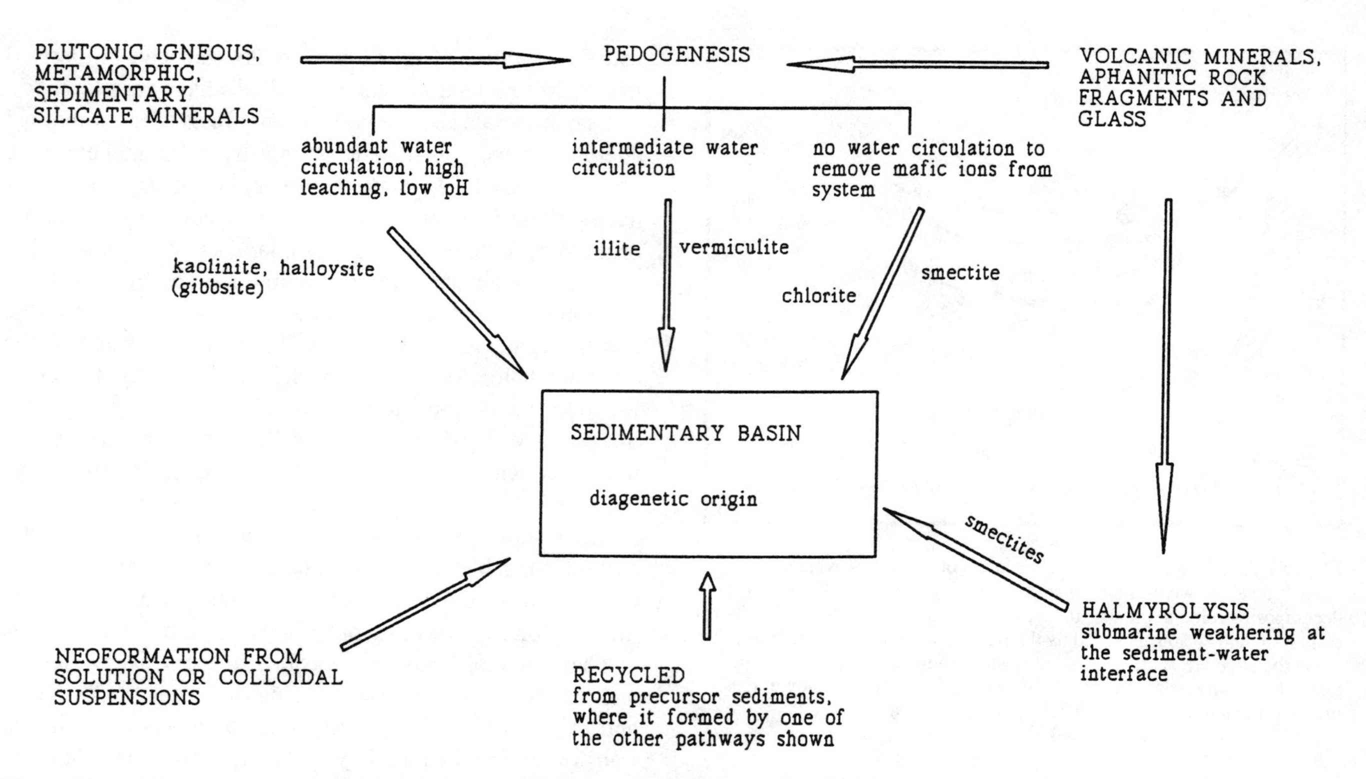

Figure 7-7. Schematic representation of the multiple potential origins of clay minerals in sedimentary deposits.

(e.g., Gibbs 1983). There are exceptions: some mineralogical changes may occur before substantial burial in sediments that have a high initial permeability and water flux; they may also be triggered by the passage through the intestinal tracts of organisms (Pryor 1975; Syvitski and Lewis 1980). Late diagenetic changes in clay mineralogy are common, as evinced by vertical changes in clay mineralogy within some stratigraphic piles wherein other effects can be discounted; particularly common are transformations of smectites through mixed-layer clays to illites and chlorites (e.g., Fig. 7-6). The transformations occur in response to temperature-pressure increases and appear to begin after burial to about 2 km (at a gradient of about 30°C/km; Weaver 1989). Smectite clays lose expandable layers and mixed-layer clays become more illitic; most clays increase their crystal ordering. Such transformations are particularly important as a late-stage means of producing water to flush out petroleum hydrocarbons (which form at intermediate burial stages).

Complicating investigations based on composition still further, clay minerals may of course be recycled from older sedimentary rocks. Whether those in the new deposit are recycled or derived from weathering of other primary minerals in the source area may be impossible to determine. Figure 7-7 provides a schematic view of the multiple possible origins for clays in sediments. Even though identification of clay minerals may require XRD or DTA, analysis with the petrographic microscope is still essential to determine textural relationships, which may be critical to the interpretation of the origin of the clay minerals (e.g., detrital vs. early diagenetic vs. late diagenetic). Under Crossed Polars (CPL), sericite (fine muscovite or coarse illite) has the highest birefringence and first reds and blues are common; illite (hydromuscovite) grades into sericite, but most commonly shows first-order whites and yellows; kaolin has a very low gray birefringence and is difficult to distinguish from chlorite (which has equally low birefringence or anomalous birefringence and takes on a greenish color in plane-polarized light if it is thick enough). The smectites also have very low birefringence, but they are the only common clay minerals with a refractive index below that of balsam; the main difficulty may be to distinguish them from authigenic interstitial chert.

COLLOIDAL SEDIMENTS

Interest in colloidal sediments is growing rapidly as their importance in many sedimentary processes, particularly those involving pollutant transport and environmental geology, becomes more apparent. Colloidal sedimentary particles are very fine grained (natural colloidal particles are generally considered to range in size between 0.1 μm and 0.1 nm; Krauskopf 1979) and show no tendency to settle out of suspension unless time and their chemical environment cause them to coagulate to form larger (noncolloidal) particles. In natural environments the most common colloidal materials are clay minerals (particularly smectites), iron oxides and iron oxyhydroxides, aluminium oxyhydroxides, and some humic substances. Less common natural colloids in modern environments include silica and various other metal oxides, metal oxyhydroxides, and metal sulfides. Studies of Precambrian sedimentation (e.g., Ewers and Morris 1981) indicate that colloidal silicon-hydroxide ($Si[OH]_4$) may have been important prior to the evolution of silica-secreting organisms.

Because colloid-size particles are so fine, they have a very high surface area/volume ratio (e.g., 1 cm^3 of material with a diameter of 10^{-5} mm has a total surface area of about 600 m^2). The high specific surface area of the colloids also means that there are many broken bonds, and both positively and negatively charged ions or molecules will be present *on* the solid surface. The net charge on the solid colloidal particle will be the sum of all these surface charges together with the contributions from the adsorbed ions. The charge will depend on the chemical and crystallographic structure of the solid and on the pH of the solution surrounding the solid (Yariv and Cross 1979; Krauskopf 1979). For each type of colloidal solid, there is a pH value at which the surface concentrations of H^+ and OH^- are equal and there is no net charge on the particle, called the point of zero charge (PZC); in the absence of ions other than H^+ and OH^-, the PZC will be at a pH referred to as the isoelectric point (IEP). At pH values above the IEP the colloidal particles will carry a net negative charge, and at pH values below the IEP the particles will have a net positive charge (see Fig. 3-5 and Table 3-2 for IEP values for some common natural materials).

Because the dipolar water molecule is attracted to charged particles, colloidal particles will not flocculate and settle from suspension unless the charge is canceled (i.e., at the PZC). Hence, flocculation can be induced by a change in electrolyte strength or pH. Monovalent cations tend to be dispersing agents whereas higher-valency ions are flocculants; however, H^+ and OH^- are particularly good coagulants. Furthermore, whereas dilute electrolytes may be dispersive, the same electrolyte at higher concentrations may cause flocculation; consequently, for example, colloidal $Fe(OH)_3$ is common in rivers but not in the sea. Colloid flocculation, induced by a rise in electrolyte concentration, is therefore an important sedimentation process in settings such as estuaries (e.g., Martin, Mouchel, and Jednacak-Biscan 1986), and has significant implications regarding the dispersion of potential pollutants that may be bound to the charged particle surface. The flocculation and precipitation of some colloids can also result in the coprecipitation of other chemical species from the water body (see Chapter 3 and Harder 1965; Yariv and Cross 1979).

Due to their high surface charge, colloids are excellent ion-exchange media (see Chapter 3) and may act as carriers for pollutants (e.g., Förstner and Wittman 1981; Balistrieri and Murray 1982; Laxen 1984; McConchie and Lawrance 1991). Hence, the discharge of colloids such as iron oxides in an effluent may lead to the deposition of polluted sediment even though the colloids themselves are nonpolluting. The very fine size and high surface charge of colloidal particles have several other effects on their physical and chemical behavior (see excellent discussion in Yariv and Cross 1979):

1. Fine colloidal iron oxides are superparamagnetic (e.g., Kündig et al. 1966; McConchie and Smith 1986), in contrast to normal coarsely crystalline hematite, which is ferromagnetic; this difference in magnetic properties is clearly evident on Mössbauer spectra (see **AS** Chapter 8). Because iron oxides precipitated in the weathering profile are initially very fine and grow with age, the ratios of superparamagnetic to ferromagnetic peak areas in their Mössbauer spectra may be useful in determining the relative age of laterite profiles (e.g., Hanstein et al. 1983) and to distinguish between geomorphologically similar profiles. Other properties of colloidal particles also change as the particles age. The recrystallization of colloidal iron oxides potentially has applications in studies of diagenetic temperatures and alterations of sediment associated with igneous activity (e.g., Johnston and Lewis 1983; McConchie and Smith 1986).

2. Colloidal particles in finely laminated sediments can have a marked influence on ion migration between layers during early diagenesis. They appear to act as a filter, allowing the passage of some ions but not others; which ions can pass through a lamination depends on the pH of interstitial solutions and on the nature of the colloidal material. This influence appears to have been highly significant as a control on early diagenetic processes during banded iron formation genesis (see Chapter 9).

3. Even when colloid-rich sediments are exposed to the action of currents strong enough to cause surface deformation, sedimentary structures formed will rapidly be obliterated by hydroplastic creep within the sediment returning it to the prestressed condition. Consequently, the absence of sedimentary structures in strata deposited as colloid-rich sediments need not imply deposition in a calm water environment.

GLAUCONITE

Glauconite is a common nondetrital component of sandstones (arenites, limestones, greensands), and is particularly useful for interpretation of geological history. Most commonly, it occurs as rounded green pellets of sand size, but also it is rarely found as a coating on grains or on broad substrates (such as carbonate hard grounds). Because the name *glauconite* has been used both as a specific mineral and in a broad sense for all green pellets of various mineralogies, Odin and Matter (1981) proposed that *glaucony* be used in the general sense for all green clay minerals, and this usage has become common (e.g., Odin 1988; Chamley 1989). The most common mineral is glauconite (discussed more fully below). Other green pellets include iron-rich smectites and mixed-layer clays, glauconite-smectite mixtures, greenalite (essentially Precambrian), and chamosite (Precambrian and Phanerozoic). The youngest form of chamosite (Pleistocene-Holocene) is called berthierine or verdine (e.g., Chamley 1989) and resembles glauconite under the microscope (some chamosite is distinctive as oolites; glauconite oolites are not known). Celadonite (an alteration product or vug filling of some basic volcanics) is indistinguishable from glauconite except by its occurrence or by complex instrumental methods such as Mössbauer analysis (see **AS** Chapter 8). Chamosite is commonly misidentified as glauconite upon superficial examination; however, its XRD pattern is different (e.g., a nearly constant 7.5-angstrom basal spacing). The origin of chamosite (particularly chamosite oolites) and most of the other less common green marine clays is still widely debated (e.g., Odin 1988), and to date there are no definitive conclusions on how they form, what subenvironments most suit their formation, or why some are most common in Precambrian strata whereas others are rarely found in sediments older than Cambrian. Some further discussion of non-glauconitic green marine clays is presented in Chapter 9, and Odin (1988) provides an excellent introduction to the study of these minerals. All of these green minerals form at the sediment-water interface, and most—especially glauconite—are restricted to the marine environment.

Glauconite minerals are iron- and potassium-rich hydrated aluminophyllosilicates with a 2:1 layer lattice. Total iron content is greater that 15% (most have 19–27%) with a Fe^{2+}:Fe^{3+} ratio of approximately 1:7; K_2O content is greater than 3% (up to 9%). Al_2O_3 generally ranges from 5% to 8%, and MgO from 3% to 5%. XRD analysis is required for identification and subdivision into varieties. There is a wide range in expandable layer content within the mineral family; according to Odin and Matter (1981), there is a fully gradational suite from glauconitic smectite (disordered, high expandable layer content) to glauconitic mica (well ordered, less than 5% expandables), reflecting progressive evolution in the process of glauconitization. Others subdivide the family into a series of varieties shown in Table 7-2 (see also Burst 1958*a*, 1958*b*; McConchie and Lewis 1980). Some grains contain both glauconitic and nonglauconitic minerals; these are distinguished morphologically as composite grains and/or mineralogically as mixed-mineral glauconite.

Table 7-2. Morphological Varieties of Glauconite

Ovoid and *spheroidal* pellets
Lobate pellets (deeply mammillated surfaces; lobes tend to break off during transport; hence presence of this morphology is best indication of authigenic glauconite)
Tabular and *discoidal* pellets
Vermicular pellets (due to replacement of micas or some types of fecal pellets)
Fossil casts and *internal molds*
Composite grains
Fragmentary grains: subclass A if more angular than 0.5 on roundness scale, probably reflecting breakage during perigenic transport; subclass B if more rounded than 0.5, probably reflecting abrasion after breakage and detrital (recycled or allogenic) origin
Spongy pellets: porous, earthy surface form. Using scanning electron microscopy (SEM), *cauliflower* vs. *serrulate* varieties can be distinguished according to whether protuberances and pores are rounded and indistinct or sharply serrated and distinct. Spongy pellets appear to reflect diagenetic corrosion of other types of glauconite pellets.
Varieties of Internal Texture (determined in thin section):
Random microcrystalline (crystallites in all orientations)
Oriented microcrystalline (crystallite orientation produces straight or wavy extinction of grain UXN)
Patch-oriented microcrystalline (patches of grains have oriented crystallites, remainder is random microcrystalline)
Micaceous
Coatings
Fibroradiated rims (coating has microflakes oriented perpendicular to rim)
Organic replacements

Note: These varieties may be distinguished under standard microscopes. Morphological varieties are determined by qualitative shape classification. Loose grains are generally required, although shapes on weathered rock surfaces may be sufficiently representative. Distinction between varieties based on internal textures of the glauconite microcrystals require the use of a polariazing microscope and thin slices of the grains (see **AS** Chapter 8).

Glauconite pellets are formed in the marine environment at the sediment-water interface, as are glauconitic films or coatings; some interstitial glauconitic cements originate below the interface. Nonmarine glauconite has been reported, but not as pellets, and it differs chemically from the marine minerals (see Kossovskaya and Drits 1971; Odin and Matter 1981). Marine glauconites are forming today and occur in rocks as old as 2×10^9 years. They are most abundant today in the outer shelf and upper slope environments (water depths of 50–500 m), but it is unlikely that water depth itself has any control, and pellets can be redeposited and may accumulate in shallower or deeper environments. Triplehorn (1966), McRae (1972), and Odin and Matter (1981) provide excellent reviews of existing knowledge (see also McConchie and Lewis 1978, 1980; McConchie et al. 1979). Pellets may be authigenic (never moved), allogenic (recycled from older sediments, in which case they are few and generally fragmental; see McConchie and Lewis 1978), or perigenic (locally redeposited; most are of this category).

Glauconite minerals form as internal molds of microfossil tests and replace a variety of mineral grains and rock fragments, most commonly fecal pellets, but also quartz, ferromagnesian minerals, volcanic fragments, bioclasts, and others. Odin and Matter (1981) recognized four evolutionary stages: (1) precipitation as (micro)pore fillings, initial substrate present and dominant; (2) initial substrate disappearing, with grains becoming green and developing paramagnetic properties of glaucony; (3) initial substrate essentially gone, with differential recrystallization and crystalline growth expanding original grain or void to produce bulbous grains with surficial cracks; (4) evolution to glauconitic mica, with new glauconite mineral forming in cracks to produce a smooth crust and rounded, subspherical shape. Others argue that glauconite forms by adsorption of iron, then potassium, into an initially disordered, swelling clay mineral lattice that may be inherited from a preexisting mineral or be an authigenic precipitate itself.

Chemically, irrespective of whether or not glauconite forms mainly by transformation of precursor clay lattices or as an entirely neoformational mineral, two processes appear to be fundamental in glauconite genesis (McConchie and Lewis 1978): first, adsorption and ion exchange, which permits a progressive rise in the iron and potassium content of the mineral, and second, the internal oxidation of Fe^{2+} to Fe^{3+}, which permits the iron content to increase without the requirement for Fe^{2+} to move up a large concentration gradient. The internal oxidation of Fe^{2+} to Fe^{3+} is facilitated by electron exchange that occurs in the glauconite crystal lattice at temperatures above 80°K (McConchie et al. 1979). Slightly reducing conditions (for supplies of Fe^{2+}) but an absence of H_2S is essential (sulfide ions preferentially scavenge iron and at the pH conditions that exist in seawater, Fe^{3+} is largely insoluble and thus iron must move to the forming glauconite in the reduced Fe^{2+} form).

Whatever the mode of origin, a slow net sediment accumulation rate is essential for glauconite formation. Odin and Matter (1981) estimate necessary exposure times at the sediment-water interface at 10^3–10^4 years for initial glauconitization, and 10^5–10^8 years for development of highly evolved glauconitic mica. Once formed, however, the pellets can be, and are, commonly redeposited rapidly in energetic, well-oxygenated settings (e.g., Ward and Lewis 1975), and they can survive in reducing surface and subsurface environments. As a result of the restricted conditions under which glauconite forms, it is widely used as a palaeoenvironmental indicator; it has also been used in radiometric dating (primarily K/Ar) studies (see Odin 1982).

SELECTED BIBLIOGRAPHY

General

Cairns-Smith, A. G., 1985, *Seven Clues to the Origin of Life.* Cambridge University Press, Cambridge, U.K.

Cairns-Smith, A. G., and H. Hartman (eds.), 1986, *Clay Minerals and the Origin of Life.* Cambridge University Press, Cambridge, U.K.

Carson, B., and N. P. Arlaro, 1983, Control of clay-mineral stratigraphy by selective transport in late Pleistocene–Holocene sediments of northern Cascadia Basin–Juan de Fuca abyssal plain: Implications for studies of clay-mineral provenance. *Journal of Sedimentary Petrology* 58:395–406.

Chamley, H., 1989, *Clay Sedimentology.* Springer-Verlag, Berlin, 623p.

Förstner, U., and G. T. W. Wittman, 1981, *Metal Pollution in the Aquatic Environment.* Springer-Verlag, Berlin/New York.

Gradusov, B. P., 1974, A tentative study of clay mineral distribution in soils of the world. *Geoderma* 12:49–55.

Millot, G., 1970, *Geology of Clays: Weathering, Sedimentology, Geochemistry.* Springer-Verlag, New York, 429p.

Singer, A., 1980, The paleoclimatic interpretation of clay minerals and weathering profiles. *Earth-Science Reviews* 15:303–26.

Weaver, C. E., 1989, *Clays, Muds, and Shales.* Developments in Sedimentology 44, Elsevier, Amsterdam, 819p.

Mineralogy

Brindley, G. W., and G. Brown, 1980, *Crystal Structures and Clay Minerals and Their X-ray Identification.* Mineralogical Society, London, 497p.

Brown, G., 1961, *The X-ray Identification and Crystal Structures of Clay Minerals.* Mineralogical Society (Clay Minerals Group), London.

Fieldes, M. 1966, The nature of allophane in soils, Part 1: Significance of structural randomness in pedogenesis. *New Zealand Journal of Science* 9:599–607.

Fieldes, M., and R. J. Furkert, 1966, The nature of allophane in soils, Part 2: Differences in composition. *New Zealand Journal of Science* 9:608–22.

Grim, R. E., 1968, *Clay Mineralogy.* McGraw-Hill, New York, 384p.

Keller, W. D., 1982, Kaolin—a most diverse rock in genesis, texture, physical properties, and uses. *Geological Society of America Bulletin* 93:27–36.

Kossovskaya, A. G., and V. A. Drits, 1971, The variability of micaceous minerals in sedimentary rocks. *Sedimentology* 15:83–101.

Spears, D. A., and H. I. Sezgin, 1985, Mineralogy and geochemistry of the Subcrenatum Marine Band and associated coal-bearing sediments, South Yorkshire. *Journal of Sedimentary Petrology* 55:570–8.

Theng, B. K. G., 1974, *The Chemistry of Clay–Organic Reactions.* John Wiley & Sons, New York.

Weaver, C. E., and L. D. Pollard, 1973, *The Chemistry of Clay Minerals.* Developments in Sedimentology 15, Elsevier, Amsterdam, 213p.

Origin and Deposition

Fieldes, M., and L. D. Swindale, 1954, Chemical weathering of silicates in soil formation. *New Zealand Journal of Science and Technology* 36B:140–54.

Gibbs, R. J., 1983, Coagulation rates of clay minerals and natural sediments. *Journal of Sedimentary Petrology* 53:1193–203.

Gibbs, R. J., D. M. Tshudy, L. Konwar, and J. M. Martin, 1989, Coagulation and transport of sediments in the Gironde Estuary. *Sedimentology* 36:987–1000.

Keller, W. D., 1970, Environmental aspects of clay minerals. *Journal of Sedimentary Petrology* 40:788–813.

Kranck, K., 1973, Flocculation of suspended sediment in the sea. *Nature* 246:348–50.

Kranck, K., 1975, Sediment deposition from flocculated suspensions. *Sedimentology* 22:111–23.

Kranck, K., 1981, Particulate matter grain-size characteristics and flocculation in a partially mixed estuary. *Sedimentology* 28:107–14.

Piper, D. J. W., and R. M. Slatt, 1977, Late Quaternary clay-mineral distribution on the eastern continental margin of Canada. *Geological Society of America Bulletin* 88:267–72.

Pryor, W. A., and W. A. Vanwie, 1971, The "Sawdust Sand"—an Eocene sediment of floccule origin. *Journal of Sedimentary Petrology* 41:763–9.

Rateev, M. A., Z. N. Gorbunova, A. P. Lizitzin, and G. I. Nosov, 1969, The distribution of clay minerals in the ocean. *Sedimentology* 13:21–43.

Rust, B., and G. C. Nanson, 1989, Bedload transport of mud as pedogenetic aggregates in modern and ancient rivers. *Sedimentology* 36:291–306. (See also 1991 Discussion by R. J. Loch, and Reply, *Sedimentology* 38:157–60.)

Shutov, V. D., A. V. Aleksandrova, and S. A. Losievskaya, 1971, Genetic interpretation of the polymorphism of the kaolinite group in sedimentary rocks. *Sedimentology* 15:69–82.

Syvitski, J. P., and A. G. Lewis, 1980, Sediment ingestion by *Tigriopus californicus* and other zooplankton: Mineral transformation and sedimentological considerations. *Journal of Sedimentary Petrology* 50:869–80.

Diagenesis

Aoyagi, K., and T. Kazama, 1980, Transformational changes of clay minerals, zeolites and silica minerals during diagenesis. *Sedimentology* 27:179–88.

de Segonzac, G. D., 1970, The transformation of clay minerals during diagenesis and low grade metamorphism: A review. *Sedimentology* 15:281–346.

Guthrie, J. M., D. W. Housekneckt, and W. D. Johns, 1986, Relationship among vitrinite reflectance, illite crystallinity, and organic geochemistry in Carboniferous strata, Ouachita Mountains, Oklahoma and Arkansas. *American Association of Petroleum Geologists Bulletin* 70:26–33.

Imam, M. B., and H. F. Shaw, 1985, The diagenesis of Neogene clastic sediments from the Bengal Basin, Bangladesh. *Journal of Sedimentary Petrology* 55:665–71.

Jennings, S., and G. R. Thompson, 1986, Diagenesis of Plio–Pleistocene sediments of the Colorado River delta, southern California. *Journal of Sedimentary Petrology* 56:89–98.

Johnsson, M. J., and R. H. Meade, 1990, Chemical weathering of fluvial sediments during alluvial storage: The Macuapanim Island point bar, Solimoes River, Brazil. *Journal of Sedimentary Petrology* 60:827–42.

Johnston, J. H., and D. G. Lewis, 1983, A detailed study of the transformation of ferrihydrite to hematite in an aqueous medium at 92°C. *Geochimica et Cosmochimica Acta* 47:1823–31.

Keller, W. D., 1963, Diagenesis in clay minerals—a review. *Clays and Clay Minerals* 13:136–57.

Matlack, K. S., D. W. Houseknecht, and K. R. Applin, 1989, Emplacement of clay into sand by infiltration. *Journal of Sedimentary Petrology* 59:77–86.

Morad, S., 1984, Diagenetic matrix in Proterozoic graywackes from Sweden. *Journal of Sedimentary Petrology* 54:1157–68.

Moraes, M. A. S., and L. F. de Ros, 1990, Infiltrated clays in fluvial Jurassic sandstones of Reconcavo Basin, northeastern Brazil. *Journal of Sedimentary Petrology* 60:809–19.

Pryor, W. A., 1975, Biogenic sedimentation and alteration of argillaceous sediments in shallow marine environments. *Geological Society of America Bulletin* 86:1244–54.

Staub, J. R., and A. D. Cohen, 1978, Kaolinite-enrichment beneath coals: A modern analogue, Snuggedy Swamp, South Carolina. *Journal of Sedimentary Petrology* 48:203–10.

Velde, B., and E. Nicot, 1985, Diagenetic clay mineral composition as a function of pressure, temperature, and chemical activity. *Journal of Sedimentary Petrology* 55:541–7.

Whetten, J. T., and J. W. Hawkins, Jr., 1970, Diagenetic origin of greywacke matrix minerals. *Sedimentology* 15:347–61.

Wilson, M. D., and E. D. Pittman, 1977, Authigenic clays in sandstones: Recognition and influence on reservoir properties and paleoenvironmental analysis. *Journal of Sedimentary Petrology* 47:3–31.

Colloids

Balistrieri, L. S., and J. W. Murray, 1982, The adsorption of Cu, Pb, Zn and Cd on goethite from major ion seawater. *Geochimica et Cosmochimica Acta*, 46:1253–65.

Ewers, W. E., and R. C. Morris, 1981, Studies on the Dales Gorge Member of the Brockman Iron Formation, Western Australia. *Economic Geology*, 76:1929–53.

Hanstein, T., U. Hauser, F. Mbesherubusa, W. Neuwirth, and H. Spath, 1983, Dating of Western Australian laterites by means of Mössbauer spectroscopy. *Zeitschrift. Geomorphologie. Neue. Folge.* 27:171–90.

Harder, H., 1965, Experimente sur "Ausfällung" der Kieselsäure. *Geochimica et Cosmochimica Acta* 29:429–42.

Krauskopf, K. B., 1979, *Introduction to Geochemistry,* 2d ed. McGraw-Hill, New York.

Kündig, W., H. Bömmel, G. Constabaris, and R. H. Lindquist, 1966, Some properties of supported small α-Fe_2O_3 particles determined with the Mössbauer effect. *Physics Review* 142(2):327–33.

Laxen, D. P. H., 1984, Adsorption of Cd, Pb and Cu during the precipitation of hydrous ferric oxide in a natural water. *Chemical Geology*, 47:321–32.

McConchie, D. M., and L. M. Lawrance, 1991, The origin of high cadmium loads in some bivalve molluscs from Shark Bay, Western Australia: A new mechanism for cadmium uptake by filter feeding organisms. *Archives of Environmental Contamination and Toxicology*, 21:303–10.

McConchie, D. M., and C. B. Smith, 1986, Iron-oxides in pisolite-like clasts in Ellendale lamproite intrusions. Fourth International Kimberlite Conference, *Geological Society of Australia Abstracts* 16:190–2.

Martin, J.-M., J.-M. Mouchel, and J. Jednacak-Biscan, 1986, Surface properties of particles at the land–sea boundary. In P. Lasserre and J.-M. Martin (eds.), *Biogeochemical Processes at the Land–Sea Boundary.* Elsevier, Amsterdam, pp. 53–71.

Yariv, S., and H. Cross, 1979, *Geochemistry of Colloid Systems for Earth Scientists.* Springer-Verlag, Berlin.

Glauconite

Bell, D. L., and H. G. Goodell, 1967, A comparative study of glauconite and the associated clay fraction in modern marine sediments. *Sedimentology* 9:169–202.

Bjerkli, K., and J. S. Ostmo-Saeter, 1973, Formation of glauconite in foraminiferal shells on the continental shelf off Norway. *Marine Geology* 14:169–78.

Burst, J. F., 1958*a*, Glauconite pellets: Their mineral nature and applications to stratigraphic interpretations. *American Association of Petroleum Geologists Bulletin* 42:310–27.

Burst, J. F., 1958*b*, Mineral heterogeneity in "glauconite" pellets. *American Mineralogist* 43:481–97.

McConchie, D. M., and D. W. Lewis, 1978, Authigenic, perigenic, and allogenic glauconites from the Castle Hill Basin, North Canterbury, New Zealand. *New Zealand Journal of Geology and Geophysics* 21:199–214.

McConchie, D. M., and D. W. Lewis, 1980, Varieties of glauconite in late Cretaceous and early Tertiary rocks of the South Island of New Zealand, and new proposals for classification. *New Zealand Journal of Geology and Geophysics* 23:413–37.

McConchie, D. M., J. B. Ward, V. H. McCann, and D. W. Lewis, 1979, A Mössbauer investigation of glauconite and its geological significance. *Clays and Clay Minerals* 27:339–48.

McRae, S. G., 1972, Glauconite. *Earth-Science Reviews* 8:397–440.

Odin, G. S. (ed.), 1982, *Numerical Dating in Stratigraphy*, 2 vols. John Wiley & Sons, Chichester, England, 1,094p.

Odin, G. S., 1988, *Green Marine Clays.* Developments in Sedimentology 45, Elsevier, Amsterdam, 445p.

Odin, G. S., and A. Matter, 1981, De glauconiarum origine. *Sedimentology* 28:611–41.

Triplehorn, D. M., 1966, Morphology, internal structure and origin of glauconite pellets. *Sedimentology* 6:247–66.

van Houten, F. B., and M. E. Purucher, 1984, Glauconitic peloids and chamositic ooids—favorable factors, constraints, and problems. *Earth-Science Reviews* 20:211–44.

Ward, D. M., and D. W. Lewis, 1975, Paleoenvironmental implications of storm-scoured, ichnofossiliferous mid-Tertiary limestones, Waihao District, South Canterbury. *New Zealand Journal of Geology and Geophysics* 18:881–908.

8
Sedimentary Carbonates

Despite the fact that they comprise a much smaller volume than detrital sediments, particularly in Holocene time, carbonate sediments have attracted much interest, only partly because of the economic importance of ancient limestones (e.g., as hydrocarbon reservoirs, sources of agricultural and industrial lime, building stone). They show a wide variety of characteristics and are mostly composed of the remains of ancient life forms, an aspect of considerable interest in itself; perhaps even more significantly, modern analogues occur in shallow, clear, warm waters and relatively easy (and comfortable) studies provide direct insights into depositional conditions of the past. In addition, because of the solubility of the carbonate minerals and their ready response to changing biological, chemical, and physical conditions, intriguing diagenetic changes in sedimentary carbonates are ubiquitous. In this book we only superficially touch on some of the major aspects of the deposits; a proportionately larger list of selected references is provided for those desiring further information (e.g., Sorby 1879; Chilingar, Bissel, and Fairbridge 1967; Folk 1973; Wilson 1975; Bathurst 1975; Scholle, Bebout, and Moore 1983; Scoffin 1987; Morse and MacKenzie 1990).

MINERALOGY

Compositional/crystallographic properties of the most common carbonate minerals are shown in Table 8-1 (see also Fig. 8-1; Goldsmith et al. 1962); Morse and MacKenzie (1990) provide thorough discussions on the origin and relationships of the various minerals. The polymorphs of $CaCO_3$—high-Mg calcite and aragonite—are common in modern warm-water sediments (aragonite comprises more than 80% of shallow-water low-latitude sediments), but are thermodynamically unstable and alter to low-Mg calcite (which initially dominates in cold and deep waters) over geologically short time spans. Hence, mineralogy is not generally used in systematic classifications of carbonate deposits. However, there are other carbonate minerals that are not $CaCO_3$ polymorphs, and because most of their properties are similar, staining techniques, X-ray diffraction, differential thermal or infrared analyses in the laboratory (see **AS** Chapter 8) usually are necessary to distinguish between them.

Most aragonite and high-Mg calcite are precipitated by organisms (see Table 8-2, Chave 1962, and Scholle 1978 for more comprehensive tables on shell mineralogies), as are most low-Mg calcites, but under certain conditions they may precipitate inorganically or with indirect organic influence (e.g., Shinn et al. 1989). Fig. 8-2 depicts major controls on the precipitation of these minerals—the major inorganic controls are pH and the abundance of dissolved CO_2 (see discussion in Chapter 3); factors such as temperature influence the amount of CO_2 that can be dissolved in water (the colder the water, the more CO_2). Siderite forms in sedimentary settings where the sulfide anion (derived from bacterial reduction of sulfate compounds and very prone to capture any ferrous iron ions) is absent; hence siderite often is characteristic of mildly reducing nonmarine environments where sulfate compounds, hence sulfide anions, are rare (see Fig. 3-8). It is unstable in the normal weathering environment and commonly changes to calcite with exsolution of iron oxides, which provide a reddish brown stain that may be the main indication of its original presence.

Dolomite is a common mineral in carbonate rocks (in contrast to modern sediments); it is more common in older rocks than more recent ones for reasons that are still debated but that include the greater thermodynamic stability of the mineral under higher pressures and temperatures than at the surface and differences in the relative activities of Ca^{2+}, Mg^{2+} and HCO_3^- in pore waters and groundwaters (and Precambrian ocean waters). Most dolomite appears to be a diagenetic replacement of aragonite or calcite (see discussion

Table 8-1. Properties of Common Sedimentary Carbonate Minerals

Calcite Group: Hexagonal system, rhombohedral-scalenohedral class. Uniaxial negative.		
Calcite $CaCO_3$	H 3 G 2.7	Low-Mg calcite—less than 4% $MgCO_3$ in solid solution the common, stable form. High-Mg calcite—more than 4% $MgCO_3$ in solid solution (up to 30% in some algal hardparts, to 43% in echinoderm teeth). Only precipitated by organisms and only stable while organic tissue is present.
Magnesite $MgCO_3$	H 3.5–4.5 G 3.0–3.5	
Siderite $FeCO_3$	H 4–4.5 G 3.9	
Rhodochrosite $MnCO_3$	H 3.5–4 G 3.2–4.0	
Smithsonite $ZnCO_3$	H 4–5 G 4.4	
Aragonite Group: Orthorhombic system. Biaxial negative.		
Aragonite $CaCO_3$	H 3.5–4 G 2.95	Metastable polymorph of calcite (hence rare in pre-Pleistocene rocks); mainly precipitated by organisms (and stable only while organic tissue is present). Rarely contains up to 1.5 mole % of magnesium.
Witherite $BaCO_3$	H 3.5 G 4.3	
Strontianite $SrCO_3$	H 3.5–4 G 3.71	
Cerussite $PbCO_3$	H 3–3.5 G 6.57	
Dolomite Group: Hexagonal system, rhombohedral-rhombohedral class. Uniaxial negative.		
Dolomite Ca $Mg(CO_3)_2$	H 3.5–4 G 2.87	Well-ordered mineral with alternating layers of $MgCO_3$ and $CaCO_3$; more stable than calcite but seems mostly of diagenetic origin in Phanerozoic rocks.
Ankerite Ca(Mg, Fe, Mn) $(CO_3)_2$	H 3.5–4 G 2.9–3.1	
Kutnohorite Ca Mn $(CO_3)_2$	(very rare)	
Uncommon but of economic importance are the monoclinic prismatic minerals.		
Azurite $Cu_3(OH)_2(CO_3)_2$	H 3.5–4 G 3.80	Biaxial negative.
Malachite $Cu_2(OH)_2$	H 6 G 4.05	Biaxial negative.

H = hardness on the Moh's scale; G = specific gravity.

below), and when present, it tends to form the majority of the carbonate rock. Such rocks are termed *dolostones*, in contrast to *limestones*, in which calcite dominates. Dolomite can often be distinguished from calcite and aragonite in the field and laboratory by its lesser reaction to dilute HCl; only the powder resulting from scratches will react, and on acid-etched surfaces, grains of this mineral will stand out in relief. It is also recognizable by its common tendency to form 1- to 2-mm subhedral-euhedral rhombic crystals (the rhombic habit of much finer crystals is apparent only with microscope examination, and coarser crystals are commonly anhedral). In thin section, dolomite is almost never twinned (whereas calcite commonly is) and it may show zoning and undulose extinction (which calcite almost never does). Because iron usually substitutes for some of the magnesium in the crystal lattice and because upon weathering that iron will commonly oxidize, dolomite often has a different color than calcite.

Carbonate Trace Elements and Isotopes

Despite the large difference in ionic radius, Mg^{2+} can substitute for the Ca^{2+} in calcite to produce Mg calcite with $MgCO_3$ contents exceeding 20 mol%; generally such substitution occurs as calcite is precipitated within organic matrixes, and when the organism dies the high-Mg calcite inverts to the low-Mg form that is stable in seawater. Calcites referred to as low-Mg calcites are generally considered to be those containing less than 5 mol% $MgCO_3$; the amount present in natural calcites depends inversely on the Ca^{2+}/Mg^{2+} ratio in the water where the crystals are growing and increases with increasing temperature (e.g., Scoffin 1987). Thus, in marine environments where the Ca^{2+}/Mg^{2+} ratio is nearly constant, carbonates formed in shallow tropical waters will generally have a higher magnesium content than carbonates formed in deeper or higher-latitude waters.

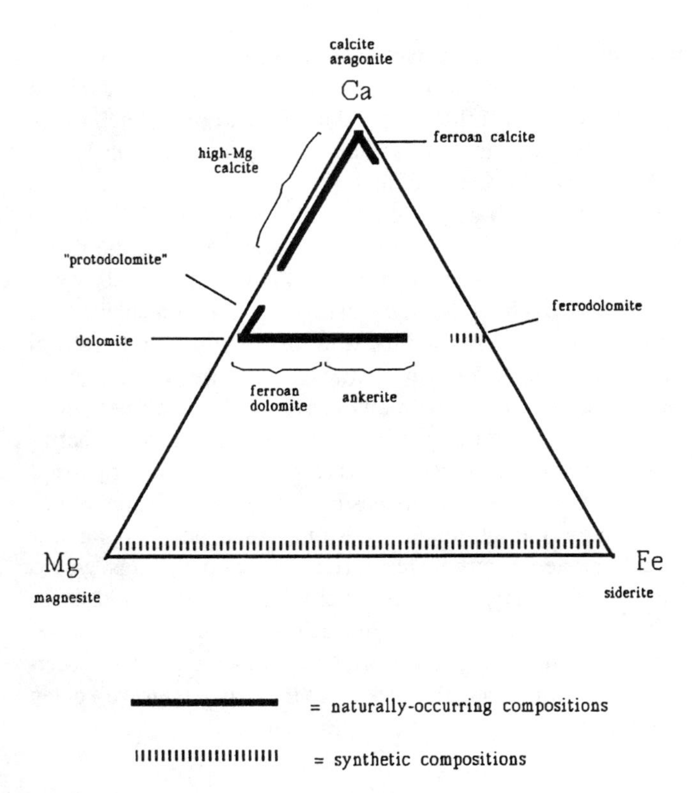

Figure 8-1. Chemical composition of the most common sedimentary carbonate minerals. (After a drawing by P. A. Scholle.)

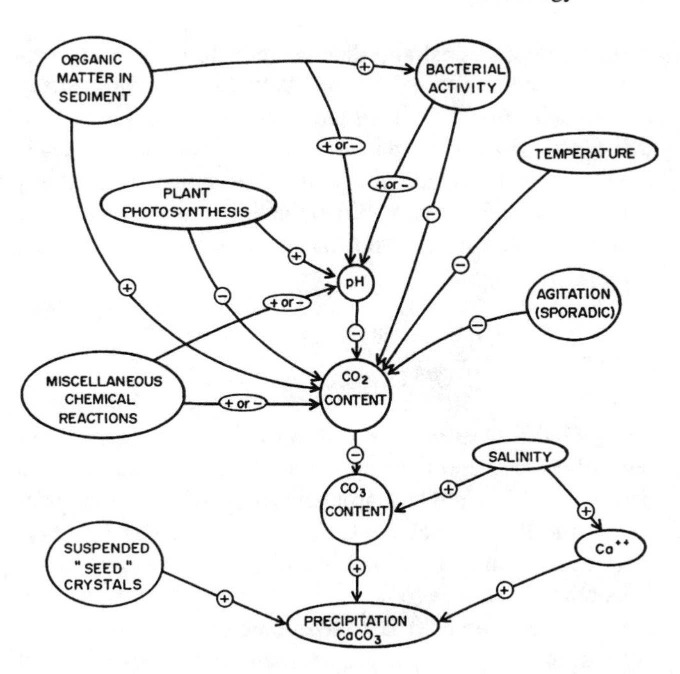

Figure 8-2. Some factors influencing precipitation of inorganic $CaCO_3$. A plus sign indicates where an increase of one factor increases the other; an antipathetic relationship is indicated by a minus sign. (After a sketch by R.L. Folk.)

Several other elements can substitute for calcium in the low-Mg calcite crystal lattice (e.g., Fe, Mn, Zn, Cu), but these are usually only present in trace concentrations. In high-Mg calcites, the concentration of trace elements is usually higher, and even large ions such as Pb^{2+} and Ba^{2+} can substitute for Ca^{2+} because the presence of the smaller Mg^{2+} ion causes sufficient lattice distortion for entry of the larger ions. When high-Mg calcite alters to low-Mg calcite, the trace elements tend to be exsolved along with the magnesium; diagenetic recrystallization also usually results in a decrease in the trace element content. Aragonite does not incorporate magnesium as readily as calcite, but its less tightly bonded crystal structure allows it to incorporate larger ions such as Sr^{2+}, Ba^{2+}, and Pb^{2+} much more readily than the better-ordered calcite. As this metastable mineral inverts to calcite, many of these ions are also exsolved, but some may remain despite the consequent distortion to the calcite lattice.

Table 8-2. Distribution of Carbonate Minerals in Organic Hardparts

High-Mg Calcite
Common in benthic foraminifera, sponges, alcyonarian corals, echinoderms, decapods, benthic red algae, some bryozoa and brachiopods
Also in a few annelid tubes, cephalopods, ostracods

Aragonite
Common in bryozoa, gastropods, bivalves, scleractinian corals, most cephalopods, pteropods, chitons, annelid tubes, benthic green algae, madreporian corals, Paleozoic stromatoporoids(?), Mesozoic/Cenozoic anthozoa
Also in a few alcyonarian corals, benthic foraminifera, sponges, bryozoa

Low-Mg Calcite
Common in bivalves, pelagic algae (coccoliths), planktic and benthic foraminifera, brachiopods, trilobites, barnacles, ostracods
Also in a few annelid tubes, bryozoa, stromatoporoids(?)

In both aragonite and calcite, the Sr/Ca ratio decreases as water temperatures increase in the environment of deposition (or diagenesis); thus, the ratio may provide a useful geothermometer and can be of assistance in deciphering diagenetic processes (e.g., Kinsman 1969). However, because the Sr/Ca ratio depends on the original mineralogy, and in biogenic carbonates is species-dependent, caution is needed when using it as a guide to mean ocean temperatures. In dolomite, significant amounts of Fe^{2+} and Mn^{2+} can substitute for Mg^{2+} (not for the Ca^{2+} in the discrete $CaCO_3$ lattice layers) to form ankerite, which is stable and shows no greater tendency to recrystallize or be replaced than does dolomite; ankerites with a low Fe/Mg ratio are commonly referred to as *ferroan dolomites* (e.g., Fig. 8-1). The relative concentrations of all of these cations in dolomite/ankerite appear to be determined by the relative concentrations of their respective ions in the diagenetic environment; thus, compositionally zoned dolomite/ankerite crystals (e.g., Katz 1971) probably record a cyclic variation in the Fe/Mg ratio in diagenetic solutions.

Stable isotope ratios in carbonate minerals (e.g., of carbon and oxygen) are increasingly used tools for deciphering the complex depositional and diagenetic histories of some carbonate sequences (e.g., Arthur 1983; Bowen 1988; Morse and MacKenzie 1990). Isotopic fractionation is any process that causes differences in isotopic ratios due to differences in their mode of formation or genetic environment. It is usually expressed in delta (δ) notation, which reports the ratio in the sample of interest relative to an international standard, for example:

$$\delta^{18}O = \frac{(^{18}O/^{16}O)_{sample} - (^{18}O/^{16}O)_{standard}}{(^{18}O/^{16}O)_{standard}} \times 1000$$

For $\delta^{18}O$, the difference in parts per thousand (referred to as per mil, ‰) is reported relative to SMOW (standard mean ocean water); the $\delta^{18}O$ in carbonates used to be reported relative to the PDB (a belemnite fossil from the Pee Dee Formation in South Carolina) standard, and it still is in paleoclimatic studies ($\delta^{13}C$ fractionation is also reported using the PDB reference). $\delta^{18}O$ fractionation in carbonates has been used for paleotemperature determination of ancient oceans, but due to uncertainty about possible changes in the isotopic composition of ocean waters over time, there remains considerable controversy about the accuracy of such determinations (e.g., see Kahn, Oba, and Ku 1981; Abell 1985; Fritz and Fontes 1986). $\delta^{18}O$ geothermometry has also been applied to a variety of other marine and nonmarine minerals (e.g., see overviews in Savin and Yeh 1981; Bowen 1988).

$\delta^{13}C$ fractionation ($^{13}C/^{12}C$ ratios) can be used to distinguish between organic carbon, biogenic carbonates, and inorganically precipitated carbonates. It is particularly useful in investigations of diagenetic processes where isotopic ratios in carbonate minerals have changed as a result of exchange with pore water or groundwater. Plots of $\delta^{18}O$ against $\delta^{13}C$ can be used in some sequences to discriminate between carbonates that have been deposited in different environmental settings or have different diagenetic histories (e.g., see Scoffin, 1987).

The radioactive ^{14}C isotope (half-life 5730 y) in biogenic carbonates has been used to date material as old as about 70,000 y, but the reliability of dates for carbonate shell material and organic matter in shells is uncertain (Bowen 1988). Dates for carbonate shell material are uncertain because the extent and significance of any gains or losses of ^{14}C due to exchange with pore waters or groundwater after the death and burial of the organism are unknown; the reliability is generally less for aragonite and high-magnesium calcite than for low-magnesium calcite. Dates for organic matter in shells are uncertain because the initial ^{14}C activity in the organic matter is not known with any certainty. In marine environments influenced by upwelling deep ocean waters, the activity of ^{14}C may have been lowered because the water carried by the upwelling currents has been isolated from the atmosphere for a long time (since it first descended in polar waters); biogenic carbonates formed in such settings would return incorrectly old ages. Shells from areas remote from upwelling currents are usually acceptable for dating, and despite the problems and uncertainties, many dates have been obtained from biogenic carbonates that agree well with dates obtained by other means (e.g., Bowen 1988).

CLASSIFICATION

Composition

Limestone classification systems are numerous (e.g., Ham 1962). The system most commonly used for composition is that of Folk (1959 and Fig. 8-3; see also Reijers and Hsu 1986 for a comprehensive glossary of carbonate sediment terminology). It is based on the presence and proportion of the following:

Micrite: carbonate mud, consisting of microcrystalline (less than 4 microns) carbonate (originally, most was probably aragonite or high-Mg calcite). Formed within the environment of sedimentation, this sediment constitutes the matrix for any grains that may be present. (Micrite is the dark

> 10% ALLOCHEMS		< 10% ALLOCHEMS		UNDISTURBED BIOGENIC ROCKS
Sparite > Micrite	Micrite > Sparite	1 - 10% Allochems	< 1% Allochems	
Intra- Oo- Pel- Bio- } sparite	Intra- Oo- Pel- Bio- } micrite	Intraclast-bearing Oolite-bearing Peloid-bearing Fossiliferous } micrite	Micrite Dismicrite Dolomicrite	Biolithite
Multiple prefixes are used if warranted, with most abundant allochem first (eg, biopelmicrite) Dominant fossil names, at the taxon level recognized, precede the bio-/fossiliferous names.				
Dolomitized or dolomitic used as preliminary term where appropriate; dolostone if fully dolomite rock.				

Figure 8-3. Simplified representation of the compositional classification of normal marine limestones. (After Folk 1959.)

"background" in thin sections; the high-power objective is necessary to see individual crystals, which are superposed in a 30-μm rock slice.) It originates by abrasion of carbonate clasts (either by inorganic processes or by organic maceration), disarticulation of skeletal parts of flora (particularly lightly calcified green and red algae and coccoliths) and some fauna (particularly thin-walled epibionts), inorganic precipitation, and inorganic recrystallization of larger crystals in the weathering or diagenetic environment (e.g., Matthews 1966; Stockman, Ginsburg, and Shinn 1967; Bathurst 1970; Perkins and Halsey 1971; Stieglitz 1973; Alexandersson 1979; Shinn et al. 1989; Reid, MacIntyre, and James 1990).

Sparite: crystalline carbonate cement comprising clear crystals larger than 10–15 microns. In the classification, only that *between* the grains of sediment is counted as sparite because the main distinction desired is the presence or absence of matrix as opposed to cement; sparry calcite *within* grains, such as that which may fill hollow shells, is not counted. Where this diagenetic component is absent (or only partially present) and the (remainder of the) intergranular pore spaces remain open, as in loose modern sediment, the same term (*-sparite*) is used in the sediment name because the important point is the absence of micrite, and because in most rocks, sparry calcite ultimately will precipitate in the water-filled pore spaces. Sparite cement is usually calcite, derived from supersaturated water moving through the sediment. However, it may also form through diagenetic recrystallization of micrite or other carbonate grains, in which cases crystal boundaries transect grain outlines and patchy replacement relationships may be evident in the field or in samples. Good evidence for a recrystallization origin is the presence of an intermediate stage of growth from micrite to sparry calcite—*microspar* (5–15-micron crystals of cement). When limestone completely recrystallizes (as when it is transformed to marble), the entire rock appears to consist of sparite, and special terminology must be applied to characterize it (e.g., Folk 1965).

Allochems (allochemical constituents): any kind of grains, generally transported locally within the environment of origin. There are four basic types:

1. *Biogenic or skeletal components, or fossil fragments*: any size or shape. Discussion of shell morphology is beyond the scope of this book: many scientific and popular books on biology and seashores depict modern shells, and books on paleontology show the morphology of fossils. However, the random two-dimensional morphologies of shells seen in a rock face, in thin section or in an acetate peel (see **AS** Chapters 5 and 8), can look very different and most workers need to refer initially to other specialized books such as those by Horowitz and Potter (1971), Bathurst (1975), and Scholle (1978) for photomicrographs of these components as seen in thin section (see also Scoffin 1987, who provides descriptions, sketches, and discussion of the internal microarchitecture of typical skeletal elements). Because the composition as well as the internal texture of bioclasts varies, breakdown in the depositional environment of these allochems is varied and can be complex (e.g., Swinchatt 1965).

2. *Ooids or oolites* (the term *oolite* is now commonly used for a rock made of ooids): spherical particles up to 2 mm in diameter, showing an internal structure of concentric or radial microcrystals of aragonite (calcite in older sediments and rocks). The nucleus is generally any kind of allochem or detrital grain. Ooids are generally formed by physiochemical precipitation about a nucleus in shallow (0–4 m), high-energy marine or lacustrine environments, such as shoals and tidal channels, where warm water is supersaturated with $CaCO_3$ (e.g., Simone 1981). *Superficial ooids* are distinguished from true ooids by having only one or two concentric coats on the nucleus. *Pisoids* or *pisolites* are similar to ooids, but they are larger than 2 mm and they form in a wider variety of ways (e.g., many originate as small concretions in calcretes and caliche soils; others can be found in the splash pools of caves). *Rhodoliths* and *oncolites* (oncoids, oncoliths) are even larger nonattached subspherical to spherical, concentrically laminated (commonly 1–10 cm) carbonate grains formed, respectively, by direct red algal precipitation (i.e., as skeletal elements) and by sediment accretion onto grains coated by nonskeletal algae or bacteria (i.e., oncoids are a type of stromatolite structure).

3. *Peloids:* sand size grains of homogeneous micrite. In rocks, these are difficult to distinguish, even under the microscope; they commonly are homogenized during compaction of the sediment. *Pellets* are round grains up to 0.2 mm diameter by the Folk classification. Most peloids originate as fecal ejecta of invertebrates such as worms, gastropods, crabs, and shrimps, but in both rocks and modern sediments it can be difficult to distinguish fecal pellets from skeletal or oolitic grains that have been converted to micrite by the boring activities of algae or fungi, and there are ways in which peloids can form inorganically (e.g., Fahraeus, Slatt, and Nowlan 1974). Some are created during diagenesis by bacterial action (*cement peloids*).

4. *Intraclasts*: irregular, rounded, or angular clasts from 0.2 mm to boulder size, consisting of homogeneous micrite or of grains (any allochem, detrital, or nondetrital noncarbonate grains) set in micrite. They represent fragments of sediment from the nearby seafloor, torn up during temporary bursts of high energy while still only partly consolidated. Intraclasts will not form even in situations in which intermittent high-energy events occur if the cohesion of earlier carbonate sediments is not sufficient; hence their absence cannot be used to infer generally quiet conditions. In a few settings, such as where beachrock has formed by very early cementation, intraclasts may lack a micritic matrix; these clasts become difficult to distinguish after subsequent lithification of the entire sediment. Intraclasts may be difficult to distinguish from lithoclasts (detrital grains), which are derived from much older, fully consolidated rocks. Distinguishing criteria are differences in allochem components (including older fossils), textures, or diagenetic fabrics (e.g., different degree of recrystallization) from the enclosing sediment; weathered rims of different color; and truncated grains on the clast margins. Intraclasts of very large size can be the dominant component of debris flows originating on bathymetric highs (e.g., Cook et al. 1972; Hiscott and James 1985).

If allochems comprise more than 10% of the rock, it is an allochem rock and the prefix *bio-*, *pel-*, *oo-*, or *intra-* is appended to the suffix *-micrite* or *-sparite* (or *-microsparite*), whichever is most abundant. If several allochems are present, several prefixes may be used (in order of abundance), for example, biopelsparite. If there is a particularly common type of fossil fragment in the rock, append the name of that fossil (at the level of recognition) to the rock type, for example, brachiopod biosparite, or *Pachymagas* biosparite. In all carbonate rocks, any detrital components that may be present can be classified and interpreted as in the detrital rocks (see Chapter 6).

Texture

Although the compositional classification of Folk (1959) is fundamentally based on textural attributes (proportions of grains to matrix and original void space), refinement of the size terminology for both sedimentary particles and diagenetic crystals is commonly desirable (e.g., Folk 1959, 1962, 1965; Fig. 8-4). Common exclusively textural terms are *calcarenite* (carbonate sediment with a dominance of sand size grains), *calcirudite* (carbonate sediment with a dominance of clasts larger than 2 mm—for consistency with detrital textures, Folk 1959 used 1 mm as the lower limit), and *calcilutite* (with a dominance of lime mud; *calcisiltite* may or may not be distinguishable as a subclass, e.g., Lindholm 1969). Whereas size and sorting terms are specified for calcarenites as they are for detrital sandstones, they have less significance than in the case of detrital sediments because the hydrodynamic behavior of skeletal carbonate grains depends as much on shape and original density (reflecting shell microstructure) as on size. Thus, bioclasts should not be subjected to the simple standard size analysis procedures such as sieving when hydrodynamic behavior is investigated. In addition, in rock faces, thin sections, and acetate peels, the maximum diameters of the inequant skeletal grains or intraclasts commonly will not be visible.

Probably the most utilized scheme for textural classification—as opposed to the compositional classification of Folk—is that of Dunham (1962); an expanded version is summarized in Table 8-3 (and see Embry and Klovan 1972). This scheme can be applied easily in both the laboratory and the field, and it has advantages in refining the texture of carbonate sediments without undue stress on the precise size and sorting of components; additional parameters can be specified to emphasize similarities and differences between deposits. Even without these additional specifications, the

mm		GRAINS	CRYSTALS
>64	calcirudite	very coarse	extremely coarsely crystalline
64–16	calcirudite	coarse	extremely coarsely crystalline
16–4	calcirudite	medium	extremely coarsely crystalline
4–2	calcirudite	fine	very coarsely crystalline
2–1.0	calcarenite	very coarse	coarsely crystalline
1.0–0.5	calcarenite	coarse	coarsely crystalline
0.5–0.25	calcarenite	medium	coarsely crystalline
0.25–0.125	calcarenite	fine	medium crystalline
0.125–0.063	calcarenite	very fine	medium crystalline
0.063–(0.03)	calcilutite	(calcisiltite)	finely crystalline
(0.03)–0.016	calcilutite	(micrite)	finely crystalline
0.016–0.004	calcilutite	(micrite)	very finely crystalline
<0.004	calcilutite	(micrite)	aphanocrystalline

Figure 8-4. Textural terminology for transported and in-place carbonate constituents in the calcilutite range, in placing the boundary between calcirudite and calcarenite at 2.0 mm (vs. 1.0 mm), and in the consequential insertion of a 1–2-mm coarse calcarenite class. The modifications are made to bring greater consistency between the textural classification of carbonates and detrital sediments.
(Modified from Folk (1962)

Dunham terms provide what to most workers is the most essential information about the sediment: the proportion of lime mud to allochems. Less easily applied is the Folk 1962 textural spectrum (Table 8-4) developed to parallel the textural maturity scheme of arenites (see Chapter 6).

In both these schemes, it is intended that the *original depositional character of the sediment* be described; diagenetic modifications must be excluded from the primary name and restricted to accessory description whenever possible. In both, the proportion of micrite is stressed because this lime mud component is thought to form abundantly in every carbonate-accumulating environment; its absence is thus a reflection of current winnowing processes. This assumption is in direct contrast to detrital sedimentation, where clays are supplied from outside the depositional basin by currents. The assumption is based on the multiple modes of origin of micrite and on field investigations in modern tropical areas (see above). However, the assumption is valid for only some of these modern tropical carbonate environments (e.g., calci-

Table 8-3. A Version of the Dunham (in Ham 1962) Textural Classification of Carbonate Sediments, with Modifications of Embry and Klovan 1972

<10% particles >2 mm	Mudstone Wackestone	<10% grains	mud-supported	with lime mud	
	Packstone Grainstone	>10% grains	grain-supported	no lime mud	append size and sorting term
>10% particles >2 mm	Floatstone	matrix-supported			
	Rudstone	grain-supported			
	Boundstone	Bafflestone Bindstone Framestone	original components respectively bound by organisms that encrust and bind, act as baffles, or build a rigid framework		

If depositional texture is not recognizable, the term *crystalline limestone* is used, with subdivisions based on crystal sizes and/or shapes (e.g., Folk 1965)

Note: In thin-section studies it can be difficult to distinguish between several classes because of the restricted two-dimensional view.

Table 8-4. The Folk (1962) Textural Classification of Carbonate Sediments

	Over 2/3 Lime Mud Matrix				Over 2/3 Spar Cement			
Percent Allochems	0–1%	1–10%	10–50%	Over 50%	Subhedral Spar and Lime Mud	Sorting Poor	Sorting Good	Rounded and Abraded
Representative rock terms	Micrite	Fossiliferous micrite	Sparse biomicrite	Packed biomicrite	Poorly washed biosparite	Unsorted biosparite	Sorted biosparite	Rounded biosparite

fied green algae are absent and there is a relative paucity of other contributing calcareous algae and microorganisms in modern temperate region shelf carbonates), and the assumption is of uncertain validity for some deposits. Even in the modern tropics, the rate of micrite production varies within subenvironments, organisms (e.g., algal mats, sea grasses) baffle and trap micrite in environments where currents would otherwise be sufficiently strong to winnow it away, and if lime mud is removed from some subenvironments it follows that adjacent subenvironments must be supplied with that lime mud. In addition, micrite can precipitate in cavities within the sediment (Reid, MacIntyre, and James 1990). Hence interpretations based on the presence or absence of micrite must be carefully evaluated.

Another aspect of texture that is important and highly variable in carbonate sediments is porosity, but since that is most commonly dictated by diagenetic modifications to the original texture and requires a separate nomenclature, it is not commonly mentioned in simple carbonate classifications and is discussed later in this chapter. A special term—*dismicrite*—was applied by Folk to calcilutites with irregular patches of sparry cement that have infilled voids created by burrowing, gas escape, or other mechanisms that disrupted the initial fabric of the lime mud.

Structure

Sedimentary structures in carbonate sediments show virtually the same range as those in detrital sediments (Chapter 4). Certain structures are more common in carbonate sediments, especially algal mat structures (e.g., stromatolites and *fenestral structures*, where parts of algal mats have decomposed during burial to leave elongate cavities later filled with cement), possibly because of a genetic relationship between them (the influence of algae on the CO_2 content of the adjacent water) or because lime mud flats are more common than detrital mud flats in the coastal regions where salinity and temperature fluctuations exclude most herbivores. Diagenetic structures (e.g., concretions) are also more common because of the greater solubility and ease of recrystallization of the carbonate minerals. A few diagenetic structures are thought by some to be unique to carbonate successions (e.g., *teepees*, formed by doming of desiccation polygons by crystallization of associated evaporite minerals in arid regions), but are almost certainly present in other sediments (e.g., evaporite-dominated sequences; see Chapter 9), albeit not recognized as such.

Full Classification

As in the case of arenites (Chapter 6), a consistent and reasonably comprehensive name should be given to each carbonate sediment in the form:

(color) (induration) (internal sedimentary structures) (sorting term) (size term from Fig. 8-4) (Dunham texture term from Table 8-3): (Folk compositional term from Fig. 8-3)

The additional information provided by the dual usage of the Folk (compositional) and Dunham (textural) terminology more than compensates for the minor duplication between them (e.g., grainstones are equivalent to [whatever]sparites).

BIOTA IN CARBONATE SEDIMENTATION

Biological factors (direct and indirect) are particularly important in carbonate sedimentation (e.g., Table 8-7 and see Chapter 3); not only do biota contribute primary sediment but they also act as secondary agents in producing sediment (eg, organisms fragment the hardparts of their prey—a single parrot fish feeding on corals can produce up to 20 kg of bioclastic sand per year). In most modern carbonate environments, fauna rework the sediment many times before it is buried to depths below the range of their action; mechanical mastication and chemical digestive effects of the sediment-ingesting infauna can be substantial. Flora not only supply much sediment (calcareous algae) and weaken skeletal grains by the microboring of encrusting algae, but unpreservable blue-green algae, sea grasses, mangroves, and other plants also are important trapping and binding agents. Types and abundance of organisms that may be present in carbonate environments depend on a plethora of ecological controls (e.g., brief summary in Chapter 2). In addition, decomposing organic matter can have substantial effects on the dissolution and other diagenetic modifications of carbonates (e.g., Reaves 1986).

Organic hardparts are particularly informative about depositional environment. Although commonly redeposited, skeletal elements are generated in the same gross environment as that in which they accumulate; most have not been transported far from their locus of origin. Hence, identification of the organic components is useful to the level (phylum, class, family, genus, species) where useful information can be derived from existing knowledge of their ecology or paleoecology. Even the presence of a single benthic fossil can provide a useful guide to paleoenvironmental conditions (e.g., the presence of a single oyster in a lime mudstone may indicate the nearby existence of a firm substrate). Modern tropical-region (20–30°C water temperatures) skeletal assemblages tend to contain or be dominated by green algal and coralline debris, together with various proportions of other organic hardparts. This *chlorozoan* association (or *chloralgal* if coralline debris is absent) differs from the *foramol* (or *bryomol*) association of temperate waters, which largely comprises bryozoa, barnacles, benthic foraminifera, molluscs, calcareous red algae, echinoderms, and lesser quantities of ostracods, sponge spicules, and serpulid worm tubes (e.g., Lees and Buller 1972). Examples of specific discussions on temperate region limestones are Nelson 1978, 1988 and other papers in that volume; James and Bone 1991; see Table 8-5 for comparison of some characteristics between cool- and warm-water carbonate systems.

Fossil constituents of limestones also provide an indication of the original mineralogical composition of carbonate

Table 8-5. Some Differences Between Tropical-Subtropical and Temperate Carbonate Sediments

	Tropical–Subtropical	Temperate
Latitude (today)	Between 30°N and 30°S	Between 60°S and 35°S
Mean annual water temperature	>23°C	<20°C
Minimum annual water temperature	About 14°C	<12°C
Dissolved $CaCO_3$	Supersaturation to saturation	Under- to supersaturation
Reef structures	Common (mainly coralgal)	Rare (mainly oyster)
Algal mats	Common	Absent or not preserved
Sedimentation Rate	10–100 cm/1000 yr	<5 cm/1000 yr
Major skeletal grains	Calcareous green algae	Bryozoans
	Calcareous red algae	Calcareous red algae
	Corals (hermatypic)	Echinoderms
	Benthic foraminifera	Benthic foraminifera
	Molluscs	Barnacles
		Bivalve molluscs, brachiopods, and serpulid worm tubes
Carbonate mud	Common (shallow and deep)	Sparse (deep)
Nonskeletal carbonate grains	Common to abundant	Rare (some intraclasts, no ooids)
Glauconite	Rare	Common
Evaporites and dolomite	Common in arid areas	Evaporites absent; dolostones absent; dolomite rare
Primary carbonate mineralogy	Aragonite>Mg-calcite>>calcite	Calcite>> Mg-calcite>aragonite
Main lithification environment	Submarine and subaerial	Shallow subsurface
Diagenesis	Commonly constructive (precipitation)	Commonly destructive (dissolution, biodegradation)
Major carbonate cement	Aragonite, Mg-calcite, calcite	Calcite
Major sources of cement	Seawater and solution aragonite grains	Selective and nonselective intergranular solution of skeletal calcite grains

Source: After Nelson 1978, 1988.

sediments, although we know little about the original mineralogy of some extinct organisms. The composition of deposits is strongly influenced by the areal distribution of skeletal organisms. Knowledge of the original mineralogy is important in interpretations of diagenetic modifications, hence is of particular importance to the petroleum industry in predictions of reservoir and trap rock histories.

Maximum information can be discovered from hardparts in life position (see discussion of reefs below). Unfortunately, the original living community in any paleoenvironment is commonly subjected to selective sorting by currents, selective destruction of remains by scavengers, selective physiochemical breakdown of hardparts composed of high Mg-calcite or aragonite; in addition more debris is contributed by short-lived organisms than long-lived ones (several generations of bryozoa, for example, may grow and die during the life span of one brachiopod). Burrowing activities of infauna or storm redeposition may also intermix populations that lived in different subenvironments. In some submarine environments, carbonate particles (including fossils) may be recycled (see MacIntyre 1970). Although death associations (*thanatocoenoses*) are most common, rare cases of life associations (*biocoenoses*) are found, and living community reconstructions can be made from these as well as from slightly transported or modified associations. Complex feedback relationships exist between organisms and the substrate in the depositional setting. For example, substrate character may change where fauna produce abundant fecal pellets and energy conditions are too low for their removal; thereafter, many organisms are excluded from inhabiting the resultant soft soupy substrate.

CARBONATE ENVIRONMENTS

Figure 8-5 illustrates most environments in which carbonate sedimentation can dominate; Table 8-5 contrasts salient characteristics of tropical- and temperate-region carbonate deposits, and Table 8-6 shows some general properties of shallow-water (platform) limestones. Brief summary characteristics of the deposits are not easy to prepare because of the wide variety of textures and components that may be present in most of the environments; as well as the general references cited in the initial paragraph of this chapter, see Logan et al. (1970); Cook and Enos (1977); Cook, Hine, and Mullins (1983) for reviews and discussions on deposits and environments. An informative specific example of depositional and diagenetic synthesis for a widespread carbonate complex is the study by Eliuk (1978).

Deposit geometries and sedimentary structures show the same range in carbonate as in detrital sediments, and are equally useful in facies recognition (e.g., Ball 1967; Schwartz 1975). Textural interpretations are not as easy as in detrital sandstones: not only the potential availability of particular allochems must be considered but also their inherent size, shape, and internal structure—for example, some fossil fragments (such as from barnacles or some bivalves) can never become rounded however much abrasion they suffer, whereas oolites and pellets are always well

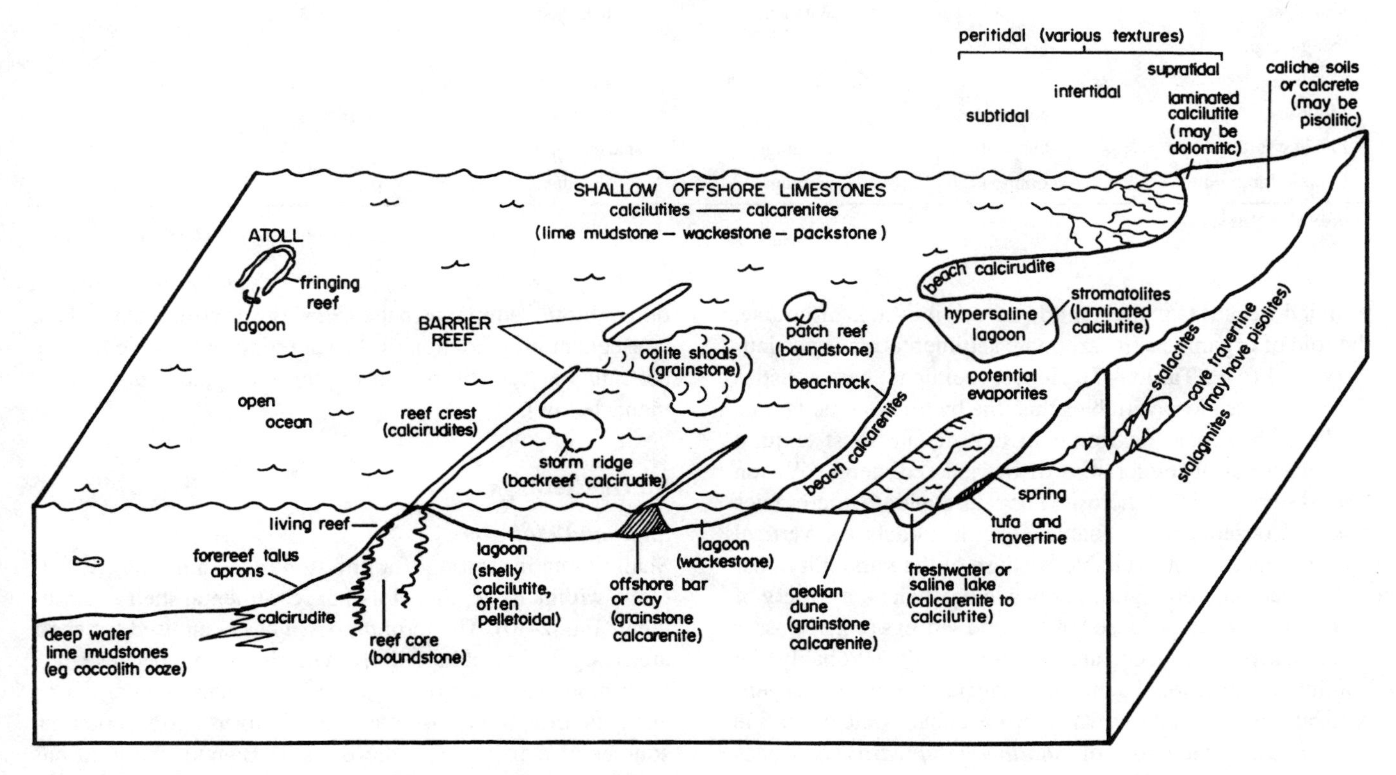

Figure 8-5. Schematic representation of carbonate depositional environments.

Table 8-6. Some Characteristics of Platform Carbonate Sediments

Morphologic Zones:	>100 km Open Sea	10–100 km? Barrier, Reef, or Shelf Edge	0 to >100 km Lagoon or Shelf Interior	<10 km Intertidal	<15 km Supratidal
Energy Sources:	Oceanic Currents	Waves and Tides		Waves, Currents, and Storms .... Tides.	...........Wind
Texture:	Muds; Sandy Muds	Sands; Gravels	(Sandy)Muds		Variable
Sediment Structures					
Lamination	abundant	rare		.... abundant	
Cross-bedding		abundant			...... possible......
Desiccation structures					common ......
Reefs		common	patchy		
Burrows	common		abundant		common
Stromatolites				....	common .....
Sediments					
Skeletal sands	abundant	abundant	abundant	abundant	common
Ooids		abundant			
Pellets				...... abundant	
Aggregates				...... abundant	
Intraclasts					..... common .....
Lime	pelagic			.... skeletal abrasion	
Mud	skeletal abrasion			.... floral disarticulation	inorganic
Evaporites					........common
Biotic Elements					
Marine grasses			abundant	common	
Blue-green algae					.... abundant
Green algae		common	abundant		
Red algae		abundant	common		
Pelagic algae	abundant		common		
Corals		.... abundant	possible		
Molluscs	present	common	abundant	present	
Planktic foraminifera	abundant	common	sparse-common		
Benthic foraminifera	common	common	abundant	present	

Source: After Nelson 1978.

rounded because of their mode of origin. And, of course, the role of organisms in carbonate sedimentation is particularly vital (e.g., Table 8-7). Hence, sediment characteristics depend on far more variables than the hydrodynamic factors of the depositional environment that dictate most textural and structural characteristics of detrital sediments (Wilson [1975] suggested that there are about 24 standard microfacies types). Consequently, reliable general models for vertical successions are not available because of the variability possible in laterally contiguous facies. Nonetheless, a variety of models have been proposed in specific settings (e.g., Strasser 1988; Cloyd, Demicco, and Spencer 1990). Probably the dominant depositional theme that can be recognized is a generalization for shallow-water tropical/subtropical marine carbonates: production of *shallowing-upward* sequences (e.g., James 1984*b*). These result from the abundant production of biotic sediments in the nearshore environment and the consequent progradation of the shoreline as these sediments accumulate (see discussion of peritidal platform environments below).

Marine Settings

Shelf and Platform

Shallow-marine settings for limestone accumulation are generally within the depth realm of the continental shelf environment (Fig. 2-30). The term *platform* has been used for shelf areas largely free of land-derived detritus either because they are remote from land or because a bathymetric low or strong currents effectively isolate the platform from potential sources of detritus (e.g., the Bahama Banks); the term has also been used in a broad sense to include all shallow-water

Table 8-7. Role of Some Organisms in Carbonate Sedimentation

Organisms	Role in Sedimentation
Blue-green algae	Trap fine sediment in mats and as stromatolites
Green algae, sponges, echinoids, barnacles, crinoids	Fragment on death to produce sand size clasts; all fragments of echinodermata comprise single calcite crystals
Planktonic foraminifera, coccoliths (post-Jurassic), benthic foraminifera, red and green algae	Form major portion of most lime muds, together with fine abrasion products of other biota
Encrusting worms, foraminifera, bryozoa, coralline algae, stromatoporoids	Encrust grains and other substrates; may cement sediments into substantial masses (e.g., modern reefs); may break into gravel or sand clasts
Bivalves, brachiopods, cephalopods, gastropods, benthic foraminifera, bryozoa, trilobites	Whole or broken, form major component of most carbonate sands and gravels
Corals, oysters, archaeocyathids, stromatoporoids, rudistid bivalves, bryozoa	Framework organism of bioherms; fragment into gravel size clasts
Worms, crustaceans, bivalves, and representatives of many other groups	Sediment ingestion and excretion modify texture and composition; burrowing modifies primary structures

Source: Modified from James 1984*b*.

limestone settings (e.g., Read 1985). Platforms (in the narrow sense) are commonly subdivided into restricted, open, reef-rimmed or sandy shoal-rimmed varieties (e.g., Wilson 1974). Sedimentary facies in the shelf settings generally are varied and complex, and most publications dealing with carbonate environments are on these settings (e.g., see review in Scholle, Bebout, and Moore 1983). Sediments can have a much wider variety of origins than the simple first-cycle allochem/micrite mixtures most expect (e.g., MacIntyre 1970). It is also in this general setting that evaporative (largely inorganic?) limestones may be deposited (e.g., Decima, McKenzie, and Schreiber 1988).

Peritidal carbonate environments (those of the shoreline) are commonly discussed separately (e.g., Wright 1984), but are an inherent part of all shelf and platform settings. Modern peritidal environments have been the subject of a plethora of studies because of their accessibility, but they are also widely represented in the geological record. Because carbonate sediments are generated in the depositional basin and because in tropical/subtropical settings the rates of biological sediment production are much more rapid than average rates of tectonic subsidence or eustatic rise (except perhaps during glaciations), the general tendency is for carbonate sediments to build up to sea level and for shorelines to prograde as a result (e.g., Strasser 1988). This shallowing-upward trend is not accompanied by a coarsening-upward trend as is the case in most prograding marine detrital successions: the nearshore gradient in most carbonate environments is very gradual, energy levels are low, and lime mud tends to be produced in much larger quantities than coarser sediments; hence muddy tidal flats are the rule.

Reefs are the prime example of life associations in limestone deposits; they are moundlike masses with the bulk of components bound together into a rigid framework with a positive bathymetric relief during growth (Figs. 2-40 and 8-6; e.g., Braithwaite 1973). *Bioherm* is the more general term for moundlike, lensoidal structures of organic origin embedded in rocks of a different character—there may not be obvious framework organisms present, and there may have been little bathymetric relief at any time. *Biostrome* is the term for a tabular, bedded accumulation of mostly sedentary organisms (e.g., oysters), with the implication of little or no bathymetric relief and without any necessary connotation of a rigid framework or of organisms being bound together. Despite these definitions, the bulk of true reefal deposits (those with framework organisms) is composed of fragmented and locally transported skeletal debris; only a very small volume is composed of framework organisms in situ. The diversity of ecological niches and the interdependence of organisms in the reefal association, as well as their influence on sedimentation, is unparalleled in sediments and is a fascinating field of study (e.g., Playford and Lowry 1966; Maxwell and Swinchatt 1970; Laporte 1974; James 1984*a*; Stanley and Fagerstrom 1988).

Framework organisms today are primarily the corals, but in the geological past, stromatoporoids, rudists, sponges, phylloid algae, bryozoa, and archaeocyathids at various times provided the framework (e.g., Rigby and Newell 1971). In modern reefs, the abundance and variety of other organisms inhabiting the reef is enormous (e.g., molluscs, bryozoa, serpulids and other worms, echinoderms, sponges, crustaceans, fish, and similar associations of "incidental" organisms, including a wide variety of soft-bodied organisms that could not leave obvious remains), and similar associations characterized most ancient reefs. Binding of the debris has most commonly been achieved by the encrusting calcareous algae. Community zonation is generally pronounced and can be interpreted as a response to the controls of light penetration (dependent on depth and turbidity), temperature, wave energy, nutrient availability, and siltation rate (whereas reefs normally develop in relatively high-energy environments, the reef-building organisms baffle the currents and micrite is common in many deposits). Because of the diversity of textures and minerals, together with the direct and indirect influence of organic matter, diagenesis of reef sediments is particularly complex (e.g., Matthews 1974).

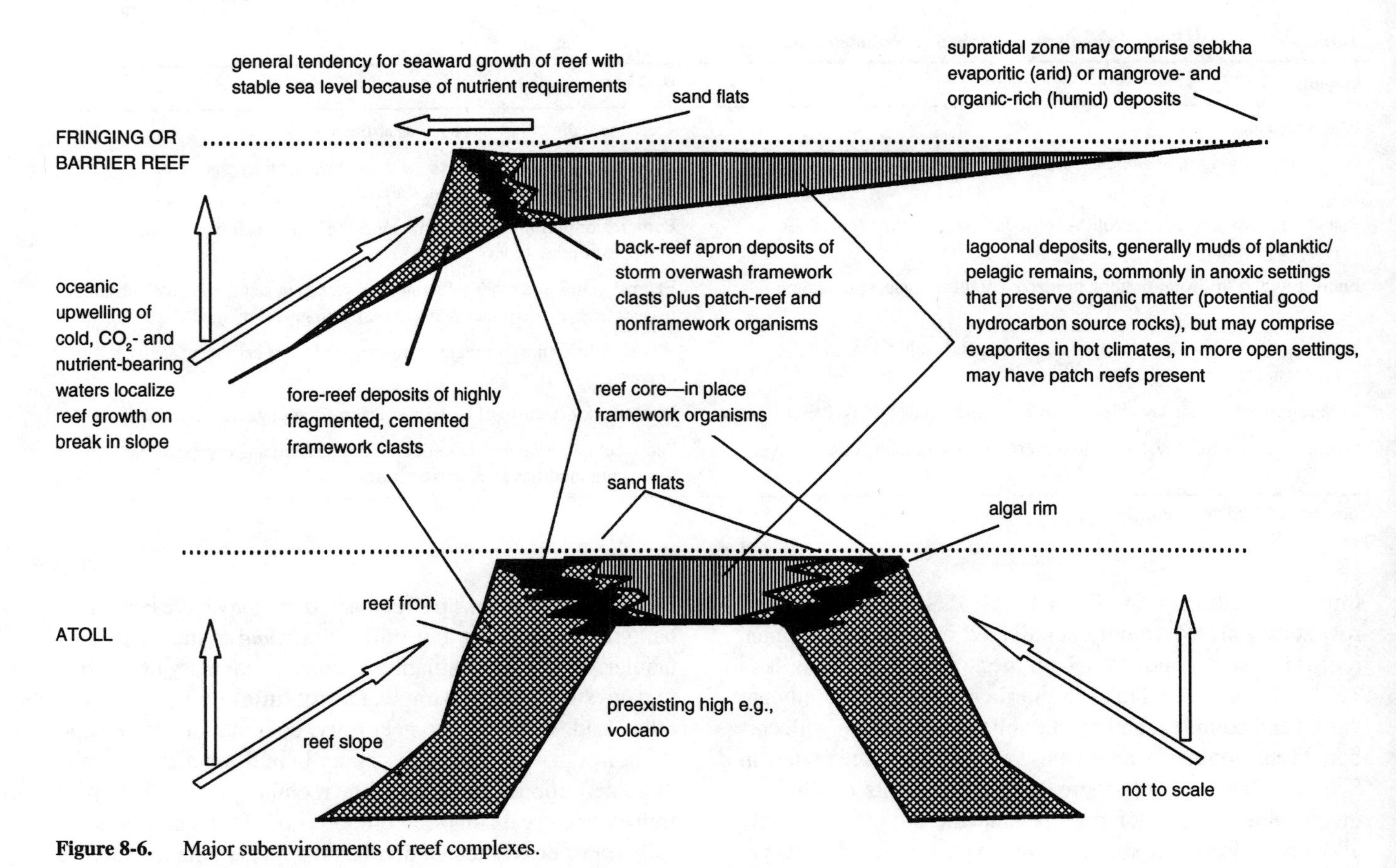

Figure 8-6. Major subenvironments of reef complexes.

Slope

Between the shelf or platform and the relatively deep ocean floor is the *slope* environment. This is the environment through which sediment gravity flows pass, and in which many may be generated together with mass movements of slumps and slides. Some deposits of the coarsest gravity-displaced sediment come to rest in this environment (e.g., reef talus), but because the slope is steeper than other environments, most sediment gravity flows pass through to the base of the slope. Background sedimentation is mainly by fallout of pelagic organic hardparts. Good examples are discussed by Cook and Mullins, and Enos and Moore in Scholle, Bebout, and Moore (1983).

Deep Marine

The deep-sea environment ranges in depth between 200 m and 4500 m (or down to the carbonate compensation level, below which no carbonates accumulate), and is the setting for deposition of relatively homogeneous and undisturbed pelagic oozes. Difficulties exist in distinguishing the deep-water lime mudstones from shelf/slope chalks in both texture and composition, since both can consist of very similar sediment (coccoliths and planktonic foraminifera); trace fossils and facies associations are the most useful distinguishing criteria (e.g., Ekdale and Bromley 1984). Sediment gravity flow deposits are common near the base of the slopes, and carbonate turbidity currents can extend for substantial distances onto the basin floor. This environment has been studied extensively during the past few decades (e.g., during the Deep Sea Drilling Project and the Ocean Drilling Program; extensive reports have been published from these research programs). Good summary and exemplary discussions of this environment and its deposits include Cook and Enos (1977), Scholle, Arthur, and Ekdale in Scholle, Bebout, and Moore (1983), and Jenkyns (1986).

The influence of temperature on the solubility of $CaCO_3$ (the colder the water, the more CO_2 can be held in solution; see discussion in Chapter 3) is the main control on the water depth (cold water is denser than warm water)) below which carbonate particles such as the tests of marine organisms begin to dissolve (the *lysocline*). At some depth below the lysocline, the rate of supply of calcareous matter is balanced by its rate of dissolution; this depth is commonly referred to as the *carbonate compensation depth*. However, because aragonite (and the even more unstable high Mg-calcite) is thermodynamically much less stable than calcite, a distinction should be made between the aragonite compensation depth and the calcite compensation depth, which will be in much deeper water. The compensation depth in the oceans will vary with the temperature profile in the water and the influence of oceanic currents; it lies at much shallower depths in cold polar waters than in warm equatorial waters. Calcite compensation depths vary from about 2.5 km to 5 km, and aragonite compensation depths vary from a few hundred metres to about 2.5 km; hence calcite is the only common mineral in deep-sea carbonate sediments. During the Precambrian, when the carbon dioxide concentration in the atmosphere was much higher than it is today, the mean pH of the oceans would have been lower (e.g., Walker 1983); consequently calcite would have been more soluble and the calcite compensation depth would have been shallower (a higher mean ocean temperature may have negated this rise).

Within the deep-marine pelagic carbonates, rhythmic alternations of intervals with more and less siliclastic mud appear to be related to regular climatic changes forced by variations in the Earth's orbital behavior; such changes on the scale of 10^4–10^5 years are called *Milankovitch cycles* in honor of the pioneering astronomical work of Milankovitch in 1930 (e.g., Berger et al. 1984; Fischer and Bottjer 1991). It is also within the deep pelagic lime mudstones that most work has been applied to diagenesis of biogenic silica that was initially dispersed as sponge, radiolarian, dinoflagellate, and diatom tests (see later discussions below and in Chapter 9).

Nonmarine Settings

In terrestrial carbonate minerals, calcium and magnesium (and other less common cations such as Fe, Mn, Sr, and Ba) are derived from igneous (particularly mafic), sedimentary, and metamorphic rocks in the catchment area of the surface drainage system. Some of the anionic carbonate (or bicarbonate, depending on pH; see Fig. 3-6) in surface and groundwaters is also a product of rock weathering (and rarely, may be derived from hydrothermal fluids), but most is present as a result of the dissolution of atmospheric carbon dioxide (see discussion in Chapter 3). Because the solubility of calcite is very strongly influenced by pH in conditions common in natural environments, small changes in environmental conditions can mean the difference between calcite dissolution and precipitation. Rainwater, some groundwaters (e.g., acid sulfate soils), and many hydrothermal fluids that reach the surface are mildly to moderately acidic (low pH); hence they tend to dissolve carbonate minerals. However, as the resulting calcite-bearing waters interact with rock materials, mix with other water bodies, or undergo evaporative concentration, they may become supersaturated with respect to calcite, which then usually precipitates. With the exception of local contributions from organisms such as charophyte algae, dissolution and precipitation reactions of this type account for most nonmarine carbonate deposits.

Limestone cave deposits (*speleothems*) form as groundwaters carrying calcium and bicarbonate ions slowly evaporate in fractures and on surfaces and precipitate calcite. In most cave systems, there is such a fine balance between calcite deposition and dissolution that even the carbon dioxide exhaled by visitors in the moist cave environment can lower the pH of water droplets and cause some dissolution of the cave formations. Fine parallel laminations in many speleothems (similar to growth rings on trees) reflect subtle seasonal variations in water composition and evaporation rates.

Laminar, finely crystalline (micritic) *tufa* masses (generally biologically mediated) form in areas where warm carbonate-rich groundwaters or hydrothermal fluids reach the surface and the solutions cool and/or evaporate (e.g., Pedley 1990); the deposits may form spectacular terraces, colored by trace elements and oxides in, or associated with, the carbonates. Laminated micritic carbonates may also form where drainage systems terminate in saline lakes in arid zones. Any carbonates remaining in solution will be among the first minerals precipitated as the lake brines evaporate (see Chapter 9); deposits are usually intimately associated with evaporite minerals (particularly gypsum). In these deposits, because they represent seasonal fluctuations (in dissolved CO_2 content due to changes in temperature, evaporation rates, and/or blooms of photosynthesizing plankton), the laminations are like the *varves* of fine detrital sediments generated by lake overturns in temperate regions. In some alkaline nonmarine lakes, the variety of allochem components may be as great as in some marine carbonates (e.g., Freytet 1973; Williamson and Picard 1974).

Of all the nonmarine carbonate deposit types, perhaps the most complex are *calcrete* (carbonate caliche) deposits, which are common in arid/semi-arid areas (Fig. 8-7; e.g., Goudie 1972; Semeniuk and Searle 1985). *Phreatic calcretes* form in paleodrainage channels and similar settings below the water table where rock/water interactions cause the pH of groundwaters to rise as they flow from one hydrogeochemical setting to another (e.g., Mann and Deutscher 1978; Mann and Horwitz 1979). *Vadose calcretes* form above the water table, or in the zone of water table fluctuation, where capillary suction combines with high soil temperatures to evaporatively concentrate carbonate-bearing waters; where capillary suction results in large volumes of fluid being drawn to the surface of sandy soils, hard calcrete surfaces may develop. Vadose and phreatic calcretes are usually found together, and because there have usually been major changes in the water table position over time, phreatic calcretes are often overprinted with vadose calcrete deposits (falling water table), or vadose calcretes are reworked as phreatic calcretes (rising water table). Calcretes also may be influenced by cyclic shifts between vadose and phreatic conditions (e.g., Arakel and McConchie 1982; Arakel 1985, 1986).

Because of the variety of changes that occur during the formation of calcretes, the resulting calcrete textures can be extremely difficult to interpret. As a general rule (with many exceptions!), phreatic calcretes are massive with coarsely crystalline calcite cement binding original sediment and fragmented carbonate clasts; concentrically layered zones formed during cycles of dissolution and precipitation may be retained. Where vadose calcretes overlie phreatic calcretes, the basal vadose zone is usually massive and is overlain by a brecciated calcrete, which shows evidence of repeated cycles of cementation, fragmentation, and dissolution. Where there has been temporary ponding of water on top of the calcrete, or where a root-mat layer develops and is calcified (e.g.,

Wright, Platt, and Wimbledon 1988), a zone of laminar calcrete is developed. Soils characterized by calcite pisolites commonly overlie vadose calcrete horizons.

Eolianites are deposits of carbonate allochems (usually bioclasts) that have been washed up on land; most are found in close proximity to beach deposits. Preservation potential is not high, particularly because of the instability of carbonates in the acidic freshwater environment.

DIAGENESIS

Carbonate sediments are subject to substantially more diagenetic modifications than detrital sediments before, during, and after burial because they are composed of minerals that can dissolve and reprecipitate readily with small changes in common surface and subsurface conditions of temperature, pressure, and water chemistry. As with other sediments, the most extensive changes take place when fluids circulate about the particles, or when temperatures and pressures rise. Common diagenetic settings are illustrated in Fig. 8-8 (and, e.g., James 1985; Simson 1985; Moore 1989). Trace element or carbon and oxygen isotope analysis (and fluid inclusion studies) may be combined with studies of crystal form and size to determine the various stages of diagenetic change and the reaction temperatures (e.g., Dickson 1985; Morse and MacKenzie 1990).

Marine Settings

Under near-surface marine conditions, diagenetic modifications reflect the roles of changing CO_2 content in the waters pervading the sediment (see discussion in Chapter 3); organic activity, pH, and both temperature and pressure changes influence the amount of CO_2 that can be held in solution.

Bioturbation and maceration of skeletal grains are particularly effective early diagenetic processes (e.g., discussion above; Alexandersson 1979); physical mass movements can also effectively influence the structure and fabric of the sediments (e.g., Bromley and Ekdale 1987). Deformation of early-formed structures such as burrows may provide clues to the extent of later compaction of the sediment (e.g., Ricken 1987).

Early submarine cementation generally requires full contact with freely circulating, warm supersaturated waters to supply sufficient carbonate for filling of the pore spaces.

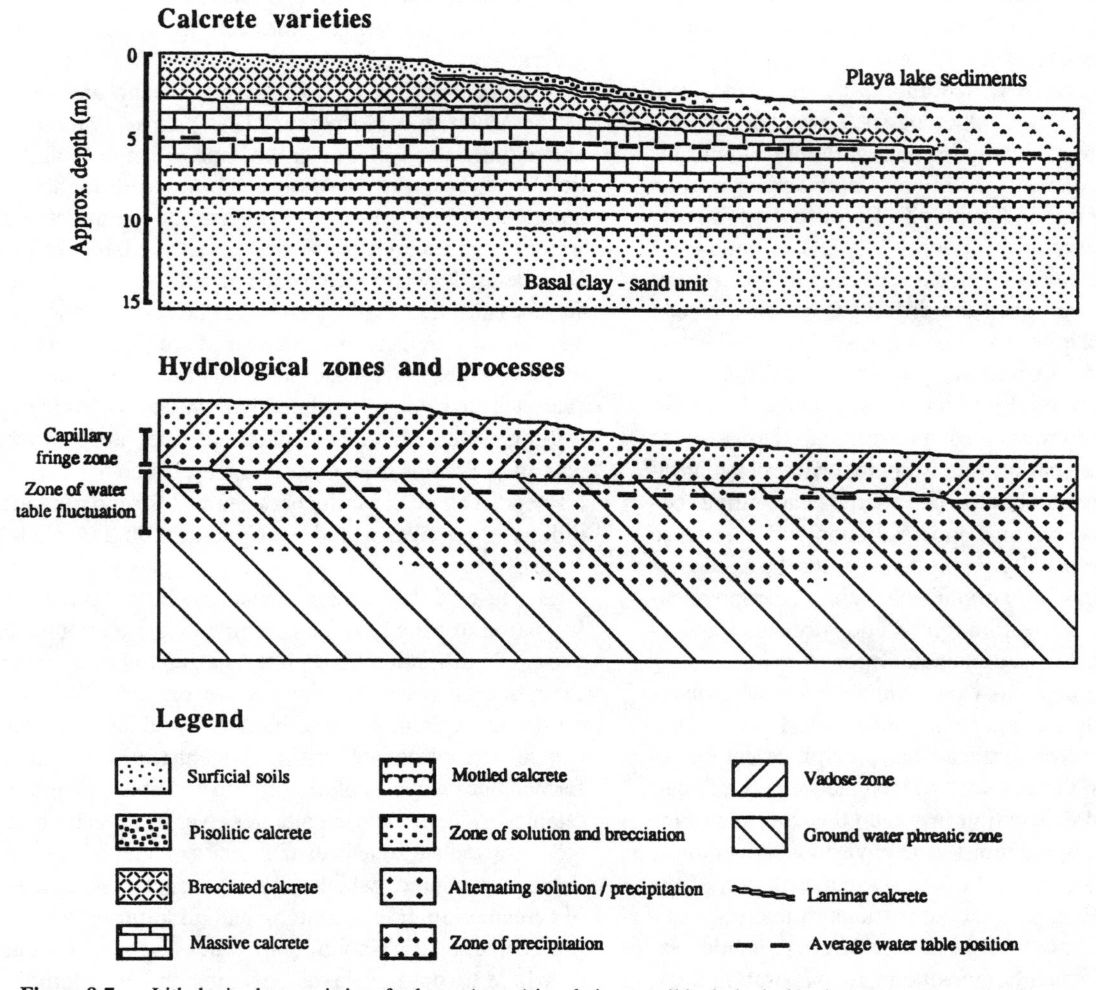

Figure 8-7. Lithologic characteristics of calcrete (upper) in relation to soil hydrological regime (lower); see also Fig. 8-8.

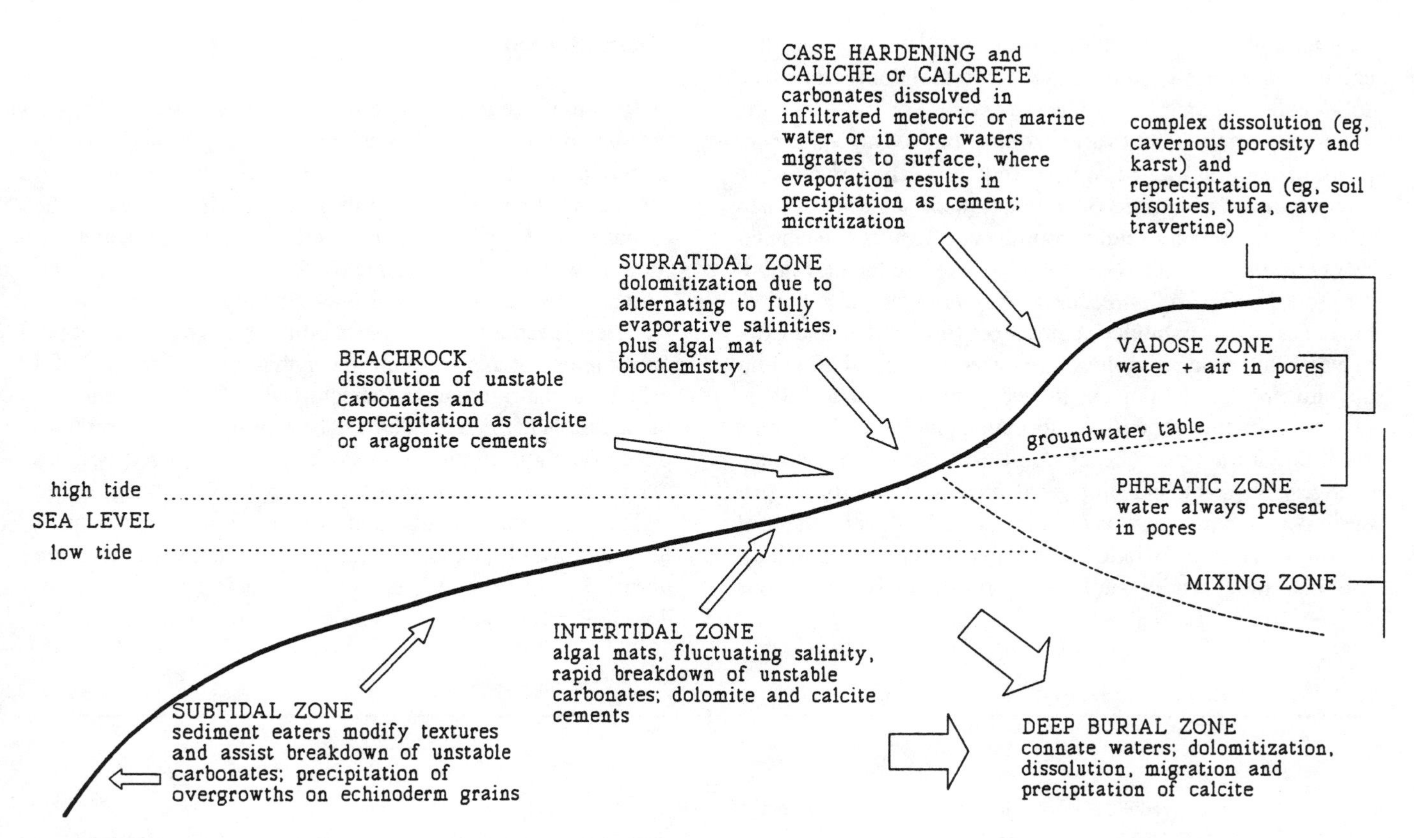

Figure 8-8. Common diagenetic environments for carbonate sediments.

Hard grounds can form at the water/sediment interface (e.g., Kennedy and Garrison 1975 and see Chapter 4) and *beachrock* at the marine+freshwater+atmosphere/sediment interface (e.g., Beier 1985). Beachrock (e.g., on a coral cay) forms where calcareous material (usually Mg-calcite or aragonite shells) is dissolved by rainwater, then reprecipitated between grains as a cement. Precipitation commonly takes place at or near the high tide level, where less dense, fresh or brackish groundwater floats to the surface on the saltwater wedge and evaporates. In some areas, deposition may be aided by partial mixing of the fresh water with higher pH sea spray. Fresh cola cans and other assorted anthropogenic debris cemented into beachrock on modern carbonate beaches in the tropics and subtropics (e.g., the Great Barrier Reef or Bahama Banks) attest to the rapid rate of beachrock development.

Cements may also be dissolved and precipitated below the sediment/water interface (e.g., where pressure-induced solution of carbonates takes place). Earliest cements in tropical areas tend to be needlelike or bladelike microcrystals of aragonite and high Mg-calcite that precipitate at grain edges and as encrustations. Later, these unstable minerals convert to low Mg-calcite, and in the process new cements and new cement morphologies are created. Cementation may take place by precipitation of calcite onto seed crystals, such as the single crystals of echinoderm fragments, whereupon the new calcite adopts the same crystallographic orientation (*syntaxial overgrowths*; e.g., Maliva 1989). Micritization of skeletal clasts and oolites is also common, principally as a result of the microboring activities of endolithic algae and fungi. At depth, where circulation is reduced and dissolved carbonates are soon exhausted, pressure-induced dissolution of parts or all of some particles is generally necessary for cementation to continue; the dissolved carbonate may be derived from subjacent grains or may travel some distance in solution from "donor" beds.

Nonmarine Settings

The most extensive and varied diagenetic modifications to carbonate sediments take place where meteoric waters are involved (e.g., Land 1970; James 1985; Quinn 1991). In subaerial environments the *phreatic* zone of permanent saturation underlies the *vadose* zone where pores are filled with alternating water and air; both of these zones may overlie sediments whose pores are filled with denser seawater, and there is a *schizohaline* boundary interval of complex chemistry between them (e.g., see discussion in Scoffin 1987). Dissolution by fresh water in the vadose zone commonly results in saturated waters from which precipitation can occur if CO_2 is released or the water starts to evaporate, as in soils of semiarid regions where caliche and calcretes form (see discussion above).

Even more extensive changes to mineralogy and texture

may take place in the phreatic zone. All unstable high Mg-calcite and aragonite convert to low Mg-carbonate either by a solid-state molecule-for-molecule process or by dissolving and reprecipitating, and various other changes take place depending on the amount of water circulation. The largest-scale nonmarine modifications of carbonate sediments are in cases where dissolution by acidic groundwaters results in deep networks of interconnected cavities—giving rise to *karst* (modern) topography and *paleokarst* (ancient—generally buried, but occasionally exhumed) complexes (e.g., James and Choquette 1988). Such dissolution features may cap shallowing-upward marine carbonate facies sequences. The calcite dissolved in this process is normally precipitated at some point down the hydraulic gradient. In addition to chemical changes, physical and biological modifications to the primary deposits may also take place (such as development of *rhizoliths* or plant rootlet structures; e.g., Klappa 1980); chemical modifications commonly follow these disruptions of the sediment fabric.

Dolomitization

Dolomite, while generally rare in modern sediments, is geochemically more stable than calcite and is next in abundance to calcite in ancient carbonate rocks. Particularly in normal marine waters where Mg^{+2}:Ca^{+2} ratios exceed 5:2, it *should* be the carbonate to form rather than calcite or aragonite. However, for a new unit cell of dolomite to form, a magnesium, a calcium, and two carbonate ions must "get together" in the same place at the same time; the probability of this event at normal ocean temperatures is much lower than the probability of a calcium and a carbonate ion meeting alone to form a unit cell of calcite. Where there is an existing carbonate mineral template, this molecularity problem is largely resolved because two of the constituents are already present; thus dolomite can replace calcite under conditions where the concentrations of magnesium and calcium ions (and the ratio between them) are unsuited to the primary precipitation of dolomite. The lack of dolomite precipitation under normal marine conditions also

Table 8-8. Possibilities for the Origin of Dolostone and Dolomite

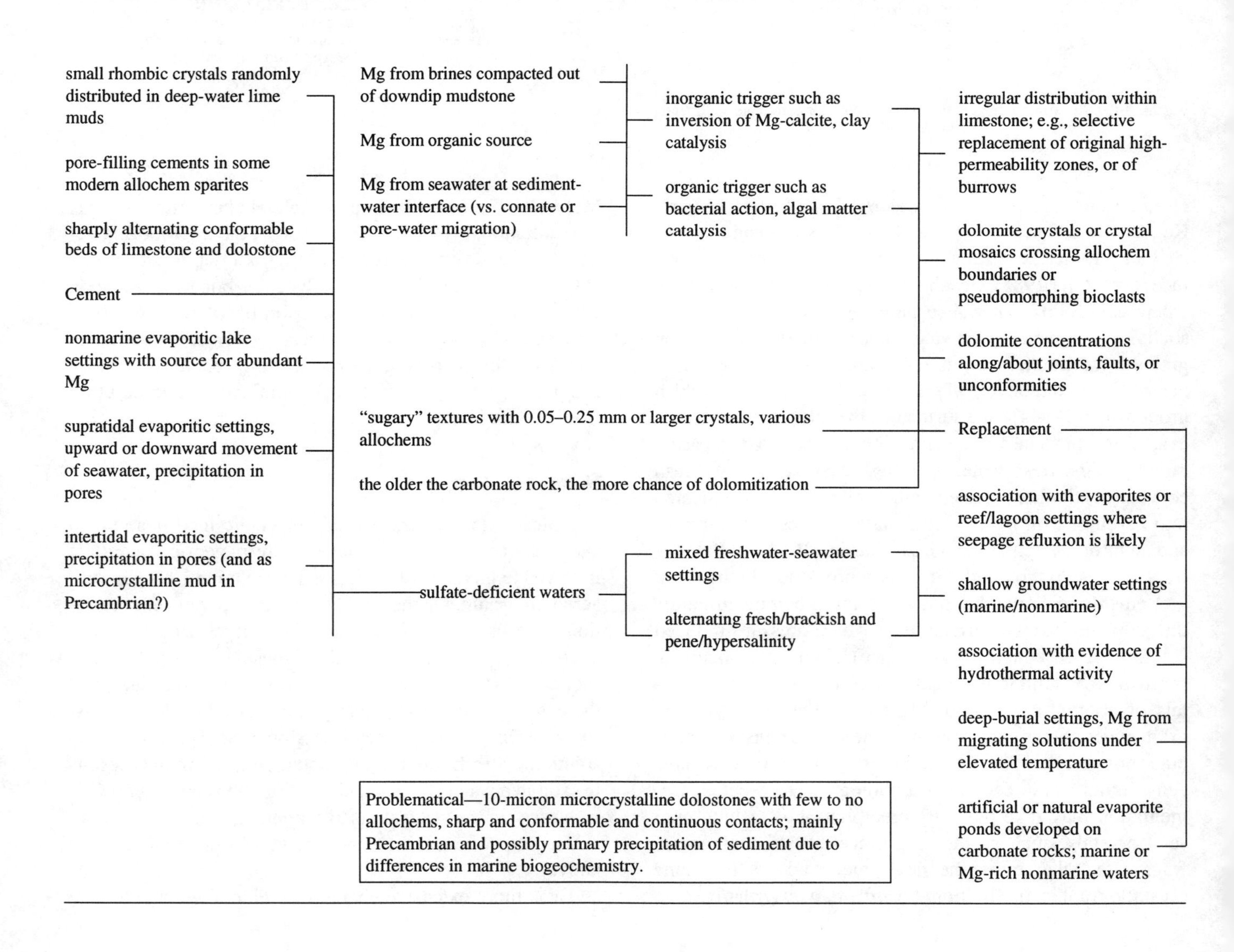

may reflect the existence of an inhibiting factor, such as sulfate ions (Baker and Kastner 1981). In nonmarine environments, under special conditions where the concentrations of Mg^{2+} and Ca^{2+} relative to other ions are high, an unstable and imperfect *protodolomite* can precipitate (e.g., Alderman and Skinner 1957; Gaines 1977; Warren 1990).

The vast majority of studies indicate that dolomite generally results from diagenetic replacement of primary carbonates, particularly high Mg-calcite and aragonite, but also low Mg-calcite. The multiple genetic possibilities for dolomite are outlined in Table 8-8. Certainly evidence is convincing that coarse-grained dolostones are replacive, and although some "ghost" outlines of allochem varieties may be distinguished locally, special terminology and treatment is necessary for most of these coarsely crystalline rocks (e.g., Sibley and Gregg 1987). Few subjects in geology have attracted as much attention as the "dolomite problem," and a multitude of models in many different settings have been proposed for diagenetic dolomite, both in very early diagenesis, as in the supratidal occurrences of the tropics, and very late in the burial history of the rock (e.g., Shinn, Ginsburg, and Lloyd 1965; Zenger 1972; Badiozamani 1973; Folk and Siedlecka 1974; Folk and Land 1975; Zenger, Dunham, and Ethington 1980; Land 1985; Machel and Mountjoy 1986; Hardie 1987; Mazzullo, Reid, and Gregg 1987; Humphrey and Quinn 1989).

Geochemical conditions for the transformation appear to be elevated temperature, a higher than normal Mg^{+2} content of the water (e.g., achieved in arid regions after some Ca^{+2} has been removed by precipitation of Ca-sulfates), or a

Table 8-9. Porosity in Carbonate Rocks

Permeability	Common Size (mm)	Shape	Selective	Timing	Relative Abundance	Relation to Permeability
Within grains, e.g., chambers of bryozoa, forams	0.01–1	Cellular	Yes	Depositional	Very rare	Poor
Between grains, normal primary porosity, includes shelter—below large irregular ones	0.05–1	Irregular between grains	Yes	Depositional	Common	Moderate
Early mechanical, burrow and gas-escape disruption	0.2–10 to vertical	Irregular, subhorizontal	No diagenetic	Early	Common	Poor to moderate
Enhanced primary, dissolution expands pores within or between grains	0.1–1	Irregular enlargements of inter- and intragranular pores	Yes	Early diagenetic	Common	Moderate
Between crystals due to recrystallization or authigenic crystal growth	0.001–0.1	Prisms, thin sheets	Yes or no	Early to late diagenetic	Rare to common	Good
Moldic, in the place of dissolved grains	0.2–10	As original grains	Yes	Early to late diagenetic	Rare to common	Poor
Vuggy or cavernous, irregular dissolution holes that cut across grains and/or cement (e.g., karst)	1–1000 vugs, 1000+ cavernous	Irregular	No	Early to late diagenetic	Common	Poor to moderate
Fractures, post-burial, due to tectonic or fluid overpressure	0.5–10	Parallel or conjugate sheets of various orientations	No	Late diagenetic	Common	Good
Stylolitic, along pressure-solution seams	0.1 wide, 100–10,000 long	Jagged seams parallel to bedding or joints	Yes or no	Late diagenetic	Very rare	Moderate to good

Pore Size Terms			
	Megapores	large	32 mm
		small	4
	Mesopores	large	0.5
		small	0.063
	Micropores		

Source: After Choquette and Pray 1970.

change in the concentration of both Mg^{+2} and Ca^{+2} ions that creates an even greater supersaturation with respect to dolomite (e.g., in brackish water or alternating saline/freshwater settings, or even evaporation of saltwater spray—e.g., Kocurko 1979). There is also some evidence for primary precipitation of dolomite (e.g., Hardie 1987), as in deep-sea sediments (e.g., Lumsden 1988) and in the pores of some supratidal sediments (e.g., Lasemi, Boardman, and Sandberg 1989), but the evidence is inconclusive. Nonetheless, many carbonate sedimentologists still suspect that even if primary dolomite precipitates cannot form in modern environments, they may have been able to precipitate from Precambrian ocean waters where ion ratios and concentrations may have differed and temperatures may have been higher; it has always been difficult to avoid an interpretation of primary deposition for the thick, continuous, nonporous dolomicrites that are particularly common in the Precambrian, and there is some suggestion that they are different from the Phanerozoic dolostones (e.g., Tucker 1982).

Porosity Changes and Stylolitization

Apart from mineralogical changes (loss of the unstable primary carbonates and transformations to dolomite), two other diagenetic effects are particularly important in carbonate sediments. One is change in porosity (e.g., Table 8-9; Choquette and Pray 1970; Friedman 1975; Moore 1989): either decreases brought about by both compaction and precipitation of cements (e.g., Wetzel 1989), or increases brought about by dissolution after burial (e.g., Giles and Marshall 1986) and/or with dolomitization (e.g., Amthor and Friedman 1991). Porosity may increase or decrease at various stages from early to late in the history of a sediment; large-scale cavernous porosity in the form of karstic pits and cave systems can develop after lithification and emergence of a limestone into the groundwater environment—paleokarst systems are recognized in limestones of many different ages and settings (e.g., James and Choquette 1988).

The other important change is a bulk loss of carbonate volume without the development of secondary porosity. This loss is accomplished by the development of *stylolites*—irregular seams (e.g., Fig. 3-16) along which pressure-assisted dissolution has resulted in the removal of carbonate (e.g., Simpson 1985). Most stylolites occur roughly along original bedding planes because the necessary pressure is most commonly provided by overlying sediment accumulations. However, stylolites also may form along joints in response to tectonically imposed pressures. The stress is transmitted along the framework of the sediment and is thus concentrated where grains touch; at these points the calcite dissolves and dissolution products are carried away in solution until the seam is clogged with insoluble residues (*stylocumulates* of noncarbonate sediments). Stylolites are thus most common in pure calcite or dolomite rocks that have some permeability. The jagged shape of the seams reflects inhomogeneities in the parent rocks, the local presence of isolated insoluble particles against which the carbonates dissolve, and the fact that shells of different structure and density dissolve at different rates; the orientation of carbonate crystals in the grains also plays a role, because dissolution is faster along the crystallographic *c*-axis than along other axes.

SELECTED BIBLIOGRAPHY

General

Bathurst, R. G. C., 1975, *Carbonate Sediments and Their Diagenesis,* 2d ed. Developments in Sedimentology 12, Elsevier, Amsterdam, 658p.

Berger, A., J. Imbrie, J. Hays, G. Kukla, and B. Saltzman (eds.), 1984, *Milankovitch and Climate: Understanding the Response to Astronomical Forcing,* 2 vols. D. Reidel Publishing Co., Dordrecht, Holland, 895p.

Chilingar, G. V., H. J. Bissell, and R. W. Fairbridge (eds.), 1967, *Carbonate Rocks,* 2 vols. Developments in Sedimentology 9A/B, Elsevier, Amsterdam, 471p, 413p.

Choquette, P. W., and L. C. Pray, 1970, Geologic nomenclature and classification of porosity in sedimentary carbonates. *American Association of Petroleum Geologists Bulletin* 54:207–50.

Cook, H. E., P. N. McDaniel, E. W. Mountjoy, and L. C. Pray, 1972, Allochthonous carbonate debris flows at Devonian bank ("reef") margins, Alberta, Canada. *Canadian Petroleum Geologists Bulletin* 20:439–97.

Dunham, R. J., 1962, Classification of carbonate rocks according to depositional texture. In W. E. Ham (ed.), *Classification of Carbonate Rocks.* American Association of Petroleum Geologists Memoir 1, Tulsa, Okla., pp. 108–21.

Embry, A. F., and E. J. Klovan, 1972, Absolute water depth limits of late Devonian paleoecological zones. *Geologische Rundschau* 61:672–86.

Fahraeus, L. E., R. M. Slatt, and G. S. Nowlan, 1974, Origin of carbonate pseudopellets. *Journal of Sedimentary Petrology* 44:27–9.

Fischer, A. G., and D. J. Bottjer, 1991, Orbital forcing and sedimentary sequences. *Journal of Sedimentary Petrology* 61:1063–9.

Folk, R. L., 1959, The practical petrographical classification of limestones. *American Association of Petroleum Geologists Bulletin* 43:1–38.

Folk, R. L., 1962, Spectral subdivision of limestone types. In E. W. Ham (ed.), *Classification of Carbonate Rocks: A Symposium.* American Association of Petroleum Geologists Memoir 1, Tulsa, Okla., pp. 62–84.

Folk, R. L., 1973, Carbonate petrography in the post-Sorbian age. In R. N. Ginsburg (ed.), *Evolving Concepts in Sedimentology.* Johns Hopkins University Press, Baltimore, pp. 118–58.

Ham, E. W. (ed.), 1962, *Classification of Carbonate Rocks: A Symposium.* American Association of Petroleum Geologists Memoir 1, Tulsa, Okla., 279p.

Hiscott, R. N., and N. P. James, 1985, Carbonate debris flows, Cow Head Group, western Newfoundland. *Journal of Sedimentary Petrology* 55:735–45.

Horowitz, A. S., and P. E. Potter, 1971, *Introductory Petrography of Fossils.* Springer-Verlag, New York, 302p.

Lees, A., and A. T. Buller, 1972, Modern temperate-water and warm-water shelf carbonate sediments contrasted. *Marine Geology* 13:M67–M73.

Lindholm, R. C., 1969, Carbonate petrology of the Onondaga limestone (Middle Devonian), New York: A case for calcisiltite. *Journal of Sedimentary Petrology* 39:260–75.

Matthews, R. K., 1966, Genesis of lime mud in Southern British Honduras. *Journal of Sedimentary Petrology* 36:428–554.

Perkins, R. D., and S. D. Halsey, 1971, Geologic significance of microboring fungi and algae in Carolina shelf sediments. *Journal of Sedimentary Petrology* 41:843–53.

Reid, R. P., I. G. MacIntyre, and N. P. James, 1990, Internal precipitation of microcrystalline carbonate: A fundamental problem for sedimentologists. *Sedimentary Geology* 68:163–70.

Reijers, T. J. A., and K. J. Hsu (eds.), 1986, *Manual of Carbonate Sedimentology.* Academic Press, New York, 302p.

Scholle, P. A., 1978, *A Colour Illustrated Guide to Carbonate Rock Constituents, Textures, Cements and Porosities.* American Association of Petroleum Geologists Memoir 27, Tulsa, Okla., 241p.

Scoffin, T. P., 1987, *An Introduction to Carbonate Sediments and Rocks.* Blackie & Son, New York, 274 p.

Shinn, E. A., R. P. Steinen, B. H. Lidz, and P. K. Swart, 1989, Whitings, a sedimentologic dilemma. *Journal of Sedimentary Petrology* 59:147–61.

Simone, L., 1981, Ooids: A review. *Earth-Science Reviews* 16:319–55.

Sorby, H. C., 1879, Structure and origin of limestones (anniversary address of the president). *Proceedings of the Geological Society of London* 35:56–95.

Stieglitz, R. D., 1973, Carbonate needles: Additional organic sources. *Geological Association of America Bulletin* 84:927–30.

Stockman, K. W., R. N. Ginsburg, and E. A. Shinn, 1967, The production of lime mud by algae in south Florida. *Journal of Sedimentary Petrology* 37:633–48.

Swinchatt, J. P., 1965, Significance of constituent composition, texture and skeletal breakdown in some recent carbonate sediments. *Journal of Sedimentary Petrology* 35:71–90.

Wilson, J. L., 1975, *Carbonate Facies in Geologic History.* Springer-Verlag, New York, 471p.

Wright, V. P., 1992, A revised classification of limestones. *Sedimentary Geology* 76:177–85.

Mineralogy, Trace Elements, and Isotopes

Abell, P. I., 1985, Oxygen isotope ratios in modern African gastropod shells: A data base for paleoclimatology. *Isotope Geoscience* 58:183–93.

Arthur, M. A., T. F. Anderson, I. R. Kaplan, J. Veizer, and L. S. Land, 1983, *Stable Isotopes in Sedimentary Geology.* Short Course 10, Society of Economic Paleontologists and Mineralogists, Tulsa, Okla., 295p.

Bowen, R., 1988, *Isotopes in the Earth Sciences.* Elsevier, Amsterdam.

Chave, K. E., 1962, Factors influencing the mineralogy of carbonate sediments. *Limnology and Oceanography* 7:218–23.

Fritz, P., and J. Ch. Fontes (eds.), 1986, *Handbook of Environmental Isotope Geochemistry,* vol. 2. Elsevier, Amsterdam, 557p.

Goldsmith, J. R., D. L. Graf, J. Witters, and D. A. Northrop, 1962, Studies in the system $CaCO_3$-$MgCO_3$-$FeCO_3$. *Journal of Geology* 70:659–88.

Hoefs, J., 1987, *Stable Isotope Geochemistry,* 3d ed. Springer-Verlag, New York, 241p.

Kahn, M. I., T. Oba, and T.-L. Ku, 1981, Paleotemperatures and the glacially induced changes in the oxygen-isotope composition of seawater during late Pleistocene and Holocene time in the Tanner Basin, California. *Geology* 9:485–90.

Kinsman, D. J. J., 1969, Interpretation of Sr(+2) concentrations in carbonate minerals and rocks. *Journal of Sedimentary Petrology* 39:486–508.

Morse, J. W., and F. T. MacKenzie, 1990, *Geochemistry of Sedimentary Carbonates.* Developments in Sedimentology 48, Elsevier, New York, 707p.

Savin, S. M., and H. W. Yeh, 1981, Stable isotopes in ocean sediments. In C. Emiliani (ed.), *The Sea,* vol. 7, Wiley, New York, pp. 1521–54.

Marine Environments

Ball, M. M., 1967, Carbonate sand bodies of Florida and the Bahamas. *Journal of Sedimentary Petrology* 37:556–91.

Cloyd, K. C., R. V. Demicco, and R. J. Spencer, 1990, Tidal channel, levee, and crevasse-splay deposits from a Cambrian tidal channel system: A new mechanism to produce shallowing-upward sequences. *Journal of Sedimentary Petrology* 60:73–83.

Cook, H. E., and P. Enos (eds.), 1977, *Deep-Water Carbonate Environments.* Society for Sedimentary Geology Special Publication 25, Tulsa, Okla., 336p.

Cook, H. E., A. C. Hine, and H. T. Mullins, 1983, *Platform Margin and Deepwater Carbonates.* Short Course 12, Society for Sedimentary Geology, Tulsa, Okla., 573p.

Decima, A., J. A. McKenzie, and B. C. Schreiber, 1988, The origin of "evaporative" limestones: An example from the Messinian of Sicily (Italy). *Journal of Sedimentary Petrology* 58:256–72.

Ekdale, A. A., and R. G. Bromley, 1984, Comparative ichnology of shelf-sea and deep-sea chalk. *Journal of Paleontology* 58:322–32.

Eliuk, L. S., 1978, The Abenaki Formation, Nova Scotia Shelf, Canada—a depositional and diagenetic model for a Mesozoic carbonate platform. *Canadian Petroleum Geology Bulletin* 26:424–514.

James, N. P., 1984*a*, Reefs. In R. G. Walker (ed.), *Facies Models,* 2d ed. Geoscience Canada Reprint Series 1, Toronto, pp. 229–44.

James, N. P., 1984*b*, Shallowing-upward sequences in carbonates. In R. G. Walker (ed.), *Facies Models,* 2d ed. Geoscience Canada Reprint Series 1, Toronto, pp. 213–228.

Jenkyns, H. C., 1986, Pelagic environments. In H. G. Reading (ed.), *Sedimentary Environments and Facies,* 2d ed. Blackwell Scientific Publications, Oxford, pp. 343–97.

Katz, A., 1971, Zoned dolomite crystals. *Journal of Geology* 79: 38–51.

Logan, B. W., G. R. Davies, J. F. Read, and D. E. Cebulski, 1970, Carbonate Sedimentation and Environments, Shark Bay, Western Australia. American Association of Petroleum Geologists Memoir 13, Tulsa, Okla., 223p.

MacIntyre, I. G., 1970, Sediments off the West Coast of Barbados: A diversity of origins. *Marine Geology* 9:5–23.

Nelson, C. S., 1978, Temperate shelf carbonate sediments in the Cenozoic of New Zealand. *Sedimentology* 25:737–71.

Nelson, C. S., 1988, An introductory perspective on non-tropical shelf carbonates. *Sedimentary Geology* 60:3–12.

Read, J. F., 1985, Carbonate platform facies models. *American Association of Petroleum Geologists Bulletin* 69:1–21.

Scholle, P. J., D. G. Bebout, and C. H. Moore (eds.), 1983, *Carbonate Depositional Environments.* American Association of Petroleum Geologists Memoir 33, Tulsa, Okla., 708p.

Schwartz, H.-H., 1975, Sedimentary structures and facies analysis of shallow marine carbonates. *Contributions to Sedimentology* 3:1–100.

Strasser, A., 1988, Shallowing-upward sequences in Purbeckian peritidal carbonates (lowermost Cretaceous, Swiss and French Jura Mountains). *Sedimentology* 35:369–83.

Walker, J. C. G., 1983, Possible limits on the composition of the Archaean Ocean. *Nature* 302:518–20.

Wilson, J. L., 1974, Characteristics of carbonate-platform margins. *American Association of Petroleum Geologists Bulletin* 58:810–24.

Wright, V. P., 1984, Peritidal carbonate facies models: A review. *Geological Journal* 19:309–25.

Reefs

Braithwaite, C. J. R., 1973, Reefs: Just a problem of semantics? *American Association of Petroleum Geologists Bulletin* 57: 1100–16.

Laporte, L. F. (ed.), 1974, *Reefs in Time and Space.* Society for Sedimentary Geology Special Publication 18, Tulsa, Okla.

Matthews, R. K., 1974, A process approach to diagenesis of reefs and reef associated limestones. In L. F. Laportee (ed.), *Reefs in Time and Space*. Society for Sedimentary Geology Special Publication 18, Tulsa, Okla., pp. 234–56.

Maxwell, W. G. H., and J. P. Swinchatt, 1970, Great Barrier Reef: Regional variation in a terrigenous-carbonate province. *Geological Association of America Bulletin* 81:691–724.

Playford, P. E., and D. C. Lowry, 1966, Devonian reef complexes of the Canning Basin, Western Australia. *Geological Survey of Western Australia Bulletin* 118:1–150.

Rigby, J. K., and N. D. Newell (eds.), 1971, Reef organisms through time. In *Symposium Volume—Proceedings of the North American Paleontological Convention, Part J.* Allen Press, Lawrence, Kansas.

Stanley, G. D., Jr., and J. A. Fagerstrom, 1988, Ancient reef ecosystems: An introduction to the volume. *Palaios* 3:110–11.

Nonmarine Carbonates

Arakel, A. V., 1985, Vadose diagenesis and multiple calcrete soil profile development in Hutt Lagoon area, western Australia. *Revue de Geologie Dynamique et de Geographie Physique* 26:243–54.

Arakel, A. V., 1986, Evolution of calcrete in palaeodrainages of the Lake Napperby area, Central Australia. *Palaeogeography, Palaeoclimatology, Palaeoecology* 54:283–303.

Arakel, A. V., and D. McConchie, 1982, Classification and genesis of calcrete and gypsite lithofacies in paleodrainage systems of inland Australia and their relationship to carnotite mineralization. *Journal of Sedimentary Petrology* 52:1149–70.

Freytet, P., 1973, Petrography and paleo-environment of continental carbonate deposits with particular reference to the Upper Cretaceous and Lower Eocene of Languedoc (southern France). *Sedimentary Geology* 10:25–60.

Goudie, A., 1972, The chemistry of world calcrete deposits. *Journal of Geology* 80:449–63.

Klappa, C. F., 1980, Rhizoliths in terrestrial carbonates: Classification, recognition, genesis and significance. *Sedimentology* 27:613–30.

Mann, A. W., and R. L. Deutscher, 1978, Hydrogeochemistry of calcrete-containing aquifer near Lake Way, Western Australia. *Journal of Hydrology* 38:357–77.

Mann, A. W., and R. C. Horwitz, 1979, Groundwater calcrete deposits in Australia: Some observations from Western Australia. *Journal of the Geological Society of Australia* 26:293–303.

Pedley, H. M., 1990, Classification and environmental models of cool freshwater tufa. *Sedimentary Geology* 68:143–54.

Semeniuk, V., and D. J. Searle, 1985, Distribution of calcrete in Holocene coastal sands in relationship to climate, southwestern Australia. *Journal of Sedimentary Petrology* 55:86–95.

Williamson, C. R., and M. D. Picard, 1974, Petrology of carbonate rocks of the Green River Formation (Eocene). *Journal of Sedimentary Petrology* 44:738–59.

Wright, V. P., N. H. Platt, and W. A. Wimbledon, 1988, Biogenic laminar calcretes: Evidence of calcified root-mat horizons in paleosols. *Sedimentology* 35:603–20.

Diagenesis

Alexandersson, E. T., 1979, Marine maceration of skeletal carbonates in the Skagerrak, North Sea. *Sedimentology* 26:845–52.

Bathurst, R. G. C., 1970, Problems of lithification in carbonate muds. *Geologists Association Proceedings* 81:429–40.

Beier, J. A., 1985, Diagenesis of Quaternary Bahamian beachrock: Petrographic and isotopic evidence. *Journal of Sedimentary Petrology* 55:755–61.

Bromley, R. G., and A. A. Ekdale, 1987, Mass transport in European Cretaceous chalk: Fabric criteria for its recognition. *Sedimentology* 34:1079–92.

Dickson, J. A. D., 1985, Diagenesis of shallow-marine carbonates. In P. J. Brenchley and B. P. J. Williams (eds.), *Sedimentology, Recent Developments and Applied Aspects.* Geological Society of London Special Publication 18, Blackwell Scientific Publications, Oxford, England, pp. 173–88.

Dickson, J. A. D. 1993, Crystal growth diagrams as an aid to interpreting the fabrics of calcite aggregates. *Journal of Sedimentary Petrology* 63:1–17.

Folk, R. L., 1965, Some aspects of recrystallization in ancient limestones. In L. C. Pray and R. C. Murray (eds.), *Dolomitization and Limestone Diagenesis.* Society for Sedimentary Geology Special Publication 13, Tulsa, Okla., pp. 14–48.

Friedman, G. M., 1975, The making and unmaking of limestones or the downs and ups of porosity. *Journal of Sedimentary Petrology* 45:379–98.

Giles, M. R., and J. D. Marshall, 1986, Constraints on the development of secondary porosity in the subsurface: Re-evaluation of processes. *Marine and Petroleum Geology* 3:243–55.

Herbert, T. D., 1993, Differential compaction in lithified deep-sea sediments is not evidence for 'diagenetic unmixing'. *Sedimentary Geology* 84:115–22.

James, N. P., and P. W. Choquette (eds.), 1988, *Paleokarst.* Springer-Verlag, New York.

James, W. C., 1985, Early diagenesis, Atherton Formation (Quaternary): A guide for understanding early cement distribution and grain modifications in nonmarine deposits. *Journal of Sedimentary Petrology* 55:135–46.

Kennedy, W. J., and R. E. Garrison, 1975, Morphology and genesis of nodular chalks and hardgrounds in the Upper Cretaceous of southern England. *Sedimentology* 22:311–86.

Land, L. S., 1970, Phreatic versus vadose meteoric diagenesis of limestones: Evidence from a fossil water table. *Sedimentology* 14:175–85.

Maliva, R. G., 1989, Displacive calcite syntaxial overgrowths in open marine limestones. *Journal of Sedimentary Petrology* 59:397–403.

Moore, C. H., 1989, *Carbonate Diagenesis and Porosity.* Developments in Sedimentology 46, Elsevier, New York, 338p.

Quinn, T. M., 1991, Meteoric diagenesis of Plio–Pleistocene limestones at Enewetak Atoll. *Journal of Sedimentary Petrology* 61:681–703.

Reaves, C. M., 1986, Organic matter metabolizability and calcium carbonate dissolution in nearshore marine muds. *Journal of Sedimentary Petrology* 56:486–94.

Ricken, W., 1987, The carbonate compaction law: A new tool. *Sedimentology* 34:571–84.

Simpson, J., 1985, Stylolite-controlled layering in an homogeneous limestone: Pseudo-bedding produced by burial diagenesis. *Sedimentology* 32:495–505.

Wetzel, A., 1989, Influence of heat flow on ooze/chalk cementation: Quantification from consolidation parameters in DSDP sites 504 and 505 sediments. *Journal of Sedimentary Petrology* 59:539–47.

Dolomite

Alderman, A. R., and H. C. Skinner, 1957, Dolomite sedimentation in the south-east of Australia and aspects of carbonate sedimentation. *American Journal of Science* 255:561–7.

Amthor, J. E., and G. M. Friedman, 1991, Dolomite-rock textures and secondary porosity development in Ellenburger Group carbonates (Lower Ordovician), west Texas and southeastern New Mexico. *Sedimentology* 38:343–62.

Badiozamani, K., 1973, The Dorag dolomitization model—application to the Middle Ordovician of Wisconsin. *Journal of Sedimentary Petrology* 43:965–84.

Baker, P. A., and M. Kastner, 1981, Constraints on the formation of sedimentary dolomite. *Science* 213:214–16.

Folk, R. L., and L. S. Land, 1975, Mg/Ca ratio and salinity: Two controls over crystallization of dolomite. *American Association of Petroleum Geologists Bulletin* 59:60–8.

Folk, R. L., and A. Siedlecka, 1974, The schizohaline environment: Its sedimentary and diagenetic fabrics as exemplified by late Paleozoic rocks of Bear Island, Svalbard. *Sedimentary Geology* 11:1–15.

Gaines, A. M., 1977, Protodolomite redefined. *Journal of Sedimentary Petrology* 47:543–6.

Given, R. K., and B. H. Wilkinson, 1987. Dolomite abundance and stratigraphic age: Constraints on rates and mechanisms of Phanerozoic dolostone formation. *Journal of Sedimentary Petrology* 57:1068–1078. (Also see discussion by D. H. Zenger, and reply, *Journal of Sedimentary Petrology* 59:162–5.)

Hardie, L. A., 1987, Dolomitization: A critical view of some current views. *Journal of Sedimentary Petrology* 57:166–83.

Humphrey, J. D., and T. M. Quinn, 1989, Coastal mixing zone dolomite, forward modeling, and massive dolomitization of platform-margin carbonates. *Journal of Sedimentary Petrology* 59:438–54.

James, N. P., and Y. Bone, 1991, Origin of a cool-water, Oligo–Miocene deep shelf limestone, Eucla Platform, southern Australia. *Sedimentology* 38:323–42.

Kocurko, M. J., 1979, Dolomitization by spray-zone brine seepage, San Andres, Colombia. *Journal of Sedimentary Petrology* 49:209–14.

Land, L. S., 1985, The origin of massice dolomite, *Journal of Geological Education* 33:112–25.

Lasemi, Z., M. R. Boardman, and P. A. Sandberg, 1989, Cement origin of supratidal dolomite, Andros Island, Bahamas. *Journal of Sedimentary Petrology* 59:249–57.

Lumsden, D. N., 1988, Characteristics of deep-marine dolomite. *Journal of Sedimentary Petrology* 58:1023–31. (See also 1989 discussion by G. M. Friedman, and reply, *Journal of Sedimentary Petrology* 59:879–81.)

Machel, H.-G., and E. W. Mountjoy, 1986, Chemistry and environments of dolomitization—a reappraisal. *Earth-Science Reviews* 23:175–222.

Mazzullo, S. J., A. M. Reid, and J. M. Gregg, 1987, Dolomitization of Holocene Mg-calcite supratidal deposits, Ambergris Cay, Belize. *Geological Society of America Bulletin* 98:224–31.

Shinn, E. A., R. N. Ginsburg, and R. M. Lloyd, 1965, Recent supratidal dolomite from Andros Island, Bahamas. In L. C. Pray and R. C. Murray (eds.), Society for Sedimentary Geology Special Publication 13, Tulsa, Okla., pp. 112–23.

Sibley, D. F., and J. M. Gregg, 1987, Classification of dolomite rock textures. *Journal of Sedimentary Petrology* 57:967–75.

Tucker, M. E., 1982, Precambrian dolomites: Petrographic and isotopic evidence that they differ from Phanerozoic dolomites. *Geology* 10:7–12.

Warren, J. K., 1990, Sedimentology and mineralogy of dolomitic Coorong lakes, South Australia. *Journal of Sedimentary Petrology* 60:843–58.

Zenger, D. H., 1972, Significance of supratidal dolomitization in the geologic record. *Geological Association of America Bulletin* 83:1–12.

Zenger, D. H., J. B. Dunham, and R. L. Ethington (eds.), 1980, *Concepts and Models of Dolomitization.* Society for Sedimentary Geology Special Publication 28, Tulsa, Okla., 320p.

9
Chemical Sediments

Strictly, chemical sediments comprise all minerals formed by inorganic processes in the sedimentary environment. In fact, biological processes are indirectly but intimately involved in the genesis of many "chemical" sediments, through their influence on chemical conditions in the depositional environment (e.g., Eh or HS^- activity). Distinction between chemical sediments and minerals formed by chemical alteration of some precursor detrital or biogenic mineral is also blurred; e.g., glauconite can form by inorganic alteration of a (often biologically degraded) clayey pellet or mineral grain as well as by direct chemical precipitation. Chapter 3 provides an introduction to processes supplemental to the following discussion.

Most chemical sediments in the geologic record are *authigenic* (formed in the place they are found, such as are all cements), but some are *perigenic* (formed nearby but locally transported before final accumulation) and in rare sediments they may be *allogenic* (derived as detrital grains from sedimentary rocks, although in that last cycle of sedimentation they are no longer classed as chemical sediments). Chemical sediments are generally formed as a result of solvent evaporation; hence they are likely to redissolve before transport in water, but there are exceptions (e.g., chemically deposited silica and apatite can survive transport as allogenic grains by virtue of their very low solubility, and chemically deposited ferric iron oxides remain insoluble wherever conditions remain oxidizing).

Chemical sediments include a wide variety of evaporite minerals as well as many iron minerals, phosphates, silica, manganese, and metal sulfide deposits. Deposition may result from a wide variety of chemical processes, including:

Evaporation of solvent (e.g., water), causing precipitation as the constituents in solution exceed saturation concentration;

The common ion effect (e.g., gypsum may precipitate if sulfate-bearing solutions derived from weathering of sulfides flow into a salt lake undersaturated with respect to gypsum);

Formation of insoluble compounds when compositionally different solutions mix (e.g., barite can deposit if solutions containing barium mix with sulfate-rich solutions);

Changes in chemical conditions (e.g., pH or Eh) in the depositional environment (e.g., iron oxyhydroxides are likely to precipitate if there is a rise in pH and/or Eh within low-pH groundwaters containing Fe^{2+});

Changes in the temperature or pressure of water bodies (e.g., precipitation from cooling exhalative hydrothermal solutions or from warming of upwelling deep ocean waters);

Coprecipitation reactions (e.g., precipitation of silica from silica-undersaturated solutions when iron or aluminium hydroxides precipitate; see Chapter 3);

Photolytic (sunlight-caused) oxidation processes (primarily applies to iron oxyhydroxides; e.g., Braterman, Cairns-Smith, and Sloper 1983);

Processes involving biological influences (e.g., sulfide mineral precipitation caused by reaction with microbially formed sulfide, precipitation due to microbially induced changes in Eh conditions, or numerous other biologically mediated reactions; e.g., Nealson 1982).

Because chemical sediments form by precipitation or alteration reactions, they provide valuable information on chemical characteristics of the depositional environment. Studies of chemical sediments also form an integral part of studies of the evolution of Earth's atmosphere and hydrosphere, together with their relationships to major economic mineral deposits (e.g., Clemmey and Badham 1982; Holland 1984; papers in Holland and Schidlowski 1982; Windley, Simpson, and Muir 1984).

Table 9-1. Common Evaporite Minerals

Borates	*Nitrates*
Boracite—$Mg(Fe,Mn)_3B_7O_{13}Cl$	Saltpeter—$NaNO_3$
Borax—$Na_2B_4O_7$	
	Sulfates
Carbonates	Anhydrite—$CaSO_4$
Aragonite—$CaCO_3$	Bassanite—$2CaSO_4 \cdot 2H_2O$
Calcite—$CaCO_3$	Celestite—$SrSO_4$
Dolomite—$CaMg(CO_3)_2$	Epsomite—$MgSO_4 \cdot 7H_2O$
Magnesite—$MgCO_3$	Glauberite—$Na_2Ca(SO_4)_2$
Siderite—$FeCO_3$	Gypsum—$CaSO_4 \cdot 2H_2O$
Strontianite—$SrCO_3$	Hexahydrite—$MgSO_4 \cdot 6H_2O$
	Kainite—$KMgClSO_4 \cdot 3H_2O$
Chlorides	Kieserite—$MgSO_4 \cdot H_2O$
Antarcticite—$CaCl_2 \cdot 6H_2O$	Langbeinite—$K_2Mg_2(SO_4)_3$
Bischofite—$MgCl_2 \cdot 6H_2O$	Mirabilite—$Na_2SO_4 \cdot 10H_2O$
Carnallite—$KMgCl_3 \cdot 6H_2O$	Polyhalite—$K_2\,Mg\,Ca_2(SO_4)_4 \cdot 2H_2O$
Halite—NaCl	Starkeyite—$MgSO_4 \cdot 4H_2O$
Sylvite—KCl	Thenardite—Na_2SO_4

EVAPORITES

Evaporite minerals form wherever evaporation is equal to or greater than the sum of precipitation, runoff, and influx from seawater or groundwater. Natural evaporite deposition occurs from lakes, springs, groundwater, and seawater (e.g., Scruton 1953; Stewart 1963; Dean and Schreiber 1978; Jankowski and Jacobson 1989; Arakel and Cohen 1991). Marine evaporites are most common and widespread in the geological column, occurring either as relatively thin and impure deposits of mainly calcium sulfate that accumulated in peritidal paleoenvironments (such as the supratidal *sebkhas*), or as thick (to over 2000 m), widespread (to thousands of km^2) basinal sequences that include monomineralic beds of the most soluble salts (such as the Permian Zechstein sequence of northern Europe, e.g., Borchert and Muir 1964). Although the dominant depositional styles and the sizes of deposits have varied over time (e.g., Muir 1987), evaporites can be found in sedimentary strata of any age (evaporite sequences have been reported in rocks as old as 3.45 Ga by Lambert et al. 1978).

The most common evaporite minerals are listed in Table 9-1, and approximate limits for precipitation of some of the common salts are shown in Table 9-2 (see also Stewart 1963; Braitsch 1971; Dean and Schreiber 1978). Table 9-3 shows examples of the succession in one of the "saline giants." In the laboratory, calcium carbonate will precipitate after about 50% of an original volume of seawater has evaporated; in nature, however, biological agencies generally influence deposition and relatively few limestones are attributed primarily to evaporation (examples are described by Bellanca and Neri 1986; Decima, McKenzie, and Schreiber 1988). Calcium sulfates precipitate after about 80% evaporation, halite after about 90%, and the most soluble potassium and mixed potassium-sodium-magnesium salts only after 98% evaporation. Natural precipitation sequences are so complex that it is difficult to account for the existence of thick or extensive monomineralic deposits of any kind more soluble than calcium sulfate. Diagenetic modifications involving dissolution, precipitation, and replacement to original textures and composition are generally substantial.

Evaporite deposits are the direct source of many economically important materials (eg, gypsum, halite, glauberite, thenardite, Mg- and K-salts [e.g., epsomite and sylvite], iodine, nitrates, potassium, magnesium, boron, and many others; see Melvin 1991). Evaporites are also closely associated with major carnotite uranium deposits in calcretes (e.g., Mann and Deutscher 1978; Arakel and McConchie 1982) and with many major sulfide ore deposits such as the Kupferschiefer, Mt. Isa, and McArthur River (e.g., Brongersma-Sanders 1968; Muir 1987). Intermediate stages in the geochemical and environmental conditions favoring evaporite deposition are not only suitable for accumulation of metal sulfides (cf. deposition in the modern Red Sea, e.g., Degens and Ross 1969; Friedman 1972) but also for accumulation of petroleum source materials (e.g., Warren 1986; Schreiber 1988). Diagenetic dissolution of evaporites can increase the petroleum reservoir potential of sediments, and impermeable evaporite strata can form effective cap rocks. The broader-scale environmental significance of salts has been reviewed by Martinez (1971).

Genetic Models

For relatively thin, impure, sulfate-dominated evaporite deposits, the model generally envisioned (and documented in

Table 9-2. Guideline Data on Precipitation of Some Common Evaporitic Salts from Seawater

		Limits for Precipitation					
			Lower Limit		Upper Limit		
Compound	Concentration in Seawater (ppt)	Volume H_2O	Density (g/cm^3)	Salinity (ppt)	Density (g/cm^3)	Salinity (ppt)	Depth of Marine Water (m) Evaporated for 1-m Precipitate
$CaCO_3$	0.12	53–19%	1.0500	72	1.264	200	25,000
$CaSO_4(2H_2O)$	1.27	19–3%	1.1264	200	1.2570	427	2,100
NaCl	27.2	9.5–1.6%	1.2138	353	?		73
$MgSO_4$	2.25	9.5–0%	1.2138	353	?		
$MgCl_2$	3.35	9.5–0%	1.2138	252	?		
KCl	0.74	1.5–0%	?		?		

Note: Precipitation in later stages of evaporation is complex and depends on temperature and other factors.

Source: After Rosler and Lange 1972; Stewart 1963; Dean and Schreiber 1978.

Table 9-3. Examples of an Evaporite Succession in the Zechstein Basin, Permian Age, Northern Europe

Germany (normal succession)		Whitby District England		
Thickness (m)	Interval	Interval		Thickness (m)
5		Upper marl		up to 183
117	Fourth evaporite	Top anhydrite		1
15		Salt clay		2–4
160–271	Third evaporite	Upper evaporites	Upper halite	15–27
			potash salts	0–9
			Lower halite	13–20
			anhydrite	5–9
			carbonate	1
8	(dolomitic top)	Carnallitic marl		9–19
49–116	Second evaporite	Middle evaporites	Upper halite	0–5
			potash salts	3–4
			Lower halite	
			halite-anhydrite	28–84
			anhydrite	15–28
9	Dolomite and other rocks	Upper magnesian limestone		36–56
57	First evaporite	Lower evaporites	Upper halite-anhydrite	12–36
			Upper anhydrite	12–183
			Lower halite-anhydrite	14–43
			Lower anhydrite	46–93
6	Chalk	Lower magnesian limestone (some anhydrite)		111
	Kupferschiefer (sulfides) conglomerate	Basal sands, breccias, marls		

Note: Thicknesses vary locally; correlations approximate; subdivisions not given for German succession.

Source: After Stewart 1963.

modern settings) is a marginal marine setting (particularly the supratidal *sebkha*, e.g., Figs. 9-1 and 9-2) where brines derived from saline groundwater or sporadic seawater influx evaporate and precipitate their least soluble salts in the pores of the sediment (e.g., Amiel and Friedman 1971; Kinsman 1976; Arakel 1980, 1981; Lowenstein and Hardie 1985; Muir 1987) . In arid regions, salts (including highly soluble varieties) precipitate on the sediment surface when playa lakes evaporate (e.g., Figs. 2-20 and 9-2), but these are ephemeral and generally redissolve in the next wet season.

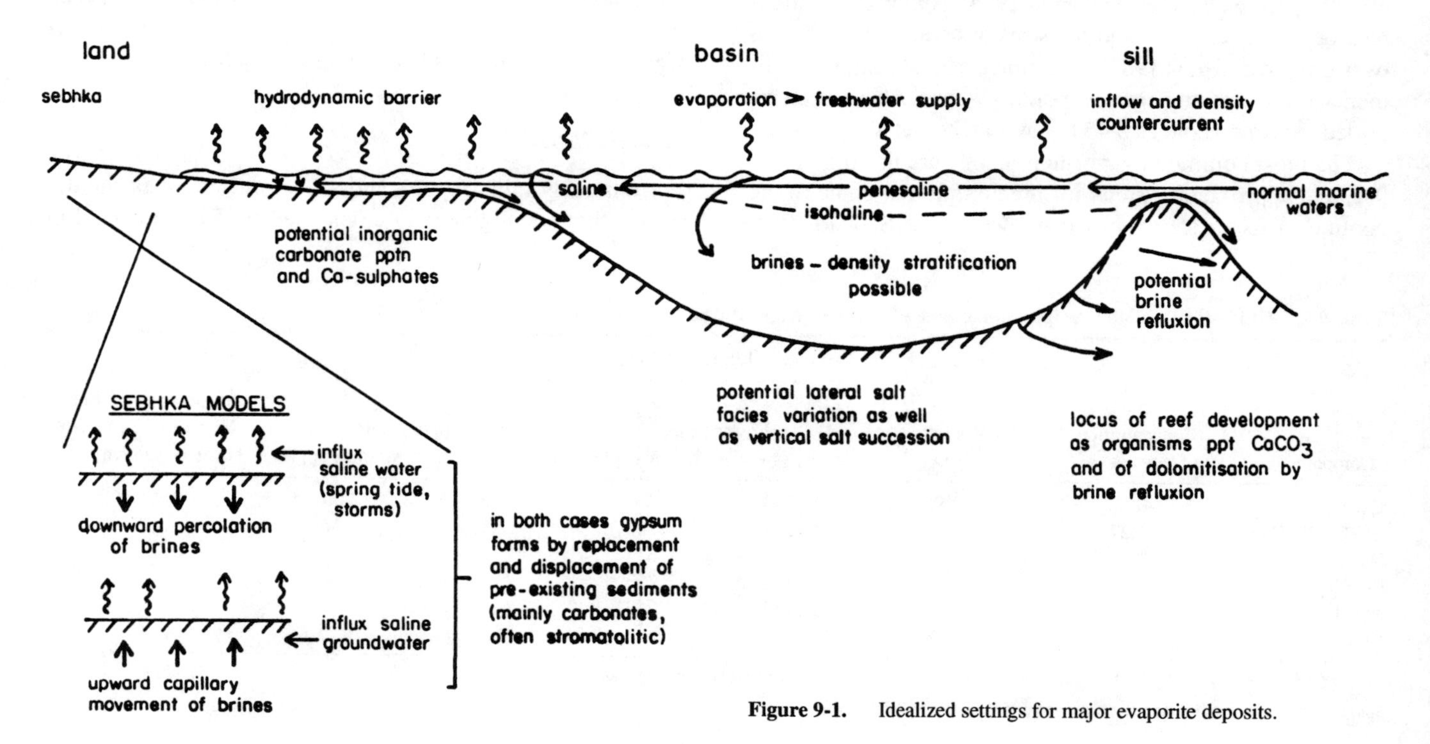

Figure 9-1. Idealized settings for major evaporite deposits.

Algal salt marsh (slightly hypersaline water)

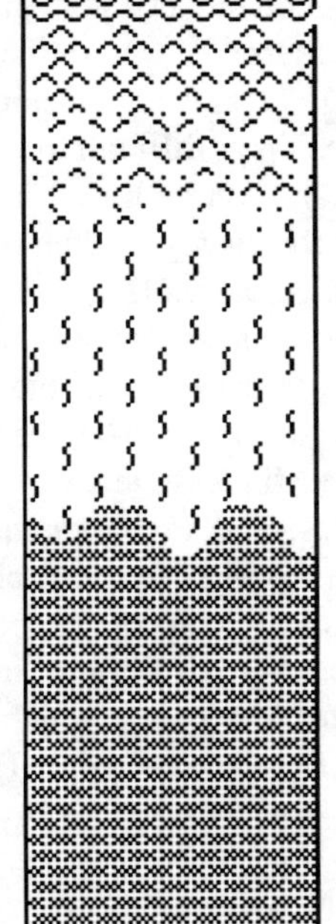

Live algal mat

Dead algal mat fragments mixed with finely laminated muds (disruption increases with depth); the sediment is commonly highly reducing

Bioturbated intertidal muds and sands

Intertidal and subtidal sediment

Coastal sebkha sequence

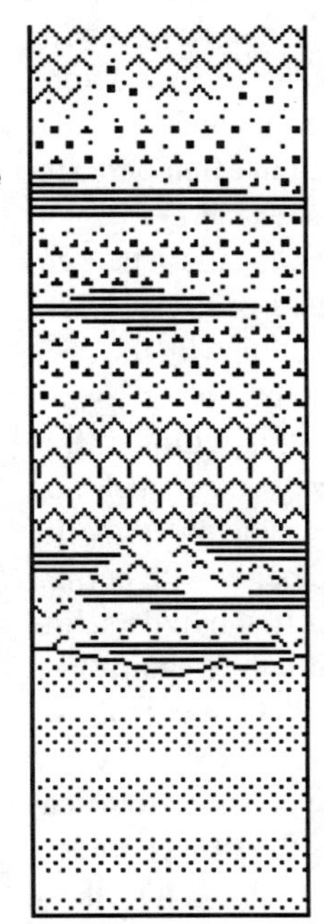

Ephemeral halite crust

Gypsarenites and laminated aragonite mud (*and detrital mineral zones?*); displacive halite crystals are abundant near the top and displacive gypsum crystals are common toward the base of the unit

Displacive prismatic gypsum

Algal mat remnants and laminated aragonitic mud

Layered intertidal and subtidal muds and sands

Seasonally dry playa lake sequence

Ephemeral bedded halite (*and bittern salts?*)

Algal mat?

Gypsarenites and gypsrudites with laminated clastic gypsite and carbonate mud (*and detrital mineral zones?*); sparse displacive diagenetic gypsum and halite crystals

Displacive prismatic gypsum

Laminated carbonate (aragonite) mud with abundant displacive diagenetic gypsum crystals

Basal clayey - sands

Playa calcrete profile

Pisolitic loose soil

Brecciated calcrete

Massive vadose calcrete

Average water table

Massive phreatic calcrete

Mottled calcrete

Basal clayey - sands

Figure 9-2. Idealized vertical facies models for evaporitic salt sequences in shallow-water settings.

Evaporite minerals precipitated in the pores of the sediment are likely to experience repeated cycles of dissolution and reprecipitation as the water table rises and falls, as well as during burial diagenesis. Sediment displacement by growing sulfate crystals often produces a "chicken-wire" structure, where the primary sediment is compressed between nodular evaporites. Coastal salt pans produce thin but often pure sulfate deposits (e.g., see Kushnir 1981; Arakel 1981)

For the thick, extensive, monomineralic deposits in the geological record, the general model imagined is a largely enclosed basin with a restriction to full circulation with the open ocean (a "sill"—as in door sill—some physical or hydrodynamic barrier system; e.g., Fig. 9-1). In this setting, influx of seawater can balance evaporation at any stage, such that precipitation of only one salt occurs. An "ideal" evaporite cycle of deposits from such a basin would begin with normal open marine sediment, gradually progress to the hypersaline stage, then "freshen" to open marine sediments again (Fig. 9-3). Both deep-water (e.g., Schmalz 1969) and shallow-water (e.g., Hardie and Eugster 1971; Hovorka 1987) models have been proposed, and the complex role of brine stratification and interaction (e.g., Raup 1970) is still not well understood. Because of the high solubility of evaporites, it is likely that many original deposits have been eliminated during the subsequent geological history of the sediment. Even in the most arid regions, groundwater dissolution prevents exposure of salts more soluble than the sulfates (however, in some areas of Australia, hypersaline groundwater brines are known with salinities exceeding 400 ppt, and they may be contributing to, rather than dissolving, the salt deposits). Clues attesting directly to the original presence of evaporite minerals may be difficult to find: indirect evidence may occur in the form of odd silica deposits or of collapse features

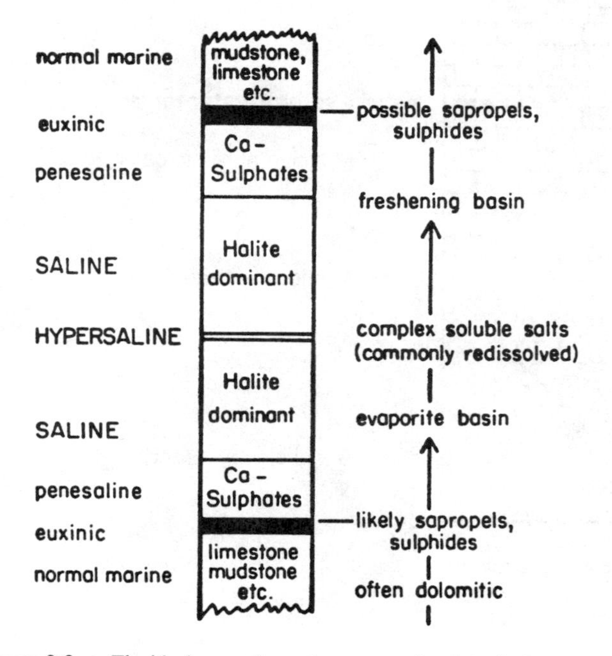

Figure 9-3. The ideal evaporite cycle, representing deposits from a basin that initially evaporates, then freshens.

where the salts have been dissolved (e.g., Folk and Pittman 1971; Muir 1987).

Many unresolved problems remain to account for the "saline giants" for which we have no good modern analogue (the Miocene evaporites in the Mediterranean basin are the youngest such deposits, e.g., Hsu 1972, and they do not attain a truly "giant" scale). For example, it is as difficult to account for consistent varve laminations that can be traced in some evaporites for over 100 km as for the analogous laminations in some banded iron formations. A particularly intriguing problem is to account for the enormous deposits of Permian evaporites around the world (e.g., in the Perm of the Commonwealth of Independent States [former USSR], the Zechstein Basin of northern Europe, the Midland-Delaware Basin of the southwestern United States, and in Saskatchewan, Canada). It is estimated that the evaporites in the Zechstein Basin alone total about 2×10^9 km^3, in contrast to the total volume of salt in the modern oceans (of similar salinity throughout the Phanerozoic, by most evidence), which is estimated at just over 2×10^{10} km^3 (Borchert and Muir 1964).

IRON IN SEDIMENTATION

Minerals and Geochemistry

Iron minerals are one of the major indicators of chemical conditions in sediments (they are also one of the main coloring agents). If detrital, they may be unmodified indicators of provenance (see Chapter 6). However, detrital iron minerals are commonly oxidized or reduced in the environment of weathering, during transport, or at the site of deposition; many iron-bearing minerals are also formed in the depositional (e.g., glauconite) or diagenetic (e.g., pyrite) environment. Hence, the assemblage of iron bearing minerals commonly reflects paleoclimatic factors and chemical and biogeochemical conditions of the depositional and diagenetic environments. The presence of organic matter (see Bass Becking and Moore 1950), sulfur compounds (e.g., Curtis and Spears 1968), and bacterial activity is particularly influential with respect to the origin and/or alteration of iron minerals.

Iron-rich sediments are those with more than 5% iron. *Iron formations* (see discussion below) are mostly Precambrian iron-rich sediments interlayered with silica-rich sediments (cherts); they are commonly hundreds of metres thick, and usually have abundant iron-oxide and iron-silicate minerals. *Blacksands*, *bog iron ores*, and *gossans* are other iron-rich sediments. Table 9-4 lists the iron-rich minerals that are commonly found in sedimentary deposits (see also James 1966). Figure 9-4 shows a simplified representation of the relationship between iron minerals and depositional setting.

Iron occurs in two valence states: Fe^{2+} (the ferrous ion), which is soluble, and Fe^{3+} (the ferric ion), which is almost insoluble. Small changes in Eh and pH conditions, within the range found in natural environments, can rapidly alter the solubility of iron and the stability of iron-bearing minerals (e.g., Fig. 3-3). Where oxygen is available, the oxidized form

Table 9-4. Common Iron Minerals and Their Characteristics

Oxides

Hematite αFe_2O_3

Hexagonal. Principal mineral in Precambrian iron formations; also common in Lower Paleozoic ores but only locally important in Mesozoic and Tertiary ironstones, except as a product of secondary enrichment. Main red coloring agent in rocks, but need not be abundant to produce distinctive color. Called *martite* when occurs as octahedral pseudomorphs after magnetite.

Geothite $\alpha FeO(OH)$ or ($\alpha Fe_2O_3 \cdot H_2O$)

Orthorhombic. Brown, submetallic, crystalline but often botryoidal. Common in soils (brown color). Absent in Precambrian rocks, but may have been present and converted to hematite. Common associate of chamosite, to which or from which it may convert.

Lepidocrocite $\gamma FeO(OH)$ or $\gamma Fe_2O_3 \cdot H_2O$

Orthorhombic. Hydrated polymorph of geothite; uncommon. Red, brown, or yellow.

Ilmenite $FeTiO_3$

Hexagonal-rhombohedral. If pure, contains 58% TiO_2; generally has less because of substitutions for Ti. Uncertain whether it forms under sedimentary conditions. May dominate blacksands.

Magnetite $FeO \cdot Fe_2O_3$

Isometric. In Phanerozoic rocks, mostly of metamorphic origin but some is diagenetic or supergene. Abundant in Precambrian, in association with Fe-silicates, apparently of early diagenetic origin. Formation requires intermediate Eh and high pH. Possibly forms from dehydration of $Fe(OH)_2$. Called *titanomagnetite* when in solid solution with ulvospinel.

Maghemite γFe_2O_3

Isometric. Magnetic dimorph of hematite; crystallography and X-ray pattern similar to magnetite, from the oxidation of which it appears to be derived. Thought to be rare in nature but is abundant in some laterites. Generally black-blue gray.

Limonite—a mixture of iron oxides (especially *goethite* and *lepidocrocite* and *turgite* ($Fe_2O_3 \cdot H_2O$)

Characteristic of bog iron ores and gossans, and locally mined. Earthy, yellow to dark brown.

Carbonates

Siderite $FeCO_3$

Hexagonal-rhombohedral. The only common Fe-carbonate, although *ankerite* (Fe-rich dolomite) is locally abundant. Easily oxidized in surficial environments, whereupon it converts to mixture of limonite and calcite. Important constituent of ironstones, where it generally is very finely crystalline and intimately intermixed with other minerals. Rarely economic because of impurities. Common accessory mineral in coal measures (e.g., as concretions and black bands).

Silicates

Glauconite $(OH)_2K_{(x+y)}(Fe^{3+}, Al, Fe^{2+}, Mg)_{\sim 2}(Si_{(4-x)}Al_x)_{\sim 4}O_{10}$ where $x = 0.2–0.6$ and $y = 0.4–0.6$

Clay mineral. Not economic to date (potential K-fertilizer if cation-exchange capability can be enhanced. Very uncommon in Precambrian (only present in Upper Proterozoic). Not a major component of iron formations or ironstones, although can comprise up to 75% of greensand beds to more than 10 m thick. Variety of parent materials; formed almost exclusively in marine environments.

Chamosite $(OH)_8(Fe_4^{+2}Al_2)(Si_2Al_2)O_{10}$

Septechlorite. Commonest Fe-silicate in post-Precambrian iron formations and ironstones. Commonly oolitic, otherwise similar occurrence and field properties to glauconite. Primary mineral, but can alter to or from goethite and hematite and chlorite. Economic ore in some places (e.g., Jurassic of Britain and Alsace-Lorraine). Commonly associated with siderite and/or clacite. Marine or nonmarine (e.g., in the lower parts of peat horizons or lagoons). *Berthierine* forming today, under conditions subtly different from glauconite.

Greenalite $Fe_3^{+2}(OH)_4Si_2O_5 \cdot H_2O$

Septechlorite analogous to antigorite. Dominantly in Precambrian (abundant); very rare in Paleozoic and Mesozoic. Typically associated with magnetite. Resembles chamosite and glauconite.

Thuringite $(OH)_{16}(Mg_{1.4}Fe_{1.5}^{+2}Al_{1.7})(Si_{4.8}Al_{3.2})O_{20}$

Fe-rich chlorite. Locally abundant in Paleozoic ironstones; may be oolitic. Diagenetic or metamorphic alteration of other Fe-silicates.

Minnesotaite $(OH)_{11}(Fe^{+2},Mg)_{11}(Si,Al,Fe^{+2})_{16}O_{37}$

Fe-analogue of talc. Greenish gray, waxy. Mostly of metamorphic origin in iron formations.

Stilpnomelane $(OH)_4(K,Na,Ca)_{0–1}(Fe,Mg,Al)_{7–8}Si_8O_{23–24}(H_2O)_{2–4}$

Brittle mica. Very similar to biotite (and probably often misidentified). Mainly metamorphic; may be primary.

Sulfides

Pyrite FeS_2

Isometric. Mainly an accessory, diagenetic mineral in sediments, disseminated or concentrated in nodular crystalline aggregates or replacing fossils. Locally to 65%+ of layers to 0.5 m thick in black shales, where it may be associated with other economic metal sulfides. Common as microspheres about or associated with bacterial sufate reducers in modern and ancient muds. Formed under reducing conditions in association generally with organic material.

Marcasite FeS_2

Orthorhombic. Dimorph of pyrite; less stable and oxidizes more readily than pyrite. Apparently formed under reducing and acidic conditions, and commonly associated with siderite in coals.

Pyrrhotite $Fe_{1-x}S$

Rare in sediments, usually then as oolites (of replacement origin?) with phosphatic nodules or siderite. In some recent muds.

Hydrotroilite amorphous $FeS \cdot nH_2O$? Highly metastable, and Melnikovite amorphous mixture containing Fe_3S_4

Both are black colloids, the latter common in Recent and Tertiary rocks and sediments. Color some sediments and may form small nodules (1 mm). Very difficult to identify, but may be common. Apparently convert readily to pyrite or other sulfides.

Phosphates

Vivianite $Fe_3P_2O_8 \cdot 8H_2O$

Bright blue colors distinctive. Occurs in bogs, swamps, and after wood fragments.

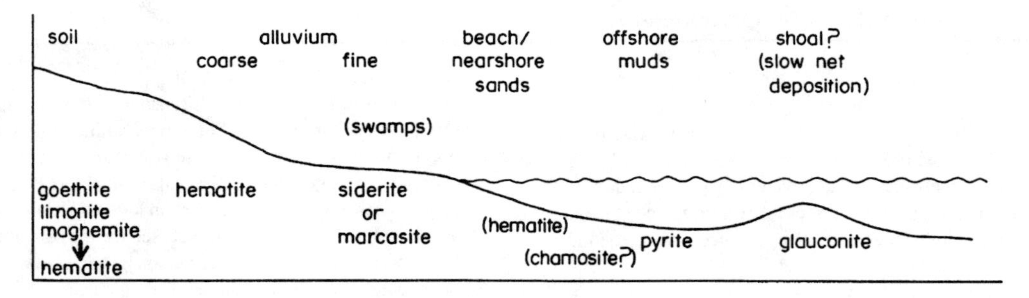

Figure 9-4. Simplistic sketch illustrating relationships between depositional environment and characteristic diagenetic iron compounds.

will be produced (and bound to oxygen) unless the iron is already held within unreactive crystals. As iron-bearing crystals weather, the iron released may oxidize, and it is common to find such minerals with oxide coatings or stains. Iron is transported in solution, in organic matter (e.g., hemoglobin and chlorophyll), within detrital (e.g., ferromagnesian) minerals, as insoluble iron oxides, or adsorbed on clay minerals. It is prone to remobilization under reducing depositional or diagenetic conditions. The influence of sulfide ions in the depositional environment is shown in Fig. 9-5. The availability of CO_2 and pH are important in the production of iron carbonates, together with an absence of sulfide ions; bacterial activity is also important in both marine and nonmarine siderite genesis (e.g., Ellwood et al. 1988; Pye et al. 1990). Iron-silicate minerals are formed under conditions not fully understood; although quite a bit is known about glauconite (see Chapter 7), conditions for its formation relative to that of chamosite are problematical (e.g., van Houten and Purucher 1984). Figure 9-6 shows an example of the diagenetic sequence of modifications that is inferred in the most complex of the iron-rich sediments—the siliceous iron formations.

Iron Formations

Banded Iron Formations

Banded iron formations (BIFs) consist of sequences of iron-rich bands (iron oxides, iron silicates, iron carbonates, and iron sulfides) and iron-poor bands (chert and some carbonates) that range in thickness from a few microns to about a metre, and commonly exhibit remarkable lateral continuity over hundreds, if not thousands of km^2 (the Brockman Iron Formation, Hamersley Ranges, Australia, covered an area of not less than 10^5 km^2, e.g., Trendall 1968; Ewers and Morris 1981). BIF sequences are often interspersed with mudstone intervals of probable volcanogenic origin. BIFs are particularly abundant in strata between 1.9 and 2.5 Ga old (James and Trendall 1982); sparse younger and older BIFs are known (e.g., Gole and Klein 1981; Appel in Appel and LaBerge 1987). Many explanations have been offered for the absence of modern BIFs (e.g., Ewers and Morris 1981; Cloud 1983; McConchie 1987) and many hypotheses have been presented to account for their abundance during the Precambrian (e.g., Trendall and Blockley 1970; Garrels, Perry, and MacKenzie 1973; Holland 1973; Drever 1974; Klein and Bricker 1977; Ewers and Morris 1981; see also papers in Trendall and Morris 1983; Appel and LaBerge 1987). Two of the few observations that almost all workers agree on are that BIFs have no modern analogues and that they formed largely as chemical precipitates.

In addition to the high lateral continuity of bands, and despite regional variations in mineralogy, chemical composition of the bands also is surprisingly constant, with the exclusion of CO_2, H_2O, and the Fe^{2+}/Fe^{3+} ratio (McConchie 1987). The regional constancy implies uniform recharge of waters in the depositional environment and regionally uniform deposition; the variation in mineralogy must reflect diagenetic/metamorphic processes rather than depositional facies, and diagenesis must have been largely isochemical (except for H_2O, H^+, CO_2, and to a lesser extent Na^+ and Mg^{2+}), even on the centimetre-band scale. Because it is very unlikely that such uniform conditions could be maintained in any enclosed or partly enclosed depositional setting, deposition in a shelf or platform setting is now favored (e.g., Morris and Horwitz 1983; McConchie 1987), with the supply of Fe and Si being maintained by upwelling oceanic currents (e.g., Button et al. 1982). The absence of current-generated sedimentary structures in BIFs has been used to argue for deposition in very quiet water settings (e.g., Trendall and Blockley 1970). However, such sedimentary structures are unlikely to form (or will be destroyed by plastic creep) in highly cohesive chemical sediments that are inferred as primary BIF precipitates (McConchie 1987).

Primary precipitation of the iron oxyhydroxides probably was controlled either by mixing of deep Fe^{2+}-rich water with oxygenated surface water (Button et al. 1982) or by direct photolytic oxidation of Fe^{2+} (e.g., Braterman, Cairns-Smith, and Sloper 1983; Braterman and Cairns-Smith 1987). The most likely mechanisms for precipitation of the silica and silicates involve polymerization of monosilicic acid and precipitation of a hydrous silica gel or coprecipitation with iron oxyhydroxides. Although there may have been fluctuations in recharge rates (e.g., Drever 1974), the most likely cyclic event that could account for the alternating pattern of banding is the year (Trendall and Blockley 1970). Several depositional models that accommodate annual cyclicity in chemical composition have been proposed (e.g., Trendall

Stability fields in marine depositional waters at pH8, solid stability at $a_{Fe^{+2}} < 10^{-6}$, as a function of Eh and
***Upper:* HS⁻ activity, with $a_{HCO_3^-} = 10^{-7}$**
***Lower:* HCO_3^- activity, with $a_{HS^-} = 10^{-7}$**

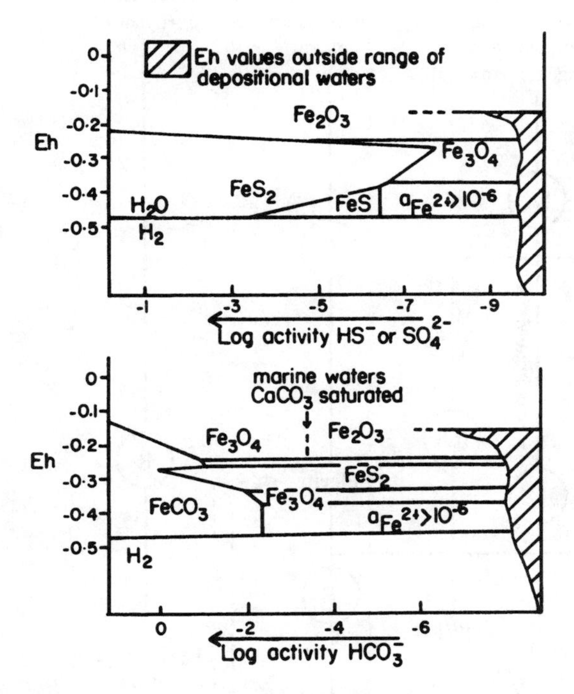

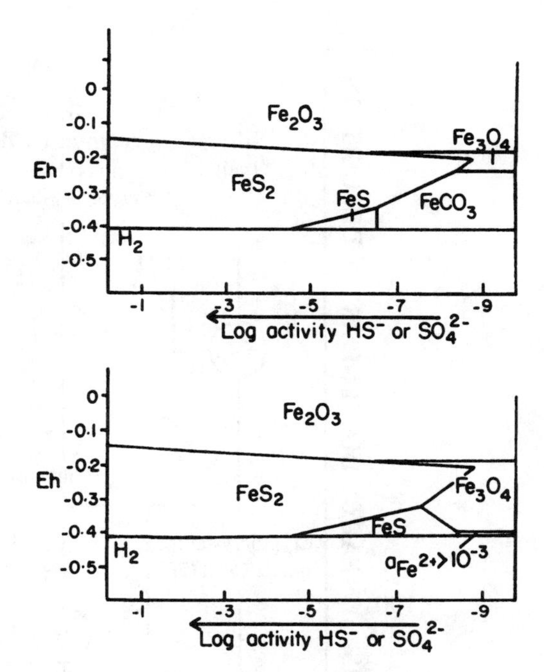

Mineral stability fields as a function of Eh and HS⁻ activity, pH7, solid stability at $a_{Fe^{+2}} < 10^{-3}$
***Upper:* probable conditions of restricted pore water circulation, $a_{HCO_3^-} = 10^{-2.5}$**
***Lower:* conditions of low carbonate activity, $a_{HCO_3^-} = 10^{-4.4}$**

Stability fields as a function of Eh and HS⁻ activity; solid stability at $a_{Fe^{+2}} < 10^{-6}$
***Upper:* anoxic marine waters, pH8, with SO_4^{-2}/HS^- equilibrium disturbed in favor of HS⁻ and $a_{HCO_3^-} = 10^{-3.5}$**
***Lower:* probable upper sediment pore waters, pH7, $a_{HCO_3^-} = 10^{-2.5}$**

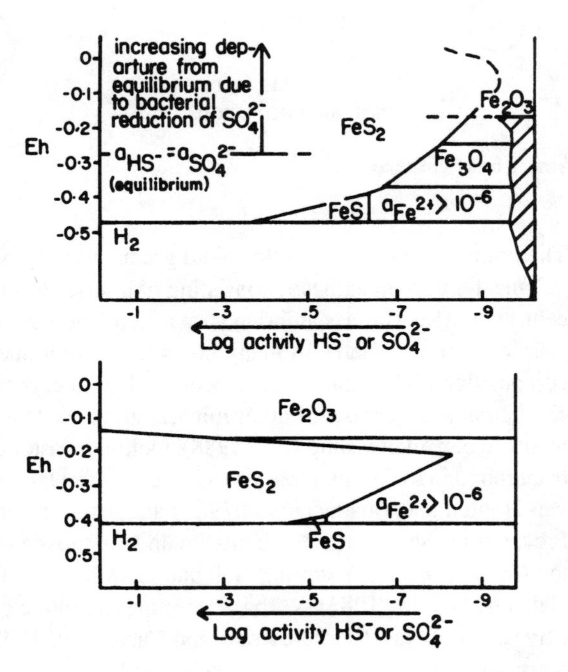

Figure 9-5. Influence of sulfide and bicarbonate ions on iron mineralogy in sedimentary environments. Major conclusions are that ferric compounds (e.g., hematite) are the only iron minerals that can exist in true equilibrium with modern depositional waters; ferrous minerals can only attain equilibrium with sediment pore waters (i.e., are only stable within sediment masses), except for pyrite, which is stable relative to all other possible phases even in the presence of low-sulfide activities. Pyrite is a metastable phase in anoxic water masses where bacteria maintain sulfide activities at nonequilibrium levels. Optimum conditions for siderite are severely restricted circulation, zero sulfide activity, and low Eh (–0.25–0.35 V); ferrous silicates are stable in similar conditions where bicarbonate activity is low and there is saturation with some active silica form. (After Curtis and Spears 1968).

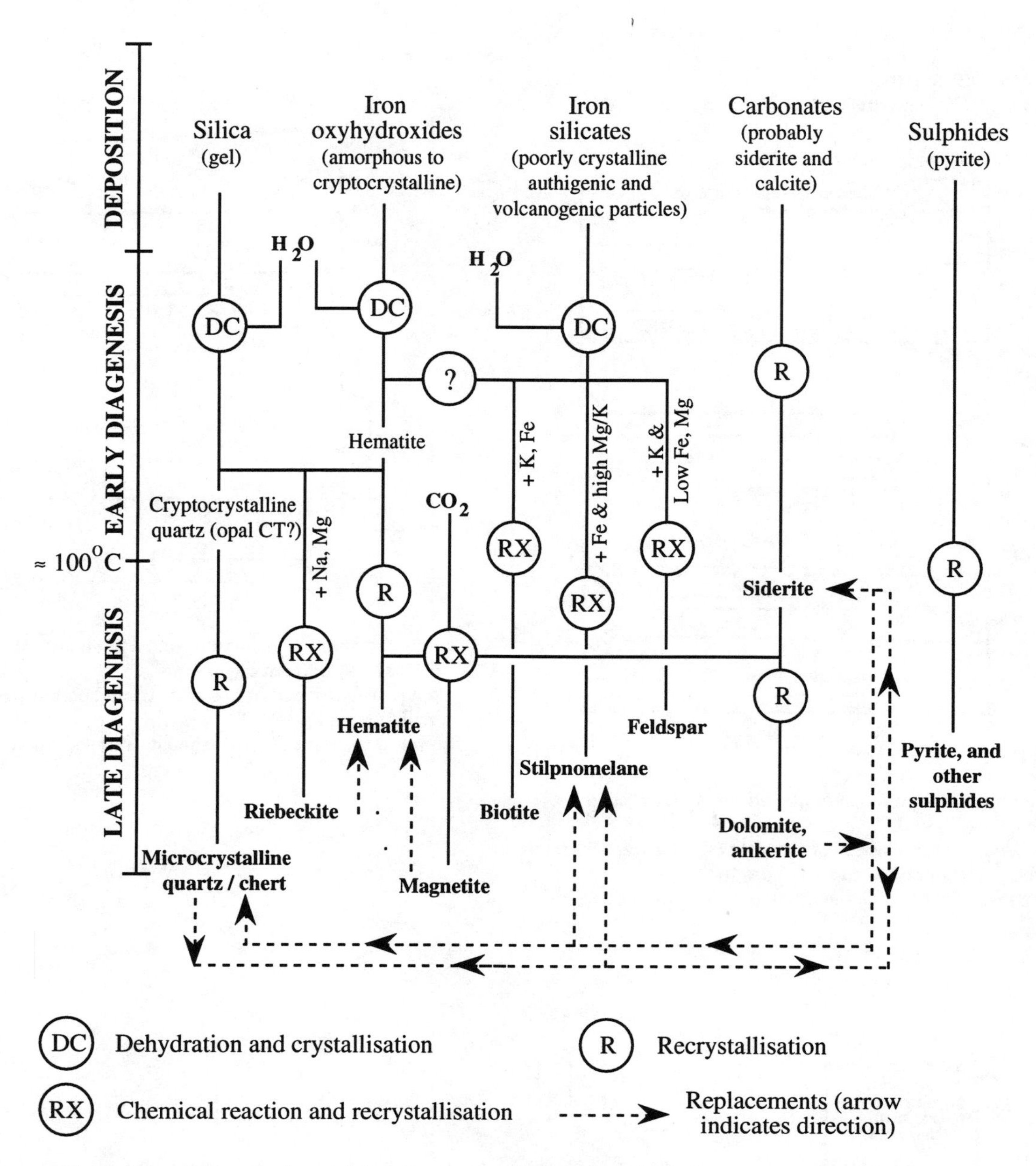

Figure 9-6. Diagenetic changes inferred to take place in banded iron formations.

and Blockley 1970; Cloud 1973; Ewers and Morris 1981; Braterman and Cairns-Smith 1987; McConchie 1987), but there is little agreement as to which is the best alternative. Probably the simplest and most flexible model involves the precipitation of Fe^{3+} oxyhydroxides as a result of major storms mixing Fe^{2+}-containing bottom waters with oxygenated surface water and the coprecipitation of silica; between storms silica may have precipitated by polymerization and flocculation. Diagenesis almost certainly played a vital role in the segregation of minerals into discrete bands.

Oolitic Ironstones

Oolitic ironstones are ferruginous sediments up to tens of metres thick that are characterized by abundant hematite, goethite, or chamosite oolites with associated chert or carbonate. The ooids are spherical to ellipsoidal grains, usually between 0.2 and 1.0 mm in diameter, consisting of a series of fine concentric laminae (in Precambrian strata, frequently alternating iron mineral and chert). In many ooids the central nucleus is compositionally similar to the concentric laminae, but in others the nucleus consists of other mineral grains or fossil fragments (e.g., Delaloye and Odin 1988). Oolitic ironstones have been found in rocks of Cambrian to Pliocene age (James 1966; van Houten and Bhattacharyya 1982) and, more recently, in Precambrian strata (e.g., the hematite and chamosite ooids in the Nabberu Basin, Australia; Hall and Goode, 1978; Goode, Hall, and Bunting 1983). With the possible exception of Loch Etive chamosites (Rohrlich, Price, and Calvert 1969), ferruginous ooids are not known from modern sediments.

There is general agreement that most ferruginous ooids

(or their precursor grains) were deposited in shallow water (less than about 30 m deep) in an environment that was periodically or persistently agitated but was not exposed to a high detrital sediment influx. A variety of mechanisms for the formation of the ferruginous ooids have been proposed (e.g., Bhattacharyya and Kakimoto 1982; Maynard 1983; Odin et al. 1988); four such mechanisms are:

Accretion of material around a nucleus either by adhesion or by direct precipitation at or above the sediment/water interface. Direct precipitation mechanisms could explain the origin of many aragonite or calcite oolites formed in carbonate-saturated water, but the origin of hematite, goethite, and chamosite oolites is more easily explained by accretion by adhesion.

Diagenetic growth as micro concretions *within* the sediment by precipitation around a nucleus.

Diagenetic replacement of preexisting aragonite and calcite (and in the case of iron silicates, hematite or goethite) ooids.

Derivation as allogenic pisoids or ooids formed as pedogenetic features in lateritic terrains, similar to bauxite (and manganese?) pisoids. This explanation is somewhat escapist, leaving the explanation to the pedologist!

All of these explanations are problematical. Using the first explanation, it is difficult to reconcile the high Fe^{2+}/Fe^{3+} ratio of chamosite ooids, normally indicative of very low-energy, oxygen-deficient water conditions, with the well-sorted occurrence of these ooids in shoals or bars, normally indicative of agitated, well-oxygenated water conditions (this difficulty does not apply to the origin of hematite, goethite, calcite, or aragonite ooids). Alternatively, if they formed by the second or third mechanisms, then it is difficult to explain the remarkably uniform size of ooids, some of their internal textures, and the common observation that adjacent sediments are not notably depleted in chemical constituents that must have migrated to form the ooids. An origin involving diagenetic replacement of aragonite oolites was strongly supported by Kimberley (1979), and synthetic replacement of a ferric iron-rich precursor by a chamosite-like mineral has been demonstrated by Harder (1978); nonetheless, despite the superficial attractiveness of diagenetic replacement models, evidence presented by James and van Houten (1979) argues against their general application. The fourth mechanism also has difficulty explaining the remarkably uniform size of ooids and some of their internal textures (e.g., nucleii of uncorroded fossil fragments are incompatible with a soil origin).

Red Beds

The origin and significance of red beds in the geological column has been the subject of much debate (e.g., Krynine 1948; Walker 1967, 1974; van Houten 1973; Ziegler and McKerrow 1975; Folk 1978); a multiplicity of origins must be accepted. The distinctive red color of these sediments is due to the presence of Fe^{+3} oxides/hydroxides, but the iron content is commonly little more than in drab beds (where the iron is mainly in the Fe^{+2} state) and most red beds are far from being iron-rich. In some red beds there is evidence of in situ alteration of ferromagnesian minerals (e.g., Czyscinski, Burnes, and Pedlow 1978), in others there is evidence that iron oxyhydroxides were authigenic chemical precipitates, and in others iron oxides are perigenic or allogenic. Because red-bed genesis requires oxic conditions, they form mainly in subaerial environments. The rarity of red beds in Proterozoic strata, relative to younger lithologies, has been used as an argument for an anoxic atmosphere prior to about 1.5 Ga (e.g., Cloud 1972). However, red beds have been found beneath some Proterozoic iron formations (eg, Dimroth and Kimberley 1976) and their rarity in very old rocks may result from the low preservation potential of subaerial deposits. The involvement of bacteria such as *Gallionella, Ferrobacillus,* and *Metallogenium* in the formation of iron oxyhydroxides in many modern environments raises the possibility that bacteria may be involved in the formation of some red beds (and would help account for the rarity of Proterozoic red beds).

OTHER CHEMICAL SEDIMENTS

Phosphates

Chemically deposited phosphates have been reported from a wide range of marine and some nonmarine environments (e.g., Southgate 1986), but they appear to be most common in shelf and platform settings (e.g., Youssef 1965; Geological Society of London 1980; Notholt 1980; papers in Bentor 1980) in the depth range from about 50 m to 500 m. Geochemical studies (Roberson 1966) suggest that seawater in warmer latitudes is nearly saturated with respect to fluorapatite in the depth range from 50 m to 200 m; in shallower water phosphate is removed by biota and in water deeper than about 200 m lower pH conditions (associated with higher dissolved CO_2 concentrations) keep concentrations below saturation. Sedimentary phosphates occur as microcrystalline phosphatic mud, phosphatic cement, sand size grains (often pelletal), replacements of shell and other biogenic fragments, phosphatic coatings on other grains, and phosphate nodules; they are usually associated with marine carbonates. Mineralogically, most marine phosphates are carbonate fluorapatite, but other varieties of apatite and X-ray amorphous phosphatic material (collophane) are common, and some marine phosphate deposits include unusual phosphate minerals. *Phosphorite* rocks of various ages contain the bulk of economic phosphate deposits (e.g., De Keyser and Cook 1972; Notholt 1980; Soudry 1987), but guano deposits have been mined on some islands for many decades, and phosphatic nodules in relatively deep oceanic waters are a prospective economic deposit when mining technologies and costs improve.

Some marine phosphates, particularly some coatings on

grains (e.g., Southgate 1986), appear to have formed as primary precipitates whereas others clearly have a replacement (commonly of carbonates) origin (e.g., d'Anglejan 1967). Upwelling of deep-sea waters was widely considered to be a prerequisite for phosphatization (e.g., McKelvey 1967), but in some large deposits such as those on the Chatham Rise East of New Zealand (Pasho 1976) there is little evidence of the involvement of upwelling waters. The importance of upwelling phosphatic waters was used to explain the abundance of marine phosphates between the fortieth parallels and on shelves on the western side of continents (e.g., Cook and McElhinny 1979), but it now appears that upwelling is only needed to counteract the biological depletion of phosphate in surface waters (Bentor in Bentor 1980). The lack of phosphate precipitation in some modern areas where there is strong upwelling and other conditions appear to be favorable suggests that other factors are involved in determining whether or not phosphatization will occur. Important controls on phosphatization are now thought to include the presence, distribution, and type of organic matter, the Mg^{2+}/Ca^{2+} ratio in interstitial and surface waters, biological activity at the site, and the sedimentation rate (pH and Eh are probably also important). For extensive phosphatization (such as the formation of large phosphorite nodules) to occur, the sedimentation rate at the site needs to be very low (or even negative) to prevent burial restricting or terminating the supply of reactants; hence, phosphatic material is often concentrated along unconformities in the stratigraphic record.

Silica

Most primary silica (as distinct from quartz) in modern sediments is biogenic (Davies and Supko 1973; Greenwood 1973), albeit that diagenetic mobilization very commonly has obscured the initial character of the biotic hardparts and has resulted in replacement of other materials. Prior to the evolution of silica-metabolizing organisms in the early Phanerozoic, substantial quantities of silica were inorganically precipitated at the sediment/water interface in both marine and nonmarine environments. Chert (microcrystalline silica) formed by recrystallization of inorganically precipitated gelatinous silica is particularly common in Precambrian sediments (e.g., banded iron formations); chert formed by diagenetic replacement reactions or by recrystallization of opalline biogenic concentrations is widely distributed but only rarely abundant in Phanerozoic sequences. Although primary silica is inorganically precipitated at the sediment/water interface in some modern environments, most inorganic silica in modern sediments (including soils) is diagenetically precipitated below the sediment surface. Were it not for the biological removal of the large quantities of dissolved silica released during rock weathering, seawater would quickly become saturated and inorganic precipitation would

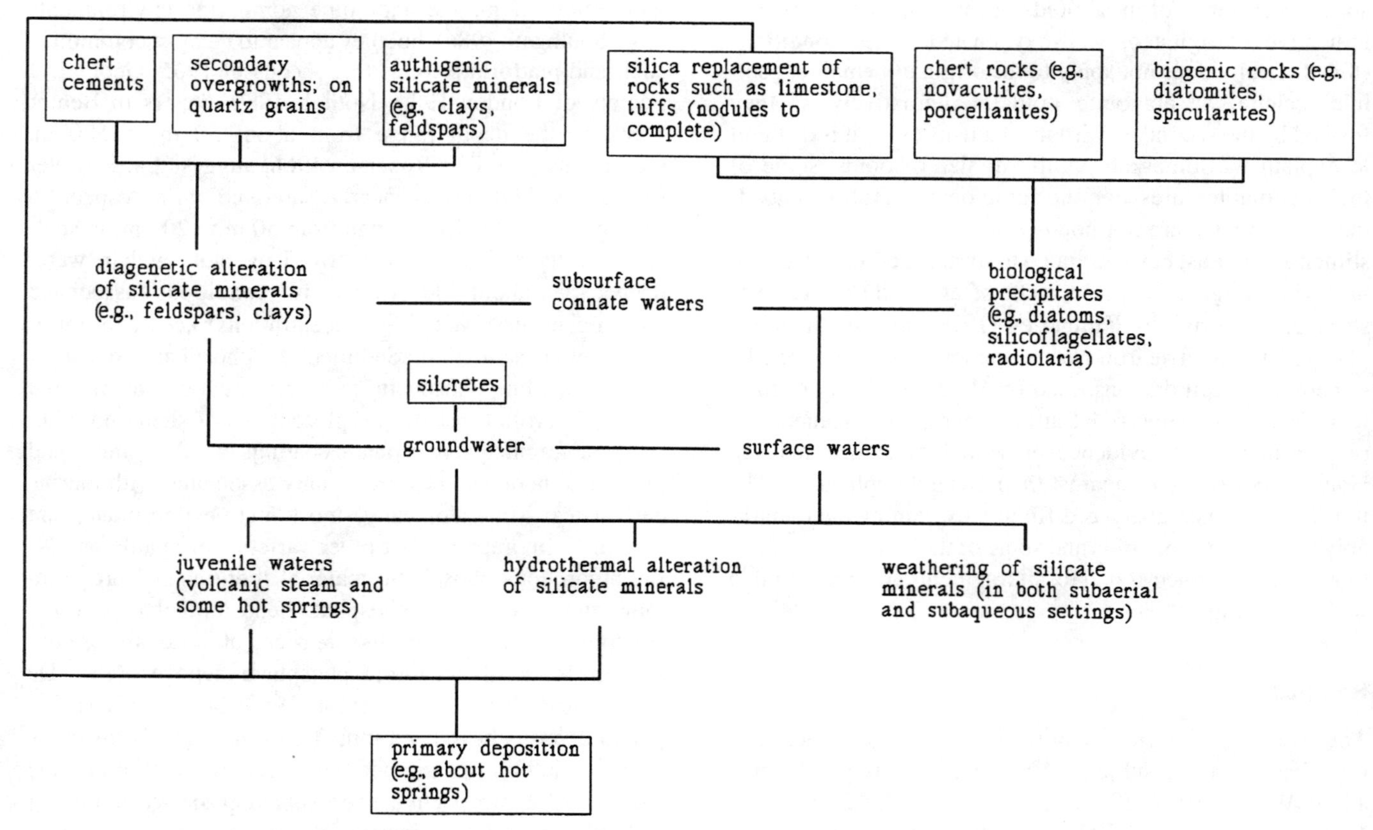

Figure 9-7. Sketch of potential routes for silica incorporation into sediments.

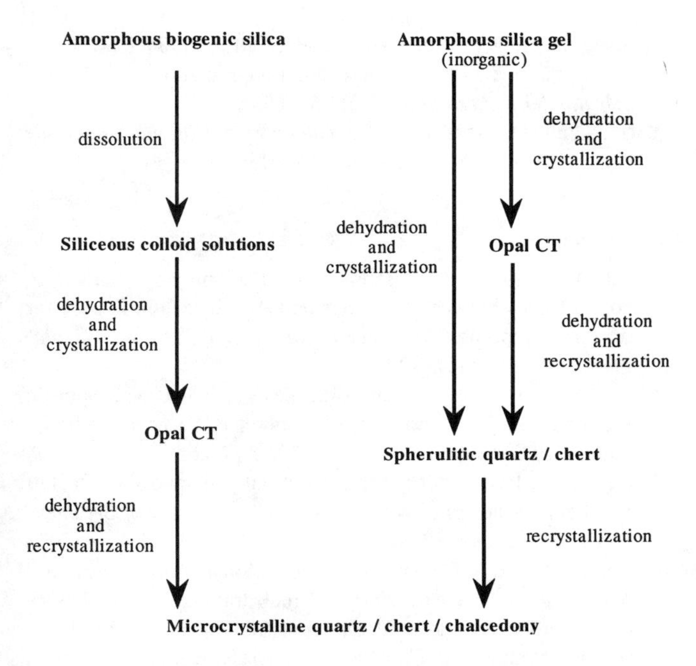

Figure 9-8. Silica recrystallization sequences.

follow. Fig. 3-19 depicts the modern silica cycle, Fig. 9-7 shows common routes for silica mobilization in sediments, and Fig. 9-8 shows common stages of diagenetic mobilization and crystallization from primary amorphous silica sediment (see also Oehler 1976).

Amorphous silica, which forms uncharged monosilicic acid [$Si(OH)_4$] in most natural waters, is only soluble at normal temperatures when the pH is greater than 9; its solubility is largely unaffected by salinity (e.g., see Fig. 3-4 and Yariv and Cross 1979). Polymerization of dissolved silica is a kinetically slow process (Iler 1979), but it can be accelerated by a change in electrolyte strength (Krauskopf 1959). Largely as a result of the influence of temperature (the solubility of amorphous silica is 60–80 mg/L at 0°C, 100–140 mg/L at 25°C, and 300–380 mg/L at 90°C) and pressure (pressure-solution effect), large volumes of silica are redistributed during diagenesis, forming overgrowths on quartz grains and silica cements and replacing other minerals such as carbonates (Fig. 9-7). In many arid zones, the precipitation of silica cements in near-surface sediments, probably stimulated by evaporative concentration of silica-bearing solutions, results in the formation of *silcretes* (e.g., Smale 1973; cf. calcretes in Chapter 8).

Manganese

Although iron and manganese are chemically similar and they are distributed uniformly together in igneous rocks, combinations of sedimentary geochemical and biogeochemical processes can promote their separation in sedimentary systems (e.g., Krauskopf 1957). Consequently, all the world's major economic manganese deposits are sedimentary (e.g., Taylor 1969).

The Nikopol deposit in the Ukraine, which contains about 75% of the world's presently commercial manganese reserves, was deposited in shallow-marine conditions as part of a transgressive sequence (Varentsov and Rakhmanov 1980). The deposit includes three mineralogically distinct facies: nearshore manganese oxides, which may be primary deposits or alteration products of precursor manganese minerals; intermediate mixed manganese oxides/carbonates; and offshore (but not deep-water) fossiliferous carbonates dominated by fine, poorly crystalline rhodochrosite ($MnCO_3$) and Mn-calcite. In contrast, the Groote Eylandt deposit in Australia is dominated by Mn-pisoids (similar to bauxite pisoids) composed mainly of cryptomelane [$K(Mn^{2+}, Mn^{4+})_8O_{16}$], with lesser pyrolusite [MnO_2] and manganite [γ-$MnO(OH)$]; the deposit also includes zones of massive fine-grained cryptomelane interbedded with detrital clays and zones where manganiferous concretions are embedded in loose sandy clay (Ostwald 1975, 1980). Nonetheless, the deposits of this major manganese field also appear to have formed under shallow-marine conditions.

Ferromanganese nodules are chemical sediments that cover large areas of the modern ocean floor, particularly in the central Pacific (e.g., Glasby 1977); they have been found concentrated along some unconformities in the stratigraphic record, but they are surprisingly rare in ancient sediments. The nodules range up to 10 cm in diameter and mainly consist of oxides of iron and manganese; most economic interest lies in the fact that they contain substantial quantities of Ni, Cu, Co, and Mo and minor quantities of other metals (e.g., McKelvey, Wright, and Rowland 1979; Heath 1981). Nodule distribution data indicate that they form only in areas where sedimentation rates are very low (less than about 1 mm/100 year), but even these low sedimentation rates are difficult to reconcile with calculations and radiometric nodule growth-rate data (Ku 1977; Heath 1979), which suggest that the average nodule spends about 1 Ma at the sediment/water interface prior to burial. There is currently no adequate explanation of how the nodules can remain at the surface so long without being buried. There is also considerable uncertainty (see discussion in Heath 1981) about the source of the metals in the nodules, why the composition of the nodules is geographically variable, the chemical processes involved in nodule growth, and whether nodule growth involves microbiological processes (particularly the activity of manganese-oxidizing bacteria).

Metal Sulfides

Many of the world's largest sulfide ore deposits (e.g., Mt. Isa, McArthur River, the Kupferschiefer, Sullivan, and even the Broken Hill deposits, although they have been extensively metamorphosed) were either originally deposited as chemical sediments or are closely associated with chemical sediments. Most metals in these deposits were transported to the site of deposition by submarine volcanic exhalations or in geothermally heated seawater brines (e.g., Degens and Ross 1969, 1976; Shanks and Bischoff 1977; Corliss et al. 1979; Hékinian et al. 1980; Haymon and Kastner 1981), where they were precipitated and/or replaced precursor sediments. Microbiological processes also may have played an important role in metal sulfide and metal oxide deposition through the activity of sulfate-reducing bacteria or metal-metabolizing bacteria such as *Metallogenium.* Discussion of these deposits and their mode of origin is beyond the scope of this book (see Amstutz and Bernard 1973; Klemm and Schneider 1977; Skinner 1981; Maynard 1983).

SELECTED BIBLIOGRAPHY

General

Amstutz, G. C., and A. J. Bernard (eds.), 1973, *Ores in Sediments.* Springer-Verlag, New York.

Berner, 1981, A new geochemical classification of sedimentary environments. *Journal of Sedimentary Petrology* 51:359–65.

Clemmey, H., and N. Badham, 1982, Oxygen in the Precambrian atmosphere: An evaluation of the geological evidence. *Geology* 10: 141–6.

Cloud, P. E., 1972, A working model of the primitive earth. *American Journal of Science* 272:537–48.

Corliss, J. B., J. Dymond, L. I. Gordon, J. M. Edmond, R. P. van Herzen, R. D. Ballard, K. Green, D. Williams, A. Bambridge, K. Crane, and T. H. van Andel, 1979, Submarine thermal springs on the Galapagos Rift. *Science* 203:1073–82.

Degens, E. T., and D. Ross, 1976, Strata-bound metalliferous deposits found in or near active rifts. In K. H. Wolfe (ed.), *Handbook of Strata-Bound and Stratiform Ore Deposits,* vol. 4. Springer-Verlag, New York, pp. 165–202.

Dimroth, E., and M. M. Kimberley, 1976, Precambrian atmospheric oxygen: Evidence in the sedimentary distributions of carbon, sulfur, uranium, and iron. *Canadian Journal of Earth Science* 13: 1161–85.

Haymon, R. M., and M. Kastner, 1981, Hot spring deposits on the East Pacific Rise at 21°N: Preliminary description of mineralogy and genesis. *Earth and Planetary Science Letters* 53:363–81.

Hékinian, R., M. Fevrier, J. L. Bischoff, P. Picot, and W. C. Shanks, 1980, Sulphide deposits from the East Pacific Rise near 21°N. *Science* 207:1433–44.

Holland, H. D., 1984, *The Chemical Evolution of the Atmosphere and Oceans.* Princeton University Press, Princeton, N.J., 582p.

Holland, H. D., and M. Schidlowski (eds.), 1982, *Mineral Deposits and the Evolution of the Biosphere.* Springer-Verlag, New York.

Klemm, D. D., and H.-J. Schneider (eds.), 1977, *Time- and Strata-Bound Ore Deposits.* Springer-Verlag, New York.

Maynard, J. B., 1983, Geochemistry of Sedimentary Ore Deposits. Springer-Verlag, New York, 305p.

Rosler, H. J., and H. Lange, 1972, *Geochemical Tables.* Elsevier, Amsterdam, 468p.

Shanks, W. C., and J. L. Bischoff, 1977, Ore transport and deposition in the Red Sea geothermal system: A geochemical model. *Geochimica et Cosmochimica Acta* 41:1507–19.

Skinner, B. J. (ed.), 1981, *Economic Geology, 75th Anniversary Volume.* Economic Geology Publishing Co., El Paso, Tex., 964p.

Windley, B. F., P. R. Simpson, and M. D. Muir, 1984, The role of atmospheric evolution in Precambrian metallogenesis. *Fortschritte der Mineralogie* 62(2):253–67.

Yariv, S., and H. Cross, 1979, *Geochemistry of Colloid Systems for Earth Scientists.* Springer-Verlag, Berlin.

Evaporites

Amiel, A. J., and G. M. Friedman, 1971, Continental sebkha in Arava Valley between Dead Sea and Red Sea: Significance for origin of evaporites. *American Association of Petroleum Geologists Bulletin* 55:581–92.

Arakel, A. V., 1980, Genesis and diagenesis of Holocene evaporitic sediments in Hutt and Leeman lagoons, Western Australia. *Journal of Sedimentary Petrology* 50:1305–26.

Arakel, A. V., 1981, Coastal sebkha and salt pan deposition in Hutt and Leeman lagoons, Western Australia, *Journal of Sedimentary Petrology* 50:1305–26.

Arakel, A. V., and A. Cohen, 1991, Deposition and early diagenesis of playa glauberite in the Karinga Creek drainage system, Northern Territory, Australia. *Sedimentary Geology* 70:41–59.

Arakel, A. V., and D. McConchie, 1982, Classification and genesis of calcrete and gypsite lithofacies in paleodrainage systems of inland Australia and their relationship to carnotite mineralization. *Journal of Sedimentary Petrology* 52:1149–70.

Bellanca, A., and R. Neri, 1986, Evaporite carbonate cycles of the Messinian, Sicily: Stable isotopes, mineralogy, textural features, and environmental implications. *Journal of Sedimentary Petrology* 56:614–21.

Borchert, T. H., and R. O. Muir, 1964, *Salt Deposits: The Origin, Metamorphism, and Deformation of Evaporites.* D. van Nostrand, Princeton and London, 338p.

Braitsch, O., 1971, *Salt Deposits: Their Origin and Composition.* Springer-Verlag, New York, 297p.

Brongersma-Sanders, M., 1968, On the geographical association of stratabound ore deposits with evaporites. *Mineralium Deposita* 3:286–91.

Dean, W. E., and B. C. Schreiber, 1978, *Marine Evaporites.* SEPM Short Course 4, Society of Economic Paleontologists and Mineralogists, Tulsa, Okla., 188p.

Decima, A., J. A. McKenzie, and B. C. Schreiber, 1988, The origin of "evaporative" limestones: An example from the Messinian of Sicily (Italy). *Journal of Sedimentary Petrology* 58:256–72.

Degens, E. T., and D. Ross (eds.), 1969, *Hot Brines and Recent Heavy Metal Deposits in the Red Sea.* Springer-Verlag, Berlin, 800p.

Folk, R. L., and J. S. Pittman, 1971, Length-slow chalcedony: A new testament for vanished evaporites. *Journal of Sedimentary Petrology* 41:1045–58.

Friedman, G. M., 1972, Significance of Red Sea in problem of evaporites and basinal limestones. *American Association of Petroleum Geologists Bulletin* 56:1072–86.

Hardie, C. A., and H. P. Eugster, 1971, The depositional environment of marine evaporites: A case for shallow clastic accumulation. *Sedimentology* 16:187–220.

Hovorka, S., 1987, Depositional environments of marine-dominated bedded halite, Permian San Andres Formation, Texas. *Sedimentology* 34:1029–54.

Hsu, K. J., 1972, Origin of saline giants: A critical review after the discovery of the Mediterranean evaporite. *Earth-Science Reviews* 8:371–96.

Jankowski, J., and G. Jacobson, 1989, Hydrochemical evolution of regional groundwaters to playa brines in Central Australia. *Journal of Hydrology* 108:123–73.

Kinsman, D. J. J., 1976, Evaporites: Relative humidity control of primary mineral facies. *Journal of Sedimentary Petrology* 46:273–99.

Kirkland, D. W., and R. Evans (eds.), 1973, *Marine Evaporites: Origin, Diagenesis and Geochemistry.* Benchmark Papers in Geology 7, Dowden, Hutchinson & Ross, Stroudsburg, Pa., 426p.

Kushnir, J., 1981, Formation and early diagenesis of varved evaporitic sediments in a coastal hypersaline pool. *Journal of Sedimentary Petrology* 51:1193–203.

Lambert, I. B., T. H. Donnelly, J. S. R. Dunlop, and D. I. Groves, 1978, Stable isotope compositions of early Archaean sulphate deposits of probable evaporitic and volcanogenic origins. *Nature* 276:808–11.

Lowenstein, T. K., and L. A. Hardie, 1985, Criteria for the recognition of salt-pan evaporites. *Sedimentology* 32:627–44.

Mann, A. W., and R. L. Deutscher, 1978, Genesis principles for the precipitation of carnotite in calcrete drainages in western Australia. *Economic Geology* 73: 1724–737.

Martinez, J. D., 1971, Environmental significance of salt. *American Association of Petroleum Geologists Bulletin* 55:810–25.

Melvin, J. L. (ed.), 1991, *Evaporites, Petroleum and Mineral Resources.* Developments in Sedimentology 50, Elsevier, Amsterdam, 556p.

Muir, M. D., 1987, Facies models for Australian Precambrian evaporites. In T. M. Peryt (ed.), *Evaporite Basins.* Lecture Notes in Earth Sciences 13, Springer-Verlag, Berlin, pp. 5–21.

Raup, O. B., 1970, Brine mixing: An additional mechanism for formation of basin evaporites. *American Association of Petroleum Geologists Bulletin* 54:2246–59.

Schmalz, R. E., 1969, Deep water evaporite deposition: A genetic model. *American Association of Petroleum Geologists Bulletin* 53:798–823.

Schreiber, B. C. (ed.), 1988, *Evaporites and Hydrocarbons.* Columbia University Press, New York, 475p.

Scruton, R C., 1953, Deposition of evaporites. *American Association of Petroleum Geologists Bulletin* 37:2498–512.

Stewart, F. H., 1963, *Marine Evaporites.* U. S. Geological Survey Professional Paper 440-Y, Washington, D.C., 53p.

Warren, J. K., 1986, Shallow-water evaporitic environments and their source rock potential. *Journal of Sedimentary Petrology* 56:442–54.

Iron

Appel, P. W. U., and G. LaBerge (eds.), 1987, *Precambrian Iron Formations.* Theophrastus Publications Co., Athens, 674p.

Bass Becking, L. G. M., and D. Moore, 1950, The relation between iron and organic matter in sediments. *Journal of Sedimentary Petrology* 29:454–8.

Bhattacharyya, D. P., and P. K. Kakimoto, 1982, Origin of ferriferous ooids: An SEM study of ironstone ooids and bauxite pisoids. *Journal of Sedimentary Petrology* 52:849–57.

Braterman, P. S., and A. G. Cairns-Smith, 1987, Iron photoprecipitation and the genesis of the banded iron formations. In P. Appel and G. LaBerge (eds.), *Precambrian Iron Formations.* Theophrastus Publications Co., Athens, pp. 215–245.

Braterman, P. S., A. G. Cairns-Smith, and R. W. Sloper, 1983, Photo-oxidation of hydrated Fe^{2+}—significance for banded iron formations. *Nature* 303:163–4.

Button, A., T. D. Brock, P. J. Cook, H. P. Eugster, A. M. Goodwin, H. L. James, L. Margulis, K. H. Nealson, J. O. Nriagu, A. F. Trendall, and M. R. Walter, 1982, Sedimentary iron deposits, evaporites and phosphorites: State of the art report. In H. D. Holland and M. Schidlowski (eds.), *Mineral Deposits and the Evolution of the Biosphere.* Springer-Verlag, New York, pp. 259–73.

Chauvel, J.-J., and E. Dimroth, 1974, Facies types and depositional environment of the Sokoman Iron Formation, Central Labrador Trough, Quebec, Canada. *Journal of Sedimentary Petrology* 44:299–327.

Cloud, P. E., 1973, Paleoecological significances of the banded iron formation. *Economic Geology* 68:1135–44.

Cloud, P. E., 1983, Banded iron formations—a gradualist's dilemma. In A. F. Trendall and R. C. Morris (eds.), *Iron Formation Facts and Problems.* Developments in Precambrian Geology 6, Elsevier, Amsterdam, pp. 401–16.

Curtis, C. D., and D. A. Spears, 1968, The formation of sedimentary iron minerals. *Economic Geology* 63:257–70.

Czyscinski, K. S., J. B. Burnes, and G. N. Pedlow, 1978, In situ red bed development by the oxidation of authigenic pyrite in a coastal depositional environment. *Palaeogeography, Palaeoclimatology, Palaeoecology* 23:239–46.

Delaloye, M. F., and G. S. Odin, 1988, Chamosite, the green marine clay from Chamoson; a study of Swiss oolitic ironstones. In G. S. Odin (ed.), *Green Marine Clays.* Developments in Sedimentology 45, Elsevier, Amsterdam, pp. 7–28.

Drever, J., 1974, Geochemical model for the origin of Precambrian banded iron formations. *Geological Society of America Bulletin* 85:1099–106.

Ellwood, B. B., T. H. Chrzanowski, F. Hrouda, G. J. Long, and M. L. Buhl, 1988, Siderite formation in anoxic deep-sea sediments: A synergetic bacterially controlled process with important implications in paleomagnetism. *Geological Association of America Bulletin* 100:980–2.

Ewers, W. E., and R. C. Morris, 1981, Studies on the Dales Gorge Member of the Brockman Iron Formation, Western Australia. *Economic Geology* 76:1929–53.

Folk, R. L., 1978, Reddening of desert sands: Simpson Desert, N.T., Australia. *Journal of Sedimentary Petrology* 46:604–15.

Garrels, R. M., E. A. Perry, and F. T. MacKenzie, 1973, Genesis of Precambrian iron formations and the development of atmospheric oxygen. *Economic Geology* 68:1173–9.

Gole, M. J., and C. Klein, 1981, Banded iron formations through much of Precambrian time. *Journal of Geology* 89:169–83.

Goode, A. D. T., W. D. M. Hall, and J. A. Bunting, 1983, The Nabberu Basin of Western Australia. In A. F. Trendall and R. C. Morris (eds.), *Iron Formation Facts and Problems.* Developments in Precambrian Geology 6, Elsevier, Amsterdam, pp. 295–323.

Hall, W. D. M., and A. D. T. Goode, 1978, The early Proterozoic Nabberu Basin and associated iron formations in western Australia. *Precambrian Research* 7:129–84.

Harder, H., 1978, Synthesis of iron layer silicate minerals under natural conditions. *Clays and Clay Minerals* 26:65–72.

Holland, H. D., 1973, The oceans: A possible source of iron in iron formations. *Economic Geology* 68:1169–73.

James, H. L., 1966, *Chemistry of the Iron-rich Sedimentary Rocks.* U.S. Geological Survey Professional Paper 440W, Washington, D.C., 61p.

James, H. L., and A. F. Trendall, 1982, Banded iron formations: Distribution in time and paleoenvironmental significance. In H.

D. Holland and M. Schidlowski (eds.), *Mineral Deposits and the Evolution of the Biosphere.* Springer-Verlag, New York, pp. 199–217.

James, H. L., and F. B. van Houten, 1979, Miocene goethitic and chamositic oolites, northeastern Colombia. *Sedimentology* 26:125–33.

Kimberley, M. M., 1979, Origin of oolitic iron formations. *Journal of Sedimentary Petrology* 49:111–32.

Klein, C., and O. P. Bricker, 1977, Some aspects of the sedimentary and diagenetic environment of Proterozoic banded iron formation. *Economic Geology* 72:1457–70.

Krynine, P. D., 1948, The origin of red beds. *New York Academy of Science Transactions* 11:60–8.

Lepp, H. (ed.), 1975, *Geochemistry of Iron.* Benchmark Papers in Geology 18, Dowden, Hutchinson & Ross, Stroudsburg, Pa., 464p.

McConchie, D. M., 1987, The geology and geochemistry of the Joffre and Whaleback shale members of the Brockman Iron Formation, western Australia. In P. Appel and G. LaBerge (eds.), Precambrian Iron Formations. Theophrastus Publications Co., City, pp. 541–601.

Morris, R. C., and R. C. Horwitz, 1983, The origin of the iron-formation-rich Hamersley Group of western Australia—deposition on a platform. *Precambrian Research* 21:273–97.

Nealson, K. H., 1982, Microbiological oxidation and reduction of iron. In H. D. Holland and M. Schidlowski (eds.), *Mineral Deposits and the Evolution of the Biosphere.* Springer-Verlag, New York, pp. 51–65.

Odin, G. S., R. W. O'B. Knox, R. A. Gygi, and S. Guerrak, 1988, Green marine clays from the oolitic ironstone facies: Habit, mineralogy, environment. In G. S. Odin (ed.), *Green Marine Clays.* Developments in Sedimentology 45, Elsevier, Amsterdam, pp. 29–52.

Pye, K., J. A. D. Dickson, N. Schiavon, M. L. Colemen, and M. Cox, 1990, Formation of siderite-Mg-calcite-iron sulphide concretions in intertidal marsh and sandflat sediments, north Norfolk, England. *Sedimentology* 37:325–43.

Rohrlich, K., N. B. Price, and S. E. Calvert, 1969, Chamosite in the recent sediments of Loch Etive, Scotland. *Journal of Sedimentary Petrology* 38:624–31.

Trendall, A. F., 1968, Three great basins of Pre-Cambrian iron formation deposition: A systematic comparison. *Geological Society of America Bulletin* 79:1527–44.

Trendall, A. F., and J. G. Blockley, 1970, *The Iron Formations of the Precambrian Hamersley Group Western Australia with Special Reference to the Associated Crocidolite.* Geological Survey of Western Australia Bulletin 119, 366p.

Trendall, A. F., and R. C. Morris (eds.), 1983, *Iron Formation Facts and Problems.* Developments in Precambrian Geology 6, Elsevier, Amsterdam, 558p.

van Houten, F. B., 1973, Origin of red beds: A review—1961–1972. *Annual Review of Earth and Planetary Science* 1:39–61.

van Houten, F. B., and D. P. Bhattacharyya, 1982, Phanerozoic oolitic ironstones—geologic record and facies. *Annual Reviews of Earth and Planetary Science* 10:441–58.

van Houten, F. B., and M. E. Purucher, 1984, Glauconitic peloids and chamositic ooids—favorable factors, constraints, and problems. *Earth-Science Reviews* 20:211–244.

Walker, T. R., 1967, Formation of red beds in modern and ancient deserts. *Geological Society of America Bulletin* 78:353–68. (See also 1968 discussion by R. E. Schmalz, and reply, *Geological Society of America Bulletin* 79:277–82.)

Walker, T. R., 1974, Formation of red beds in moist tropical climates: A hypothesis. *Geological Society of America Bulletin* 85:633–8.

Ziegler, A. M., and W. S. McKerrow, 1975, Silurian marine red beds. *American Journal of Science* 275:31–57.

Phosphates

Bentor, Y. K., 1980, *Marine Phosphorites—Geochemistry, Occurrence, Genesis.* Society of Economic Paleontologists and Mineralogists Special Publication 29, Tulsa, Okla., 249p.

Cook, P. J., and M. W. McElhinny, 1979, A reevaluation of the spatial and temporal distribution of sedimentary phosphate deposits in the light of plate tectonics. *Economic Geology* 74:315–30.

d'Anglejan, B. F., 1967, Origin of marine phosphorites off Baja California, Mexico. *Marine Geology* 5:15–44.

De Keyser, F., and P. J. Cook, 1972, *Geology of the Middle Cambrian Phosphorites and Associated Sediments in Northwestern Queensland.* Bureau of Mineral Resources, Geology and Geophysics Bulletin 138, 79p.

Geological Society of London, 1980, Phosphatic and glauconitic sediments. *Journal of the Geology Society of London* 137:657–805.

McKelvey, V. E., 1967, *Phosphate Deposits.* U.S. Geological Survey Bulletin 1252-D, Washington, D.C., 21p.

Notholt, A. J. G., 1980, Economic phosphatic sediments: Mode of occurrence and stratigraphical distribution. *Journal of the Geological Society of London* 137:793–805.

Pasho, D. W., 1976, *Distribution and Morphology of Chatham Rise Phosphorites.* New Zealand Oceanographic Institute Memoir 77, Wellington, N.Z., 27p.

Roberson, C. E., 1966, *Solubility Implications of Apatite in Sea Water.* U.S. Geological Survey Professional Paper 500-D, Washington, D.C., pp. 178–85.

Soudry, D., 1987, Ultra-fine structures and genesis of the Campanian Negev high-grade phosphorites (southern Israel). *Sedimentology* 34:641–60.

Southgate, P. N., 1986, Cambrian phoscrete profiles, coated grains, and microbial processes in phosphogenesis: Georgina Basin, Australia. *Journal of Sedimentary Petrology* 56:429–41.

Youssef, M. I., 1965, Genesis of bedded phosphates. *Economic Geology* 60:590–600.

Silica

Davies, T. A., and P. R. Supko, 1973, Oceanic sediments and their diagenesis: Some examples from deep-sea drilling. *Journal of Sedimentary Petrology* 43:381–90.

Greenwood, R., 1973, Cristobalite: Its relationship to chert formation in selected samples from the Deep Sea Drilling Project. *Journal of Sedimentary Petrology* 43:700–8.

Iler, R. K., 1979, *The Chemistry of Silica.* Wiley-Interscience, New York, 866p.

Krauskopf, K. B., 1959, The geochemistry of silica in sedimentary environments. In H. A. Ireland (ed.), *Silica in Sediments.* Society of Economic Paleontologists and Mineralogists Special Publication 7, pp. 4–19.

Oehler, J. H., 1976, Hydrothermal crystallization of silica gel. *Geological Society of America Bulletin* 87:1143–52.

Smale, D., 1973, Silcretes and associated silica diagenesis in southern Africa and Australia. *Journal of Sedimentary Petrology* 43:1077–89.

Manganese

Glasby, G. P. (ed.) 1977, *Marine Manganese Deposits.* Elsevier, Amsterdam, 523p.

Heath, G. R., 1979, Burial rates, growth rates and size distributions of deep-sea manganese nodules. *Science* 205:903–4.

Heath, G. R., 1981, Ferromanganese nodules of the deep sea. In B. J. Skinner (ed.), *Economic Geology, 75th Anniversary Volume.* Economic Geology Publishing Co., El Paso, Tex., pp. 736–65.

Krauskopf, K. B., 1957, Separation of manganese from iron in sedimentary processes. *Geochimica et Cosmochimica Acta* 12:61–84.

Ku, T. L., 1977, Rates of accretion. In G. P. Glasby (ed.), *Marine Manganese Deposits.* Elsevier, Amsterdam, pp. 249–67.

McKelvey, V. E., N. A. Wright, and R. W. Rowland, 1979, Manganese nodule resources in the northeastern equatorial Pacific. In J. L. Bischoff and D. Z. Piper (eds.), *Marine Geology and Oceanography of the Pacific Manganese Nodule Province.* Plenum Press, New York, 842p.

Ostwald, J., 1975, Mineralogy of manganese oxides from Groote Eylandt. *Mineralium Depositum* 10:1–12.

Ostwald, J., 1980, Aspects of the mineralogy, petrology, and genesis of the Groote Eylandt manganese ores. In I. M. Varentsov and Gy. Grasselly (eds.), *Geology and Geochemistry of Manganese,* vol. 2. E. Schweizerbart'sche Verlagsbuchhandlung, Stuttgart, pp. 149–82.

Taylor, J. H., 1969, Sedimentary ores of iron and manganese and their origin. *Proceedings 15th Inter-University Geological Congress, 1967—Sedimentary Ores: Ancient and Modern.* University of Leicester, pp. 171–86.

Varentsov, I. M., and V. P. Rakhmanov, 1980, Manganese deposits of the USSR (a review). In I. M. Varentsov and Gy. Grasselly (eds.), *Geology and Geochemistry of Manganese,* vol. 2. E. Schweizerbart'sche Verlagsbuchhandlung, Stuttgart, pp. 319–92.

10
Environmental and Scientific Reporting

SEDIMENTOLOGISTS AND ENVIRONMENTAL IMPACT STUDIES

Over the past few decades there has been a rapid increase in awareness of humanity's interaction with natural environmental processes and properties and of the consequences of humanity's impact on the environment (e.g., Archer, Lüttig, and Snezhko 1987). In most countries, awareness of the interaction between humanity and the environment has led governments to enact legislation designed to minimize the impact, and to establish ministries, departments, and various other statutory authorities to monitor and police compliance with the relevant legislation. Consequently, development and implementation of environmental management and protection programs have become significant economic considerations in the planning of major public and private enterprises. The resulting need for sound advice means that scientists from many disciplines are increasingly being required by both government and industry to investigate and evaluate environmental issues and to suggest appropriate courses of action. Environmental issues are regularly raised in scientific books and journals of many disciplines and several journals (e.g., *Archives of Environmental Contamination and Toxicology; Marine Pollution Bulletin*). Humanity's activities impact most directly and obviously on sediments (including soils), particularly since sediments also reflect interaction with the hydrosphere, atmosphere, and biosphere. Hence, sedimentologists are being asked ever more frequently to contribute to a wide range of environmental studies (e.g., see Chapter 1), and it is useful to consider some of the new types of sedimentological investigations that are required and the approaches to reporting the findings.

In the feasibility study and planning phases of major development projects, sedimentological input is required to identify environmental conditions that may affect the viability of the project or may impose particular design or operation constraints. At this early stage, sedimentological investigations are carried out in much the same way as they are in more traditional research projects, but objectives are more tightly defined. Sedimentological information required for development planning commonly includes:

- Identification of target strata (e.g., for mining, hydrocarbon extraction, or waste disposal);
- Detailed description (physical and chemical) of sediments at the site and in nearby areas that may influence, or be affected by, the development;
- Description of sedimentary processes (physical, chemical, and biological) acting at the site, or capable of influencing the development;
- Prediction of any likely changes in sediments or sedimentary processes due to construction or operation of the proposed development;
- Prediction of the likely effect of long-term sedimentation trends near the site;
- Assessment of the way in which any sedimentary characteristics of the site need to be accommodated in the design of the development.

Reports resulting from these studies generally take a similar form to traditional scientific papers or reports, with the following salient differences: the *abstract* is replaced by an *executive summary* (a little more detailed than a conventional abstract and written with a minimum of jargon so that the findings are comprehensible to nonscientists); the report states who the work was commissioned by and how it relates to the planned development; the conclusions focus on answers to questions specified in the commissioning brief.

Once the technical and economic viability of a development has been established, the next stage involves preparing an environmental impact statement (EIS) to assist in obtaining the necessary government and local authority development permits. The type of information to be included in an environmental impact statement varies substantially because the requirements for each EIS are dictated by applicable legislation and the nature of the proposed development (e.g., cf. Warden and Dagodag 1976; Baker, Kaming, and Morrison 1977; Porter 1985; Lee 1987); some of the major considerations to be addressed are summarized in Table 10-1. EIS's have a different format from that of traditional scientific papers or reports primarily because they are prepared for a rather different readership (in some places the format is specified by legislation). They must be comprehensive and objective, but they do not usually include all the details and analytical discussion that normally form a major part of scientific papers. Most scientists contributing to EIS's find it a difficult task to write in a style that is comprehensible to nonspecialist readers (free of jargon) and to produce a report that requires a definitive conclusion for which they may be liable; in contrast, lawyers, engineers, and policy administrators usually want definite answers and have great difficulty accepting the degree of uncertainty with which many scientists must qualify their interpretations.

During the construction phase, and often after many larger development projects have been completed, there can be further work for sedimentologists in monitoring environmental impacts and preparing environmental audits and reviews. The granting of development consent is often conditional on agreement to undertake these types of studies, which are designed to detect any unanticipated adverse environmental impact before it causes irreversible change, so that corrective measures can be instituted, and to review the accuracy of the predictions contained in the feasibility studies and the EIS.

Table 10-1. General Requirements for Environmental Impact Statements

1. *Description of the proposed development*
   Including the location, extent, form, purpose, current status, and life expectancy of the proposed development, ownership of all affected lands, identification of individuals or organizations responsible for implementation of the project, identification of legislative requirements applicable to the project, and any other land use constraints. A summary of contacts made with relevant governmental authorities in relation to the proposed development should be included; the correspondence is placed in an appendix. Prospective plans for site rehabilitation, maintenance of the aesthetic value of the site, noise control, the number of persons to be employed (during construction and operation), provisions for staff accommodation, transport routes and traffic densities, provision of infrastructure, and the main proposals for minimizing environmental impacts are summarized.

2. *Description of the existing environment that may be affected by the proposed development*
   Including the physical (geomorphological, geological, hydrological, land stability, sediment movement, etc.), biological and ecological (terrestrial and aquatic fauna and flora with particular reference to endangered species), social, archeological, and economic aspects of the proposed development site and of adjacent land. The description should be sufficiently comprehensive to provide a basis for assessing the environmental impact of the proposed development at and near the site and in any other areas that could be influenced by the development.

3. *Identification of likely interactions between the development and the environment*
   All significant interactions between the proposed development and the environment described in section 2 must be clearly flagged; where the proposed development may have specific impacts on individuals or organizations, it may be necessary to identify those individuals or organizations.

4. *Assessment of likely environmental impacts or consequences of the proposed development*
   This assessment needs to be thorough and objective because the conclusions are likely to be the most contentious and to be those most likely to be challenged if the proposed development becomes the subject of litigation. The assessment must evaluate long-term and short-term impacts of the proposed development on the physical, biological, social, archeological, and economic status of the proposed development site and of adjacent land; the impact of the environment on project design, life expectancy, the project staff, and operation strategy must also be addressed. Specific impacts commonly addressed include those affecting terrestrial and aquatic fauna (e.g., feeding, breeding, and other behavior patterns), terrestrial and aquatic vegetation (e.g., distribution, diversity, and density), drainage and rivers (e.g., hydrodynamics, siltation, flooding, bank stability), land stability and erosion rates, coastal processes and geomorphology, air and water quality (leachates, surface runoff, point source discharges), waste management, traffic and transport (land and water based as appropriate), the aesthetic value of the site, land use practices in the area, social conditions, recreational activities, economic activity (local, regional, and national), and any areas with high preservation value (environmental, cultural, archeological, or scientific). The assessment must also address differences in construction phase and postconstruction impacts, constraints imposed on future site use options, the reversibility of the various impacts, flow-on impacts, and cumulative impacts.

5. *Justification of the proposed development in relation to environmental, economic, and social considerations*
   Assessment of the benefits of the project (social, economic, etc.) in relation to anticipated environmental impacts and the expected effectiveness of proposed impact mitigation measures.

6. *Description of measures to be taken in conjunction with the proposed development to protect the environment and assessment of their likely effectiveness*
   Evaluation of the likely effectiveness of measures designed to minimize the impacts described in section 4.

7. *Evaluation of any feasible alternatives to the proposed development*
   Assessment of the advantages and disadvantages of reducing the scale of the proposed development, relocating the development to another site, or modifying the proposed design or development strategies.

8. *Evaluation of the consequences of not carrying out the proposed development*
   Assessment of the social, economic, and environmental consequences (local, regional, and national) of canceling the proposed development and continuing existing land use practices at the site.

PLANNING AND WRITING A REPORT

Whether preparing a report for a client, an EIS, or a scientific paper, writing will benefit greatly from prior planning, clear and concise presentation, and logical and consistent presentation of data and development of arguments (some common faults are listed in Table 10-2). Effective communication is essential; the most brilliant studies and profound conclusions count for little if they cannot be clearly communicated to others. Concise writing is also essential because most readers grow bored with excessive verbiage and because with the ever increasing volume of information published each year, no scientist or administrator has the time to struggle through a jungle to discover some well-hidden item of interest. Some books to peruse before writing are Lester (1976); Barrass (1978); Day (1983); Turabian (1987); Forbes (1991); Klein and Klein (1991); Eisenberg (1992); Locke (1992).

Plan to have a colleague critically review your work at penultimate draft stage. Science progresses as faults or inadequacies are found in existing hypotheses, forcing construction of new hypotheses that overcome the faults (Table 10-3). Thus, constructive criticism should always be welcomed; it provides a chance to make improvements and indicates areas warranting further attention. If a colleague can prove that your hypothesis is defective or your conclusions are inconsistent with your data, you should be grateful, but do not abandon your ideas without an argument—you may find that defense of your hypothesis produces new arguments to strengthen it!

There are many approaches to planning a report once the data have been accumulated and the literature has been surveyed; the following are considerations we have found helpful in our work:

1. *Discover the formal requirements of the client, editors, or conference organizers* (e.g., page layout, length, citation format, whether an abstract or summary is required; the particular requirements of journals are generally published on the inside cover but may have to be requested).

Table 10-2. Common Faults in Developing Arguments and Reaching Conclusions

Begging the question: basing a conclusion on an assumption that requires as much validation as the conclusion itself.
Inadequate sampling: making a generalization based on too small a sample.
Undocumented assertion: giving no supporting evidence.
False disjunction: polarizing the argument by assuming there are only two alternative hypotheses from which to choose.
Post hoc fallacy: assuming that the first event causes the second when there are two sequential events.
Appeal to authority: giving weight to an argument by using the name of a supposed authority.
Circular reasoning: using conclusions based on assumptions to show that the assumptions were valid.

Table 10-3. The Scientific Method

Scientific research generally proceeds in a sequence of steps ideally represented as follows:

1. Selection and definition of the problem.
2. Planning methods of attacking the problem and collecting data.
3. Collection of data.
4. Interpretation of data, and comparison with previous knowledge.
5. Construction of working hypotheses to explain the observations.
6. Prediction of anticipated observations in unstudied situations.
7. Testing the predictions, and modification of the hypothesis.
8. Repetition of the cycle and selection of the best hypothesis.
9. Communication of the results in a publication or at a conference.
10. Discussion and criticism by the readers or audience.
11. Further refinement of the hypotheses.

2. *Carefully define the scope of the report*, ensuring that it is neither too narrow nor too broad with respect to its context and targeting the characteristics of the audience who will read it.

3. *Plan the overall structure of the report*. Prepare a preliminary table of contents, whether or not one is required (i.e., an outline that shows the main points to be discussed as headings of appropriate rank). Although this organization will probably be modified and refined later, it will help establish an initial plan and a logical progression of ideas, and it will help avoid repetition.

4. *Do a literature search for published (and unpublished theses) works of relevance*. Start with the most recent publications and their reference lists. Search the indexing and abstracting journals (e.g., *Australian Science Index; Bibliography and Index of Geology; British Geological Literature; Index New Zealand; Science Citation Index*) and/or undertake a computer search through the DIALOG system of such data bases as *Geoarchive* and *Georef*.

5. *Edit your notes*. Assemble data and ideas relevant to each subtopic from all sources. Consider whether there are sufficient data for each point you wish to cover; if not, decide whether to seek more or to refer the reader to a "need for further work." Find and evaluate apparent contradictions between different sources; decide whether they can be reconciled and whether they should be noted in your text with or without your solution to the conflict. Delete data that are excessive or not directly relevant to your topic.

6. *Compile a detailed table of contents showing all headings and subheadings you plan to use*. Most scientific reports have abundant subheadings (rarely are there more than two full pages of text in one subsection). Order the headings and subheadings into a hierarchical series; distinguish the relative importance of each subheading clearly and consistently (e.g., major headings may be capitalized, boldface, and centered; second-level headings may be capital/lowercase, italicized, and flush left; etc.).

7. *Write a first draft, preferably in one pass.* Construct separate paragraphs for each major point or idea; arrange them in a logical progression under each subheading.

8. *Construct and assemble illustrations and tables and write captions or titles for them.* Illustrations are almost as important in reports as in verbal presentations; they save words and impart a message much more effectively (e.g., Rodolfo 1979). Insert them where they are most useful in the text (readers are irritated when they must seek them in a separate compilation, although excessively long data tables or large figures may be appropriately placed in an appendix). Ensure that all figures and tables are numbered in consecutive order as they are used in the text, and that each is mentioned in the text. Next, compile the reference list. A useful way to begin is to read through the text, noting on a separate page in alphabetical order each citation. Bibliographic computer programs may undertake this step for you, but remember that sources of previously published figures and tables must also be cited as references.

9. *Do not read the first draft for at least a few days. Then re-read and revise it,* concentrating on coherence, logical progression of ideas, convincing development of the major theme and of each minor theme in each subsection. Another read is then warranted for clear expression and correct usage of punctuation, spelling, length of sentences, and other aspects of grammar; also replace overused words or phrases with alternatives to enliven readability. Reorganize as necessary, recast ambiguous sentences or phrases, and delete all repetitious or irrelevant material to produce a concise text.

When re-reading, imagine that you are seeing the report for the first time. Has the writer provided a clear statement of intent, problem, and scope? Are there unwarranted assumptions concerning the reader's familiarity with the material presented (e.g., is there abstruse jargon)? Is a specific feature mentioned in relation to others that have not been introduced previously (e.g., is a unit discussed before its position in a group is established)? Has the writer considered alternatives to the hypothesis or conclusion? Are unresolved problems clearly indicated? If based on fieldwork, has the report presented a clear picture of the field area size, topography, abundance and quality of accessible exposures? Has the writer fallen victim to the "Agatha Christie syndrome" (all will be revealed on the last page)? Subvocalization of each sentence is useful for checking clarity, and the pauses (of appropriate length) that would normally be made when reading aloud indicate the kind of punctuation marks that are required. After all, writing is a substitute for good dialogue (see Murray 1968).

10. *Finally, if at all possible, have a colleague read the report critically, and revise it a final time* (checklist in Table 10-4). The rechecking and refining of your text may involve time and effort, but it will be worthwhile if it means that you can look back on your report with pride, knowing that it is scientifically sound, grammatically correct, well presented, and easy to read.

STRUCTURE AND COMPONENTS OF A PAPER OR REPORT

The structure of a report is designed to allow a reader easy and rapid access to the items of interest for various purposes; information is grouped into sections under suitable headings, arranged in an appropriate sequence. Different organizations may have slightly different guidelines for their own publications (e.g., cf. Blackadar, Dumych, and Griffin 1980; N.Z. Government Printing Office 1972; Royal Society, London 1974; De Bakey 1976; USGS 1991; Council of Biology Editors 1983), but there are also general rules and guidelines (present in virtually all guides such as those cited above, and see Rubens 1992). The conventional headings and sequence of presentation are listed below, but it is often necessary to have supplemental headings in large reports.

Title
Abstract (or Executive Summary)
Table of Contents (for large reports and some papers)
Introduction
Methodology or Procedures (may form a subheading of Introduction)
Observations (or Results)
Interpretation (or Discussion)
Conclusions and/or Summary
Acknowledgments
Appendixes
References

Table 10-4. Effective Communication Checklist

Clarity:	Will the paper be readily understood by the range of likely readers?
Association:	Is the paper in a language and style with which both the writer and reader are at ease?
Association:	Does it use concepts with which the readers are familiar to construct the arguments?
Relevance:	Does the paper keep to the point, or does it digress into confusing side issues?
Structure:	Does the paper have a logical structure and sequence?
Priorities:	Does the writing style emphasize the most important aspects of the report?
Interest:	Will the structure and style of the paper interest the reader?
Unity:	Is the message complete?
Accuracy:	Is the content of the paper both accurate and precise?
Brevity:	Can the report be shortened by rephrasing or deleting words, clauses, or sentences?

The title (read by all readers) and abstract (about 40%) are by far the most commonly read parts of a report; therefore, they should be written with the greatest care.

Title

Brief titles that inform about the contents attract readers. Try to use fewer than 100 letter spaces and make every word count; there is no room for insignificant words. Titles should contain key words that provide precise information on the scope and subject matter of the report; in scientific reports, the title also usually indicates the location of the study.

The title is normally written on a separate page together with the name of the author; this page may also contain the date submitted, address of the author, and his or her company or organization. In a growing number of cases, a list of key words suitable for use in electronic information-retrieval systems may be appended to the page.

Abstract

The abstract of a report must state the main results and conclusions of the study; whereas it may contain a short comment on the purpose of the study and the investigative procedures employed, it should not discuss how the results were obtained unless a radically new investigative strategy was used. The abstract is *the* most important part of a report and is commonly read by many more readers than the report itself. Thus, it must convey the essentials of the report without the support of other material and must not contain information absent in the body of the report. Use a clear, concise, and forceful style (use strong verbs in the active voice); rewrite at least three times. For detailed advice on writing an abstract, consult Landes (1966).

Executive Summary

Reports prepared for organizations (companies and government departments) commonly contain an executive summary *instead* of an abstract. Executive summaries are similar to abstracts, but they usually present more detail and place greater emphasis on the implications of the study; because executive summaries are likely to be read by people who are not specialists in the same discipline (some may not be scientists), the use of specialist technical terms is minimized.

Table of Contents

The table of contents should contain a list of headings (with italics, capitalization, and other formatting as used in the text) with page numbers; separate lists of tables and figures (with page numbers) may be required/desirable with long reports. Headings not only assist the reader in locating items but also provide a summary of the sections in a report and indicate the logical pattern and priorities adopted by the author.

Introduction

The introduction sets the scene for the contribution and puts it in perspective. Different types of information can be given in different order in the introduction, depending on the particular study. Subheadings may be useful, particularly if the introduction covers more than a page of text. The introduction should contain the following:

General aim, scope, and specific objectives of the investigation.
Background and reasons for carrying out the investigation and reporting on it.
Significance and relevance of the study to the fields of inquiry (commercial, EIS, science in general, and/or the discipine in particular).
Plan of presentation. The proposed method of treatment should be stated in general terms; do not repeat the headings used in the report. If there is a table of contents, this step may be unnecessary.
The study area and a locality map. Assume the reader is from overseas and lacks local knowledge. If the report involves the geology of an area, provide a simplified map of regional geological setting and a summary stratigraphic column.
A sample location map with sample numbers at the sites for all samples used in the study (may be combined with preceding locality map; grid reference locations for sample sites are frequently included separately in an appendix).
Review of current knowledge. Briefly summarize the main contributions of relevance by other workers. In theses and some kinds of major scientific or commercial reports, this section may warrant a heading of equal rank to the introduction.
Present nomenclature and definitions for uncommon technical terms, new terms or nomenclature, and nonstandard abbreviations used in the report.

Methodology (or Procedures)

Description of the procedures used in the study may be given in a subsection of the introduction, but in the natural sciences this section is often large enough to warrant a higher ranking. Readers need to know the methodology used both to understand the nature of the study and to evaluate the quality of the results; if future studies are contemplated, the same procedures must be used to obtain comparable results. This section may include some or all of the following (see **AS** for discussion of categories):

Sampling strategies: describe the sampling strategies used in the study and justify their selection in terms of the aims of the study.
Analytical methods: describe the analytical methods used in the study and note the accuracy, precision, and sensitivity achieved. Specify the type of analytical equipment used.

Warn the reader of dangerous compounds and procedures in case an inexperienced worker attempts to duplicate/extend your methods.

Quality (error) control: describe any strategies adopted to ensure quality control during sampling, sample pretreatment, and analysis (keep description brief, but ensure all likely sources of error are considered).

Statistical analysis: describe any statistical procedures used in sampling and data evaluation (keep this section brief, except where nonstandard methods were selected).

Careful consideration should be given to the selection of material for inclusion in a methodology section and to how it is written up. Because it is difficult to avoid using the passive voice when describing the procedures used, lengthy descriptions can be very boring to read; keep the descriptions short by referring to strategies and techniques used in other published studies wherever they are the same or similar, and tabulate procedures (or use flowcharts) when possible.

Observations (or Results)

Observation (yours and others) and interpretation (yours and others) must be clearly distinguished. The observations section should state what work was actually performed and what results were obtained; leave interpretations to the discussion section. Data from previously published studies may be included in the results section for reference purposes, but evaluation of any similarities or differences between these data and those from the current investigation should be reserved for the discussion section.

Tables, figures, and graphs are particularly useful for presenting data clearly and concisely (especially numerical data), but they must be designed carefully; a well-prepared diagram or table can be much more effective than many words, but a bad one may obfuscate and irritate the reader. Tables and figures must be directly relevant to the text and all must be cited in the text; they are numbered in the order of mention in the text. Place tables and figures as close as possible to the relevant section in the text. Some general suggestions for presentation of figures, tables, photographs and plates, maps, and cross-sections are given in Table 10-5. All figures and tables require a brief caption to explain what each is about; the caption must be complete by itself—some readers will examine the figures and tables without reading the entire text. Captions do not necessarily require complete sentence construction. Other information may be given in captions, such as explanations of symbols used, sources of data given for comparison, the analyst or observer, and the apparatus used.

The results section commonly includes biological (or paleontological) names, stratigraphic terms, chemical data, and statistical information. Where biological names are used, ensure they are correct and are written in the correct Linnaean format. Guides for stratigraphic nomenclature are provided in the *International Stratigraphic Guide* (Hedberg 1976) and in other guides for specific countries. Observe conventions for the representation of chemical elements, ions, and formulae and check that all subscripts and superscripts in chemical symbols and formulae are correctly represented (e.g., K^+, Ca^{2+}, Ti^{4+}, CO_3^{2-}, Mg_2SiO_4). Use standard units and conventional symbols for presentation of all numerical data (SI units rather than those archaic imperial units; e.g., Table 10-6), and do not report data with unrealistic numbers of significant figures (e.g., 22.1751 mg/L in a chemical analysis with a standard error of $\pm$ 5% should be given as 22 mg/L).

Interpretation (or Discussion)

The discussion section must lead to the conclusions, and careful planning is required to avoid unnecessary repetition. The discussion should evaluate the data presented in the results section, compare these data with the conclusions of other workers, evaluate the merits of alternative explanatory hypotheses, document the preferred hypothesis (giving reasons and outlining any limitations), and recommend suitable approaches for future investigations. Generally, discussion progresses from the more specific to the less specific, e.g., from the known to the unknown in a series of logical steps (avoid faults in logic listed in Table 10-2). All arguments must be clearly developed and all assumptions involved in each step should be justified. State the sources of all information or ideas used in the discussion, especially when they do not originate from your own work; statements based on the work of other authors must be accurate, and reference citations must be complete and clear to the reader.

Summary and Conclusions

A summary states the main results of the work; conclusions are statements drawn from the results. In neither case should new material be introduced (rare exceptions are when your conclusions are compared with those of others.) This part of a report is vital to the reader who does not wish to follow all the discussion but requires more information than the abstract provides. Most authors find it difficult to cast a summary and conclusions section that is substantially different from an abstract: the difference should be that an abstract simply and briefly states the results whereas synthesis and reasoning are presented in the former.

Acknowledgments

Acknowledgments are necessary when assistance, such as access to material, equipment, or information, was provided by an organization or individual (in the case of information, citations of published work in the text and references section serve the same purpose). Acknowledge all sources of finance, and the institution or company that gave permission to publish the whole or any part of a report. Acknowledge the source of all personal communications (pers. comm.) used in the text and the assistance of any critical reviewer.

Table 10-5. Suggestions for Presentation of Nontext Components of Reports or Papers

Tables

Follow conventional formats for data presentation (e.g., the order of listing oxides in whole rock analyses).
Check that all units and symbols are correct.
Tabulate data neatly; do not overcrowd, but try to avoid splitting tables (e.g., change font size in preference).
Explain all symbols used with a key (either in the table or in a caption).
Report numerical data in metric units.
A zero should precede the decimal point in numbers less than one (e.g., 0.5 rather than .5).
Use exponents rather than writing large numbers in full (e.g., write 3.92×10^7 rather than 39,200,000).

Figures and Graphs

Emphasize the main theme of the figure and exclude nonessential material, which may reduce clarity.
Consider aesthetic balance when drafting the figure and avoid overcrowding.
Ensure that all lettering is readable (minimum letter height in the final printed copy = 1 mm).
Label all essential features (use a key if necessary).
If a scale is necessary and the figure is likely to be reduced, use a graphical scale (i.e., a labeled scale bar).
For figures depicting three-dimensional features, indicate the spatial orientation for each section you draw, e.g., horizontal section (or plan), vertical section and viewer's perspective of the section, or some other orientation; if necessary, include a reference sketch to show the spatial relationships of the different sections.
Label the axes of graphs clearly and specify the units used; in triangular diagrams, label all poles and state the units. Ensure that the scale is appropriately marked on all axes that represent a continuum.
Do not waste space on graphs by providing for nonexistent points, but ensure that truncating the axes does not reduce clarity; show a break in the axes if they are not continuous.
Label multiple curves or trends on graphs (or use different kinds of lines and provide a legend).
Choose a graphical form appropriate to the type of data plotted. Because very few natural variables change suddenly at specific values, it is usually advisable to fit smooth curves or a single straight line to data points on a graph; the probability that natural conditions will be better reflected by a series of straight lines connecting successive data points is extremely remote. For many graphs, additional information is available that gives some guidance on the shape of the curve that should be fitted. If it is unclear what sort of curve should be fitted, it is better to avoid connecting the points on the graph, even if both axes represent continuous series.

Photographs and Plates

Scale (graduated centimeter or meter scale, preferably with alternating black and white divisions; pencils, lens covers, or rock hammers are commonly used, but coins should be avoided because they mean little to readers unfamiliar with the currency used). Photomicrographs should include a bar scale (use 0.1 or 1.0 mm rather than odd sizes such as 0.15 or 1.3 mm), or state the magnification (e.g., ×100) if the photograph is not going to be reduced or enlarged (e.g., on a projected slide).
Field location data and orientation: specify map reference coordinates and indicate the "way up" for close-up photographs of sections where the orientation may be unclear; it may also be necessary to indicate the compass bearing on which the photograph was taken.
For petrographic photomicrographs, indicate whether the photograph was taken using plane-polarized light or crossed polars.
Specify the official collection number (and the name of the collector) if the photograph is of a sample retained in an official collection.

Maps and Cross-Sections

A map or section must have a title, a legend (ensure that all symbols, colors or shade patterns, etc. used are explained in the legend; use standard symbols for geological and geographic features), a scale (use a metric scale *bar* because the map may be reduced) and a north direction indicator.
Cross-sections must show bearings or other indicators of orientation. East-west cross-sections are usually drawn so that the observer faces north, and north-south sections so that the observer looks west.
If a map and a cross-section apply to the same area, a section line should be marked on the map; label the points at each end of the section and ensure that the labels on the map match those on the section.
Maps may need to show coordinates to reference grids (i.e., the origin of the map grid and relation to other grids), air photograph runs and centers.
Maps may require a location sketch showing the relationship of the map to surrounding areas or adjacent maps and figures. Insert a small outline sketch of the larger recognizable area(s) in a corner of the main map that is unoccupied by essential detail, with a highlight box showing the locality of the main map.
Indicate which data on the maps or sections are factual and which are interpreted; differentiate between factual, interpretive, and approximate boundaries. Maps may also require a reliability diagram that indicated the methods used (e.g., air photograph interpretation and transverse, detailed mapping, etc.).
Mark all contours and contour intervals clearly, and specify a reference datum.
If vertical exaggeration is used on cross-sections, indicate clearly the degree of exaggeration.
Geological maps require the inclusion of a legend in the form of a stratigraphic column (even if a column is shown as a separate figure), with rock types of formations listed in order from youngest at the top to oldest at the bottom.

Table 10-6. Standard Units and Symbols Used in Quantitative Measurement

	Unit	Symbol	Definition
Length	metre	m	S.I. base unit
	micrometre	μm	10^{-6} m (formerly micron, μ)
	nanometre	nm	10^{-9} m (formerly millimicron, mμ)
Volume	cubic metre	m^3	10^3 L
	cubic centimetre	cm^3	1 ml
	litre	L	S.I. base unit
Area	square metre	m^2	
	hectare	ha	100 m^2
Mass (m)	kilogram	kg	S.I. base unit
	gram	g	10^{-3} kg
Time (t)	second	s	S.I. base unit
	hour	hr	3600 s
Temperature (T)	degree Celsius	°C or deg C	
	degree Kelvin	°K	(°C + 273.15) (S.I. base unit)
Force	Newton	N	$m \cdot kg. s^{-2}$
Pressure	Pascal	Pa	$m^{-1.} kg. s^{-2}$
	kilobar	kb	(1 bar = 10^8 Pa)
Energy	Joule	J	$m^2 \cdot kg. s^{-2}$
Power	Watt	W	$m^2 \cdot kg. s^{-3}$
Frequency	Hertz	Hz	s^{-1}
Magnetic flux	Tesla	T	10^4 Gauss
	1 nanotesla	nT	10^{-5} Gauss = 1 gamma
Gravity	milligal	mGal	10 $\mu m.s^{-2}$

Prefixes for Multiples of Units

10^9	giga	G	10^1	deca	da	10^{-6}	micro	μ
10^6	mega	M	10^{-1}	deci	d	10^{-9}	nano	n
10^3	kilo	k	10^{-2}	centi	c	10^{-12}	pico	p
10^2	hecto	h	10^{-3}	milli	m			

Symbols for Other Physical Quantities

angle	α, β, γ, etc.	viscosity	η
radius	r	normal stress	σ
rectangular coordinates	x, y, z	refractive index	n
electric current	A	frequency	ν, f
sediment grain size	φ	wavelength	λ
conductivity (siemens)	S	density	r
magnetic field strength	H	fugacity	*f*
pressure	P	partial pressure	p
electrical resistance	Ω	voltage, potential difference	V
velocity of electromagnetic radiation in a vacuum	c		
activity of radioactive substances (Becquerel)	Bq		

Appendixes

Information relevant to the report but which is lengthy and interrupts the readability of the text is best placed in an appendix. Appendixes should be cited in the relevant section of the text. Examples of the types of information that may be presented in appendixes are:

Details of apparatus and procedures, where these differ from standard methods. If the method is part of the subject matter of the report, these details must be given in the text.

Lists of specimens studied or cited (e.g., reference collections—note the relevant specimen numbers and the location of the reference collection).

Detailed descriptive information (e.g., detailed petrography of rock samples).

Tables of detailed numerical data (e.g., analytical results that are not essential to the text). Particularly extensive tables will seldom be published but may be referred to as "available from the author."

CITATIONS, REFERENCES, AND BIBLIOGRAPHIES

Citations are made in the text of all relevant sources of information used in the report; scientific reputations (and jobs) are lost when plagiarism occurs so ensure comprehensiveness.

Leeway is permitted when the information can be judged to be "general knowledge" within the discipline—all science builds on previous results; however, it is better to err on the side of caution. Do not cite articles that have not been directly consulted. Where possible, consult and cite the *original* references; if the original is unobtainable, use the form of referencing "Smith (1968) in Brown (1975) stated . . ." Conventions for in-text citations are:

Single authors: "author (date) stated . . . " or "the calcrete is silicified (author date)" with or without a comma between the author and date depending on the publisher. If more than one citation is made at the same position, semicolons separate the citations (e.g., "Jones 1983; Smith 1989"). If quoting, the page reference should be given.

Multiple authors: for two or three authors (depending on the publisher), cite them all. Where there are more than two or three authors, use the first author's surname and "et al." (e.g., "Brown et al. 1973"). Some publishers permit the full English equivalent "and others" instead of "et al."; some editors require "et al." to be italicized.

Personally supplied information (verbal or via letter): cite the supplier and "(pers. comm. year)" or "(written comm. year)" is sometimes cited.

Accepted for publication, but not in print: cite author "(in press)". Some publishers allow "(in prep.)" for work in preparation (but since there is no guarantee of publication, many publishers do not!).

Unpublished information provided in written form to someone else: (e.g., theses deposited in a university library)—cite "Smith (unpublished information)" or "Smith (1976)" and list the unpublished article in the references. *You should obtain permission to use such unpublished information.*

Reference to previously cited references: the format "Smith (op. cit.)" can be used, but it is preferable to restate the date of the reference, particularly if other citations have been made in between or if there are several publications by Smith.

References are lists of *all* published works cited in the text and *no others;* they are presented in a separate section at the end of a report. All citations made in captions, tables, diagrams, and appendixes must be included. *Bibliographies* are comprehensive lists of *all* relevant publications, cited or not; in science none can be safely eliminated, hence major research is involved in compiling and checking bibliographies (computer searches are not infallible). *Selected bibliographies* are what most science writers provide if they go beyond the simple list of references cited in the text; their "selection" reflects a personal choice of what they consider "best" or most relevant.

The presentation of references must adhere strictly to a consistent format and order, which may be dictated by the publisher or employer. The general format for geological publications is given here. References are listed in the reference list alphabetically for single and first of multiple authors (works by single authors precede works by multiple authors where the same person is first author). Where there is more than one reference for the same author or group of authors, chronological order is used. Where there are two or more references in the same year for the same author or authors, the references are differentiated by attaching a letter to the year (e.g., 1973*a*, 1973*b*); the letters (*a*) or (*b*) should reflect the month of publication, but more commonly they reflect the order of citation in the text.

Author last names are cited first for the first author and either also for all authors or with preceding initials for all subsequent authors. Year is given next (logically—readers wish to know the relative dates of the authors' work first; rarely it is given at the end of the reference). Title is followed by the journal name, or in the case of a book, by publisher and place of publication. Volume number and page ranges are given for journal papers (total pages may or may not be given for books). Because of inconsistencies between publishers, it is wise to acquire a computer software package for bibliographic information that permits automatic translation to the different formats. See your favorite journal for their usage, and compile your own selected bibliography according to their style.

WRITING STYLE

A good writing style should be clear, concise, and direct. Good guides to correct grammatical use of words, punctuation, and construction of sentences and paragraphs, together with other aspects of good writing, are Strunk and White (1959); Tufte (1971); O'Connor and Woodford (1977); Style Manual Committee (1978); Fowler (1983); Cochran, Fenner, and Hill (1984); Gowers (1987); The University of Chicago Press (1993). Choose words carefully to avoid complexity, ambiguity, and imprecision (cf. Partridge 1976). Keep a good dictionary and a thesaurus handy (e.g., Roget 1984)—and use them! (Most computer software spellcheckers and thesauri do not have adequate scientific coverage—and be careful when adding a word to the user dictionary that it is correctly spelled!)

Ensure that the only abbreviations you use without specific in-text definition are internationally accepted standards; where nonstandard abbreviations are necessary (e.g., to avoid excessive repetition of lengthy words or phrases), clearly define the abbreviation in the text the first time you use it (or in a special subsection of an introduction) (see Table 10-7). Do not overuse abbreviations: they can distract from smooth reading. Underline or italicize foreign words and genus or species names. Make sure you use technical and foreign words correctly (for geological terms, consult the *Glossary of Geology* by Bates and Jackson 1987).

Sentence Construction

Good sentence construction is vital. As a general rule (with exceptions), the use of simple short sentences, familiar words

Table 10-7. Some Common Abbreviations

aq.	aqueous
av. (or ave)	average
BP	before present
b.y. (or Ga)	billion years
ca.	about
cf.	compare with
conc.	concentrated
ed.(s)	editor(s)
edn(s)	edition(s)
e.g. (or eg,)	for example
et al.	and others
etc.	and so on
et seq.	and what follows
Fig.	figure
gen. nov.	new genus
ibid.	in the same book
i.e. (or ie,)	that is
IR (or i.r.)	infrared
loc. cit.	in the passage already cited
max.	maximum
min.	minimum
m.p.	melting point
MS(s)	manuscript(s)
m.y. (or Ma)	million years
N	north
NE	north-east
NNE	north-north-east
No(s).	number(s)
op. cit.	in the work already cited
p(p).	page(s)
ppb	parts per billion
ppm	parts per million
ppt	parts per thousand
R.I.	refractive index
SI (or S.I.)	System Internationale
S.G. (or sp. gr.)	specific gravity
sp(p).	species (plural form)
sp. nov. (or n.sp.)	new species
stet	leave unchanged
UV (or u.v.)	ultra violet
viz.	namely

(balance against necessary use of jargon), and variety of expression makes text more readable. Properly constructed sentences usually have a subject, verb, and object, with or without modifying phrases and clauses that either help develop the central idea in the sentence or should be omitted. Ensure that the principal clause contains the key idea and is distinct from the subclauses. The subject is the key word or idea in a sentence, so make the subject clause strong. Avoid excessive use of indefinite pronouns. Be careful when using "this" and "these" as pronouns. Difficulties can arise when what is actually being referred to is ambiguous. Clarity can better be served by considering them adjectives that require accompanying nouns, for example, "This situation encouraged further investigations." Check that the verb matches the subject. If connectives are used in a sentence, use an appropriate connective to indicate the relative value of clauses (e.g., "and" connects clauses of equal rank, whereas subordinating connectives such as "however" introduces clauses of unequal rank).

Many writers of scientific reports feel compelled to establish their objectivity by writing in the third-person passive voice, but this strategy is neither necessary nor desirable: exclusive use of the passive voice is an effective way to ensure that readers become bored and disinterested as quickly as possible (e.g., Kirkman 1975). Of course there are exceptions and the passive voice should not be excluded, but the active voice is generally more direct, clear, and conducive to shorter sentences (cf. the passive "the area was mapped by Smith" and "Smith mapped the area"). Sparse use of the first person (I, we) is not only permissible but desirable to maintain the active voice and to distinguish your own contributions from those of others; the reason that many consider such usage anathema is because *excessive* use was made of the first person in scientific writings of the 1800s and early 1900s.

Paragraph Construction

Paragraphs are the basic units of any written report. Each paragraph should have a single theme; sentences in the paragraph should relate to the theme and develop it in a logical order. The first and last sentences in a paragraph are the most important to a reader. Paragraphs begin with a topic sentence that introduces the key idea; the topic sentence resembles an abstract for a paragraph. The last sentence is commonly used to emphasize or summarize, or it presents an important point in the theme. Try to use a word, phrase, or clause in each sentence to provide a link with adjacent sentences; a link between ideas can often be established by using connectives such as "similarly," "in contrast," "alternatively." Avoid excessively long paragraphs because solid blocks of print cause the reader to lose concentration; sequences of unusually short paragraphs are also distracting.

Footnotes

A footnote is like a small appendix in the text; it is used as an aside, an elaboration, or an addendum. Footnotes must be brief. They are unacceptable to many journals because they can be distracting and interrupt the flow of thought (and are costly to typeset!).

VERBAL PRESENTATION OF REPORTS

In addition to writing reports, all scientists must at least occasionally present information effectively to a live audience. Clarity and simplicity are even more important in verbal than in written communication because the audience cannot pause, return to, or reexamine any points they have missed in a talk. A few publications with useful general discussions of do's and don'ts for talks are Clifton (1978); Booth (1981); Bryant and Wallace (1982). Particularly germane considerations when preparing and delivering a verbal presentation are:

Know your audience: make the talk appropriate to their interests, needs, and background; aim for a balance between overestimating and underestimating their knowledge. Avoid using terminology that is unfamiliar to the average member of the audience, and avoid unnecessary jargon. Some of the word choices and grammatical rules important in writing are less so when talking, but be careful not to follow the politician's diction and style too closely!

State the purpose and objectives at both the beginning and the end, and give an outline of the paper at the beginning. During the course of the talk, review what you have said and indicate the objective at least every 10 minutes.

Show the audience how the particular topic relates to broader fields of inquiry, at both the beginning and the end.

Use a logical, clear, stepwise sequence to develop your ideas around a central concept. Maintain continuity by connecting different ideas and showing how they are related.

Concentrate on the essentials of the topic; do not waste time on the finer details. Try to achieve a suitable balance between too much information and too little.

Relate new information to knowledge already available to the audience (go from the known to the unknown, from the simple to the complex, and from the concrete to the abstract).

Illustrate the main points with clear examples and the use of simple visual aids (see AAPG 1970; Shinn 1981).

Leave time for questions and discussion. An important part of the talk is to obtain feedback. Rehearse the talk to ensure staying within the time limit!

Make the introduction stimulating and try to capture the interest of the audience; once you have their attention, try to hold it with brief and entertaining delivery.

Speak clearly, loudly enough for those at the rear of the room to hear, and not too quickly. Vary the tone of your voice to avoid monotony. Slow down and repeat yourself when presenting the most complex points. During your rehearsal, have your listener sit at the back of the room and take notes on these aspects of your talk.

Keep your eye on the audience as much as possible, and try to establish eye contact with random individuals in the audience—once you break this bond their attention can stray. (This guideline is difficult for a new speaker to follow, and there is a danger of reacting to an individual, but it is an important skill to develop.) Look at the visual aids no more than is necessary; never talk to a visual aid.

Do not read the talk. It is very difficult to avoid a monotonous and boring presentation; audiences give considerable leeway to free-flow speakers even if they forget words. Use brief notes clearly printed on small cards as a framework. Center the presentation around diagrams or slides to reduce the need for notes.

Use hand and body movements, but do not pace and avoid monotonous and distracting gestures or mannerisms (jingling coins in a pocket or swaying back and forth). Have a friend check you out initially and during the talk. Watch out particularly for the "mouth open, mind in neutral" syndrome that produces the dreaded "Ahhhh . . ." or "Ummmmm . . ."!

Be enthusiastic and confident, and try to stimulate participation, thought, and discussion. Try to become aware if the audience is following or not, and adjust your pace and presentation to stimulate the audience. (This is where eye contact can be informative—if no one meets your eye, you've lost them!) Spontaneous humor can help generate enthusiasm in the audience, but don't overdo it!

Vary the form of communication from verbal to visual. Visual aids of any kind greatly assist in getting a message across and stimulate the interest of the audience. While audiences generally accept on-the-spot sketching on a pad or board, they allow less leeway for poorly prepared slides or overhead transparencies. Prepared materials must clearly illustrate what you intend to convey in a simple and legible (even at the back of the room) fashion. Be aware that an audience needs some time to comprehend visual aids.

Organize the verbal and visual components for maximum continuity and minimum interruption. Avoid long gaps in talking and delays while sorting through notes and diagrams or operating equipment—such interruptions annoy the audience and waste time. Check the sequence of presentation of visuals immediately before the talk, and number them clearly so that time is not wasted sorting through them if they get out of order. Check the facilities for, and operation of, visual aids (switches for lights and projectors, projector controls) before the talk to avoid delays and interruptions.

Do not apologize—it wastes time and does not win any sympathy.

Do not panic. Remember what your reaction is to speakers who try hard, and assume the audience will react the same way to you.

SELECTED BIBLIOGRAPHY

AAPG, 1970–, *Slide Manual.* American Association of Petroleum Geologists, Tulsa, Okla.

Archer, A. A., G. W. Lüttig, and I. I. Snezhko (eds.), 1987, *Man's Dependence on the Earth: The Role of Geosciences in the Environment.* UNESCO, E. Schweizerbart'sche Verlagsbuchhandlung, Stuttgart, 216p.

Baker, M. S., J. S. Kaming, and R. E. Morrison, 1977, *Environmental Impact Statements: A Guide to Preparation and Review.* Practicing Law Institute, New York, 334p.

Barrass, R., 1978, *Scientists Must Write.* Chapman & Hall, London.

Bates, R. L., and J. A. Jackson (eds.), 1987, *Glossary of Geology,* 3d ed. American Geological Institute, Alexandria, Va., 788p.

Blackadar, R. G., H. Dumych, and P. J. Griffin, 1980, *Guide to Authors: A Guide for the Preparation of Geological Maps and Reports,* 2d revised reprinting, Canada Geological Survey Miscellaneous Report 29, Ottawa, 66p.

Booth, V., 1981, *Writing a Scientific Paper and Speaking at Scientific Meetings,* 5th ed. The Biochemical Society, London, 48p.

Bryant, D. C., and K. R. Wallace, 1982, *Oral Communication,* 5th ed. Prentice-Hall, New York.

Clifton, H. E., 1978, Tips on talks and how to keep an audience attentive, alert, and around for the conclusions at a scientific meeting. *Journal of Sedimentary Petrology* 48:1–5.

Cochran, W., P. Fenner, and M. Hill (eds.), 1984, *Geowriting,* 4th ed. American Geological Institute, Alexandria, Va., 80p.

Council of Biology Editors, 1983, *Style Manual: A Guide for Authors, Editors, and Publishers in the Biological Sciences,* 5th ed. Council of Biology Editors, Inc., Bethesda, Md., 324p.

Day, R. A., 1983, *How to Write and Publish a Scientific Paper.* ISI Press, Philadelphia, 160p.

De Bakey, L., 1976, *The Scientific Journal, Editorial Policies and Practices.* C. V. Mosby Co., St. Louis.

Devereaux, P., 1991, *New Zealand Style Book.* Government Printer, Wellington, N.Z.

Eisenberg, A., 1992, *Effective Technical Communication,* 2nd ed. McGraw-Hill, New York.

Forbes, M., 1991, *Writing Technical Articles, Speeches, and Manuals.* Krieger Publishing Co., Malabar, Fla.

Fowler, H. W., 1983, *A Dictionary of Modern English Usage,* 2d ed., revised by E. Gowers. Oxford University Press, Oxford, 725p.

Gowers, E., 1987, *The Complete Plain Words.* Penguin Books, Harmondsworth, 288p.

Hedberg, H. D. (ed.), 1976–, *International Stratigraphic Guide.* International Subcommission on Stratigraphic Classification of IUGS Commission on Stratigraphy, John Wiley & Sons, New York, 200p.

Kirkman, J., 1975, That pernicious passive voice. *Physics in Technology* 5:197–200.

Klein, D. H., and J. H. Klein, 1991, *Write for Success: A Guide to Effective Technical and Professional Writing.* Kendall/Hunt Publishing Co., Dubuque, Ia., 174p.

Landes, K. K., 1966, A scrutiny of the abstract, II. *American Association of Petroleum Geologists Bulletin* 50:1992.

Lee, N., 1987, *Environmental Impact Assessment: A Training Guide.* Department of Town and Country Planning Occasional Paper 18, University of Manchester, Manchester, 168p.

Lester, J. D., 1976, *Writing Research Papers: A Complete Guide,* 2d ed. Scott, Foresman, Dallas, Tex., 196p.

Locke, D. M., 1992, *Science as Writing.* Yale University Press, New Haven, Conn.

Murray, M. W., 1968, Written communication—a substitute for good dialogue. *American Association of Petroleum Geologists Bulletin* 52:2092–7.

North American Commission on Stratigraphic Nomenclature, 1982, North American Stratigraphic Code. *American Association of Petroleum Geologists Bulletin* 67:841–75.

O'Connor, M., and F. P. Woodford, 1977, *Writing Scientific Papers in English: An ELSE–Ciba Foundation Guide for Authors.* Elsevier/Exerpta Medica, North Holland, Amsterdam.

Partridge, E., 1976, Usage and Abusage. Penguin Books, Harmondsworth.

Porter, C., 1985, *Environmental Impact Assessment: A Practical Guide.* University of Queensland Press, St. Lucia, Queensland, 269p.

Rodolfo, K. S., 1979, One picture is worth more than ten thousand words: How to illustrate a paper for the Journal of Sedimentary Petrology. *Journal of Sedimentary Petrology* 49:1053–60.

Roget, P. M., 1984, *Roget's Thesaurus of English Words and Phrases.* Penguin, Harmondsworth, 1350p.

Royal Society, London, 1966, Guide for preparation and publication of abstracts. *American Association of Petroleum Geologists Bulletin* 50:1993.

Royal Society, London, 1974, *General Notes on the Preparation of Scientific Papers.* The Royal Society, Cambridge University Press, London.

Rubens, P., 1992, *Science and Technical Writing: A Manual of Style.* H. Holt, New York.

Shinn, E. A., 1981, Make the last slide first. *Journal of Sedimentary Petrology* 51:1–6.

Strunk, W., Jr., and E. B. White, 1959, *The Elements of Style.* Macmillan, New York.

Style Manual Committee, 1978, *Style Manual.* Australian Government Publishing Service, Canberra, Australia.

The University of Chicago Press, 1993, *The Chicago Manual of Style,* 14th ed. Chicago.

Tufte, V., 1971, *Grammar as Style.* Holt Reinhart & Winston, New York, 280p.

Turabian, K. L., 1987, *A Manual for Writers of Research Papers, Theses, and Dissertations,* 5th ed. Heinemann, London.

USGS, 1991, *Suggestions to Authors of the Reports of the United States Geological Survey,* 7th ed. U.S. Government Printing Office, Washington, D.C., 289p.

Warden, R. E., and W. T. Dagodag, 1976, *A Guide to the Preparation and Review of Environmental Impact Reports.* Security World Publishing Co., Los Angeles, 138p.